TOXICITY OF HEAVY METALS
IN THE ENVIRONMENT

HAZARDOUS AND TOXIC SUBSTANCES
A Series of Reference Books and Textbooks

Series Editor: Seymour S. Block
University of Florida
Department of Chemical Engineering
Gainesville, Florida

1. Highly Hazardous Materials Spills and Emergency Planning, J. E. Zajic and W. A. Himmelman

2. Toxicity of Heavy Metals in the Environment (in two parts), edited by Frederick W. Oehme

Other volumes in preparation

TOXICITY OF HEAVY METALS IN THE ENVIRONMENT

IN TWO PARTS

Part 1

EDITED BY

Frederick W. Oehme

Comparative Toxicology Laboratory
Kansas State University
Manhattan, Kansas

MARCEL DEKKER, INC. New York and Basel

Library of Congress Cataloging in Publication Data

Main entry under title:

Toxicity of heavy metals in the environment.

   (Hazardous and toxic substances ; 2)
    1. Heavy metals--Toxicology. 2. Heavy metals--
Environmental aspects. 3. Veterinary toxicology.
I. Oehme, Frederick W. II. Series.
RA1231.M52T69   615.9'2   78-14758
ISBN 0-8247-6718-7 (p. 1)

Chapters 4 and 20 appeared in Clinical Toxicology 5(2) June 1972.
Chapters 1, 5, 10, 11, and 19 appeared in Clinical Toxicology 6(3)
1973. Chapter 7 appeared in Clinical Toxicology 9(1) 1976.

MARCEL DEKKER, INC.
270 Madison Avenue, New York, New York 10016

Current printing (last digit):
10 9 8 7 6 5 4 3 2

PRINTED IN THE UNITED STATES OF AMERICA

The environment is here to stay and, chemically and biologically, the heavy metals fall in the same category. Their persistence and their hazard and toxicity to human and animal life has been one of the prime impulses behind the development of this publication. Through a careful selection of authors, each with a specific expertise and knowledge, an effort has been made to present an overview of the toxicology associated with heavy metal chemicals as they occur in the environment of all animals, including humans. Hence, most of the contributors have adopted a comparative approach in discussing their assigned topics. This was deliberately requested since much of the information available has been derived from animals; occasionally, human instances of environmental toxicity have stimulated the conduct of experiments in laboratory or domestic animals.

This book has been organized by first discussing the basic concepts and principles of heavy metal pollution, how the heavy metals enter the environment and animal or human food chain, and the fundamental principles and mechanisms of toxicity due to heavy metal chemicals. The more common toxic heavy metals are then described together with their biochemistry and clinical syndromes. This is followed by an excursion into the area of trace heavy metals and the interactions of specific metallic compounds. The concluding chapters deal with quantitative assay of environmental metallic contaminants and some concepts of chelation therapy. Although printing time introduces a lag factor, every attempt has been made to provide the most current material available. The diligence of the contributors in maintaining this objective is greatly appreciated.

It is hoped that this publication will serve two functions: (1) To provide a review and discussion of the current information on comparative heavy metal toxicity; and (2) to further demonstrate that comparative

studies involving scientists with diverse backgrounds, training, and spe-
cialty-disciplines provide a well-balanced fuel with which a specific effort can rapidly and accurately arrive at its objective. The contributors
are commended for their cooperation and enthusiasm in responding to this
undertaking. Appreciation is given to the publisher for his interest and
continual guidance in seeing this book through to its fulfillment. The
product is one which the readers can best evaluate; their comments and
suggestions are welcomed.

Frederick W. Oehme

CONTENTS

Cumulative Index to appear in Part 2

ARONSON, ARTHUR L., D.V.M., M.S., Ph.D., Professor, Department of Physiology, Biochemistry, and Pharmacology, Cornell University New York State College of Veterinary Medicine, Ithaca, New York

BUCK, WILLIAM B.,[1] D.V.M., M.S., Professor of Veterinary Toxicology, Veterinary Diagnostic Laboratory, Iowa State University College of Veterinary Medicine, Ames, Iowa

CALLENBACH, JOHN CORRIE, M.D., F.A.A.P., Pediatrician, Departments of Pediatrics and Neonatology, The Children's Mercy Hospital, Kansas City, Missouri

CASSIDY, DELMAR RONALD, D.V.M., M.S., Ph.D., Chief, Pathology, Toxicology, and Parasitology Laboratory, National Veterinary Services Laboratories, Animal and Plant Health Inspection Service, U.S. Department of Agriculture, Ames, Iowa

ERDMAN, JAMES A., M.A., Ph.D., Botanist, U.S. Department of the Interior, Geological Survey, Denver, Colorado

EWAN, RICHARD C., Ph.D., Professor, Department of Animal Science, Iowa State University of Science and Technology, Ames, Iowa

EXON, JERRY H., B.S., Researcher, Oregon State University School of Veterinary Medicine, Corvallis, Oregon

FURR, ALLAN, D.V.M., M.S., Assistant Professor, Toxicology, Veterinary Diagnostic Laboratory, Iowa State University, Ames, Iowa

GREEN, VERNON A., Ph.D., Chief of Toxicology Services, The Children's Mercy Hospital, Kansas City, Missouri

HAMMOND, PAUL B.,[2] D.V.M., Ph.D., Professor, Department of Veterinary Physiology and Pharmacology, University of Minnesota College of Veterinary Medicine, St. Paul, Minnesota

HARADA, MASAZUMI, M.D., Associate Professor, Department of Neuropsychiatry, Institute of Constitutional Medicine, Kumamoto University, Kumamoto, Japan

Present Affiliation:

[1]Professor, Division of Pharmacology and Toxicology, Department of Veterinary Biosciences, University of Illinois College of Veterinary Medicine, Urbana, Illinois

[2]Professor of Environmental Health, Department of Environmental Health, University of Cincinnati Medical Center, Cincinnati, Ohio

HARMS, THELMA F., B.S., Research Chemist, U.S. Department of the Interior, Geological Survey, Denver, Colorado

HARR, JAMES R., D.V.M., M.S., Director, Department of Toxicology, Pennwalt Corporation, Rochester, New York

HAUFLER, MAURICE,[3] B.A., Biological Lab Technician, Biochemistry, U.S. Livestock Insects Laboratory, Agricultural Research Service, U.S. Department of Agriculture, Kerrville, Texas

KOBAYASHI, JUN,[4] Sc.D., Professor, Okayama University Institute for Agricultural and Biological Sciences, Kurashiki, Japan

LEDET, ARLO E., D.V.M., M.S., Ph.D., Professor, Department of Veterinary Pathology, Iowa State University College of Veterinary Medicine, Ames, Iowa

OEHME, FREDERICK W., D.V.M., Ph.D., Professor of Toxicology, Medicine, and Physiology, and Director, Comparative Toxicology Laboratory, Kansas State University, Manhattan, Kansas

OSWEILER, GARY D., D.V.M., M.S., Ph.D., Associate Professor, Veterinary Anatomy-Physiology, University of Missouri – Columbia, Columbia, Missouri

PALMER, J. S., D.V.M., M.P.H., Veterinary Medical Officer, Veterinary Toxicology and Entomology Research Laboratory, Science and Education Administration, U.S. Department of Agriculture, College Station, Texas

PAPP, CLARA S. E., B.S., M.S., Research Chemist, U.S. Department of the Interior, Geological Survey, Denver, Colorado

RINER, JAYME C.,[5] Physical Science Technician, U.S. Livestock Insects Laboratory, Agricultural Research Service, U.S. Department of Agriculture, Kerrville, Texas

RUSSELL, LEON H., JR., B.S., D.V.M., M.P.H., Ph.D., Professor, Department of Veterinary Public Health, College of Veterinary Medicine, Texas A & M University, College Station, Texas

SCHARDING, NANCY N.,[6] D.V.M., M.S., Department of Surgery and Medicine, College of Veterinary Medicine, Kansas State University, Manhattan, Kansas

SHACKLETTE, HANSFORD T., M.S., Ph.D., Research Botanist, U.S. Department of the Interior, Geological Survey, Denver, Colorado

VAN GELDER, GARY A., D.V.M., M.S., Ph.D., Professor and Chairman, Veterinary Anatomy-Physiology, University of Missouri College of Veterinary Medicine, Columbia, Missouri

---

Present Affiliation:

[3]U.S. Livestock Insects Laboratory, Science and Education Administration, U.S. Department of Agriculture, Kerrville, Texas

[4]Professor Emeritus, Okayama University Institute for Agricultural and Biological Sciences, Kurashiki, Japan

[5]U.S. Livestock Insects Laboratory, Science and Education Administration, U.S. Department of Agriculture, Kerrville, Texas

Present Address:

[6]Box 840, Memphis, Tennessee

WESWIG, PAUL H., Ph.D., Professor, Department of Agricultural Chemistry, Oregon State University, Corvallis, Oregon

WHANGER, PHILIP D., Ph.D., Associate Professor, Department of Agricultural Chemistry, Oregon State University, Corvallis, Oregon

WISE, GEORGE W., M.D., F.A.A.P., Director of Poison Control, Pediatrics, The Children's Mercy Hospital, Kansas City, Missouri

WRIGHT, FRED C.,[7] B.S., M.A., Research Chemist, U.S. Livestock Insects Laboratory, Agricultural Research Service, U.S. Department of Agriculture, Kerrville, Texas

YOUNGER, R. L.,* D.V.M., Veterinary Medical Officer, Agricultural Research Service, U.S. Department of Agriculture, Veterinary Toxicology and Entomology Research Laboratory, College Station, Texas

ZOOK, BERNARD C., D.V.M., Associate Professor of Pathology, The George Washington University Medical Center, Washington, D.C.

*Deceased

Present Affiliation:

[7]U.S. Livestock Insects Laboratory, Science and Education Administration, U.S. Department of Agriculture, Kerrville, Texas

CONTENTS OF PART 2

PERSISTENCE VERSUS PERSEVERANCE

Frederick W. Oehme
Kansas State University
Manhattan, Kansas

A special characteristic of heavy metal chemicals is their strong attraction to biological tissues and, in general, the slow elimination of these chemicals from biological systems. These same characteristics are apparent in the environment, where pollution with heavy metals has produced great enthusiasm. In total, heavy metals are persistent. Once in a system, they remain for relatively long periods. Whether their presence produces good, evil, or no effect frequently depends upon the speaker or author, the audience being addressed, and the particular study involved. While the great cry of "pollution" has tended to label the presence of any foreign chemical as "evil," the impression is gained that studies documenting the evil effect of heavy metals sometimes employ levels that exceed physiological concentrations of the particular chemical. On the other hand, it is becoming increasingly apparent, due to more sophisticated methods of clinical and biochemical evaluation, that low environmental levels of heavy metals may produce subtle and chronic effects previously unrecognized. It is obvious that health scientists and toxicologists are used to working with relatively massive toxicoses, producing dynamic clinical signs. The casualness with which environmental pollutants may produce their effects raises the problem of identifying these changes, characterizing them for clinical diagnoses, developing treatment procedures, and finally disseminating this information as promptly as possible to interested and involved medical personnel.

With this basic stimulus, and recognizing that fundamental studies of human health problems are of necessity initially investigated in animals, the American College of Veterinary Toxicologists co-sponsored, together with the American Veterinary Medical Association, the Symposium of the Biological Effects of Heavy Metal Pollutants during the 108th Annual Meeting of the American Veterinary Medical Association. Several of the papers included in this printing were presented at the ACVT-AVMA co-sponsored session. Some have been modified and updated by material generated since the papers were initially given. Other papers, providing additional relevant information and contributing to the overall topic, have also been included. "Toxicology and Adverse Effects of Mineral Imbalance" was presented as an invited paper at the 138th meeting of the American Association for the Advancement of Science, December 28, 1971, in Philadelphia. "Comparative Studies of Lead Poisoning" is based upon material given at the First Annual Meeting of the Institutional Consortium on Endemic Lead Poisoning, held May 3-4, 1972, in Chicago, sponsored by the American Academy of Clinical Toxicology.

As can be seen by readers of these papers, the biological effects of these chemicals are at once diverse and significant. Heavy metals are environmentally stable, and as long as the chemicals are present they may induce their toxic effect. However, the medical scientist must also be persistent in his efforts to understand the mechanisms of the toxic effect, to recognize the clinical signs (even though the subtleness of the interaction may belie its importance), and to attain a clinical response to therapy. Toxicologists must maintain perseverance to continue their efforts to understand and combat the effects of these pollutants.

It is hoped that publication of these papers will fulfill the purpose of providing further insight into the biological interactions of the common heavy metal pollutants. The authors have been most cooperative in providing reviews of the pertinent literature and documentation of their research efforts. We hope these contributions will add to the base of information continually being generated and will eventually develop an increased recognition and effectiveness of treatment for these all-too-common environmental hazards.

Chemical **persistence** in biological systems can produce undesirable effects. **Perseverance** in efforts to identify and solve problems produced by these chemicals will be rewarding. Scientific perseverance can effectively combat chemical persistence.

HEAVY METALS IN FOODS OF ANIMAL ORIGIN

Leon H. Russell, Jr.
Texas A & M University
College Station, Texas

Heavy metals have long been associated with acute and subacute food-borne intoxications. However, with the exception of mercury poisoning from aquatic foods, little attention has been given to the significance of animal food products as a source of intoxicating levels of heavy metals. With the recognition of the hazardous bioaccumulation of organic chemicals [28,62], such as chlorinated hydrocarbons, it is also imperative that the significance of food chain accumulation of heavy metals be clarified [94,113].

Heavy metals, unlike the synthetic organic chemical compounds, are not man-made. To the contrary, the environmental pollution problem associated with metals is one of redistribution by man's industrial and agricultural societies [112]. During the redistribution process, three natural resources (water, land, and air) are utilized for the dilution and deposition of waste heavy metal products [56]. Once a heavy metal is released into the natural environment, it is possible for contaminants to enter man's food through various routes, including fish, poultry, and livestock food products [84].

The objective of this paper is to review current knowledge on the occurrence and safety of heavy metals in foods of animal origin.

ANALYTICAL METHODS

A heavy metal, as defined by the VanNostrand's Chemist Dictionary [114], is a "metal of high specific gravity"; whereas, a metal is "characterized by luster, ductility, malleability, high electric and thermal conductivity, and chemically, of forming bases which can react with acids." There is no sharp demarcation between metals and nonmetals; in fact, it may be more appropriate to classify some elements, such as arsenic and selenium, as metalloids since they have characteristics of both groups of elements [114].

Historically, understanding the effects of metals on biological systems has been intimately associated with the development of analytical chemistry [85]. Sensitive and specific methods are basic for assessing minute biological concentrations of heavy metals. As the sensitivity of analytical procedures has increased, so has our understanding of metals and their action on biological systems. Therefore, it is imperative that the period of time during which analyses are reported should be considered when evaluating the occurrence and safety of chemicals in food.

The methods commonly utilized for the detection of heavy metal contaminants in food are the atomic absorption spectrophotometry and neutron activation analysis [8,29,52]. The method choice, and its modifications, limit the degree of detection as indicated in Table 1. Neutron analysis is more sensitive for arsenic, cadmium, mercury, and selenium; whereas atomic absorption techniques are preferred for lead and zinc [85]. To determine methyl mercury, gas-liquid chromatographic analysis is required [100].

TABLE 1. Analytical Methods (Pier, 1975) [81]

| | Limits of Detection | |
Element	AAS[a] (ppm)	NAA[b] (μg)
Arsenic	0.2	$1\text{-}3 \times 10^{-4}$
Beryllium	0.002	Not detected
Boron	6.0	Not detected
Cadmium	0.005	4.9×10^{-3}
Lead	0.01	1-3
Mercury	0.5	$1\text{-}3 \times 10^{-3}$
Nickel	0.005	$4\text{-}9 \times 10^{-3}$
Selenium	0.5	$1\text{-}3 \times 10^{-3}$
Thallium	Not detected	$4\text{-}9 \times 10^{-2}$
Titanium	0.2	$1\text{-}3 \times 10^{-2}$
Vanadium	0.04	$1\text{-}3 \times 10^{-4}$

[a]Atomic absorption spectrophotometry.
[b]Neutron activation analysis.

EPIDEMIOLOGY

The epidemiology of heavy metal food intoxication generally has been associated with one of three patterns of occurrence, namely (1) environmental pollution, (2) accidental inclusion during processing, and (3) contamination from containers during processing or storage of foods. Most epidemiologic investigations of heavy metal food-borne intoxications have been the retrospective type of procedure. Very little has been done with the prospective type of long-term exposure, or cohort study, to low concentrations of heavy metals in human food [10,50].

An epidemiological model for the flow of low levels of heavy metals into human food was given by MacGregor in his review on mercury and cadmium pollution (Fig. 1) [70]. This model can be utilized for better understanding of the epidemiology and control of all heavy metal problems associated with man's food.

GUIDELINES

In evaluating the concentrations of the various metals in foods, World Health Organization/Food Agriculture Organization (WHO/FAO) and Food and Drug Administration (FDA) provisional recommendations or guidelines are given wherever available (Table 2). Also, the estimated dietary intakes of heavy metals by class is given according to Mahaffey et al. (Table 3) [70].

ACTIONS OF HEAVY METALS

Metals constitute an important fraction of the biosphere [42] and many are required for the life processes of animals, including man. A classification of the requirements of metals by man has been given by Pier [84]: (1) metals required in substantial amounts, for which the body has a rather wide tolerance, e.g., sodium and iron, (2) metals required in much smaller amounts, and for which the body has a quite narrow tolerance, e.g., the "micronutrients" copper and zinc, and (3) metals for which we have not determined an essential role in the life processes and are toxic at low levels, e.g., lead and mercury.

It is beyond the limits of this chapter to discuss the specific toxicity mechanism of each heavy metal, which is influenced by the various characteristics of every metal, namely molecular configuration, solubility, particle size, and various other physicochemical characteristics [1]. In some cases the metals attack nervous tissue by enzymatic blocking of critical

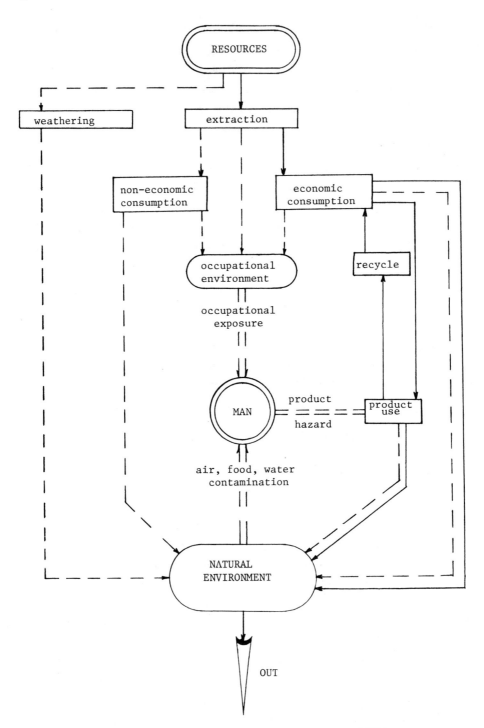

FIG. 1. Schematic metal flow (MacGregor [68]).

TABLE 2. Dietary Intakes of Heavy Metals

	WHO/FAO Provisional Tolerated Weekly Intakes for Man[a]		Transformation of WHO/FAO Limits to Daily Basis		FY 1973 Total Diet Findings		% of Tolerable Daily Intake
	(mg/person /week)	(mg/kg body wt/week)	(μg/person /day)	(μg/kg body wt/day)	(μg/person /day)	(μg/kg body wt/day)	
Mercury							
Total mercury	0.3	0.005	42.9	0.71	2.89[b]	0.041	6.7
Methyl mercury expressed as mercury	0.2	0.0033	28.6	0.48	--	--	--
Lead	3.0[c]	0.05	429.0	7.15	60.4	0.87	14.1
Cadmium	0.4-0.5	0.0067-0.0083	57.1-71.4	0.95-71.4	51.2	0.73	89.7-71.7
Selenium	Not established		--	--	149.7	2.15	--
Arsenic	Not established		--	--	110.1 (as As2O3)	0.15 (as As2O3)	--

[a]From 16th Report of Joint FAO/WHO Expert Committee on Food Additives, Mahaffey et al. [70].

[b]FDA determines total mercury in the total diet study.

[c]WHO/FAO intakes of lead were specifically stated as not being applicable to infants and children.

TABLE 3. Estimated Dietary Intakes of Heavy Metals by Food Class[a]

		Lead		Cadmium		Zinc		Mercury		Arsenic		Selenium	
		(μg /day)	(% of total)	(μg /day)	(% of total)	(μg /day)	(% of total)	(μg /day)	(% of total)	(μg /day)	(% of total)	(μg /day)	(% of total)
I.	Dairy products	0.0	0.0	3.94	7.7	3837	21.4	0.0	0.0	2.34	23.1	0.0	0.0
II.	Meat, fish, and poultry	4.00	6.6	2.49	4.9	6660	37.2	2.89	100.0	5.64	55.6	56.3	37.6
III.	Grain and cereal	4.16	6.9	11.66	22.8	3370	18.8	0.0	0.0	1.35	13.7	92.5	61.8
IV.	Potatoes	0.70	1.2	9.11	17.8	1198	6.7	0.0	0.0	0.64	6.3	0.65	0.4
V.	Leafy vegetables	3.03	5.0	3.18	6.2	136	0.8	0.0	0.0	0.0	0.0	0.0	0.0
VI.	Legume vegetables	18.80	31.1	0.42	0.8	542	3.0	0.0	0.0	0.0	0.0	0.0	0.0
VII.	Root vegetables	3.83	6.4	0.76	1.5	80	0.5	0.0	0.0	0.0	0.0	0.25	0.2
VIII.	Garden fruits	11.36	18.8	1.71	3.4	267	1.5	0.0	0.0	0.0	0.0	0.0	0.0
IX.	Fruits	9.49	15.7	9.38	18.3	194	1.1	0.0	0.0	0.0	0.0	0.0	0.0
X.	Oil and fats	0.67	1.1	1.36	2.7	314	1.8	0.0	0.0	0.17	1.7	0.0	0.0
XI.	Sugars and adjuncts	0.55	0.9	0.68	1.3	254	1.4	0.0	0.0	0.0	0.0	0.0	0.0
XII.	Beverages	3.81	6.3	6.49	12.7	1066	6.0	0.0	0.0	0.0	0.0	0.0	0.0
	Totals	60.4		51.2		17918 (17.9 mg)		2.89		10.1		149.7	

[a]Based on FY 1973 Total Diet Survey data of the Food and Drug Administration, Mahaffey et al. [70].

biochemical reactions [81]. More **frequently**, heavy metals may exert their action indirectly through destruction of detoxifying and excreting organs such as the liver and kidneys [81]. These latter organs should be closely monitored as indicators of heavy metals in the food animal's body (and also be critically evaluated as to their wholesomeness for food, even in the absence of clinical toxicity [78,79]).

TOXIC METALS

The reported occurrence, toxicity, detection, and safety of arsenic, cadmium, mercury, lead, selenium, and zinc will be summarized for the red meats, poultry, aquatic foods, and dairy products. The concentrations reported will be by those procedures currently believed to be most reliable. A group summary of these surveys is given (Table 4).

Arsenic

Although arsenic intoxication has been associated with occupational and criminal exposures, only one outbreak of acute disease has been reportedly associated with the ingestion of arsenic in food or beverage. Reynolds reported arsenical beer poisoning in 1902 [88]. Acute poisoning can result in severe gastroenteritis, but chronic intoxication is characterized by loss of weight and hair and skin lesions. The most recent concern is related to the carcinogenic property of arsenic, as observed from occupational exposure to arsenical compounds [119]; however, with the exception of water, carcinogenesis from ingestion of arsenic-contaminated foods is unknown [111].

TABLE 4. Group Means of Heavy Metal Contamination

Food Group	Heavy Metal (ppm)					
	Arsenic	Cadmium	Lead	Mercury	Selenium	Zinc
Red meat	0.02	0.0093	0.015	0.0093	0.3	24.5
Poultry	0.02	0.0093	0.015	0.0093	0.1	24.5
Eggs	0.02	0.02	0.005	0.02	0.1	24.5
Dairy products	0.0033	0.005	0.04	0.005	0.1	5.0
Aquatic foods	2.6	0.02	0.6	0.02	--	24.5

Artesian well water with high concentration of arsenic has been associated with chronic arsenicism and high prevalence of skin cancer in Taiwan [111].

Arsenic, which has no known vital function, is ubiquitous in the biosphere. Most foods contained minute amounts, averaging 0.02 ppm, including meats, fish, and poultry [70]. A recent survey indicated that beef should have less than 0.5 ppm in the muscle, liver, and kidneys [122]. Seafoods generally were higher in arsenic concentration. Zook et al. reported an aquatic food mean of 2.6 ppm with a range of 0.1 to 6.0 ppm for catfish and shrimp, respectively [121]. Fowler reported concentrations of up to 51 ppm in some bottom-dwelling marine organisms [32].

The dairy products were much lower, averaging 0.0033 ppm, but accounted for 31% of the dietary intake according to Mahaffey et al. [70].

Arsenic levels may be determined by several techniques and are often reported as As_2O_3. However, neutron activation and x-ray fluorescent procedures are generally considered most accurate [67]. In addition, the methylation of inorganic arsenic by bacteria and mammals should be considered during analysis [63].

The WHO/FAO recommends a maximum daily intake of 0.05 mg/kg of body weight. The FDA requires that liver and edible animal by-products contain less than 2 ppm and that muscle tissue contain less than 0.5 ppm [96].

Two potential problem areas may develop in the near future. One may expect to see an increase of arsenicals in all types of foods grown in areas where arsenicals are commonly used as herbicides and defoliants; e.g., foods associated with cottonseed by-products [120]. It should be noted that arsenic is a metalloid close to selenium on the periodic table of elements [114]. It is possible for selenium-accumulator plants to concentrate arsenic from the soil and pass the element on to food animals. Other possible hazards could develop around areas using coal as a source fuel in which the fly ash could contaminate the surrounding soil [65]. Also, the emission of arsenic into runoff water from geothermal power plants has been reported to result in the subsequent contamination of rivers receiving such discharges (up to 500 ppm in Ohaki, New Zealand [91]). This has also been reported in water contaminated with gold mining wastes [5].

Cadmium

Acute cadmium toxicity, characterized by severe gastroenteritis, has been associated with acid foods prepared or stored in cadmium-plated utensils. Chronic cadmium toxicity has been reported to be the cause of

Japanese itai-itai disease [33]. This was apparently acquired through ingestion of rice containing 0.6 to 1.1 ppm cadmium; but fish that were harvested from cadmium-contaminated water may have contributed to the malady
[116].

Cadmium has produced several types of chronic lesions in laboratory
animals [102]. This includes testicular necrosis [102], fetal malformations [102], and cardiovascular hypertension [18]. Their significance to
similar diseases in man is unknown at the present time.

Several methods can be used for the detection of cadmium in foods,
but atomic absorption spectrophotometry is the most commonly used technique
[116]. Because this method may give erroneous results, due to interference
by salts such as NaCl, Pier suggested that atomic fluorescence spectroscopy
should be considered as a method of detection [84].

Cadmium, which has no known vital function, is widely distributed in
the environment [113], as well as commonly used in industrial procedures
and metallic products [68]. Consequently, this heavy metal was commonly
present at low levels in most foods but was found at high concentrations in
foods associated with a locality where environmental pollution with cadmium
is common [27,68].

Livestock and poultry have rarely, if ever, been recognized as being
poisoned by cadmium [78]. The mean concentrations for dairy products
(0.005 ppm), red meats, and poultry (0.0093 ppm) were generally low [76];
however, the kidneys of animals were generally much higher than other parts
of a carcass, averaging 0.05 ppm [27,78,79]. Aquatic foods had the highest
mean concentration with 0.02 ppm, but some oysters were reported as high as
2.0 ppm [19,21]. The brown body meat of some crabs has been reported to
contain cadmium at levels of up to 12 ppm [121].

The WHO/FAO recommends dietary limits of 0.95 to 1.19 mg/kg of body
weight. The FDA has not published any guidelines for the allowable content
of cadmium in foods.

Future problems could be associated with increased use of cadmium in
metal alloys [69]. For example, cadmium has generally appeared as a contaminant in any use of zinc [12]. Recent agriculture developments may
indirectly result in increased cadmium intake, such as sewage-sludge-
fertilized corn. No clinical or pathological signs were observed in swine
fed sewage-sludge-fertilized corn; however, determinations were not reported on the cadmium concentrations in pork from swine fed the grain,
which had five times the normal cadmium content [46].

Lead

This heavy metal, which has been mined and used since ancient times, is present in most living creatures and plants [44,64]. It has no known essential function but can accumulate in many biological systems until it reaches toxic levels [2,6].

Lead exerts its most significant toxic effects on the nervous system, the kidney, and the hematopoietic system [17,22]. The intoxication is progressive and may lead to acute encephalopathy, anemia, and nephrosis [14]. The literature has reports concerning the effects of low-level dietary lead (5 to 50 ppm) on behavioral changes within laboratory animals [87,95]. Others have investigated the effects of lead on the brain metabolism [4,16], hyperactivity [25,99], and even the increased susceptibility to infectious agents [35].

The detection of lead in foods has been done by atomic absorption spectroscopy, neutron activation analysis, and optical emission spectroscopy. The most common technique, atomic absorption spectroscopy, gave a very wide range of values when over 60 laboratories participated in a sensitivity study [60]. The flameless atomic absorption method has been reported to be a more accurate method of determination and probably should be used as the standard technique [53,74]. However, many variables, especially the hydrogen ion concentration, must be controlled when extracting lead and other metals with organic chelating agents [30].

It has been estimated that in the United States man has a daily lead intake of approximately 300 μg [64], of which about 60% [70] comes from the ingestion of contaminated foods. The lead content of these foods has been largely influenced by environmental contamination. This may have been through surface contamination as well as by plant-tissue incorporation of lead that was deposited on soil or crops grown in areas near main highways or lead smelters [27]. When such crops were used as forage for farm animals, the lead was transported into milk or meat [27,66].

Meat and poultry products averaged 0.015 ppm [60] whereas fish have been reported to average 0.6 ppm [121], with oysters up to 1.0 ppm, from environments not considered to be polluted with lead [121].

The lead content varied widely in fresh milk with a mean of 0.04 ppm, but canned, evaporated milk generally had much higher values, up to 0.2 ppm [5,6]. In fact, the concentration of lead in most canned foods was found to be higher (0.17 ppm) than in similar products packaged in glass

containers (0.04 ppm) [76]. Kirkpatrick [58], in a survey of most types of
cured meats, found that the ready-to-eat canned, shelf-stable products had
the highest concentration of lead (0.16 ppm). This has been associated
with the use of lead in solder used to seal the seams on cans [76].

However, no foods of animal origin have been shown to possess excessively high levels of lead unless they leach the lead from a container,
such as a metallic or lead-glazed utensil [74], or the ink from printed
paper containers [48] and polyethylene bags [45]. According to Mahaffey et
al. [68], meat, poultry, and fish accounted for only 6.6% of the estimated
dietary intake of lead.

The WHO/FAO recommended limits for daily intake of lead of 7.15 mg/kg
of body weight [70]. The FDA guideline for milk is 0.3 ppm and 0.5 ppm for
whole milk and evaporated milk, respectively [76]. There are no guidelines
published for other foods such as meat and poultry.

The future will probably see more concentrated efforts to survey the
lead content of all food products harvested from livestock raised in areas
which are considered to have a high level of possible exposure, such as
smeltering regions. Solder will be replaced by some type of epoxy or other
adhesive, and the use of lead-containing inks should be recognized as a
hazard and eliminated as a major source of food contamination.

Mercury

Most of the mercury which man acquires in his diet comes from foods of
animal origin, especially fish, which act as bioaccumulators of this heavy
metal [34,104]. Mercury has no known beneficial biological function. Because of its increased economic consumption in electrical apparatus, paints,
fungicides, and plastic catalysts, mercury has become a serious environmental pollutant [54,68]. When considering that mercury wastes eventually
flow back into the aquatic environment, highest concentrations are expected
to occur in water and aquatic life of circumscribed areas [31,63].

Organic mercury, especially methyl mercury [34], is more toxic and
accumulates in the muscle of fish, swine [11,13,43,54,63,83], and bovine
[79]. Recent studies have confirmed that inorganic mercury may be converted into methyl mercury by microbes in some aquatic environments [51,
80,92]; therefore, any degree of mercurial pollution should be considered
hazardous [38,39,47].

Acute mercury intoxication, generally from inorganic compounds, results in severe gastroenteritis and nephritis [55,79]. Chronic mercurial-

ism, especially from organomercurial compounds, affects the brain, causing
serious and irreversible damage [69]. This condition has been reported in
individuals who ate fish containing high levels of methyl mercury, above
6.7 ppm, as occurred at Minamata Bay [34]. However, one outbreak occurred
from the consumption of pork that had been fed organomercurial-treated seed
grain [69]. The selenium/methyl mercury ratio should be considered as an
indicator of toxicity such as the 1:10 ratio in fish foods products of the
Minamata Bay disaster [7,103]. However a low ratio, 1:1, is indicative of
a low potential for toxicity from the ingestion of such fish food products
[33,37,104].

Reports indicated that, except for fish, mercury levels in foods were
generally so low as to be undetected by the usual flameless atomic absorption techniques [75,93,107]; therefore, neutron activation analysis should
be used to determine the mercury content of foods other than fish [49,105].

Extensive surveys have been conducted on the mercury content of foods
[100,101,121]. A recent survey, including 3000 samples of canned tuna,
indicated that more than 96% of seafoods were below the FDA "guideline" of
0.5 ppm mercury, with a mean level of 0.13 ppm [121]. Only large halibut,
red snappers, and tuna exceeded the guideline [83].

Milk was found to have mercury levels of about 0.01 ppm [42,79]. Meat
and poultry as a group ranged from 0.001 to 0.04 ppm. Beef and pork liver,
raw ground beef, and chicken breast were 0.03 ppm [72], whereas pork had
only 0.005 ppm [42]. In bovine dosed with methyl mercury chloride, beef
kidneys have been found to have the highest concentration, 4.4 ppm [79,97].
Chickens fed 10 ppm of methyl mercury chloride for 10 days produced eggs
with 10 and 5 ppm in egg whites and yolks, respectively [71,98]. However,
in a study in Michigan it was reported that the survey of market eggs revealed only 0.03 ppm [42].

The WHO/FAO recommended maximum daily levels of total mercury for
adults is 0.71 mg/kg body weight and that for methyl mercury is 0.48 mg/kg
body weight or 0.02 to 0.05 ppm of the total diet. The FDA guideline, as
mentioned above, is 0.5 ppm. This has eliminated the use of some fish at
the top of the food chain, namely swordfish and large tuna.

Future considerations should include the problem of the increasing
number of fishing areas condemned because of mercury pollution. If additional fishing areas and lakes are considered to be mercury polluted,
activation of demethylation procedures in sediment will have to be developed [104]. Also, more knowledge is needed of the Se/Hg ratio and its
relationship to toxicity and possible use as an index of safety [36].

Selenium

This essential element or micronutrient has caused poisoning when ingested in excessive amounts by animals but not in humans [3,26]. However,
it has been suspected as being a carcinogen in man and lower animals [119].
Selenium's juxtaposition to arsenic on periodic tables of elements should
be considered when evaluating the distribution of these elements in the
environment.

The selenium content of beef averaged 0.3 ppm, while pork, chicken,
and dairy products averaged 0.1 ppm. Animal food products accounted for
38% of the total dietary intake of selenium [70].

There have been no guidelines established by either the FDA or the
WHO/FAO.

Zinc

Although zinc is an essential element or micronutrient, excessive
amounts have caused acute gastroenteritis upon ingestion [15]. Outbreaks
of zinc poisoning have followed the ingestion of acid foods, e.g., curried
chicken, stored in galvanized utensils [15]. These **toxic** foods contained
over 1000 ppm zinc. Since cadmium is also present in galvanized metal
containers, the role of zinc as a cause of food-borne intoxication should
be confirmed. Some investigators have speculated that zinc may be associated with cancer; however, this has not been confirmed [77].

Foods, especially seafoods, varied considerably in their zinc content.
Oysters from certain areas have been found to contain up to 1500 ppm zinc
[94]. Meats, fish, and poultry averaged 24.5 ppm, while dairy products
averaged 5 ppm [70]. In a smeltering area of Missouri, Dorn [27] reported
milk levels up to 375.9 ppm, which varied with the seasons of the year as
well as the degree of environmental contamination.

Since it is not commonly considered a toxic metal, neither the WHO/FAO
nor the FDA gives guidelines for zinc other than as an essential micronutrient.

Other Metals Suspected as Toxic

Numerous heavy metals have been suspected of being toxic; however, the
evidence of common food-borne transmission is lacking. Chromium [106],
manganese [20], nickel [24], silver [24], tin [9,118], and vanadium [95]
are all fairly common in foods and some have been found in high concentrations in foods of animal origin, such as titanium at levels of over 1:1000
in canned tuna [85].

Copper, an essential metal that is found in all living organisms, has been the cause of some food-borne intoxications. All of the outbreaks have involved acid foods prepared or stored in copper vessels. Of the foods with the greatest natural concentration of copper, oysters have been reported to contain up to 137 ppm, and beef liver contained 11 ppm [85]. Most all other foods of animal origin contained between 2 and 4 ppm. Chronic copper poisoning has been recognized in sheep [109] and suspected in humans [40,109].

Since their introduction with catalytic converters on many of the 1975 and later automobiles [57,110], platinum [115] and palladium [117] have been investigated as potential environmental pollutants. However, the concentration of these metals in foods has not reached a detectable level. Low contamination of the environment from beryllium related to missile propellants did not appear to be an extensive problem [89].

Safety Measures

Surveillance. This is the backbone of any effective control program and probably more so with a potential problem such as food-borne heavy metal intoxications. A heavy metal toxicant surveillance program has three essential factors that must be discerned before undertaking any other control measures. These factors are (1) the determination of levels of heavy metal toxicants in foods based on standard methods of sampling analysis, (2) the determination of food consumption patterns and the typical total diet, and (3) the estimation of total load of the heavy metal contaminant from all sources of exposure, including air, water, and occupational sources [41].

Once a proficient surveillance program is accomplished, it should not include common faults often found in such programs. Firstly, avoidance of the appearance of a "witch hunt" [94]. Heavy metals are present in most all foods, including those of animal origin. It should be realized that, within accepted levels of risk [86], the purpose of the program is to watch the total sum of heavy metal concentration for all types of food.

Secondly, the surveillance program should select critical "key" points in the food chain that are the best early indicators of contamination. Since the bioaccumulation in a food chain is generally not as noticeable as with other chemicals, such as DDT, it should be realized that localized and insidious hazards exist. Also, although increased environmental concentrations always indicate a potentially critical food-borne intoxication problem,

e.g., lead, inorganic mercury, and cadmium may be ingested by some food
animals, their muscle and milk will contain insignificant amounts if the
animals have developed no clinical signs.

Thirdly, select sentinel animals and accumulator tissues for analysis,
for example, the bovine kidney for cadmium, the large-sized fish for methyl
mercury, and oysters for zinc. The FDA total diet program is more than
adequate for pesticides that tend to be spread over large regions; however,
heavy metal pollutions and accidents tend to be more localized and generally
depend upon a characteristic geosphere or industry group. Therefore, the
generalized type of surveillance may overlook problem areas. A diversified
program could be installed by being more selective of the potential problem
foods in each area. A total diet-type surveillance could be expanded where
the expense and work load are reduced by freezing representative diets for
later analysis, should the need arise as signaled by critical point or
sentinel animal indicators.

Control of Exposed Animals. Livestock should be quarantined whenever
a severe heavy metal exposure occurs. Problem areas, such as lead mining
and smeltering regions, should have a mandatory animal identification procedure. This would be feasible with a national livestock identification
program. Infectious diseases can be eradicated, but toxicity problems must
be controlled continuously. This includes both surveillance and identification of animals and their food products that may be potential sources of
toxic materials in man's diet.

For animals recovering from clinical intoxications, firm guidelines
on their use as food are needed. Selby et al. [96] recommended that after
exposure to arsenic which results in clinical illness, cattle should not
be slaughtered for at least 6 weeks.

Decontamination of Problem Areas. Many lakes have been rendered useless for fishing because of mercury pollution. Scientific methods should
be applied to decontaminate water and sediment. No purpose would be served
if starvation resulted from avoiding food sources which could be salvaged
or recycled.

Removal of Toxicants from Food. Methods have been developed which will
remove mercury from milk [90] and fish products [23,108]. These methods
could be utilized for producing protein supplements for livestock, if not
food for humans. There is a paucity of information concerning the removal
of other heavy metals from other animal food products.

<u>Removal of Hazardous Containers and Inks Used in Packaging</u>. Lead solder in food containers and lead colors in inks on printed plastic and paper containers should be eliminated as soon as feasible.

<u>Epidemiological Investigation</u>. A thorough epidemiological study should be made for all outbreaks. This should enable one to determine the factors responsible for the intoxication so as to prevent future occurrences. Also, it would probably help get a better grasp on the dose-response relationship of man for ingested heavy metals [50].

CONCLUSIONS

The best safety program will be a mixture of several strategics, which will be flexible and change with man's food habits and his disposal of waste heavy metals. This can best be visualized by utilization of flow diagrams (Figs. 2 and 3) where the critical control points are the following:

Mercury

FIG. 2. Food chain flow diagram of the heavy metals most likely to cause clinical disease in humans. Control: (1) environmental control, (2) animal surveillance, and (3) food product surveillance.

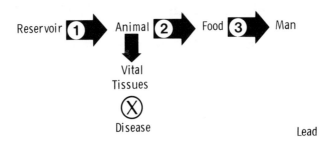

Lead

FIG. 3. Food chain flow diagram of the heavy metals most likely to cause disease in animals (sentinels). Control: (1) environmental control, (2) animal quarantine, and (3) food product surveillance.

1. Avoid polluted animal foodstuffs and polluted environmental sources.

2. Avoid the slaughter or harvest of clinically ill animals; quarantine and surveillance of the exposed.

3. Surveillance of the animal food products at key food service or distribution points.

   Hopefully, this will prevent any extensive and severe incidence of food-borne heavy metal intoxications.

REFERENCES

1. A. Albert, Selective Toxicity, 5th ed., Chapman and Hall, London, 1973.

2. F. W. Alexander, Environ. Health Perspect., Suppl., 7, 155 (1974).

3. W. H. Allaway, Cornell Vet., 63, 151 (1973).

4. T. R. Allen and P. J. MeWey, Environ. Health Perspect., Suppl., 7, 239 (1974).

5. S. K. Amasa, Environ. Health Perspect., 12, 131 (1975).

6. D. G. Anderson and J. L. Clark, Environ. Health Perspect., Suppl., 7, 3 (1974).

7. M. S. Ansari and W. M. Britton, Poultry Sci., 53, 1134 (1974).

8. R. A. Baetz and C. T. Kenner, J. Agr. Food Chem., 21, 436 (1973).

9. W. H. Barker and V. Runte, Amer. J. Epidemiology, 96, 657 (1972).

10. J. M. Barnes, Brit. Med. Bull., 31, 196 (1975).

11. E. G. Bligh, Can. Inst. Food Sci. and Technol., 5, A6 (1972).

12. J. D. Bogden, N. P. Singh, and M. M. Toselow, Environ. Sci. Technol., 8, 740 (1974).

13. G. Birke and A. G. Johnels, Arch. Environ. Health, 25, 77 (1972).

14. T. W. Bouldin, P. Mushak, and L. A. O'Tuama, Environ. Health Perspect., 12, 81 (1975).

15. M. A. Brown, J. V. Thom, G. L. Orth, P. Cova, and J. Juarez, Arch. Environ. Health, 8, 657 (1964).

16. R. J. Bull, P. M. Stanaszek, J. J. O'Neill, and S. D. Lutkenhaff, Environ. Health Perspect., 12, 89 (1975).

17. S. J. Carpenter, Environ. Health Perspect., Suppl., 7, 129 (1974).

18. R. E. Carroll, J. A. M. A., 198, 177 (1966).

19. J. E. Cearley and R. L. Coleman, Bull. Environ. Contamn. and Toxicol., 11, 146 (1974).

20. S. V. Chandra, P. K. Seth, and J. K. Mankeshwar, Environ. Res., 7, 374 (1974).

21. E. A. Childs and J. N. Caffke, J. Food Sci., 39, 453 (1974).

22. B. E. Clayton, Brit. Med. Bull., 31, 236 (1975).

23. G. B. Cohen and E. E. Schrier, J. Agr. Food Chem., 23, 661 (1975).

24. R. L. Coleman and J. E. Cearley, Bull. Environ. Contamn. and Toxicol.,
     12, 53 (1974).

25. O. J. David, Environ. Health Perspect., Suppl., 7, 17 (1974).

26. A. T. Diplock, Critical Review in Toxicology, 4, 271 (1976).

27. C. R. Dorn, J. O. Pierce, G. R. Chase, and P. E. Phillips, Environ.
     Res., 9, 159 (1975).

28. R. E. Duggan and P. E. Corneliussen, Pesticides Monitor, 5, 331 (1972).

29. H. Egan and A. W. Hubbard, Brit. Med. Bull., 31, 201 (1975).

30. R. J. Everson and H. E. Parker, Anal. Chem., 46, 1966 (1974).

31. R. Finch, Fishing Bull., 71, 615 (1973).

32. B. A. Fowler, R. C. Fay, R. L. Walter, R. D. Willis, and W. F.
     Cutknecht, Environ. Health Perspect., 12, 71 (1975).

33. L. Friberg, M. Piscato, C. G. Nordbert, and T. Kjellstrom, Cadmium in
     the Environment, 2nd ed., CRC Press, Cleveland, 1971.

34. L. Friberg and J. Vostal, Mercury in the Environment, CRC Press,
     Cleveland, 1972.

35. J. H. Gainer, Environ. Health Perspect., Suppl., 7, 113 (1974).

36. H. E. Ganther and M. L. Sunde, J. Food Sci., 39, 1 (1974).

37. H. E. Ganther, C. Goude, M. L. Sunde, and W. C. Hoekstra, Science, 175,
     1122 (1972).

38. M. Gilbertson, Bull. Environ. Contamn. and Toxicol., 12, 726 (1974).

39. M. Gilmartin and N. Revelante, Fishery Bull., 73, 193 (1975).

40. M. W. Gleason, R. E. Gosselin, H. C. Hodge, and R. P. Smith, Clinical
     Toxicology of Commercial Products, 3rd ed., Williams and Wilkins,
     Baltimore, 1969.

41. A. Goldberg, Environ. Health Perspect., Suppl., 7, 103 (1974).

42. M. I. Gomez and P. Markakis, J. Food Sci., 39, 673 (1974).

43. D. A. Hancock, Australian Fisheries, 4 (1976).

44. L. Hankin, J. Milk Food Technol., 35, 86 (1972).

45. L. Hankin, G. H. Heichel, and R. A. Botsford, Bull. Environ. Contamn.
     and Toxicol., 12, 645 (1974).

46. L. G. Hansen, J. L. Dorner, C. S. Byerly, R. P. Tarara, and T. D.
     Hinesly, Amer. J. Vet. Res., 37, 711 (1976).

47. J. E. Hardcastle and N. Mavichakana, Bull. Environ. Contamn. and
     Toxicol., 11, 456 (1974).

48. G. H. Heichel, L. Hankin, and R. A. Botsford, J. Milk and Food
     Technol., 37, 499 (1974).

49. M. W. Heitzman and R. E. Simpson, J. Ass. Offic. Anal. Chem., 55,
     960 (1972).

50. I. T. T. Higgins, Brit. Med. Bull., 31, 230 (1975).

51. H. W. Holm and M. F. Cox, Appl. Microbiol., 29, 491 (1975).

52. W. Horwitz (ed.), Official Methods of Analysis of the Association of
     Official Analytical Chemists, Ass. Offic. Anal. Chem., Washington,
     D.C., 1975.

53. A. W. Huffman and J. A. Caruso, J. Agr. Food Chem., 22, 824 (1974).

54. A. G. Hugunin and R. L. Bradley, Jr., J. Milk Food Technol., 38, 285 (1975).

55. A. G. Hugunin and R. L. Bradley, Jr., J. Milk Food Technol., 38, 354 (1975).

56. J. B. Johnson, C. R. Hoglund, and B. Buxton, J. Dairy Sci., 56, 1354 (1973).

57. D. E. Johnson, J. B. Tillery, and R. T. Prevost, Environ. Health Perspect., 12, 27 (1975).

58. D. C. Kirkpatrick and D. E. Coffin, J. Sci. Food Agr., 24, 1595 (1973).

59. D. S. Klauder and H. G. Petering, Environ. Health Perspect., 12, 77 (1975).

60. A. C. Kolbye, K. R. Mahaffey, and J. A. Fiorino, Environ. Health Perspect., Suppl., 7, 65 (1974).

61. F. C. Kopfler, Bull. Environ. Contamn. and Toxicol., 11, 275 (1974).

62. H. F. Kraybill, Can. Med. Ass. J., 100, 204 (1969).

63. J. H. Lakso and S. A. Peoples, J. Agr. Food Chem., 23, 674 (1975).

64. Lead, National Academy of Science, ANON, Washington, D. C., 1972.

65. S. E. Lindberg, A. W. Andren, and R. J. Rairdon, Environ. Health Perspect., 12, 9 (1975).

66. D. J. Lisk, A. K. Furr, D. R. Mertens, W. H. Gutenmann, and C. A. Bache, J. Agr. Food Chem., 22, 954 (1974).

67. G. Lunde, T. Sci. Food Agr., 24, 1021 (1973).

68. A. MacGregor, Environ. Health Perspect., 12, 137 (1975).

69. L. Magos, Brit. Med. Bull., 31, 241 (1975).

70. K. R. Mahaffey, P. E. Corneliussen, C. F. Jelinek, and T. A. Fiorino, Environ. Health Perspect., 12, 63 (1975).

71. B. E. March, R. Soong, E. Bilinski, and R. E. E. Jonas, Poultry Sci., 53, 2175 (1974).

72. B. E. March, R. Soong, E. Bilinski, and R. E. E. Jonas, Poultry Sci., 53, 2181 (1974).

73. R. Meers and L. K. Graham, Environ. Res., 8, 1 (1974).

74. T. C. Meranger, Can. J. Public Health, 64, 472 (1973).

75. R. K. Munns and D. C. Holland, J. Ass. Offic. Anal. Chem., 54, 202 (1971).

76. D. G. Mitchell and K. M. Aldous, Environ. Health Perspect., Suppl., 7, 59 (1974).

77. N. D. McGlasham, Lancet, 1, 578 (1967).

78. M. W. Neathery and W. J. Miller, J. Dairy Sci., 58, 1767 (1975).

79. M. W. Neathery, W. J. Miller, R. P. Gentry, P. E. Strake, and D. M. Blackman, J. Dairy Sci., 57, 1177 (1974).

80. G. A. Neville and M. Berlin, Environ. Res., 7, 75 (1974).

81. C. A. Oehme, Clin. Toxicol., 5, 151 (1972).

82. C. A. Owen, Mayo Clin. Perspect., 49, 368 (1974).

83. C. L. Peterson, W. L. Klawe, and G. D. Sharp, Fishery Bull., 71, 603 (1973).

84. S. M. Pier, Texas Rep. Biol. Med., 33, 85 (1975).

85. S. M. Pier and T. L. Valentine, Contract No. NAS9-11701, The University of Texas Health Science Center at Houston School of Public Health (1975).

86. E. E. Pochin, Brit. Med. Bull., 31, 184 (1975).

87. L. W. Reiter, G. E. Anderson, and T. W. Laskey, Environ. Health Perspect., 12, 119 (1975).

88. E. S. Reynolds, Lancet, 1, 167 (1970).

89. F. R. Robinson, Clin. Toxicol., 6, 497 (1973).

90. J. K. Roh, R. L. Bradley, T. Richardson, and K. G. Weckel, J. Dairy Sci., 58, 1782 (1975).

91. J. E. Sabadel and R. C. Axtman, Environ. Health Perspect., 12, 1 (1975).

92. G. S. Sayler, J. D. Nelson, and R. R. Colwell, Appl. Microbiol., 30, 91 (1975).

93. M. L. Schafer, U. Rhea, J. T. Peeler, C. H. Hamilton, and J. E. Campbell, J. Agr. Food Chem., 23, 1079 (1975).

94. H. A. Schroeder, The Poisons Around Us, Indiana University Press, Bloomington, 1974.

95. H. A. Schroeder, M. Mitchener, and A. P. Nason, J. Nutr., 100, 59 (1969).

96. L. A. Selby, A. A. Case, C. R. Dorn, and D. J. Waystaff, J. Amer. Vet. Med. Ass., 165, 1010 (1974).

97. J. L. Sell and K. L. Davison, J. Agr. Food Chem., 23, 803 (1975).

98. J. L. Sell, W. Guenter, and M. Sifri, J. Agr. Food Chem., 22, 248 (1974).

99. E. K. Silbergeld and A. M. Goldberg, Environ. Health Perspect., Suppl., 7, 227 (1974).

100. R. E. Simpson, W. Horwitz, and C. A. Roy, Pesticides Monitor, 7, 127 (1974).

101. F. A. Smith, R. P. Sharma, and R. I. Lynn, Bull. Environ. Contamin. and Toxicol., 12, 153 (1974).

102. B. R. Sonawane, M. Nordberg, G. Nordberg, and G. W. Lucier, Environ. Health Perspect., 12, 97 (1975).

103. G. S. Stoewsand, C. A. Bache, and D. J. Lisk, Bull. Environ. Contamin. and Toxicol., 11, 152 (1974).

104. W. Stopford and L. J. Goldwater, Environ. Health Perspect., 12, 115 (1975).

105. J. T. Tanner, M. H. Friedman, D. N. Lincoln, and L. A. Ford, Science, 177, 1102 (1972).

106. I. J. Tartakow, Amer. J. Public Health, 59, 1674 (1969).

107. F. M. Teeny, J. Agr. Food Chem., 23, 668 (1975).

108. F. M. Teeny, Marine Fisheries Rev., 36, 15 (1974).

109. R. H. Thompson and J. R. Todd, Res. Vet. Sci., 16, 97 (1974).

110. J. B. Tillery and D. E. Johnson, Environ. Health Perspect., 12, 19 (1975).

111. W. P. Tseng, H. M. Chu, S. W. How, J. M. Fong, C. S. Lin, and S. Yeh, J. Nat. Cancer Inst., 40, 453 (1968).

112. D. D. Ulmer, Fed. Proc., 32, 1758 (1971).

113. R. I. Vanhook and A. J. Yates, Environ. Res., 9, 76 (1975).

114. Chemist Dictionary, Van Nostrand, New York, 1973.

115. M. D. Waters, T. O. Vaughan, D. J. Abernethy, H. R. Garland, C. C. Cox, and D. L. Coffin, Environ. Health Perspect., 12, 45 (1975).

116. M. Webb, Brit. Med. Bull., 31, 246 (1973).

117. M. J. Weiester, Environ. Health Perspect., 12, 41 (1975).

118. S. Warburton, W. Udler, R. M. Ewert, and W. S. Haynes, Public Health Reports, 77, 798 (1942).

119. A. H. Wolff and F. W. Oehme, J. Amer. Vet. Med. Ass., 164, 623 (1974).

120. E. A. Woolson, J. Agr. Food Chem., 23, 677 (1975).

121. E. G. Zook, J. J. Powell, B. M. Hackley, J. A. Emerson, J. R. Brooker, and G. M. Knobl, J. Agr. Food Chem., 24, 47 (1976).

122. E. M. Bailey, Texas A. & M. University, personal communication.

3

TRACE ELEMENTS IN PLANT FOODSTUFFS

Hansford T. Shacklette, James A. Erdman, Thelma F. Harms,
and Clara S. E. Papp

U.S. Department of the Interior, Geological Surve
Denver, Colorado

INTRODUCTION

Food and feed plants constitute the principal ultimate source of trace elements for man and other animals [1]; lesser amounts are derived from water, earth materials, and the air. These elements in plant foodstuffs are commonly present as components of the plant tissue, having been absorbed through cell membranes from soil solutions or the atmosphere. They may also be present as particulates on the external or internal surfaces of plants. The surficial deposits on plant foodstuffs commonly are removed or reduced in concentration by washing or other methods of food preparation in which the surface layer of the food material is removed and discarded. The deposits on feed plants (those consumed by domestic animals) commonly are not removed by washing or other processes, except in milled feeds, and are therefore ingested along with the plant tissue.

The intake of trace elements by man in industrial societies is generally much less closely related to local geochemical environments than is the intake by domestic animals. Foodstuffs that are consumed by man originate in many different places, and commonly are subjected to a great variety of processing techniques including cooking, which may either increase or decrease the content of particular trace elements. Farm animals are much less dependent on imported feedstuffs, commonly consuming as the major part of their ration locally produced feeds that have undergone little, or no, processing prior to being consumed. Many examples could be given of nutritional deficiencies and toxicities in farm animals that relate directly to natural geochemical environments in particular localities, whereas similar

25

problems in human nutrition in industrial societies are much less frequently
reported; in fact, Underwood [2] suggested that the relationship between
iodine and goiter is the only association between trace element levels in
soils and human health that is completely convincing. Allaway [3], in
discussing the associations between human health and trace element composition of plants in certain areas in contrast to these associations with
domestic animals, stated, "Unquestionably, the fact that domestic animals
frequently derive all their diet from a specific environment, in contrast
to the multisource diets common to people, has been a major factor in the
success of soil-plant-livestock studies."

In evaluating the trace element composition of plant foodstuffs it is
important to differentiate the amounts of these elements that were contained
in the plants as they grew and the amounts that were added, incidentally or
intentionally, to the products in processing for consumption. If these
amounts are to be modified, the point in the production and preparation of
the foodstuffs at which trace elements are added or removed must be known.

Naturally occurring amounts of these elements may be modified by
changing the amounts or availability of these elements in the plant's environment and, possibly, by genetically altering the plant so that it absorbs more, or less, of a particular element. The relationships of trace
element concentration in plants and their supporting soils are generally
complex and indirect, as were admirably discussed by Hodgson [4]. The
available amounts, rather than only total amounts, determine the uptake of
an element by a plant from the soil, and available amounts change as the
chemical environment changes. Scientists in agricultural research have
extensively studied the factors of availability for elements that are necessary for optimum growth and development of the plant. Less research has
been applied to agricultural practices that would enhance the ability of
plants to serve the trace element nutritional requirements of humans and
farm animals, although Beeson [5] in 1957 emphasized the importance of the
nutritional quality of feedstuffs. He wrote, "The term crop quality means
both marketable quality and nutritional quality of a crop...Nature has not
always combined two aspects of crop quality in one package, and man has
seldom improved matters in his efforts to breed plants and manage soil so
as to produce crops that are both attractive and high yielding."

The control of trace element composition of food plants by genetic
modification has received less attention than has the genetic control of
organic compounds in these plants. Selection and breeding of crop plants

for greater sugar, starch, and oil production or for reduction of organic
toxins is an ancient practice; more recently, other organic constituents
have been considered in plant breeding programs to the neglect of the trace
elements, as for example, the high-lysine strains of corn. A task force of
the Food and Drug Administration [6] has recently proposed monitoring the
content of two elements, magnesium and calcium (along with protein, vitamins
A, B6, and C, thiamin, riboflavin, and niacin), in nine food crops. This
action may portend an awareness of the importance of the nutritive elements
in foodstuffs. Recommended standards for allowable toxic element concentrations in the diet have been available for several years [7].

A discussion of the introduction or removal of trace elements by processing and preparing foods and feeds for consumption is outside the scope
of this chapter. The concentration or dilution of certain elements in
foods by cooking, and in feeds by milling grains, is well documented in
the literature. However, before an assessment can be made of this important
aspect of the nutritional value of foods as they are consumed, data on the
trace element composition of the unprocessed food and feed plants must be
available.

The object of this chapter is to present data on the trace element
composition of plants and plant parts that are derived from reports in the
literature as well as from our studies and those of our colleagues in the
U.S. Geological Survey. Evaluations of the nutritive importance or toxicity of these elements to man and other animals will not be attempted, except
as statements of researchers considered competent to make these judgments.

OCCURRENCE AND DISTRIBUTION OF TRACE ELEMENTS
IN PLANTS AND PLANT PRODUCTS

The term trace element cannot be precisely defined; it originated in
chemical laboratories to designate a group of elements commonly found in
concentrations too low to be quantitatively evaluated by the analytical
methods then available; therefore they were reported as found in "trace"
amounts. In plant materials we sometimes include in this category the
elements that occur in concentrations of less than one percent by weight
in the plant ash. If this concept is followed, a large number of naturally
occurring trace elements may be found among the many kinds of plant tissues
that are analyzed. However, there is wide variation in element concentrations in plants, both among and within species and organs; therefore on
this basis an objective designation of certain elements as trace elements

is not always possible. These elements may be more easily defined by ex-
clusion, that is, trace elements include all naturally occurring elements
in organisms except the macronutrients carbon, hydrogen, oxygen, nitrogen,
phosphorus, potassium, calcium, magnesium, and sulfur which are essential
for all organisms.

The trace elements can be grouped into nutritional elements or micronutrients, toxic elements, and elements having no known nutritional or toxic
qualities at the concentrations ordinarily found in organisms.

The micronutrient elements that are generally accepted as essential
for the higher plants (practically speaking, the seed plants) are boron,
chlorine, copper, iron, manganese, molybdenum, and zinc. Other trace elements may be essential for certain other plants, or for all plants, but in
concentrations so low that their essentiality has not been established.
The micronutrients essential for mammals are chromium, cobalt, copper, iron,
iodine, molybdenum, manganese, selenium, tin, and zinc. Arsenic, barium,
fluorine, silicon, and strontium are beneficial to some animals under certain conditions but are not recognized as generally essential [1]. The
content of trace elements in food plants is important not only for promoting
proper growth and development of the plants but also for supplying the
additional elements cobalt, chromium, iodine, selenium, and tin that are
required by animals. This latter aspect of trace elements in plants has
not been of primary importance to nutritionists concerned with human and
animal health, probably because either no widespread deficiencies in these
elements have been noted or the fact that these deficiencies can easily be
therapeutically corrected, as with iodine for humans and cobalt for domestic
animals.

With all micronutrient elements the difference between their nutritive
function and their toxicity to both plants and animals is only a matter of
concentration of the elements in the organism. For most of these elements
this difference is so great that generally it is of no practical concern.
However, for some micronutrients the difference between amounts required
for nutritional sufficiency and for toxicity is very small, for example,
boron in plants and selenium in animals.

Another group of trace elements commonly detected in foodstuffs includes those that have no known metabolic function in either plants or
animals. Most of these elements (e.g., bismuth, lithium, titanium, and
zirconium) also have no recognized toxic properties at concentrations
usually found in natural organic materials. These elements are sometimes

referred to as "ballast" elements to express their inactivity in organisms.
Two elements, cadmium and mercury, have no known metabolic function and are
universally recognized as toxic elements that are undesirable in any concentration in foods.

The largest group of trace elements in foods of plant origin are those
generally occurring in such minute amounts that they are infrequently detected by commonly used analytical methods. They have no known metabolic
effects on either plants or animals in concentrations at which they occur,
and except for their known or potential use in biogeochemical prospecting
for minerals, their occurrence is at present of little practical concern.
Yet we must consider that some of these elements are radioactive as they
occur in nature, and some can have artificially produced radioactive iso-
topes that may be absorbed to the same extent as the forms that are commonly
found in plants.

A complete list of elements that have been determined to occur in
plants probably has not been compiled, but quantitative data on at least
71 of the 94 elements of the periodic table have been reported. A few atoms
of all naturally occurring stable elements seem likely to be present in
plant tissues. The occurrence of all these elements in whole human tissue
was reported by Hamilton and Minski [8], and if this report is correct, the
same elements may be assumed to be present also in tissues of domestic
animals. The problem of determining the presence of some of these elements
is approached by use of increasingly sensitive analytical methods. In
recent years these methods have been greatly improved in their ability to
detect and quantify extremely minute concentrations of elements to the extent that sample contamination has become a major problem in both accuracy
and precision of the methods.

Notes on the occurrence of the trace elements (defined as all elements
except the macronutrients) in plant tissue follow. This list includes food
and feed plants, as well as other species not cultivated for consumption by
domestic animals or man, because it is axiomatic that each plant species
enters the food chain of at least one kind of animal. Furthermore, there
is no reason to believe that commercial food and feed plants do not, as a
group, absorb the same elements as do other plants, although quantitative
data on the former may not be available. This list includes readily available data that may suggest the occurrence and concentration of many trace
elements in plant tissues in general. There is, of course, an important
species effect on element absorption; therefore extrapolation of these data

to other species must be done cautiously. Concentrations of some of the
more uncommon elements, usually detected only by neutron activation or mass
spectrographic methods, are represented by a single analysis; this fact
does not necessarily indicate that other analyses do not exist, but only
that they were not available to us at the time of writing this report. The
values for some elements in plants given by Brooks [9] are presumed to represent "typical" or "average" concentrations with particular reference to
biogeochemical prospecting and were derived largely from Bowen [10] and
Cannon [11]. These concentrations may be greatly different in the various
organs of a plant. The toxicity of elements to plants refers to element
occurrence in anomalous concentrations in the substrate. It should be
noted that tolerance to these elements varies greatly among plant species
and biotypes; therefore the evaluations of toxicity given, which are those
reported by Brooks [9] and McMurtrey and Robinson [12], are only general-
izations.

Ag: Range, if detected, 0.4-20 ppm in ash [13]; most values are in
the 1-5 ppm range, but usually reported in no more than 5% of the samples
analyzed [14]. Highest value found in food plants: tomato fruits, 7 ppm
in ash [15]. Brooks [9] gave 1 ppm as the concentration in plant ash;
toxicity to plants, severe.

Al: Range, 150 to >10,000 ppm in ash [14]. Usually determined in all
samples. Most food plants (edible parts) range from 200 to 1500 ppm in ash
(highest value found was in asparagus stems, 7000 ppm) [13]. The element
is toxic to plants at low concentrations in water culture [12].

As: In 193 samples of Rocky Mountain trees, As ranged from 1 to 50 ppm
in ash [16]. Less than 0.25 ppm in dry material was found in all but two of
200 samples of trees and shrubs of Missouri [17]. Warren et al. [18] reported 3-8200 ppm in ash of Douglas fir; other associated plants had lower
amounts, or As was not detected. Williams and Whetstone [19] reported the
following amounts (ppm in dry material): beet roots, 1; tomato fruits, 1;
rutabaga roots, 0.5; corn grains, 0.3; onion bulbs, 0.2; apple fruits, 0.1;
wheat grains, 0.2; other food plants were within the range given here, but
vegetation growing on As-contaminated soils contained <1 to 38 ppm in dry
material. Beeson [20] reported As content of some vegetables to range from
0.1 to 3.87 ppm in dry material of edible parts, with leafy vegetables being
in the upper range. Duke [21] reported 0.2 ppm in dry matter of taro.
Brooks [9] gave 4 ppm as the concentration in plant ash; toxicity to plants,
severe. Arsenic compounds have been found to prevent growth of legumes in
sandy soil [12].

Au: Jones [22] reported an average concentration in plant ash of 7 ppm, with a maximum of 36 ppm. Warren and Delavault [23] gave a high value of 1.02 ppm in ash of willow, or 0.023 ppm in dry material. In gold absorption experiments using a radioactive gold tracer, a maximum of 320 ppm in dry material of impatiens stems was reported [24]. Wormwood growing in mineralized zones in the Soviet Union was reported to contain up to 125 ppm Au in ash [25]. Brooks [9] gave 0.005 ppm as the concentration in plant ash; toxicity to plants, slight.

B: Range, <30-3000 ppm in ash of a variety of plant species from throughout the United States [14]. Trees and shrubs (range, 50-500 ppm in ash) generally contain 2 to 10 times as much B as do vegetables (range, <30 to 1000 ppm in ash), with the exception of asparagus [13]. Beeson [20] reported edible parts of fruits and vegetables to range from 66 to 515 ppm in dry material. Soybean seeds were stated to range from 100 to 300 ppm in ash, whereas corn grains from the same areas in Missouri ranged from 50 to 100 ppm [26]. Brooks gave 700 ppm as the concentration in plant ash. Toxicity to plants, slight [9]; however, McMurtrey and Robinson [12] wrote, "...very small quantities are necessary for normal growth of many, if not all, plants, and only slightly higher concentrations can cause injury. With a number of plants the range between these two levels in water culture is only a few parts per million."

Ba: Range, 2-70,000 ppm in ash of a variety of plant species from throughout the United States [14]. Trees and shrubs (range, 30 to >10,000 ppm in ash) generally contain much higher Ba concentrations than do the edible parts of vegetables (range, 5-1500 ppm in ash) [13]. Soybean seeds were reported to range from 30 to 1500 ppm in ash, whereas corn from the same areas in Missouri ranged from 5 to 100 ppm [26]. Robinson et al. [27] reported BaO in a variety of fruits and vegetables (dry matter, edible parts) to range from 3 ppm in apple to 80 ppm in lettuce, and noted the unusually high values in Brazil nuts (range, 700-3200 ppm). Brooks [9] gave 220 ppm as the concentration in plant ash; toxicity to plants, moderate. The quantities of Ba taken up by plants ordinarily are so low that there is little likelihood of animals being poisoned by eating the plant [12].

Be: Range, <2 to 100 ppm in ash of a variety of native species from throughout the United States, but found in measurable amounts in only eight of 1153 samples [14]. Seventy samples of Rocky Mountain trees ranged from 0.1 to 1 ppm Be in ash [16]. Hickory trees were reported to contain 2-7 ppm in ash [13]. Duke [21] reported 0.001 ppm in dry matter of yam and cassava. Brooks [9] gave 0.7 ppm as the concentration in plant ash; toxicity to plants, severe. Very low concentrations are toxic to citrus cuttings [12].

Bi: Range, 1-15 ppm in ash in 37 of 226 samples of Rocky Mountain trees [16]. Brooks [9] gave 1.2 ppm as the concentration in plant ash; toxicity to plants, moderate. We have no other reports of Bi in plants.

Br: Beeson [20] reported concentrations in dry matter of 35 fruits and vegetables (edible portion) to range generally from about 1 to 10 ppm. Exceptions to this range included asparagus (20.2 ppm), cantaloupe (94.5 ppm), potato tubers (14.3 ppm), and watermelon fruit (262 ppm). Br in some field crops (ppm, dry matter) was given as follows: alfalfa hay, 6.4; vetch, 2.1; barley grain, 5.5; corn grain, 1.7; millet grain, 3.8; oats grain, 3.1; and rye grain, 1.9. Duke [21] reported the following Br concentrations (ppm, dry matter): breadfruit, 1.0; coconut, 0.2; yam, 1.0; cassava, 0.3; plantain fruit, 0.3; banana, 0.8; and dry beans, 15.0. As sodium bromide, this element may accumulate to levels that are toxic to animals as the plants age [12].

Cd: Shacklette [28] reported concentrations in a wide variety of plants, which are summarized as follows (ppm, dry matter): vegetables, 0.1-1.2, with a high value of 50 ppm for leafy vegetables grown in polluted areas; grasses and alfalfa, 0.02-0.03; grains, 0.1 to a high of 2 in polluted areas; tree leaves, 0.1 to a high of 17 in polluted areas; mosses, 0.7 to a high of 340 in polluted areas. Schroeder and Balassa [29] found lettuce, turnips, and radishes grown on soil fertilized by superphosphate to contain 0.5-14.0 ppm (wet-weight basis), whereas Cd was not detected in these plants in control plots. Kropf and Geldmacher-von Mallinckrodt [30] reported Cd content of many kinds of foodstuffs; the edible parts of vegetables ranged from 0.004 to 282 ppm, wet-weight basis. Warren et al. [31] suggested "normal" Cd values in lettuce, potato, cabbage, and beet of 1 ppm in ash, whereas values indicative of pollution ranged from 3.9 in cabbage to 87 in lettuce. Brooks [9] gave 0.1 ppm as the concentration in plant ash; toxicity to plants, moderate.

Ce: Duke [21] gave the following ppm in dry material: breadfruit, 0.1; taro, 0.5; and cassava, 0.05. An extreme value of 300 ppm in ash of tomato fruits grown in Georgia was reported by Connor and Shacklette [13]. Found in only six of 912 samples of native plants from throughout the United States in amounts ranging from 300 to 700 ppm in ash [14].

Cl: Beeson [20] reported average Cl contents of many fruits and vegetables (percent in dry matter) as follows: apple, 0.04; apricot, 0.09;
and mandarin orange, 0.36; artichoke, 0.38; beans, snap, 0.09; beet roots,
0.03; broccoli, 0.85; cabbage, 0.42-0.85; carrot roots, 0.52; cauliflower,
0.48-0.57; celery, 1.92-2.28; Jerusalem artichoke, 0.23; kale, 0.99-1.17;
lettuce, 1.24-1.30; mangel-wurzel roots, 1.38; onion bulbs, 0.12; pea seed,
0.03; and potato tubers, 0.13-0.58. Cl in forage plants (alfalfa, alsike
clover, crimson clover, red clover, and white clover) ranged from 0.20 to
0.79%. Grains (barley, corn, and oats) averaged 0.17, 0.59, and 0.13, respectively. Duke [21] reported (ppm dry material) 18 in cassava and 300 in
avocado. Excessive amounts are harmful to plant growth [12].

Co: Range, <5 to 300 ppm in ash of a variety of native species from
throughout the United States; detected in 232 of 1119 samples [14]. Co in
16 of 226 samples of Rocky Mountain trees ranged from 5 to 20 ppm in ash
[16]. Beeson [20] reported Co (ppm, dry material) in vegetables to range
from 0.02 in carrot roots to 0.13 in onion bulbs; in fruits (apricots and
pears), 0.18 and 0.032, respectively; and in grains (buckwheat, corn, oats,
rice, and wheat), 0.36, 0.11, trace, 0.006, and 0.012, respectively. The
extremely wide range of Co concentrations among samples of a species is
noteworthy; 30 samples of snap bean fruits from Georgia ranged from <7 to
70 ppm in ash, and 30 samples of black gum stems ranged from <7 to 10,000
ppm [15]. Black gum and persimmon trees are notable Co accumulators, yet
some samples will have very low values [13]. Duke [9] gave the following
Co values (ppm, dry material): breadfruit, 0.2; coconut, 0.01; taro, 0.25;
yam, 0.04; cassava, 0.1; avocado, 0.05; dry beans, 0.1; and corn, 0.02.
Beeson [32] discussed Co deficiency in ruminants and gave the Co content
of some forage plants. Brooks [9] gave 9 ppm as the concentration in plant
ash; toxicity to plants, severe.

Cr: Range, <1 to 700 ppm in ash of a variety of native species from
throughout the United States; detected in 1096 of 1139 samples [14]. Cr in
vegetables from Georgia (ppm in ash) ranged from <7 to 30, and was found
at the 7 ppm level or higher in about 75% of the samples [15]. Black gum
appears to be a Cr accumulator, the values ranging from 2 to 100 ppm in
ash. Duke [21] reported the following Cr values (ppm in dry material):
breadfruit, 2.0; coconut, 0.15; taro, 8.0; yam, 0.2; cassava, 0.15; plantain
fruit, 0.5; banana, 0.5; rice grain, 0.6; avocado, 0.03; dry beans, 0.05;
and corn grains, 0.25. Brooks [9] gave 9 ppm as the concentration in plant

ash; toxicity to plants, severe. Cr salts, particularly chromates, in very
low concentrations, are toxic to plants [12].

    Cs: Duke [21] reported the following values (ppm) in dry edible parts:
breadfruit, 0.011; pineapple, 0.003; papaya, 0.35; coconut, 0.004; yam,
0.007; cassava, 0.007; plantain fruit, 0.005; avocado, 0.011; dry beans,
0.005; corn, 0.002; and taro, 0.2. We have no other reports of naturally
occurring Cs in plants; however, the ready uptake of ^{134}Cs isotope was
demonstrated in broomsedge [33] and tulip tree [34] in experimental studies.

    Cu: Range, 5-1500 ppm in ash of a variety of native species from
throughout the United States; detected in all 1153 samples [14]. Connor
and Shacklette [13] reported Cu (ppm in ash) to range from 10 to 300 in
most vegetable samples, with tomato fruits tending to be somewhat higher
(30-500 ppm). Cu in ash of **corn** grains was found to range from 30 to 220
ppm, whereas in soybeans the range was 100 to 300 ppm. Black cherry, black
gum, and sweet gum appear to be Cu accumulators, with upper values ranging
from 2000 to 7000 ppm in ash. Duke [21] reported the following Cu concentrations (ppm in dry matter): pineapple, 17; coconut, 4; cassava, 5-12;
rice, 9-27; and corn, 3-5. Brooks [9] gave 180 ppm as the concentration in
plant ash; toxicity to plants, severe.

    Dy: Found in eight samples of the Georgia study [15] as follows (ppm
in ash): hickory leaves, 50-300; dogwood stems, < 100 to 150. Duke [21]
reported 0.08 ppm in dry material of taro. Erämetsä and Yliruokanen [35]
found Dy in lichens to range from 2.5 to 18 ppm in ash, and in mosses, 1.3
to 26 ppm. The Dy in ash of a fern, Lastrea dryopteris, was reported to
vary with the season from 0.6 to 3.6 ppm in ash [36], and in three species
of horsetail Dy ranged from 2 to 9 ppm in ash [37]. Robinson et al. [38]
reported Dy to average 2.4 atomic % of total rare earths in hickory.

    Er: Found in eight samples in the Georgia study [15] as follows (ppm
in ash): hickory leaves, 150-300; dogwood stems, < 100 to 150. Duke [21]
reported 0.08 ppm in dry material of taro. Erämetsä and Yliruokanen [35]
reported Er in lichens to range from 0.62 to 9.3 ppm in ash, and in mosses
from 0.63 to 13 ppm. The Er in ash of three species of horsetail was found
to range from 2 to 7 ppm [37]. Robinson et al. [38] reported Er to average
2.4 atomic % of total rare earths in hickory.

    Eu: Duke [24] reported 0.08 ppm in dry material of taro. Erämetsä
and Yliruokanen [35] found Eu in lichens to range from 1.1 to 8.7 ppm in
ash, and in mosses, from 0.70 to 5.0 ppm. The Eu in three species of

horsetail was reported to range from 1 to 2 ppm in ash [37]. Robinson et
al. [38] reported Eu to average 0.8 atomic % of total rare earths in hickory.

F: Garber [39] gave the F content of a wide variety of plants and
plant parts, both in normal and in polluted areas. Food-plant data from
normal areas (ppm in dry material) follows: alfalfa, 2-4; red clover, 3-9;
oat grains, 0.5; wheat grains, 1; rye grains, 1.53; barley grains, 1.74;
corn grains, 0.15-0.38; rice grains, 0.76; spinach, 1.3-28.3; lettuce, 4.4-
11.3; cabbage, 1.5; cauliflower, 0.9; parsley, 9; parsnip root, 0.4-8.4;
celery, 5.7; radish, 1; carrot root, 2; beet root, 4-7; potato tubers, 1.5-
3; sugar beets, 2.3-6; onion bulbs, 3.04; apple, 1.3-5.7; pear, 2.1-4.4;
and peach, 0.21. Duke [21] reported F concentrations (ppm in dry material)
as follows: coconut, 0.02; taro, 2.0; yam, 0.15; and cassava, 0.1. The F
concentration (ppm in ash) of a variety of native trees in Missouri ranged
from < 0.5 to 3 [26].

Ga: Range, < 3 to 30 ppm in ash of a variety of native species from
throughout the United States; reported in 139 of 912 samples [14]. Connor
and Shacklette [13] reported Ga in tree and shrub samples to commonly range
from < 5 to 10 ppm in ash, with black gum having extreme values up to 200
ppm. We have no reports of Ga in food plants. Brooks [9] gave 1 ppm as
the concentration in plant ash; toxicity to plants, slight.

Gd: Found in eight samples in the Georgia study [58] as follows (ppm
in ash): hickory leaves, 150-300, and dogwood stems, < 100 to 150. Duke
[21] reported 0.08 ppm in dry material of taro; we have no other reports of
Gd in food plants. Robinson et al. [38] reported Gd to average 3.1 atomic %
of total rare earths in hickory.

Ge: Shacklette and Connor [40] reported 15 ppm in ash of one Spanish
moss sample of 124 samples that were analyzed. Paul [41] reported 5% Ge
in ash of one sample of lettuce; most values in samples of lettuce, sugar
beet, corn, and deciduous and coniferous trees ranged from 2% to not detected, the values commonly being in the hundredths of a percent range.
Five ppm were found in dry material of alpine fir needles [16]. Brooks
[9] gave 5 ppm as the concentration in plant ash; toxicity to plants,
slight.

Hf: Duke [21] reported 0.08 ppm in dry material of taro. We have no
other records of Hf in plants.

Hg: Shacklette [42] reported Hg (ppm in dry material) in trees and
shrubs that grew over a cinnabar deposit in Alaska to range from 0.5 to 3.5.
Connor and Shacklette [13] reported maximum Hg values (ppm in dry material)

in trees and shrubs of Missouri as follows: buckbush, 0.02; cedar, 0.025; post oak, 0.025; and shortleaf pine, 0.025; however, the Hg content was above the 0.025 detection limit in only nine of the 500 samples analyzed. Tanner et al. [43] analyzed 10 common foods for Hg; among these foods, potatoes were reported to range from <1 to 15 ppb in fresh ("wet") material, with a median value of 3 ppb. They concluded, "With the exception of certain fish, the major foods in the United States are essentially free of mercury." Brooks [9] gave 0.01 ppm as the concentration in plant ash; toxicity to plants, severe.

Ho: Found in one sample in the Georgia study [13]: in hickory leaves, with 150 ppm in ash. Duke [21] reported 0.08 ppm in dry material of taro. Erämetsä and Yliruokanen [35] found Ho in lichens to range from 0.20 to 3.5 ppm in ash. The Ho in three species of horsetail was reported to range from 1 to 2 ppm in ash [37].

I: Of 362 samples of a variety of native plants from Missouri, I ranged from 2 to 10 ppm in dry material [26]; a similar range occurred in 38 samples of corn from Missouri [26]. Soybeans from Missouri, however, contained over twice the amount in corn, ranging from 10 to 16 ppm in dry material. Shacklette and Cuthbert [44] gave average I amounts (ppm in dry material) as follows: deciduous tree stems, 2.7; herbaceous plants, 6.9; and marine brown algae, 2488. They gave the following I values (ppm in dry material) in food and feed plants: onion bulbs, 7.8-10.4; asparagus stems, 5.6-5.9; cabbage, 9.0-10; carrot roots, 6.0-8.0; snap beans, 5.7-9.5; potato tubers, 2.8-4.9; bluegrass stems and leaves, 4.3-7.1; and corn grains, 3.4-3.9.

La: Range, <50 to 1500 ppm in ash of a variety of species from throughout the United States; reported in 46 of 1123 samples [14]. Connor and Shacklette [13] showed the La content of woody plants (ppm in ash) to generally range from <30 to 300, except for hickory which had maximum concentrations of 2000. Duke [21] reported 0.2 ppm in dry material of taro. We have no other records of La in food plants.

Li: In samples of a variety of native plants from Missouri and Kentucky Li content (ppm in ash) ranged from <4 to 130 [13]. Borovik-Romanova [45] reported the Li concentration in many plants from the Soviet Union to range from 0.15 to 55 ppm in dry material; she reported Li in food plants as follows (ppm in dry material): tomato, 0.4; rye, 0.17; oats, 0.55; wheat, 0.85; and rice, 9.8. Duke [21] reported the following Li concentrations (ppm in dry material): breadfruit, 2; coconut, 0.4; taro, 0.25; yam, 0.4; cassava, 0.6; plantain fruits, 0.5; banana, 0.01; rice, 0.3; avocado, 0.01;

dry beans, 0.5; and corn grains, 0.05. Brooks [9] gave 2 ppm as the concentration in plant ash; toxicity to plants, slight. Excessive soluble
salts of Li in soil have caused injury to tobacco [12]. A recent paper
[74] summarizes information on Li concentration in plants and reports the
stimulating effect of Li as a fertilizer for certain species, especially
those in the Solanaceae family.

Lu: Duke [21] reported 0.08 ppm in dry material of taro. Erämetsä
and Yliruokanen [35] found Lu in lichens to range from 0.05 to 1.0 ppm in
ash; in mosses the range was 0.12-2.2 ppm in ash. We have no other reports
of Lu in plants.

Mn: Range, 30-70,000 ppm in ash of a variety of native species from
throughout the United States; reported in 881 of 912 samples [14]. The Mn
concentrations in stems of trees commonly is an order of magnitude greater
than that of herbaceous plants: tree stems range from about 150 to 50,000
ppm in ash and of herbs from about 15 to 2000 ppm [13]. Vegetables ordinarily contain about 300-400 ppm in ash and no strong species effects on
this concentration were noted [15]. Duke [21] reported the following Mn
values (ppm, dry material): breadfruit, 19; yam, 11-30; cassava, 45; plantain fruits, 10-18; banana, 14-36; rice, 17-35; avocado, 8-17; and corn
grains, 80. Brooks [9] gave 4800 ppm as the concentration in plant ash;
toxicity moderate. The Mn in organic soils, if the soils are submerged
for short periods, becomes soluble in concentrations greatly exceeding the
Mn tolerance of plants [12].

Mo: Range, <2 to 500 ppm in ash of a variety of native species from
throughout the United States; reported in 335 of 912 samples [14]. Warren
and Delavault [46] reported concentrations (ppm in ash) as high as 2900 in
alpine fir needles and 5400 in Sitka alder leaves from trees growing at
mine sites. Mo in vegetables commonly ranges from <5 to 70 ppm in ash
[13]. Duke [21] reported the following Mo concentrations (ppm in dry material): coconut, 0.06; taro, 0.5; yam, 0.25; cassava, 0.2; rice, 0.4;
avocado, 0.1; dry beans, 0.3; and corn grains, 0.1. The availability of
Mo to plants is strongly controlled by soil pH, being favored by alkaline
conditions. Many studies have been made of the Mo content of pasture plants
related to molybdenosis in livestock; Dye and O'Harra [47] wrote, "With
respect to forage plants growing in these areas, one finds that legumes
usually collect more Mo than grasses; of the legumes, clover contains
higher concentrations of Mo than alfalfa; strawberry, Sierra, and birdsfoot
trefoil (Lotus sp.) clovers are particularly bad in this respect. From 5
to 30 ppm Mo [in dry material] are common concentrations in toxic areas.

The record concentration analyzed by the Experiment Station Laboratory at Nevada was 372 ppm Mo [in dry material] in a sample of Black Medic (Medicago lupulina) from Carson Valley." Brooks [9] gave 13 ppm as the concentration in plant ash; toxicity to plants, moderate. McMurtrey and Robinson [12] wrote, "Mo has been found toxic to some higher plants when present in any considerable concentration."

Na: Range, 400-360,000 ppm in ash of a variety of native species from throughout the United States [14]. Desert shrubs may be high Na accumulators; Rickard [48] reported 40,000-200,000 ppm in dry material of greasewood. Beeson [20] gave the mean Na content (ppm in dry material) of some food and feed plants as follows: potato tubers, 400-1900; prunes, 1500; rice, 600-1000; rutabaga roots, 3500, soybean seed, 2400; spinach, 8400-19,900; sweet potato, 600; tomato fruit, 2200-2600; turnip root, 15-100; rye grain, 400; and white sweet clover, 700. The ranges of Na in ash of corn grains and soybean seeds from Missouri were <25 to 70 ppm [13]. Practically all agricultural crops are deficient in Na for the nutrition of herbivorous animals [12].

Nb: Of 1155 samples of a variety of native species from throughout the United States, only one was reported to contain Nb (30 ppm in ash) [14]. Six of 38 Wisconsin moss samples contained Nb in amounts ranging from 15 to 20 ppm in ash [49]. Brooks [9] gave 0.3 as the concentration in plant ash; toxicity to plants, not known but probably slight. We have no other reports of Nb in plants.

Nd: Range, 150-1500 ppm in ash of a variety of native species from throughout the United States; reported in 11 of 1155 samples [14]. Reported in 23 of 1564 Georgia plant samples, ranging from 30 to 700 ppm in ash [15]. Found in one out of 60 tomato fruit samples, 150 ppm in ash [13]. Reported by Duke [21] in taro at a concentration (ppm) of 0.08 in dry material. Erämetsä and Yliruokanen [35] found Nd in lichens to range from 12 to 150 ppm in ash; in mosses the range was 7.8 to 100 ppm. Nd in the fern L. dryopteris was reported to have a seasonal range of 4-35 ppm in ash [36]; in three species of horsetail Nd ranged from 3 to 50 ppm in dry material [37]; and in three species of lycopodium, 5 to 40 ppm in ash [50].

Ni: Range, <5 to 500 ppm in ash of a variety of plants from throughout the United States; reported in 822 of 912 samples [14]. Ni occurs in most samples of food plants in concentrations greater than 5 ppm in ash, and rarely exceeds 100 ppm [13]; an exception is provided by soybean seeds, samples of which usually range from 30 to 500 ppm in ash, which contrasts

with corn grains from the same area in which Ni ranged from <5 to 70 ppm
in ash [26]. Beeson [20] reported Ni concentrations (ppm in dry material)
in 21 kinds of food plants including apricots, 0.64; buckwheat grain, 1.34;
cabbage, 3.3; carrot roots, 0.29; corn grain, 0.14; figs, 1.20; oat grains,
0.45; onion bulbs, 0.16; pear fruits, 1.30; peas, 2.25; potato tubers, 0.25;
spinach, 2.37; tomato fruit, 0.15; and wheat, 0.35. Duke [21] gave Ni concentrations (ppm in dry material) as follows: breadfruit, 0.5; coconut,
1.0; taro, 8.0; yam, 1.0; cassava, 1.5; plantain fruits, 0.25; banana, 0.25;
rice, 3.0; avocado, 10.0; dry beans, 1.0; and corn, 1.0. Brooks [9] gave
65 ppm as the concentration in plant ash; toxicity to plants, severe. In
all but the most minute concentrations Ni is toxic to plants [12].

Pb: Range, <7 to 7000 ppm in ash of a variety of plants from throughout the United States; reported in 761 of 912 samples [14]. In contaminated
areas, the Pb content may be much greater; cedar trees in Missouri on a
roadside contaminated with lead ore dust contained as much as 20,000 ppm
in ash [51], and Spanish moss near heavily traveled roads contained as much
as 50,000 ppm in ash [40]. Vegetables growing in a normal environment seldom exceed 70 ppm in ash, but even in these plants Pb is usually less than
10 ppm in ash [13]. In contrast, leaves and stems of woody plants nearly
always have greater than 10 ppm in ash, typical values ranging from about
100 to 400 ppm, with extreme values of about 2000 ppm when growing in a
normal environment [13]. Many studies of lead-contaminated vegetation have
been made since the report of Cannon and Bowles [52] on Pb in grasses along
highways was published. A survey of the possible physiological significance
of Pb in the environment was made by Warren [53] who concluded, "Because
food supplies come from a wide variety of sources it seems reasonable to
assume that under normal conditions humans must anticipate ingesting some
400 µg of lead daily of which some 20-40 µg will be absorbed. However,
rural persons and individuals deriving a significant portion of their food,
and more particularly their vegetables, from an area anomalously high in
lead may ingest and absorb from three to five times this amount." Duke's
study [21] of food sources of the Chocó Indians in Central America provides
data on elements in food plants that grew in an environment probably free
of industrial and vehicular pollution. He reported the following lead concentrations (ppm in dry material): breadfruit, 0.1; coconut, 0.015; taro,
0.25; yam, 0.03; cassava, 0.1; plantain fruits, 0.1; banana, 0.02; rice,
0.02; avocado, 1.0; dry beans, 0.02; and corn, <0.02. Brooks [9] gave
70 ppm as the concentration in plant ash; toxicity to plants, severe.

Soluble Pb compounds, more likely to occur in acid soils, are toxic to
plants except at very low concentrations. Lead from spray applications in
orchard soils has prevented normal plant growth [12].

Pm: Duke [21] reported 0.08 ppm in dry material of taro; however,
because Pm does not occur naturally in the earth's crust, this report may
be in error. We have no other reports of Pm in plants.

Po: The naturally occurring radioelement ^{210}Po was reported in tobacco
leaves by Tso et al. [54]. They stated, "Polonium-210 in tobacco plants is
derived from **either** the soil or the air. It may be taken up directly from
the soil or may result from radioactive decay of lead-210 or radium-226
taken up from the soil. It may also result from radioactive decay of the
daughters of radon-222 deposited on the leaves." They reported the range
of this element to be 0.15-0.48 pc/g in tobacco leaves.

Pr: Found in one sample of hickory leaves from Georgia, 700 ppm in
ash [15]. Range in values in ash of lichens, 3.8-31 ppm; of mosses, 1.2-25
ppm [35]; of horsetails, 10.7-30.2 ppm [37]; and of lycopodium, <1 to 16
ppm [50]. Pr averaged 1.8 atomic % of total rare earths in hickory [38].

Pt: Reported in a mixed sample of mountain avens and Labrador tea
that grew on platinum-bearing ultrabasic rock at a concentration of 6.6 ppm
(presumably) in ash; samples of the same species growing on metamorphic
rock contained 3.5 ppm [55]. We have no other reports of Pt in plants.

Ra: The naturally occurring radioelement ^{226}Ra was reported in tobacco
leaves in concentrations ranging from 0.059 to 0.39 pc/g [54]. Brooks [9]
gave 2×10^{-8} ppm as the concentration in plant ash; toxicity severe due to
radioactivity.

Rb: Ranges in concentration (ppm in ash) were reported as follows [9]:
hickory stems, 18-192; black oak stems, 37-210; red oak stems, 37-260; and
white oak stems, 37-400. Robinson et al. [56] reported 2.7 ppm in dry material of wheat plants. Duke [21] gave Rb concentrations (ppm in dry material) as follows: breadfruit, 2-10; coconut, 3.0; taro, 50.0; yam, 8.0;
cassava, 10.0; plantain fruit, 3.0; banana, 1.0; avocado, 20.0; dry beans,
3.0; and corn grains, 3.0. Brooks [9] gave 2 ppm as the concentration in
plant ash.

Re: Reported for the first time in plants by Myers and Hamilton [57];
the following values (ppm in ash) were found: locoweed, 70-300; desert
trumpet, 70-150; gumweed, 150; poisonvetch, 150; Mormon tea, 150; shadscale
saltbush, 300; and evening primrose, 150. Found in American elm tree leaves
from a tree growing in an ore mill deposit in Wisconsin, 150 ppm in ash [58].

Brooks [9] gave 0.005 ppm as the concentration in plant ash; toxicity to plants, slight.

Ru: Duke [21] reported the following concentrations in food plants (ppm in dry material): breadfruit, 0.0004; papaya, 0.002; yam, 0.004; cassava, 0.0002; plantain fruit, 0.0001; and avocado, 0.0004. We have no other reports of Ru in plants.

Sb: Shacklette [58] found 27 samples of trees and shrubs growing over cinnabar deposits in Alaska to contain from 7 to 50 ppm Sb in the dry material. Samples of woody plants from the Rocky Mountains contained the following concentrations (ppm in ash): alpine fir stems, 30-50; Engelmann's spruce stems, 20-30; and myrtle blueberry stems, 50-100 [16]. Twinpod collected by Chaffee [16] that grew at a mineralized site contained 100 ppm in ash. Brooks [9] gave 1 ppm as the concentration in plant ash; toxicity to plants, moderate.

Sc: Range, < 10 to 30 ppm in ash of a variety of species from throughout the United States; reported in 39 of 1153 samples [14]. Reported in ash of black gum leaves, sassafras stems and leaves, and winged sumac, in single samples of each, at 10 ppm [13]. Duke [21] gave the following Sc concentrations (ppm in dry material): breadfruit, 0.1; coconut, 0.005; yam, 0.07; cassava, 0.06; plantain fruit, 0.05; rice, 0.1; avocado, 0.1; dry beans, 0.05; and corn grains, 0.1.

Se: Many reports of Se in plants have been published because of its toxicity to animals and its use in biogeochemical prospecting. Some native species, especially of the genus Astragalus, concentrate Se to extremely high levels [11]. Data on Se in food and feed plants are given in tables of this report. Selenium in crops in relation to Se-responsive livestock diseases was reviewed by Kubota et al. [59]. Se in the environment was discussed by Lakin [60], and the toxic effects on animals were presented by Underwood [61]. Selenium concentrations (ppm in dry material) in native plants usually ranged from < 0.01 to 0.08 in samples of some native U.S. plants [13]. Brooks [9] gave 1 ppm as the concentration in plant ash; toxicity to plants, moderate. Lakin [60] stated, "Selenium is required in the diet of animals at a minimum level of 0.04 ppm and is beneficial to 0.1 ppm; at levels above 4 ppm it becomes toxic to animals. In fact, farmers have suffered greater losses because of selenium deficiencies than selenium toxicities."

Si: Scouring rush (horsetail) plants contain as much as 33% Si in ash (as SiO_2) and grasses commonly have about 24% [62]. Lovering [63] gave

the silica contents (here converted to silicon) in percent of dry weight
for several species as follows: giant reed, 4.45; common gromwell, 3.8;
water chestnut, 2.6; giant horsetail, 2.53; English ryegrass, 1.64; and a
palm (Calamus rotang), 1.01. Silicon in ash of beech leaves was reported
to have a seasonal range of 0.56-26.7%. The Si content of hickory and oak
stem ash ranged from <0.34 to 3.8% [13]. McMurtrey and Robinson [12]
stated, "While silica is not poisonous, cases have been reported where
cattle have died as a result of lacerations of the walls of the digestive
tract from the sharp siliceous spikes of rice hulls." Brooks [9] gave
1500 ppm as the average Si content of plant ash.

Sm: Reported in five samples in the Georgia study [13] as follows
(ppm in ash): hickory leaves, 200-700; and dogwood stems, 200. Range of
values in ash of lichens (ppm), 2.6-40 [35]; of mosses, 1.8-23 [35]; of
horsetail, 102 [37]; and of the fern L. dryopteris, 1.1-5.1 [36]. Samarium
was reported [38] to average 1.4 atomic % of total rare earths in hickory.
Duke [21] reported 0.08 ppm Sm in dry material of taro. We have no other
record of Sm in food plants.

Sn: Connor and Shacklette [13] reported Sn in only one sample each
of three kinds of vegetables as follows (ppm in ash): carrot, 20; corn
grains, 30; and beet root, 20. Tin was but rarely detected in samples of
buckbush from Missouri and ranged from 20 to 30 ppm in ash [17]. Duke [21]
reported 0.10 ppm Sn in dry material of avocado. Brooks [9] gave 1 ppm as
the concentration in plant ash; toxicity to plants, severe.

Sr: Range, <10 to 15,000 ppm in ash of a variety of plants from
throughout the United States; detected in 880 of 912 samples [14]. Strontium is commonly reported in all vegetables, but concentrations usually
vary widely both among and within species. Some selected values (ppm in
ash) follow [13]: asparagus stems, 300-2000; lima bean seeds, 30-1000;
snap bean fruits, 50-700; beet root, 300-500; cabbage leaves, 100-3000;
corn grains, <7 to 150; potato, 30-700; and tomato fruits, <7 to 700.
Duke [21] reported the following concentrations (ppm in dry material):
breadfruit, 7.0; coconut, 0.15; taro, 40.0; yam, 2.0; cassava, 3.0; plantain fruits, 1.0; banana, 1.5; rice, 0.2; avocado, 1.0; dry beans, 1.5;
and corn, 0.06. Brooks [9] gave 30 ppm as the concentration in plant ash;
toxicity to plants, slight.

Tb: Duke [21] reported 0.08 ppm in dry material of taro. Range of
values in ash (ppm) of lichens, 0.90-3.3 [35]; of mosses, 0.40-3.3 [35];

and of horsetail, 1-2 [37]. Average in hickory, 0.6 atomic % of total rare earths [38].

Ti: Range, < 10 to 15,000 ppm in ash of a variety of plants throughout the United States; reported in 880 of 912 samples [14]. Titanium is commonly detected in vegetable samples, but concentrations usually vary widely both among and within species. Some selected values (ppm in ash) follow [13]: asparagus stems, 150-200; lima bean seeds, < 2 to 200; snap bean fruits, 3-700; beet root, 10-70; cabbage leaves, 15-1500; corn grains, < 2 to 700; potato, 7-200; and tomato fruits, < 2 to 700. Duke [21] reported the following concentrations (ppm in dry material): breadfruit, 6.0; coconut, 0.3; taro, 80.0; yam, 15.0; cassava, 6.0; plantain fruits, 0.25; banana, 0.2; rice, 2.0; avocado, 1.0; dry beans, 2.0; and corn, 2.0.

Tl: Found by Curtin and King, as analyzed by Mosier [16], in trees and shrubs from the Rocky Mountain region as follows (ppm in ash): alpine fir, needles 2-100, stems 2-70; limber pine, needles 2-5, stems 3-5; lodgepole pine, needles 2-5, stems 3-7; Engelmann's spruce, needles 2-10, stems 15; myrtle blueberry stems and leaves, 2-7; phyllodoce stems, 2; and ponderosa pine stems, 15. Zýka [64], in studying vegetation of a region in Yugoslavia remarkable for high levels of Tl, found Tl in ash of herbaceous plants to range from 10 ppm in eryngium to 17,000 ppm in bedstraw; toadflax contents ranged from 3000 to 3800 ppm. He wrote, "The high thallium contents in the plants explain their toxic effects on cattle." He also noted zonation of species related to soil concentrations of Tl, which indicated the degree of toxicity of the element. McMurtrey and Robinson [12] stated that Tl is poisonous to both plants and animals, that as little as 35 ppm in sandy soil has practically prevented the growth of plants.

Tm: Duke [21] reported 0.08 ppm Tm in dry material of taro. Erämetsä and Yliruokanen [35] found 0.078-1.3 ppm in ash of lichens and 0.15-2.2 ppm in ash of mosses. Three species of horsetail contained 1 ppm Tm in ash [37]. Thulium was reported [38] to average 0.2 atomic % of total rare earths in hickory.

U: Cannon [65] reported average ppm U in ash of trees on mineralized ground as follows: juniper, 1.74; pinyon, 1.31; fir, 2.18; and ponderosa pine, 1.28. On "barren" ground the same species contained 0.33, 0.56, 0.35, and 0.63 ppm, respectively. Whitehead and Brooks [66] gave the range in U concentrations (ppm in ash) in aquatic bryophytes (mosses and liverworts) as 0.7-86.0. Malyuga [67] reported 16 ppm U in ash of the moss Scorpidium

scorpioides, and 40 ppm in ash of Norway spruce. Brooks [9] gave 0.6 ppm
as the concentration in plant ash; toxicity to plants, moderate. We have
no reports of U in food plants.

V: Range, <7 to 300 ppm in ash of a variety of plants from throughout
the United States; reported in 526 of 912 samples [14]. Vanadium is infrequently found in ash of food plants; it ranges in concentration from <5 to
50 ppm in ash of most vegetables, except snap beans which may have values of
as much as 700 ppm in ash [13]. This element also was not commonly reported
in ash of woody plants, except for buckbush (5-50 ppm in ash) and shortleaf
pine (<5 to 30 ppm in ash). Vanadium in some food plants (ppb, wet weight)
was reported [68] as follows: lettuce, 1080; radish, 3020; apple, 330; tomato, 0.027; potato, 1490; pear, 50; carrot, 990; beet, 880; and pea, 460.
Duke [21] reported the following values (ppm in dry material): breadfruit,
0.1; coconut, 0.004; taro, 0.4; yam, 0.4; cassava, 0.3; plantain fruits,
0.05; banana, 0.01; rice, 0.2; avocado, 0.1; dry beans, 0.5; and corn, 0.07.
Brooks [9] gave 22 ppm as the concentration in plant ash; toxicity to plants,
moderate.

W: Found by Curtin and King, as analyzed by Mosier and Nishi [16],
(ppm in dry material) in samples of Rocky Mountain trees and a shrub as
follows: alpine fir stems, 5-50; common juniper, needles 50, and stems
100; and Douglas fir stems, 50. Brooks [9] gave 0.5 as the concentration
in plant ash; toxicity to plants, moderate. We have no reports of W in
food plants.

Y: Range, <20 to 700 ppm in ash of a variety of plants from throughout the United States; found in 158 of 1128 samples [14]. Infrequently
detected in samples of vegetables; the following values (ppm in ash) were
reported [13]: snap bean, 30; blackeyed pea, 20; cabbage, 30-100; and tomato, 30. Duke [21] reported 0.4 ppm in dry material of taro. The following values were given (ppm in ash): lichens, 11-82 [35]; mosses, 2.1-200
[35]; and lycopodium, 1-20. Concentrated in ash of hickory to as much as
500 ppm in ash [13].

Yb: Range, <1 to 70 ppm in ash of a variety of plants from throughout the United States; found in 101 of 1108 samples [14]. Reported in ash
of four samples of cabbage (5-30 ppm) and one sample of corn grains (500
ppm) from Georgia [13]. Duke [21] reported 0.08 ppm in dry material of
taro. Concentrated in ash of winged sumac and sweetgum to as much as 300
ppm [13].

Zn: Range, <25 to 5800 ppm in ash of a variety of plants from throughout the United States; detected in all but one of 699 samples [14]. Examples of concentration (ppm in ash) of vegetables follow [13]: asparagus,
200-400; lima bean, 200-1000; snap bean, 200-1000; beet root, 300-500;
blackeyed pea, 400-1200; cabbage, 100-6000; carrot, 100-300; corn, 500-2800;
sweet pepper, 200-400; soybean, 700-1500; and tomato, 100-600. Duke [21]
reported the following concentrations (ppm in dry material): pineapple, 20;
coconut, 16; cassava, 16-27; rice, 33-38; and corn, 31-36. Brooks [9] gave
1400 as the concentration in plant ash; toxicity to plants, moderate.

Zr: Range, <20 to 500 ppm in ash of a variety of plants from throughout the United States; detected in 470 of 1152 samples [14]. Infrequently
detected in food plants; examples of the few concentrations (ppm in ash)
reported in vegetables follow [13]: lima bean, 50-70; snap bean, 30-150;
blackeyed pea, 20; cabbage, 500-700; corn grains, 20; and tomato, 20. Much
more commonly detected in ash of woody plants, especially in buckbush which
usually contains 20-500 ppm in ash [13]. Duke [21] reported the following
concentrations (ppm in dry material): pineapple, 0.005; breadfruit, 0.055;
papaya, 0.015; coconut, 0.15; taro, 0.10; yam, 0.035-0.065; cassava, 0.015-
0.020; banana, 0.015-0.2; rice, 0.075; avocado, 0.005-0.010; and corn grains,
0.20.

CONCENTRATION OF SELECTED TRACE ELEMENTS
IN FOOD AND FEED PLANTS

The preceding list of trace elements reported in certain plants gives
only suggestions of the concentrations that may be expected. Several
methods of expressing these values were used, according to the data that
were available, such as single analyses, ranges in concentration, or averages. The reliability of most of these values as given remains unknown;
however, for some citations more information was provided in the original
publication which does provide a basis for estimating levels of confidence
in the data. The purpose of this section is to present some original data
on several trace elements in important food and feed plants to which described sampling, analytical, and statistical methods were applied in order
to enhance the reliability of the values that are given.

Reliability of Trace Element Data

The usefulness of trace element analyses is determined by their reliability. If sufficiently reliable, the analyses may be used to establish

bases for judging the relevance of trace element concentrations in food and
feed plants to nutritional or toxicity problems. Reliability of data seldom,
or never, reaches the absolute level; there are various levels each of which
may have valid application to specific questions or may be generally useful
in the absence of data that are more adequately supported. A discussion
follows of the factors contributing to the reliability of trace element
data which should be known, but which are often omitted in published reports.

Accurate identification of the material that is analyzed is the primary requirement for deriving reliable data. Identification generally
presents no problems with most food and feed plants because they are well
known by their common names in the language that is used. However, there
may be uncertainties in the use of common names, for colloquial names may
differ or improper synonyms may be used. For example, the names "yam" and
"sweet potato" in some American usage are considered synonyms, or of varietal difference only, whereas more properly yam is applied to a genus of
plants (Dioscorea) largely grown in tropical countries, while sweet potato
(genus Ipomea) is an entirely different plant in another family. Orchardgrass of American usage is known as cocksfoot in England; many such examples
could be cited. For noncultivated plants common names are generally a
source of much more confusion. Therefore, if the possibility of confusion
exists, the scientific name should also be given, especially for the reader
who may not be fluent in the language that is used. In this report, the
common names as used in the data source (or their English translation) and
their equivalent scientific names are given in the Appendix.

The designation of the plant part that is analyzed may be as important
as giving the correct name of the plant, because commonly the trace element
concentration varies among plant organs such as root, stem, leaf, fruit,
and seed. Furthermore, procedures of preparation before analysis must be
described because different tissues of a plant organ may have different
trace element concentrations or may be subject to different degrees of
contamination. For example, paring, washing, cooking, grinding, and drying
may have important effects on analytical values.

The reliability of data is enhanced by use of proper procedures for
obtaining the sample suite that is to be analyzed. The basic aim in sampling is to obtain a set of samples that faithfully represent the compositional variation present in the entire quantity of interest. Each individual sample, of course, must be adequate for the analytical method used.

However, the extent to which a sample represents the whole cannot be subjectively determined; one can only attempt to eliminate or reduce the
effects of bias in sample selection by dividing the whole into a convenient
number of equally sized units, then randomly selecting the units from which
the samples are collected. By this objective method each part of the whole
has an equal chance of being selected for sampling.

        For comparing analyses of a foodstuff with those of other workers or
with an established toxic or nutritional level, it is important to know
which analytical method was used. For the various elements these methods
may differ in their accuracy and precision, as well as in their upper and
lower limits of determination as commonly calibrated. A useful, but seldom
reported, statistic is a term expressing the analytical error of a method.
The "best" estimate of error is that obtained by duplicate analyses using
a particular subset of samples unknown to the analyst.

        Several facts are essential in giving trace element analyses of food-
stuffs, some of which are often omitted or are only implied in reports.
The units used to express ratios of concentration of an element, such as
percent, parts per million, or micrograms per milligram, and the basis of
the ratios (whether fresh or "wet" material, dry material, or ash) must be
stated if the results are to be useful. The results may be expressed as
single analyses, as an average or median value, or as a range in values.
If average (mean) values are given, the number of samples included in the
average, whether the average is arithmetic or geometric, and some measure
of the scatter of values (variation) and the observed range are necessary
for optimum utilization of the data. In trace element analyses the limits
of determination may commonly be exceeded for certain elements, resulting
in nonnumeric expressions such as "trace," or "less than" a stated limit.
The handling of these censored data must be explained if summary statistics
for these elements are presented.

        Statistical Analysis of Data

        The frequency distribution of concentrations of trace elements in
plants and other natural materials tends to exhibit positive skewness [69,
70]; therefore we logarithmically transform the original analysis values,
which results in a more nearly normal distribution. Log transformations
also tend to overcome the adverse effects of the geometric brackets in
which spectrographic analyses are commonly reported [71]. In the work reported here the means have been estimated as the antilog of the arithmetic

mean of the logs and are, therefore, geometric. The geometric mean (GM) is
the appropriate measure of central tendency for a log-normal distribution
and provides a "typical" or "characteristic" value for the element in
samples of the various plant materials. A measure of variation in element
concentrations is provided by the geometric deviation (GD), which is the
antilog of the standard deviation of the logs. About two-thirds of the
area under the log-normal distribution curve lies in the range GM ÷ GD to
GM X GD. Some properties of plant material, such as the dry matter and
ash content, may tend to exhibit a normal frequency distribution. These
properties were not logarithmically transformed in this study, and the
arithmetic mean and the standard deviation express the central tendency
and the variation, respectively.

Another characteristic of trace element analyses is the common occurrence of qualified or censored values because although the element in
question is presumably present in the material, its concentration is below
the sensitivity of the analytical method and can be reported only as "less
than" the sensitivity limit. In deriving means of data sets containing a
rather small proportion of these censored values, the technique devised by
Cohen [72] and applied to geochemical studies by Miesch [71] may be used
to estimate the censored values, as was generally done in the studies that
follow. If the data sets contain a large proportion of censored values,
or if the frequency distributions are severely skewed or bimodal, the Cohen
technique may not be applicable, and the best estimate of the mean that can
be given may, in fact, be a median value that represents the total distribution, or a less than value that expresses the limit of determination of
the analytical method.

Variation is a common property of natural materials. Part of the
measured compositional variation in a suite of samples lies in the inherent
properties of the material, and contributes the natural variation to the
total that is observed in reports of analyses. But there is error in all
geochemical data that may result from sampling bias and analytical imprecision which accounts for a part of the total observed variation. Sampling
bias can be reduced by objective methods, including randomizing procedures
in sample selection, but its contribution to the total error is difficult
to estimate [73]. Laboratory error is introduced by deficiencies in precision (reproducibility) of the analytical method, by operator bias, or by

other procedures such as sample preparation or sample "splitting." The
error due to analytical imprecision can be estimated from results of rep-
licate analysis and the use of analysis of variance techniques [73]. When
identified, this amount of variation can be subtracted from the total observed variation in element concentrations to produce a better estimate of
the "natural" range of concentrations; but such error, if randomly distributed, has little effect on the validity of the geometric means that are
calculated. Laboratory error, expressed as percent of the total variation,
was calculated for Study 1 which follows.

Study 1: Samples of Wheat Grains from the Northern Great Plains

Eleven samples of wheat, including hard red winter and spring bearded
and beardless (Triticum aestivum) and one sample of durum (T. durum), were
obtained in the field at harvest time (1974) in an area including southeastern Wyoming, eastern Montana, western North Dakota, and northwestern
South Dakota. The heads of wheat were clipped from the straws, placed in
cloth bags, and oven dried at approximately $50°$ C. The grains were then
threshed from the heads with an Almaco thresher and cleaned of chaff with
a grain cleaner at the Colorado State University experimental farm through
the courtesy of Dr. Burns R. Sabey. Each of the 12 samples was divided
into two equal parts (splits) and the resulting 24 samples were submitted
to the laboratory in a randomized sequence where a weighed portion of each
sample was burned to ash in a muffle furnace in which the heat was increased
$50°$ C/hr to a temperature of $550°$ C and held at this temperature for 14 hr.
The resulting ash was then weighed to determine the ash yield of the dry
grains. The methods of analysis and the elements determined in ash follow:
atomic absorption (chromium, cobalt, copper, lead, lithium, nickel, and
zinc), colorimetric (molybdenum), and catalytic (vanadium). For determining
concentrations of some elements that would be volatilized and lost by burning, weighed aliquots of the dried grains were "wet ashed" as follows: arsenic (nitric, sulfuric, and perchloric acids), mercury (nitric and sulfuric
acids), and selenium (nitric and perchloric acids and hydrogen peroxide).
For determining fluorine the wheat grains were pulverized in a blender and
fused with sodium hydroxide. Methods of analysis used for these volatile
elements were atomic absorption for arsenic, selective ion electrode for
fluorine, flameless atomic absorption for mercury, and fluorimetric for
selenium. Results of these determinations are given in Table 1.

TABLE 1. Selected Trace Elements in Wheat Grains[a]

Element	Ratio	GM	GD	Observed Range	E
Arsenic	0:24	<0.05	--	--	--
Cadmium	24:24	0.047	1.63	0.022-0.15	2
Chromium	3:24	<0.034	--	<0.034-0.034	--
Cobalt	24:24	0.024	1.78	0.016-0.13	13
Copper	24:24	4.4	1.13	3.5-5.8	9
Fluorine	10:24	<1	--	<1-1	--
Lead	22:22	0.083	1.39	0.051-0.15	73
Lithium	12:24	~0.068	--	<0.068-0.25	--
Mercury	1:24	<0.01	--	<0.01-0.01	--
Molybdenum	24:24	0.95	1.93	0.32-2.3	3
Nickel	24:24	0.31	1.57	0.13-0.56	5
Selenium	24:24	0.45	2.16	0.10-2.0	3
Vanadium	2:24	<0.017	--	<0.017-0.017	--
Zinc	24:24	30	1.32	20-47	4

[a]Concentrations reported as parts per million in dry material. Ratio, number of samples in which the element was found in measurable concentrations to number of samples analyzed. GM, geometric mean; GD, geometric deviations; E, percent of the total log variance attributed to laboratory procedure; --, no data available.

The arithmetic mean ash yield of the dry wheat grains is 1.7%. The concentrations of elements expressed as means in dry weight (Table 1) can be converted to approximate concentrations in ash as follows:

$$M_A = \frac{100 \ M_D}{M_P} \tag{1}$$

where M_A approximates the mean in ash weight, M_D is the mean in dry weight, and M_P is the mean of the percent ash (for wheat grains, 1.7%).

Study 2: Samples of Fruits and Vegetables
from Retail Markets in 11 Metropolitan
Areas of the United States

Fruit and vegetable samples were purchased in summer (1974) from 11 retail markets in metropolitan areas that were selected on the basis of wide geographic coverage and large size as follows: Atlanta, Georgia;

Bismarck, North Dakota; Chicago, Illinois; Dallas, Texas; Denver, Colorado;
New Orleans, Louisiana; Phoenix, Arizona; Portland, Maine; San Francisco,
California; Seattle, Washington; and Washington, District of Columbia.
The markets were all large enterprises that obtained their produce largely,
if not entirely, from commercial wholesale firms; therefore the fruits and
vegetables came from many parts of the United States. No foreign produce
was sampled. The markets were selected on the basis of their large size
and for convenience in sampling and are not purported to be representative
of markets in the area. Samples were subjectively chosen from the display
of produce at a market, with the intent to obtain "typical" examples of the
particular produce that was offered. Varietal names for most produce could
not be determined and therefore were ignored in the sampling. The samples
as purchased were sealed in plastic bags and shipped to the U.S. Geological
Survey laboratory in Denver, where they were prepared as for eating (but
not cooked), weighed, then dried to approximate constant weight in an electric oven with circulating air at a temperature of 45-50° C. Before submit-
ting the samples to the laboratory for analysis, 29 samples were selected
at random, without regard to kind of produce, from the suite of 132 samples.
These selections were then split to provide duplicate samples for assessing
the degree of precision of the laboratory procedures including chemical
analyses. In estimating the error by analysis of variance techniques, if
censored values constituted one-third or less of the total number of analytical values, we replaced the censored data with a value equal to 0.7 of
the lower limit of sensitivity. If censored values exceeded one-third of
the total values, no error estimate was made. The precision of the analytical methods was found to be satisfactory, as was demonstrated by the low
laboratory error for all elements except lead.

The methods used for chemical analysis in this study are the same as
those given in Study 1. The data for produce samples are given in Table 2.

The means in Table 2 are given as parts per million in the dried material. For comparison with much data in the literature the mean concentrations expressed on an ash weight basis may be desired, or for some purposes
the mean concentrations may be wanted in the "fresh" or "wet" material.
These approximate conversions can be made by using Eq. (1) for changing
dry weight to ash value, and Eq. (2) for changing dry weight to wet weight
values, using the means provided in Tables 2 and 3.

$$M_W = \frac{M_D \times M_{PD}}{100} \qquad (2)$$

TABLE 2. Selected Trace Elements in Fruits and Vegetables Obtained from Retail Stores Throughout the Conterminous United States[a]

Element	Fruits			Major Vegetables					Leafy Green and Salad Vegetables			
	Apple[b]	Orange[c]	Potato[d]	Dry Bean[e]	Sweet Corn[f]	Carrot[d]	Snap Bean[g]	Cabbage[h]	Head Lettuce[h]	Tomato[i]	Bulb Onion[j]	Cucumber[h]
Arsenic												
Ratio	3:10	0:11	1:11	2:11	2:10	8:11	5:11	2:11	4:11	1:11	10:11	9:11
GM	<0.05	<0.05	<0.05	<0.05	<0.05	~0.05	0.049	<0.05	<0.05	<0.05	~0.05	0.20
GD	--	--	--	--	--	--	1.56	--	--	--	--	2.65
Range	<0.05-0.20	--	<0.05-0.05	<0.05-0.06	<0.05-0.05	<0.05-0.08	<0.05-0.10	<0.05-0.05	<0.05-0.25	<0.05-0.12	<0.05-0.12	<0.05-0.50
Cadmium												
Ratio	9:10	2:11	11:11	10:11	10:11	11:11	10:11	11:11	11:11	11:11	11:11	10:11
GM	0.0082	<0.013	0.12	0.023	0.018	0.29	0.045	0.068	0.72	0.26	0.093	0.15
GD	1.47	--	1.60	1.60	1.93	2.39	1.60	1.59	2.23	2.19	1.81	2.31
Range	<0.0064-0.013	<0.013-0.020	0.050-0.26	<0.016-0.076	<0.008-0.049	0.032-0.82	<0.030-0.14	0.029-0.11	0.15-2.4	0.13-2.0	0.022-0.17	0.045-0.48
Chromium												
Ratio	6:10	5:11	5:11	9:11	2:11	6:11	8:11	1:11	5:11	3:11	3:11	4:11
GM	~0.032	<0.090	<0.096	0.11	<0.062	~0.16	0.23	<0.18	<0.30	<0.30	<0.098	<0.30
GD	--	--	--	1.59	--	--	2.05	--	--	--	--	--
Range	<0.032-0.051	<0.090-0.18	<0.096-0.17	<0.082-0.31	<0.062-0.072	<0.16-0.23	<0.16-1.5	<0.18-0.36	<0.30-2.9	<0.30-0.41	<0.098-0.12	<0.30-0.81
Cobalt												
Ratio	5:10	1:11	11:11	10:11	5:11	9:11	11:11	9:11	10:11	9:11	11:11	9:11
GM	~0.016	<0.045	0.16	0.18	<0.031	0.12	0.20	0.16	0.21	0.20	0.080	0.17
GD	--	--	1.95	2.07	--	1.53	1.70	1.93	1.66	1.71	1.74	1.48
Range	<0.016-0.021	<0.045-0.047	0.045-0.48	<0.051-0.36	<0.031-0.037	<0.082-0.19	0.081-0.42	<0.072-0.44	<0.18-0.54	<0.15-0.49	0.036-0.16	<0.14-0.32
Copper												
Ratio	10:10	11:11	11:11	11:11	11:11	11:11	11:11	11:11	11:11	11:11	11:11	11:11
GM	1.3	3.4	3.8	7.3	2.0	5.2	6.6	2.7	7.3	9.8	5.3	10
GD	1.28	1.27	1.46	1.36	1.45	1.34	1.48	1.39	1.31	1.50	1.22	1.26
Range	0.80-1.8	2.2-5.0	2.4-9.4	3.1-9.0	1.1-3.3	3.9-11	3.0-11	1.7-4.5	4.6-11	4.9-16	4.0-8.7	8.5-15
Fluorine												
Ratio	6:10	3:11	1:11	5:11	1:10	7:11	10:11	3:11	10:11	7:11	6:11	7:11
GM	~1	<1	<1	<1	<1	~1	1.6	<1	~1	~1	~1	~1
GD	--	--	--	--	--	--	1.72	--	--	--	--	--
Range	<1-3	<1-2	<1-2	<1-1	<1-1	<1-3	<1-4	<1-1	<1-11	<1-2	<1-2	<1-3
Lead												
Ratio	10:10	11:11	10:11	10:11	11:11	11:11	11:11	10:11	11:11	11:11	11:11	11:11
GM	0.12	0.28	0.13	0.13	0.11	0.36	0.49	0.31	1.0	0.86	0.32	0.66

	1	2	3	4	5	6	7	8	9	10	11	12
GD	1.69	1.97	1.58	1.81	1.51	1.51	1.38	1.60	2.27	1.58	1.83	1.57
Range	<0.50-2.6	0.11-0.77	<0.090-0.22	<0.092-0.33	0.052-0.21	0.14-0.59	0.30-0.75	<0.17-0.58	0.44-3.8	0.49-2.1	0.15-0.63	0.42-1.6
Mercury												
Ratio	0:10	3:11	1:11	0:11	0:10	3:11	3:11	3:11	2:11	3:11	1:11	5:11
GM	<0.01	<0.01	<0.01	<0.01	<0.01	<0.01	<0.01	<0.01	<0.01	<0.01	<0.01	<0.01
GD	--	--	--	--	--	--	--	--	--	--	--	--
Range	--	<0.01-0.01	<0.01-0.01	--	--	<0.01-0.01	<0.01-0.02	<0.01-0.02	<0.01-0.01	<0.01-0.02	<0.01-0.01	<0.01-0.02
Molybdenum												
Ratio	11:11	11:11	11:11	11:11	11:11	11:11	11:11	11:11	11:11	10:11	11:11	11:11
GM	0.13	0.26	0.40	2.4	0.30	0.52	2.5	0.91	0.64	1.1	0.28	2.0
GD	1.69	1.65	1.59	1.23	1.66	1.65	2.68	1.81	1.48	1.69	1.89	2.41
Range	0.064-0.32	0.14-0.66	0.19-0.90	1.8-3.8	0.08-0.62	0.26-0.97	0.65-13	0.35-2.9	0.42-1.8	<0.62-1.9	0.14-0.74	0.52-6.4
Nickel												
Ratio	11:11	11:11	11:11	11:11	11:11	11:11	11:11	11:11	11:11	11:11	11:11	11:11
GM	0.059	0.39	0.48	1.3	0.34	0.98	3.7	0.99	1.8	1.0	0.84	2.0
GD	2.10	1.78	1.61	1.57	2.10	1.91	1.58	2.10	2.20	1.69	2.12	1.86
Range	0.028-0.30	0.18-0.99	0.25-1.3	0.50-2.3	0.10-1.1	0.26-2.1	1.6-7.4	0.36-3.1	0.37-5.2	0.46-2.6	0.34-4.0	0.53-4.8
Selenium												
Ratio	1:11	10:11	11:11	11:11	8:11	11:11	11:11	11:11	11:11	11:11	11:11	11:11
GM	<0.01	0.020	0.065	0.068	0.023	0.080	0.075	0.078	0.057	0.054	0.080	0.088
GD	--	1.91	2.44	3.07	3.07	2.38	2.66	2.91	1.53	2.22	2.64	2.23
Range	<0.01-0.02	<0.01-0.06	0.02-0.30	0.02-0.35	<0.01-0.10	0.01-0.20	0.02-0.30	0.02-0.50	0.04-0.15	0.02-0.35	0.02-0.35	0.04-0.40
Vanadium												
Ratio	0:10	0:11	1:11	2:11	1:10	0:11	9:11	2:11	3:11	0:11	2:11	0:11
GM	<0.016	<0.045	<0.048	<0.046	<0.031	<0.080	0.11	<0.091	<0.15	<0.15	<0.049	<0.15
GD	--	--	--	--	--	--	1.94	--	--	--	--	--
Range	--	--	<0.048-0.048	<0.046-0.051	<0.031-0.031	--	<0.069-0.50	<0.091-0.14	<0.15-1.1	--	<0.049-0.079	--
Zinc												
Ratio	10:11	11:11	11:11	11:11	11:11	11:11	11:11	11:11	11:11	11:11	11:11	11:11
GM	1.2	6.9	14	25	25	22	38	23	49	36	17	56
GD	1.99	1.34	1.30	1.14	1.37	1.48	1.25	1.33	1.54	1.36	1.41	1.49
Range	<0.50-2.6	4.1-10	10-27	21-33	12-37	14-58	27-56	14-34	25-83	24-60	10-32	23-110

[a] Concentrations reported in parts per million in dry material. Ratio, number of samples in which the element was found in measurable concentrations to number of samples analyzed. GM, geometric mean; GD, geometric deviation; --, no data available.
[b] Washed, cored, and sliced.
[c] Peeled, seeded, and cut up.
[d] Peeled and sliced.
[e] No preparation.
[f] Grains cut from cob.
[g] Stems removed, broken into pieces.
[h] Washed and sliced.
[i] Stems removed, washed, and sliced.
[j] Peeled, sliced, and divided into rings.

TABLE 3. Yields of Dry Material Obtained from Fresh Produce and of Ash
Obtained from Dry Material, Expressed as Weight Percent[a]

Kind of Produce	Dry Material			Ash		
	AM	SD	Observed Range	AM	SD	Observed Range
Apple	16	0.71	14.3-16.6	1.6	0.26	1.3-2.1
Orange	12	2.04	8.5-15.7	4.5	1.09	3.2-6.3
Potato	20	1.75	17.2-23.7	4.8	0.38	4.2-5.5
Dry bean	91	2.12	86.3-94.1	4.6	0.37	4.1-5.2
Sweet corn	24	3.61	18.4-30.0	3.1	0.64	2.0-4.0
Carrot	12	2.0	8.8-14.4	8.0	0.83	6.5-9.7
Snap bean	9.0	1.50	6.4-10.9	8.2	1.56	6.5-12.4
Cabbage	7.2	1.01	6.0-8.6	9.1	2.16	6.8-14.4
Head lettuce	3.3	0.64	2.5-4.8	15	2.93	10.4-18.0
Tomato	4.8	1.29	3.6-7.3	15	1.83	12.3-19.0
Bulb onion	8.6	2.41	5.6-14.0	4.9	1.42	3.6-7.9
Cucumber	3.3	0.63	2.6-4.4	15	4.49	10.5-26.9

[a]AM, arithmetic mean; SD, standard deviation.

where M_W approximates the mean in wet weight, M_D is the mean in dry weight,
and M_{PD} is the percent yield of dry material (from Table 3).

The percentage of water in the produce as prepared for esting can be
calculated by subtracting the mean yield of dry material (Table 3) from
100.

Study 3: Samples of Grain, Seed, and Forage Plants
from Various Locations in the United States

This study is a compilation of data from our files, some of which have
been published, and the references are given in footnotes to Table 4. The
study of corn grains and soybeans from Missouri included an objective sampling design based on a hierarchical arrangement of geographic areas, and
estimates of laboratory error were given for the mean element concentrations
[26]. Laboratory error was not calculated for other data in Table 4. The
sampling plans and laboratory methods used in studies of corn grains in
Wisconsin and forage plants sampled in various parts of the United States
were described by Connor and Shacklette [13].

The mean concentrations of elements in Table 4 are given as parts per
million in ash of the samples. These values can be converted to approximate concentrations in dry material by using the mean weight percent ash
given in Table 5 and the mean concentration in dry material from Table 4
in the following equation:

$$M_D = \frac{M_A \, M_{PA}}{100} \tag{3}$$

where M_D approximates the mean dry weight, M_A is the mean in ash (Table 4),
and M_{PA} is the percent ash in the dry material (Table 5).

CONCLUDING REMARKS

The data presented in this chapter should serve to convey some appreciation of the abundances and occurrences of the trace elements in foods
and feeds. The question of the relevance of the data to current or future
toxicity problems is intentionally left unanswered. It is the function of
specialists in this field to demonstrate whether the usual levels of these
elements in the foods and feeds as commonly consumed present a real or potential threat to the health of man and livestock. The data in this report
have been largely based on samples of plant products that were considered
to have grown in a "normal" geochemical environment (not necessarily in a
"natural" one, for the total effects of man's activities on the natural
geochemical environment cannot be fully evaluated). Underwood [61, p. 44]
gave perspective to the question of toxicity in foodstuffs as follows:

> Some of the trace elements, such as arsenic, lead, cadmium, and
> mercury, are frequently classified as toxic elements because their
> toxicity to man and animals is relatively high and their biological
> activity is largely confined to toxic reactions. However, all the
> trace elements can be toxic if consumed in large-enough quantities
> or for long-enough periods. With some, such as fluorine in man and
> copper in sheep, the margins between beneficial and toxic intakes
> are quite small. Furthermore, the toxicity of a particular element
> can be greatly influenced by the extent to which other elements or
> compounds affecting its absorption, excretion, or metabolism are
> present in the diet. Interactions of this type are of such importance
> with some elements, notably iron, copper, molybdenum, and zinc, that
> there is no single maximum "safe" level in foods, or no single
> minimum toxic level. There is a series of such levels, depending
> upon the chemical form, the duration and continuity of intake of the
> element, and the nature of the rest of the diet, including the
> amounts and proportions of various other elements and organic com-
> pounds. These facts should constantly be borne in mind in any con-
> sideration of the toxic potential of trace elements occurring in
> foods.

TABLE 4. Selected Trace Elements in Samples of Grain, Seed, and Forage Plants[a]

| Element | Corn Grains | | | Soybean Seeds[b] | Alfalfa Stems and Leaves[c] | Yellow Sweetclover Stems and Leaves[c] |
	Missouri[b]	Wisconsin[b]	United States[c]			
Boron						
Ratio	38:38	19:27	36:35	37:37	25:25	10:10
GM	64	37	120	210	22	160
GD	1.35	1.76	1.97	1.38	1.04	1.56
Range	50-100	<30-100	100-700	100-300	100-700	100-300
Chromium						
Ratio	2:38	11:27	23:38	0:37	23:25	9:10
GM	<2	1.5	<1	<2	5.5	4.6
GD	--	2.34	--	--	2.69	2.40
Range	<2-5	<2-6.9	<1-150	--	<2-50	<1-30
Cobalt						
Ratio	13:38[e]	0:27	1:39	29:37[e]	0:25	9:10
GM	<1	--	<7	1.7	--	<3
GD	--	--	--	2.63	--	--
Range	<1-9	--	<7-7	<1-13	--	<3-10
Copper						
Ratio	38:38	27:27	39:39	37:37	25:25	10:10
GM	79	150	120	190	94	100
GD	1.42	1.37	1.56	1.32	1.49	1.63
Range	30-150	70-300	50-300	100-300	30-150	50-200
Lead						
Ratio	2:38	10:27	23:35	0:37	13:25	8:10
GM	<20	<20	14	<20	20	34
GD	--	--	2.56	--	3.71	2.22
Range	<20-50	<20-50	<10-150	--	<20-150	<10-150
Molybdenum						
Ratio	32:38	22:27	35:37	30:37	20:25	8:10
GM	11	16	18	14	18	20
GD	2.07	1.94	2.15	2.89	3.32	3.40
Range	<7-30	<7-70	<7-70	<7-70	<7-150	<3-70

Bluegrass Stems and Leaves[c]	Tall Fescue Stems and Leaves[d]	Johnson Grass Stems and Leaves[c]	Wheat Stems and Leaves[c]	Crested Wheatgrass Stems and Leaves[c]	Western Wheatgrass Stems and Leaves[c]
16:16	14:18	10:12	6:11	16:16	13:13
100	48	81	43	160	170
1.32	1.24	1.78	2.11	1.64	1.76
70-200	< 50-70	< 50-300	< 50-100	70-300	70-700
16:16	18:18	10:12	7:11	16:16	13:13
6.7	8.6	4.5	2.9	19	17
2.36	2.28	2.92	2.97	2.09	1.58
< 2-20	< 2-30	< 2-30	< 2-15	3-70	10-50
0:16	0:18	0:12	0:1	5:16	1:13
--	--	--	--	7.5	< 10
--	--	--	--	1.27	--
--	--	--	--	< 10-10	< 10-10
16:16	18:18	12:12	11:11	16:16	13:13
110	59	110	86	42	50
1.39	1.56	1.64	1.46	1.38	1.31
50-150	30-150	50-200	50-150	30-70	30-70
14:16	18:18	8:12	5:11	4:16	4:13
89	43	32	16	11	7.8
3.11	1.59	3.15	2.73	2.04	5.81
< 20-300	20-150	< 20-200	< 20-70	< 20-30	< 20-150
5:16	10:18	6:12	8:11	3:16	9:13
3.3	4.8	6.0	12	2.1	7.3
3.39	2.47	2.85	2.52	3.06	2.56
< 7-20	< 5-20	< 7-30	< 7-30	< 7-15	< 5-30

TABLE 4 (Continued)

Element	Corn Grains			Soybean Seeds[b]	Alfalfa Stems and Leaves[c]	Yellow Sweetclover Stems and Leaves[c]
	Missouri[b]	Wisconsin[b]	United States[c]			
Nickel						
Ratio	37:38	27:27	37:38	37:37	23:25	10:10
GM	23	14	20	100	15	12
GD	2.14	1.94	2.16	1.93	2.01	1.63
Range	<5-70	7-50	<5-70	30-500	<5-50	7-30
Strontium						
Ratio	34:38	21:27	35:39	37:37	25:25	10:10
GM	16	21	20	290	1700	880
GD	1.88	2.02	2.04	1.64	2.19	1.89
Range	<7-50	<15-70	<10-70	70-700	500-5000	500-3000
Vanadium						
Ratio	1:38	0:27	1:39	0:37	8:25	3:10
GM	<10	--	--	<10	4.1	<7
GD	--	--	--	--	4.68	--
Range	<10-115	--	<10-10	--	<10-50	<7-100
Zinc						
Ratio	38:38	25:27	38:38	37:37	12:12	--
GM	1800	970	1700	970	220	--
GD	1.22	1.34	1.26	1.19	1.25	--
Range	1200-2800	<500-1500	1100-2800	700-1500	150-300	--

[a] Concentrations of elements, reported as parts per million in ash, were obtained by emission spectrography except as noted. Ratio, number of samples in which the element was detected in measurable concentrations to number of samples analyzed. GM, geometric mean; GD, geometric deviation; --, no data available.

[b] Connor and Shacklette [13].

[c] Samples collected in a geochemical survey of U.S. soils and plants [14].

[d] Samples collected in a geochemical study in Missouri [26].

[e] Analysis by the atomic-absorption method.

Bluegrass Stems and Leaves[c]	Tall Fescue Stems and Leaves[d]	Johnson Grass Stems and Leaves[c]	Wheat Stems and Leaves[c]	Crested Wheatgrass Stems and Leaves[c]	Western Wheatgrass Stems and Leaves[c]
16:16	12:18	9:12	4:11	15:16	10:13
14	7.1	9.7	3.2	9.8	6.3
1.54	2.56	3.38	1.98	1.72	1.64
7-30	< 5-30	< 5-50	< 5-7	< 5-20	< 5-10
16:16	18:18	12:12	11:11	16:16	13:13
330	280	310	170	350	410
1.68	1.56	2.10	1.85	1.66	1.84
150-1000	150-700	100-1000	50-500	200-1000	150-1500
5:16	6:18	3:12	0:11	14:16	10:13
7.7	< 5	< 15	--	42	24
3.04	--	--	--	2.23	1.90
< 15-30	< 5-30	< 15-50	--	< 15-100	< 15-50
--	--	7:7	3:3	--	--
--	--	520	1000	--	--
--	--	1.62	1.82	--	--
--	--	400-600	680-2000	--	--

The question of nutritional sufficiency of trace elements in these foods and feedstuffs is also left unanswered in this chapter. Not only the concentration and chemical form of a trace element in foods, but also the total amount consumed in relation to body weight and the duration of this intake, must be used in assessing, in general terms, the nutritional property of a food. General statements that a certain food is "rich" in zinc or some other element should be accepted only if supporting explanations are given. For example, some element concentrations in foods, as given in this report, could lead to inaccurate conclusions about their

TABLE 5. Weight Percent Ash in Dry Material
of Grain, Seed, and Forage Plants[a]

Plant Material	GM	GD	Observed Range
Corn grains			
Missouri[b]	1.6	1.20	1.0-2.2
Wisconsin[b]	1.8	1.44	1.0-6.3
United States[b]	1.4	1.31	1.0-4.0
Soybean seeds[b]	5.3	1.08	4.3-6.0
Alfalfa stems and leaves[c]	11.7	1.24	8.0-20
Yellow sweetclover stems and leaves[c]	7.5	--	--
Tall fescue stems and leaves[c]	11.5	1.23	7.8-15
Johnson grass stems and leaves[d]	6.2	1.57	2.9-10
Wheat stems and leaves[d]	9.0	1.84	3.0-16
Crested wheatgrass stems and leaves[d]	8.6	1.29	5.8-13
Western wheatgrass stems and leaves[d]	7.7	1.35	5.3-14
Bluegrass stems and leaves[d]	6.8	1.38	3.9-11

[a]GM, geometric mean; GD, geometric deviation; --, no data available.

[b]Connor and Shacklette [13].

[c]Samples collected in a geochemical study in Missouri [26].

[d]Samples collected in a geochemical survey of United States soils and plants [14].

TABLE 6. Relationship of Concentrations of Zinc and Copper Expressed as
Parts per Million in "Fresh," Dry, and Ash Weights in Three Vegetables

	Zinc			Copper		
	"Fresh" Material	Dry Material	Ash	"Fresh" Material	Dry Material	Ash
Sweet corn	6	25	820	0.48	2.0	64
Lettuce	1.6	49	330	0.24	7.3	49
Dry beans	23	25	550	6.6	7.3	160

nutritional significance. The effect of changing the basis of reporting
the same concentrations of two elements in three vegetables is illustrated
in Table 6.

If only the concentration in ash is considered, sweet corn seems to
be almost twice as rich in zinc as are dry beans. But on a dry weight
basis the zinc concentration is the same for both vegetables. However, on
a fresh weight basis, the zinc concentration in dry beans is about four
times as high as in sweet corn. Lettuce has a greater zinc concentration
in dry material than have either sweet corn or dry beans, but the concentrations in "fresh" material and ash are much lower. Copper is higher on
all three bases in dry beans than in sweet corn and lettuce; therefore dry
beans could be characterized as rich in copper compared to the other two
vegetables. Yet dry beans have water added when cooked for eating, and
this addition would decrease their apparent richness in copper compared to
that of corn and lettuce.

These examples illustrate some of the difficulties in evaluating the
contribution of the various essential trace elements in the food and feed
plants to dietary sufficiency, and emphasize again the importance of reliable data concerning the trace element composition of foods and feedstuffs.

APPENDIX: COMMON AND SCIENTIFIC NAMES
USED IN THIS REPORT

Plant names given in some of the literature cited include only the
genus or a common name referring to the genus or species; for these we
supply a species name only if we have no doubt as to the intention of the
writers. Some of the cited references give reports of several species
which, for presenting ranges in element concentrations, we have grouped in
a genus or higher category. The authority for a scientific name is included
if we are reasonably certain of the identity of the plant mentioned or if
the authority is given by the writer. The common names are listed alpha-
betically in the form that is used in the text; for example, "lodgepole
pine" is listed rather than "pine, lodgepole."

Alfalfa, *Medicago sativa* L.

Alpine fir, *Abies lasiocarpa* Nutt.

Alsike clover, *Trifolium hybridum* L.

American elm, *Ulmus americana* L.

Apple, *Pyrus malus* L.

Apricot, *Prunus armeniaca* L.

Artichoke, *Cynara scolymus* L.

Asparagus, *Asparagus officinale* L.

Avocado, *Persea gratissima* Gaertn. f.

Banana, *Musa paradisiaca* var. *sapientum* Kuntze

Barley, *Hordeum vulgare* L.

Bedstraw, *Galium* sp.

Beet, *Beta vulgaris* L.

Black cherry, *Prunus serotina* Ehrh.

Blackeyed pea, *Vigna sinensis* Endl.

Black gum, *Nyssa sylvatica* Marsh.

Black oak, *Quercus velutina* Lam.

Bluegrass, *Poa pratensis* L.

Brazil nut, *Bertholletia excelsa* H. and B.

Breadfruit, *Artocarpus altilis* (Parkinson) Fosberg

Broccoli, *Brassica oleracea* var. *italica* Plenck

Broomsedge, *Andropogon scoparius* Michx.

Buckbush, *Symphoricarpos orbiculatus* Moench

Buckwheat, *Fagopyrum sagittatum* Gilib.

Bulb onion, *Allium cepa* L.

Cabbage, *Brassica oleracea* var. *capitata* L.

Cantaloupe, *Cucumis melo* L.

Carrot, *Daucus carota* var. *sativa* DC.

Cassava, *Manihot esculenta* Crantz

Cauliflower, *Brassica oleracea* var. *botrytis* L.

Cedar, *Juniperus virginiana* L.

Celery, *Apium graveolens* var. *dulce* DC.

Coconut, *Cocos nucifera* L.

Common gromwell, *Lithospermum officinale* L.

Common juniper, *Juniperus communis* L.

Corn, _Zea_ _mays_ L.

Crested wheatgrass, _Agropyron_ _desertorum_ (Fisch.) Schult.

Crimson clover, _Trifolium_ _incarnatum_ L.

Cucumber, _Cucumis_ _sativus_ L.

Desert trumpet, _Eriogonum_ _inflatum_ Torr. and Frem.

Dogwood, _Cornus_ _florida_ L.

Douglas fir, _Pseudotsuga_ _menziesii_ (Mirb.) Franco

Dry bean, _Phaseolus_ _vulgaris_ L.

Durum wheat, _Triticum_ _durum_ Desf.

Engelman spruce, _Picea_ _engelmannii_ Parry

English ryegrass, _Lolium_ _perenne_ L.

Eryngium, _Eryngium_ sp.

Evening primrose, _Oenothera_ _caespitosa_ Nutt.

Fig, _Ficus_ _carica_ L.

Giant horsetail, _Equisetum_ _telmateia_ Ehrh.

Giant reed, _Arundo_ _donax_ L.

Greasewood, _Sarcobatus_ _vermiculatus_ (Hook.) Torr.

Gumweed, _Grindelia_ _fastigiata_ Greene

Head lettuce, _Lactuca_ _sativa_ L.

Hickory, _Carya_ sp.

Horsetail, _Equisetum_ sp.

Impatiens, _Impatiens_ _holstii_ Engler and Warb.

Jerusalem artichoke, _Helianthus_ _tuberosus_ L.

Johnson grass, _Sorghum_ _halepense_ (L.) Pers.

Juniper, _Juniperus_ _monosperma_ (Engelm.) Sarg.

Kale, _Brassica_ _oleracea_ var. _acephala_ DC.

Labrador tea, _Ledum_ _decumbens_ (Ait.) Lodds

Lettuce, _Lactuca_ _sativa_ L.

Lima bean, _Phaseolus_ _limensis_ Macf.

Limber pine, _Pinus_ _flexilis_ James

Locoweed, _Astragalus_ _pattersoni_ Rydb.

Lodgepole pine, _Pinus_ _contorta_ var. _latifolia_ Engelm.

Lycopodium, _Lycopodium_ sp.

Mandarin orange, _Citrus_ _nobilis_ var. _deliciosa_ Swingle

Mangel-wurzel, _Beta_ _vulgaris_ L.

Millet, _Setaria_ _italica_ (L.) Beauv.

Mormon tea, _Ephedra_ _viridis_ Coville

Mountain avens, _Dryas_ _octopetala_ L.

Myrtle blueberry, _Vaccinium_ _myrtillus_ L.

Norway spruce, _Picea_ _abies_ (L.) Karst.

Oak, _Quercus_ sp.

Oats, _Avena_ _sativa_ L.

Onion, _Allium_ _cepa_ L.

Orange, _Citrus_ _sinensis_ Osbeck

Orchardgrass, _Dactylis_ _glomerata_ L.

Papaya, _Carica_ _papaya_ L.

Parsley, _Petroselinum_ _crispum_ Mansf.

Parsnip, _Pastinaca_ _sativa_ L.

Pea, _Pisum_ _sativum_ L.

Peach, _Prunus_ _persica_ L.

Pear, _Pyrus_ _communis_ L.

Persimmon, _Diospyros_ _virginiana_ L.

Phyllodoce, _Phyllodoce_ _empetriformis_ (Smith) D. Don

Pineapple, _Ananas_ _comosus_ (L.) Merr.

Pinyon, _Pinus_ _cembroides_ var. _edulis_ Zucc.

Plantain, _Musa_ _paradisiaca_ L.

Poisonvetch, _Astragalus_ _preussii_ Gray

Ponderosa pine, _Pinus_ _ponderosa_ Laws.

Post oak, _Quercus_ _stellata_ Wang.

Potato, _Solanum_ _tuberosum_ L.

Prune, _Prunus_ _domestica_ L.

Radish, _Raphanus_ _sativus_ L.

Red clover, _Trifolium_ _pratense_ L.

Red oak, _Quercus_ _rubra_ L.

Rice, _Oryza_ _sativa_ L.

Rutabaga, _Brassica_ _napobrassica_ Mill.

Rye, _Secale_ _cereale_ L.

Sassafras, _Sassafras_ _albidum_ (Nutt.) Nees

Shadscale saltbush, _Atriplex_ _confertifolia_ (Torr. and Frem.) Wats.

Shortleaf pine, _Pinus_ _echinata_ Mill.

Sitka alder, _Alnus_ _crispa_ subsp. _sinuata_ (Regel) Hult.

Snap bean, _Phaseolus_ _vulgaris_ L.

Soybean, _Glycine max_ Merr.

Spanish moss, _Tillandsia usneoides_ L.

Spinach, _Spinacia oleracea_ L.

Sugar beet, _Beta vulgaris_ L.

Sweet corn, _Zea mays_ var. _rugosa_ Bonaf.

Sweet gum, **Liquidambar** _styraciflua_ L.

Sweet pepper, _Capsicum frutescens_ var. _grossum_ Bailey

Sweet potato, _Ipomea batatas_ L.

Tall fescue, _Festuca elatior_ L.

Taro, _Colocasia esculenta_ Schott.

Toadflax, _Linaria triphylla_ Jacq.

Tobacco, _Nicotiana tabacum_ L.

Tomato, _Lycopersicum esculentum_ Mill.

Tuliptree, _Liriodendron tulipfera_ L.

Turnip, _Brassica napus_ L.

Twinpod, _Physaria_ sp.

Vetch, _Vicia_ sp.

Water chestnut, _Trapa natans_ L.

Watermelon, _Citrullus vulgaris_ Schrad.

Western wheatgrass, _Agropyron smithii_ Rydb.

Wheat, _Triticum aestivum_ L.

White clover, _Trifolium repens_ L.

White oak, _Quercus alba_ L.

White sweetclover, _Melilotus alba_ Desr.

Willow, _Salix_ sp.

Winged sumac, _Rhus copallina_ L.

Wormwood, _Artemisia terres-alba_ Krasch.

Yam, _Dioscorea alata_ L.

Yellow sweetclover, _Melilotus officinalis_ (L.) Lam.

REFERENCES

1. E. J. Underwood, Trace Elements in Human and Animal Nutrition, 3rd Ed., Academic, New York, 1971, p.543.

2. E. J. Underwood, in Trace Substances in Environmental Health IV (D. D. Hemphill, ed.), Missouri University, Columbia, 1971, pp. 3-11.

3. W. H. Allaway, in Geochemical Environments in Relation to Health and Disease (H. C. Hopps and H. L. Cannon, eds.), Vol. 199, Ann. New York Acad. Sci., New York, 1972, pp. 17-25.

4. J. F. Hodgson, in Trace Substances in Environmental Health III (D. D. Hemphill, ed.), Missouri University, Columbia, 1970, pp. 45-58.

5. K. C. Beeson, in Yearbook of Agriculture 1957, U.S. Dept. Agr., 1957, pp. 258-267.

6. J. Miller, Science, 185, 240 (1974).

7. National Research Council, Food and Nutrition Board, Recommended Dietary Allowances, 7th Ed., National Acad. Sci., Washington, 1965.

8. E. I. Hamilton and M. J. Minski, Sci. Total Environ., 1, 104 (1972).

9. R. R. Brooks, Geobotany and Biogeochemistry in Mineral Exploration, Harper and Row, New York, 1972, p. 290.

10. H. J. M. Bowen, Trace Elements in Biochemistry, Macmillan, London, 1966, p. 241.

11. H. L. Cannon, Science, 132, 591-598 (1960).

12. J. E. McMurtrey, Jr., and W. O. Robinson, in Yearbook of Agriculture 1938, U.S. Dept. Agr., 1938, pp. 807-829.

13. J. J. Connor and H. T. Shacklette, U.S. Geol. Survey Prof. Paper 574-F, 1975, p. 168.

14. Unpublished work on analyses of plants collected throughout the conterminous United States at intervals of about 50 miles, in connection with soil study described by H. T. Shacklette, J. C. Hamilton, J. G. Boerngen, and J. M. Bowles, U.S. Geol. Survey Prof. Paper 574-D, 1971, p. 71.

15. H. T. Shacklette, H. I. Sauer, and A. T. Miesch, U.S. Geol. Survey Prof. Paper 574-C, 1970, p. 39.

16. Unpublished data on samples collected by G. C. Curtin and H. D. King or by M. A. Chaffee (twinpod samples only); analyses by E. L. Mosier or J. M. Nishi, using a method described by Mosier in Appl. Spectrosc., 26, 636-640 (1972).

17. J. A. Erdman, H. T. Shacklette, and J. R. Keith, U.S. Geol. Survey Prof. Paper 954-D, 1976, p. 23.

18. H. V. Warren, R. E. Delavault, and J. Barakso, Can. Mining Metallurg. Bull., July, 860-866 (1968).

19. K. T. Williams and R. R. Whetstone, U.S. Dept. Agr. Tech. Bull. 732, 1940, p. 20.

20. K. C. Beeson, U.S. Dept. Agr. Misc. Publ. 369, 1941, p. 164.

21. J. A. Duke, Econ. Botany, 24(3), 344-366 (1970).

22. R. S. Jones, U.S. Geol. Survey Circ. 625, 1970, p. 15.

23. H. V. Warren and R. E. Delavault, Bull. Geol. Soc. America, 61, 123-128 (1950).

24. H. T. Shacklette, H. W. Lakin, A. E. Hubert, and G. C. Curtin, U.S. Geol. Survey Bull. 1314-B, 1970, p. 23.

25. Kh. Aripove and R. M. Talipov, Uzb. Geol. Zh., Akad. Nauk Uzb. SSSR, 3, 45-51 (1966).

26. J. A. Erdman, H. T. Shacklette, and J. R. Keith, U.S. Geol. Survey Prof. Paper 954-C, 1976, p. 87.

27. W. O. Robinson, R. R. Whetstone, and Glen Edgington, U.S. Dept. Agr. Tech. Bull. 1013, 1950.

28. H. T. Shacklette, U.S. Geol. Survey Bull. 1314-G, 1972, p. 28.

29. H. A. Schroeder and J. J. Balassa, Science, 140, 819-820 (1963).

30. R. Kropf and M. Geldmacher-von Mallinckrodt, Archiv. Hygiene, 152, 218-224 (1968).

31. H. V. Warren, R. E. Delavault, and K. W. Fletcher, Can. Min. Metallurg. Bull., July, 1-12 (1971).

32. K. C. Beeson, U.S. Dept. Agr. Misc. Publ. 369, 1941.

33. J. D. Dodd and G. L. Van Amburg, Ecology, 51(4), 685-689 (1970).

34. G. N. Brown, Science, 143(3604), 368-369 (1964).

35. O. Erämetsä and I. Yliruokanen, Suomen Kemistilehti B, 44, 121-128 (1971).

36. O. Erämetsä and M. Haukka, Suomen Kemistilehti B, 43, 189-193 (1970).

37. O. Erämetsä, A-R. Haarala, and I. Yliruokanen, Suomen Kemistilehti B, 46 (1973).

38. W. O. Robinson, H. Bastron, and K. J. Murata, Geochim. Cosmochim. Acta, 14, 55-67 (1958).

39. Von K. Garber, in Deutsche Forschungsgemeinschaft, Forschungsberichte, 14, Franz Steiner, Wiesbaden, 1968, pp. 42-48 (in German).

40. H. T. Shacklette and J. J. Connor, U.S. Geol. Survey Prof. Paper 574-E, 1960, p. 46.

41. P. F. M. Paul, unpublished manuscript, Germanium in Contemporary Vegetation, notes of a discussion held with Dr. Hans Brauchli, Johns Hopkins University, Baltimore, August 25, 1953.

42. H. T. Shacklette, in U.S. Geol. Survey Prof. Paper 713, 1970, pp. 35-36.

43. J. T. Tanner, M. H. Friedman, D. N. Lincoln, and L. A. Ford, Science, 177(4054), 1102-1103 (1972).

44. H. T. Shacklette and M. E. Cuthbert, in Geol. Soc. America Special Paper 90 (H. L. Cannon and D. D. Davidson, eds.), 1967, pp. 31-46.

45. T. F. Borovik-Romanova, in Problems of Geochemistry (N. I. Khitaron, ed.), Acad. Sci. USSR, 1965, pp. 675-682.

46. H. V. Warren and R. E. Delavault, Western Miner, 38, 64-72 (1965).

47. W. B. Dye and J. L. O'Harra, Nevada Univ. Agr. Exp. Station Bull. 208, 1959, p. 32.

48. W. H. Rickard, Bot. Gaz., 126(2), 116-119 (1965).

49. H. T. Shacklette, U.S. Geol. Survey Bull. 1198-D, 1965, p. 21.

50. O. Erämetsä, R. Viinanen, and I. Yliruokanen, Suomen Kemistilehti B, 46, 355-358 (1973).

51. J. J. Connor, H. T. Shacklette, and J. A. Erdman, in U.S. Geol. Survey Prof. Paper 750-B, 1971, pp. 151-156.

52. H. L. Cannon and J. M. Bowles, Science, 137, 765 (1962).

53. H. V. Warren, J. Biosoc. Sci., 6, 223-238 (1974).

54. T. C. Tso, N. A. Hallden, and L. T. Alexander, Science, 146(3647), 1043-1045 (1964).

55. W. W. Rudolph and J. R. Moore, Alaska Construction and Oil Report, February, 41-42 (1972).

56. W. O. Robinson, L. A. Steinkoenig, and C. F. Miller, U.S. Dept. Agr. Bull. 600, 1917, p. 25.

57. A. T. Myers and J. G. Hamilton, in U.S. Geol. Survey Prof. Paper 424-B, 1961, pp. 286-288.

58. H. T. Shacklette, unpublished data.

59. J. Kubota, W. H. Allaway, D. L. Carter, E. E. Cary, and V. A. Lazar, J. Agr. Food Chem., 15, 448-453 (1967).

60. H. W. Lakin, in Trace Elements in the Environment (E. L. Kothny, ed.), Adv. in Chem. Ser. 123, Amer. Chem. Soc., Washington, 1973, pp. 96-111.

61. E. J. Underwood, in Toxicants Occurring Naturally in Foods, 2nd Ed., National Acad. Sci., Washington, 1973, pp. 43-87.

62. W. O. Robinson and G. Edgington, Soil Sci., 60, 15-28 (1945).

63. T. S. Lovering, Bull. Geol. Soc. America, 70, 781-800 (1959).

64. V. Zýka, Sbornik Geologických Věd, Těch. Geochem., 10, 91-95 (1972).

65. H. L. Cannon, in Symposium de Exploración Geoquímica, XX Int. Geol. Cong., 2, 235-241 (1959).

66. N. E. Whitehead and R. R. Brooks, The Bryologist, 72, 501-507 (1969).

67. D. P. Malyuga, Biogeochemical Methods of Prospecting, Consultants Bureau, New York, 1964, p. 205.

68. Committee on Biologic Effects of Atmospheric Pollutants, Vanadium, National Academy of Science, Washington, 1974, p. 117.

69. J. S. Duval, Jr., T. Schwarzer, and J. A. S. Adams, in Trace Substances in Environmental Health IV (D. D. Hemphill, ed.), Missouri University, Columbia, 1971, pp. 120-131.

70. L. H. Ahrens, Geochim. Cosmochim. Acta, 5, 49-84 (1954); 6, 121-131 (1954); 11, 205-212 (1957).

71. A. T. Miesch, U.S. Geol. Survey Prof. Paper 574-B, 1967, p. 15.

72. A. C. Cohen, Jr., Technometrics, 1, 217-237 (1959).

73. A. T. Miesch, U.S. Geol. Survey Prof. Paper 574-A, 1967, p. 17.

74. E. E. Angino, H. L. Cannon, K. M. Hambridge, and A. W. Voors, in Geochemistry and the Environment (Walter Mertz, ed.), Vol. 1, National Academy of Science, 1974, pp. 36-42.

4

MECHANISMS OF HEAVY METAL INORGANIC TOXICITIES

Frederick W. Oehme
Kansas State University
Manhattan, Kansas

Although heavy metals were formerly important therapeutic agents, present interest lies primarily with the toxic reactions they are capable of producing. The problems created by water and air pollution, food contamination, and the widespread use of agricultural chemicals are largely concerned with these toxicants. Humans, as well as domestic and wild animals, are vulnerable.

In order to recognize, study, and treat the conditions caused by these compounds, the biological distribution and fate of heavy metals must be understood. It is generally recognized that the absorption of inorganic compounds depends heavily upon the compounds' solubility. Metals coming in contact with the body in elemental form are usually poorly absorbed. However, elements placed on the skin in lipid suspensions may be more readily utilized, and the ingestion of metals by carnivors with highly acid digestive tracts provides increased opportunity for absorption. Finely powdered metals are more soluble than large pieces of the same compound.

Once absorbed into the body, inorganic metals are capable of reacting with a variety of binding sites; indeed, their action is spoken of as that of "protein precipitation." Their specific toxic effect in a biological system, however, depends upon reactions with ligands that are essential for the normal function of that system. Metal complexes (coordination compounds) are formed with sulfhydryl groups and to lesser degrees with amino, phosphate, carboxylate, imidazole, and hydroxyl radicals of enzymes and other essential biological proteins. More stable complexes are formed with the sulfur and nitrogen than with oxygen. This specific area is ripe for biochemically oriented research.

The sensitivity of the particular system attacked by the metal and the degree of cellular-activity interference caused by the metal-protein complex determines the clinical effects and course. In general, the digestive system shows involvement due to its part in absorption, the liver is affected by its role as body filter and detoxifier, and the kidney glomeruli and tubular cells are damaged through their function in excretion. The disease state may be altered by the introduction of artificial ligands having equal or greater affinity for the toxic material than do the biological proteins. The use of these antidotes has been remarkably effective, together with general supportive care, in reversing the toxicosis.

As examples of the information currently available about the fate of heavy metals in the body, and to see how this material may be applied to toxic states produced by these compounds, the absorption, distribution, biotransformation, and excretion of arsenic, lead, mercury, selenium, and thallium are discussed.

ARSENIC

Absorption

Inorganic arsenic is absorbed significantly through the intact skin. When applied in a lipid-soluble ointment, the absorption is increased. Necrosis and ulcerations of the area in contact with the arsenic may occur; cell division is inhibited and nuclear abnormalities are seen [13]. Absorption from the digestive tract is dependent upon solubility. Finely divided powders are more completely absorbed than coarse materials. Soluble compounds of arsenic are well absorbed through the digestive tract and from all mucous surfaces, including the lung. Parenteral administration results in complete removal from intramuscular and subcutaneous sites within 24 hr [46].

Distribution

Following absorption from any surface, 95-99% of the inorganic arsenic is located in the red blood cells in combination with the globin of hemoglobin. The arsenic in serum is bound to proteins [46]. The metal rapidly leaves the blood in 24 hr and distributes to the liver, kidney, lung, wall of the gastrointestinal tract, and spleen. Smaller amounts are found in the muscle and nervous tissue. After 2 weeks, or with continuous arsenic intake, the skin, hair, and bones begin to accumulate the metal. Dogs chronically poisoned with arsenic have four to six times as great an amount of arsenic in the hair per unit weight as in the liver [3]. The arsenic is bound tightly to the sulfhydryl groups of the protein fraction of the respective tissues and is only slowly released. Arsenic deposition in the hair and bone is fixed at these sites for years.

Biotransformation

Inorganic arsenic causes widespread damage by combining with the readily available sulfhydryl groups of proteins. The sulfhydryl enzyme systems essential to cellular metabolism are therefore inhibited and permeability of capillaries and small arterioles with vascular dilation occurs [48]. It has been suggested that the first point of attack in the inactivation of the essential sulfhydryl enzymes is pyruvate oxidation [36]. Rather than reacting with the sulfhydryls of two separate molecules, arsenic usually reacts with both sulfhydryls of one molecule, forming a ring compound more stable than noncyclic thioarsenicals.

Arsenic is also capable of uncoupling oxidative phosphorylation in the liver mitochondria [52]. The metal substitutes for inorganic phosphorus in the reaction: $X \sim Y + Pi \rightleftarrows Y \sim P + X$. This yields an unstable arsenic ester intermediate which hydrolyzes spontaneously. Energy, that is ATP formation, does not result. Other enzyme systems (alpha-glycerol phosphate dehydrogenase, lactic acid dehydrogenase, and cytochrome oxidate) are also affected by arsenic.

The thioarsenic compounds are relatively stable, but eventually may be metabolized further by being oxidized to active arsenoxides and then to additional oxidation compounds, as follows: $RAs-AsR \rightarrow RAsO \rightarrow RAsOH$ [38,51].

Excretion

Single doses of arsenic leave the body slowly, principally via the kidneys and urine. The greater portion of the arsenic excreted in the urine appears in the first 4 days [46]. Excretion starts 2-8 hr after administration, but small concentrations are still found in the urine of man after 10 days [25]. Up to 70 days may be required for complete elimination after repeated administration of arsenic [17]. Increased amounts of urinary coproporphyrin are of diagnostic value in arsenic toxicosis [47].

Following administration, arsenic is also eliminated in the feces. The excretion of the metal into the gastrointestinal tract is minor, however [12]. Small amounts are excreted by the sweat glands. The absorption of arsenic does not result in a significant increase in excretion in the milk [35].

The administration of dimercaprol greatly increases urinary excretion of arsenic by removing the metal from the tissues into a thiolarsenic complex which is readily excreted by the kidneys. Maximal urinary excretion occurs 3-4 hr after the antidote is given. Fecal arsenic concentrations are not affected.

LEAD

Absorption

Metallic lead is slowly, but constantly, absorbed by most routes except the skin. Skin abrasions and lesions, however, will allow significant absorption. Lead particles buried subcutaneously or intramuscularly are frequently absorbed in sufficient concentration to cause poisoning within 1 month.

Respiratory tract absorption of lead dust is commonly the cause of industrial poisoning. Lead is absorbed from all portions of the respiratory tract, including the nasal passages, and indeed this absorption is more complete and rapid than by any other route [17].

Lead absorption from the gastrointestinal tract is slow and dependent upon the species involved. In general, about 10% of ingested inorganic lead is absorbed. Increased motor activity of the digestive tract decreases the absorption of lead; decreased activity causes increased lead absorption. Lead absorption in man occurs mainly from the small intestine, to a lesser extent from the colon, and not at all from the stomach [22-24].

In domestic animals, the alimentary tract absorption of lead proceeds slowly over a period of hours. The lead salts, insoluble in water, are dissolved readily by the gastric juices and absorbed from the small intestine [7].

Distribution

Following absorption, lead is rapidly removed from the plasma to combine with the blood cellular elements. Nearly all of the circulating inorganic lead is associated with the erythrocytes, chiefly in the membrane stroma, and only when lead is present in large amounts does the plasma contain significant quantities [17].

The pattern of distribution in the tissue varies with the route of administration. Following intravenous injection, the highest quantities

occur in the liver and bone marrow; smaller amounts are found in the kidneys, spleen, lungs, and bone. This probably reflects initial distribution
and the formation of colloidal lead phosphate in the plasma [1]. After oral
ingestion of inorganic lead, 60% is distributed to the bone, 25% is found
in the liver, 4% is in the kidney (but in high concentration per gram of
tissue), 3% is deposited in the intestinal wall, another 3% is in the reticuloendothelial system, and 4% is spread throughout the remaining tissues
of the body, including the teeth and hair. The slower absorption by this
route probably allows for uniform distribution and redistribution [7,22-24].

The tissue concentrations in the kidney cortex provide higher levels
for a longer period of time than do any other tissues in the body. Kidney
thus provides the best tissue for analysis [7]. Lead concentrations in
hair gradually increase until 2 months after the last lead administration.
The degree of exposure to lead may be fairly evaluated by measuring the
hair lead content [49].

Biotransformation

Despite the frequency of lead poisoning, little work has been done on
the metabolic action and biotransformation of inorganic lead. Although
generally assumed to be a protoplasmic poison and to combine with sulfhydryl groups (resulting in enzyme inhibition), the relative ineffectiveness
of dimercaprol treatment and the affinity of lead for bone suggest that
dithiol binding is the only mechanism involved in vivo and that other interactions should also be investigated.

Complete inhibition of the sulfhydryl-containing enzyme succinoxidase
has been demonstrated in the dog [9]. Other studies have shown that lead
produces an inhibition of active potassium transport in the red blood cell
membrane, a finding consistent with the occurrence of anemia and increased
urinary coproporphyrin excretion in lead poisoning cases [17,53].

Porphyrinuria is one of the first signs of lead poisoning. Abnormal
amounts of coproporphyrin III (5 to 50 times increase over normal) and coproporphyrin I (accounting for 30% of porphyrin excretion) appear in the
urine within a day [18,47].

Excretion

Inorganic lead is excreted from the body chiefly in the feces and
urine, the concentrations being dependent upon the duration of exposure to
the metal. Ninety percent of orally ingested lead is unabsorbed and passed
in the feces. In addition, the small quantities of lead excreted in the

bile and from the intestinal mucosa are present [46]. Bile excretion may
represent a major pathway of elimination, however, in cases of heavy lead
absorption. After ingestion of lead is stopped, fecal excretion of lead
ceases shortly.

Most of the absorbed lead is excreted by the kidneys at rates directly
proportional to the rate of absorption. Industrial workers in lead smelter
plants have excretion values of 5-736 µg of lead per 100 ml of urine [47].
The urinary excretion of lead in sheep is dependent upon the quantity of
lead absorbed, but does not exceed 0.8 mg per day [8]. After discontinuance
of lead exposure, the concentration excreted in the urine at first decreases
rapidly and then more evenly. The normal range of urine lead is reached in
a period from a few weeks to 18 months or more. The concentrations in the
tissues diminish in accordance with urinary excretion and reach normal
levels in 12-18 months. The skeleton is the last tissue to return to normal
[17,22].

Lead is eliminated through the milk in proportion to the concentration
found in the blood cells, but probably never exceeds 1.0 ppm. The concentration of lead in the sweat is about the same as that in urine, but the
total quantity excreted is small [17].

The excretion of lead is enhanced by conditions that favor mobilization
of bone and soft tissue reserves. Dietary imbalances, vitamin D injections,
parathyroid hormone, acidosis, and iodide and bicarbonate administration
increase urinary levels of lead. Exogenous chelating agents such as calcium
disodium-edetate, penicillamine, and dimercaprol also serve to dissociate
lead from bone deposits and to hasten its urinary excretion.

MERCURY

Absorption

Absorption of mercury varies considerably with the chemical form of
the metal. Elemental mercury is probably not absorbed; its solubility depends upon the oxidation of mercury or the formation of sulfides. In the
presence of atmosphere or water, particularly if chloride is available,
mercury will be slowly oxidized to the mercurous form, which is capable of
being still further oxided to mercuric ions. Except where long-time exposure to the body is allowed, this oxidation and solution in body fluids
does not usually proceed at a rapid enough rate to permit significant absorption.

If elemental mercury is applied to the skin in suitable vehicles, it may readily be absorbed, especially if oxidation occurs. Inhaled mercury is usually oxidized rapidly enough to assure sufficient exposure for intoxication [21]. The ionic mercury passes into the blood with a half-life of 7 hr. When applied to the skin as an ointment, amounts of mercury up to 31% of that applied were excreted [46].

Orally ingested mercury may be absorbed if oxides or sulfides are formed. Insoluble inorganic mercurous compounds, such as calomel, undergo oxidation to soluble absorbable compounds. The inorganic soluble mercurials (mercuric compounds) readily gain access to the circulation, although considerable portions remain fixed to the alimentary mucosa and intestinal contents [17].

Mercury compounds present in vaginal jellies are easily absorbed and retained in the body.

Distribution

As the absorbed mercury passes into the blood, one-half is firmly bound to the albumin of plasma (in combination with sulfhydryl groups) and the other half is associated with the red cells [6]. It is, therefore, readily redistributed to the tissues and in a few hours is found in highest concentration in the kidneys. The liver, blood, spleen, respiratory mucosa, intestinal and colon walls, skin, salivary glands, heart, skeletal muscle, brain, and lung contain decreasing amounts. Mercury is stored temporarily in the bone and bone marrow, but after 24 hr very small amounts remain [47].

By the end of 1 week, 85-95% of all the mercury in the body is stored in the kidneys [42] and is virtually limited to the proximal convoluted tubules. With increasing dosages the lesion and concentration progress proximally [40]. Increasing the quantity of mercury absorbed raises the concentration found in other body tissues but has no effect on kidney levels [15]. The kidney has the highest concentration of mercury and retains 60-80% of a given amount after one week.

Biotransformation

Mercury ions are general protein precipitants and cause severe necrosis at the points of direct contact with tissue. This is especially common at the sites of entry (mouth, esophagus, skin, conjunctiva, cornea, and the cells lining the gastrointestinal tract) and the excretory routes (kidney and large bowel).

The ionic metal readily and with preference forms covalent linkage to the sulfur of sulfhydryl groups. Even in low concentrations this affinity for mercury capture is great and gives the -SH group the name mercaptans. Ring compounds, however, are not formed with mercury because the bond angle of mercury is 180 degrees. Mercury also combines readily with other important ligands: amide, amine, carboxyl, and phosphoryl groups [16,21].

The specific and varied biochemical mechanisms of the many chemical mercurial states are still unknown, as are the relationships of the mercury-binding radicals to the various protein-precipitant, corrosive, and enzyme-inhibitory effects of mercury. Some information about tissue and cellular interactions has been reported, however [10,34]. Mercury appears to first combine with the functional groups on the cell membrane, probably the thiol (sulfhydryl) groups because of preferential interaction. The ion then slowly enters the cell. The uptake is rapid for 20 min and then proceeds at a much slower rate. The cell uptake of glucose is inhibited during the first, slow phase; in the second phase, cellular respiration is progressively inhibited. When the cell membrane is broken, the inhibition of respiration occurs immediately upon the addition of mercuric ion, suggesting that the first phase is primarily to allow membrane permeability, while the direct inhibition of cellular respiratory enzymes is the definitive interaction. A decrease in the electrical potential across the cell membrane and loss of cellular potassium is also observed. Inhibition of glucose uptake by the cell membrane is reversed by the addition of extraneous mono- and dithiol groups. The effect on cellular respiration is not reversible.

Mercury has also produced red blood cell hemolysis and porphyria. The metallic ion enters into a chelation or coordination complex with the erythrocytes, which causes the red cells to clump [5].

Excretion

Absorbed mercury is excreted chiefly in the urine, but considerable quantities are also passed in the feces through secretion into the gastrointestinal lumen (primarily the colon), bile, saliva, and gastric and intestinal juices. Excretion begins immediately upon absorption and continues at a rapid rate, the tissue concentrations falling at different schedules. The mercury content of the brain falls especially slowly [17]. Most of the mercury is excreted within a week, but low levels may be found in the urine and feces for months.

The kidney tissue retains its mercury content with great tenacity. One week after exposure, 60-80% of the mercury originally stored is still present. Because of its storage in that organ, the concentration of mercury in the kidney, and to a lesser degree in the urine, is an indication of the severity of exposure. Increased dosages result in a proportionate increase in the urinary output of mercury [49]. Sixty to 70% of the mercuric ions are excreted in the urine as sulfhydryl-mercury compounds of the type RHg-SR'. One of the common thiols to combine with mercury is cysteine; another is n-acetylcysteine. No inorganic mercury is excreted in the urine [32].

The kinetics of mercury clearance from the body has recently been reported by Rothstein and Hayes [42]. Using radioactive mercury, three distinct phases were shown. The first involved 35% of the absorbed mercury. A half-time of 3-4 days was noted and represented rapid buildup in the liver with subsequent excretion into the feces (via the bile and other secretions) and translocation into the kidneys. The second component involved 50% of the absorbed dose and had a half-time of 30 days. This was the result of accumulation of the ion in the kidneys and excretion into the urine. The third phase involved 15% of the dose and was accounted for by excretion from the kidneys; the half-time was 100 days.

The urinary excretion of mercury may be materially hastened by the administration of external sources of dithiols and other chelating agents. Dimercaprol, d,1-penicillamine, and n-acetyl-d,1-penicillamine are commonly and successfully employed in poisoning instances [17].

SELENIUM

Absorption

Selenium occurs in nature as the toxic selenite ion or in some organic combination. In commercial compounds, the metal is present as the ionic selenium or as the sodium salts, sodium selenate, and sodium selenite. All selenium compounds are readily absorbed from the gastrointestinal tract and through the lungs. Virtually none is absorbed when applied to the skin. After subcutaneous injection, the maximum blood levels are reached within 15 min [46].

Distribution

Following absorption, selenium is carried fixed in the red blood cells and is found associated with plasma albumin and globulin. The albumin is the immediate receptor and is directly or indirectly involved in the transport of selenium to more stable binding sites in blood and tissues [49]. It is distributed to all soft tissues, but the concentration is especially

high in the liver and kidneys. The lungs and milk contain increased amounts
of selenium following its administration. The selenite ion has a greater
affinity for tissues than does selenate [41].

Single doses cause blood levels of 10-25 ppm. An hour after administration nearly half the absorbed selenium is found in the liver. Significant amounts are found in the hair and hooves of acutely poisoned animals
[20]. Selenium is also transported to the fetus through the placenta [49].

Continuous administration of selenium compounds causes greater levels
to accumulate in red blood cells than in plasma. The largest tissue concentrations are in liver and kidney, but greater amounts are deposited in
hair, nails, and hooves following multiple doses. These high concentrations
in the latter structures persist for at least 6 weeks [46].

Biotransformation

Much of the fundamental biochemistry and toxicology of selenium compounds is still being developed. Inorganic selenium salts are converted to
organic selenium-protein complexes by the liver, muscle, hemoglobin, and
plasma. Selenate salts are reduced enzymatically by the liver, spleen,
whole blood, and plasma to selenite and eventually to selenium, which is
then tied in protein complexes. Sodium selenate is not toxic until it is
reduced to selenite [41].

Portions of selenium are converted to volatile methyl selenide, which
produces a garlic-like odor in the breath.

The mechanism of selenium methylation is not known, but the changes
that occur are apparently as follows: $Na_2SeO_4 \rightarrow Na_2SeO_3 \rightarrow Se(CH_3)_2$ [29].

Selenium replaces sulfur in plant cystein and methionine. In sucking
insects, selenium acts as a metabolite analog to sulfur compounds necessary
to insect metabolism [2]. The exact role of selenium in producing toxicity
in higher animals, however, is still not clear.

Scott has suggested that the toxic mechanism of selenium is that of
competition with sulfur for sites at which sulfur normally plays a role in
cellular metabolism [43]. The administration of sulfur concurrently with
selenium produces no specific antidotal action, nor does selenium accom-
plish a direct substitution for sulfur in vivo. The mixing of selenium-
sulfur solutions (1:0.4) in vitro produces a precipitate which might also
occur in the biological system [54].

Dog liver proteins labeled in vivo with radioactive selenium contain
activity in three different compounds. The greatest concentration of

activity is associated with cystine and selenocystine; the second greatest
concentration appears with methionine and selenomethionine; the third area
of activity is associated with leucine [30].

Selenium may take the place of sulfur in the synthesis of mercapturic
acids, since the feeding of bromobenzene to steers with selenium poisoning
causes a reduction in the blood selenium level and a rise in urinary selenium excretion, probably as selenium-mercapturic acid [31]. Westfall and
Smith [55], however, found no effect of bromobenzene in increasing selenium
excretion.

The capability of selenium to inhibit certain enzyme systems is widely
reported, particularly its effect upon those enzymes requiring sulfhydryl
groups. The administration of dithiol groups (as dimercaprol) has no effect
upon selenium toxicity, however [55]. Very little effect is shown by selenite on the combination of several sulfhydryl enzymes [5]. Mercaptide
formation is thus an unlikely mechanism to explain the effects of selenium.

More likely the toxicity develops from a selenium-catalyzed oxidation
of such cofactors as glutathione, coenzyme A, and dihydrolipoic acid and
the resultant disturbance in intermediary metabolism. The metal may also
exert its effect by inhibiting enzymes involved in cellular oxidation.
Oxygen uptake of yeast decreases 80% in the presence of selenite. The de-
hydrogenating enzymes which play essential roles in tissue oxidation are
particularly sensitive to selenium [37].

Thus, while the evidence would point toward enzyme inhibition, especially of oxidative processes, no clear-cut explanation is yet available
to explain the metabolic action of selenium.

Excretion

The elimination of selenium is slow. Selenium retention is greater
after feeding natural seleniferous foods than after the administration of
selenite. Selenium given orally is one-third lost during the first week;
traces still remain after a month. After the injection of selenite, however, urinary excretion accounts for more than half the dose in 24 hr, with
small quantities passed on subsequent days [29]. In chronically poisoned
animals, the elimination of selenium is small until after the tissues become saturated. Thereafter, elimination increases until severe poisoning
depresses body functions.

Selenium excretion occurs mainly in the urine. Some is found in the
feces through elimination in the bile, and sizable portions are excreted

through the lungs and in the milk. Perspiration contains minor quantities. Two-thirds of radioactive selenium is excreted in the urine, 10% is found in the feces, and 7.5% is recovered in the expired air in 48 hr [20]. The urine of animals fed seleniferous wheat contains selenium primarily in organic combinations [55], while the greater part of the selenium administered as selenite is excreted in the urine in the inorganic form [49].

The portion excreted in the breath is passed as dimethyl selenide and may account for as much as 30% of the ingested selenite. Selenate produces far less dimethyl selenide than does selenite [29]. The concentration of selenium in the milk of poisoned cows reaches 3 ppm, and enough of the toxic metal may be eliminated in this way to produce toxicity in nursing calves [30]. Approximately 54% of the selenium in milk is in the casein fraction, 40% is present in the albumin fraction, and 7% is in globulin binding [28].

In general, the distribution among the various excretory pathways is rather variable and depends in part on the rate and form of selenium absorption.

There is no current specific antidote that may be given to greatly alter the excretion of seleniferous compounds. Application of the previously mentioned factors may indirectly increase elimination. Sodium sulfate may be administered and will increase the rate of urinary selenium excretion without altering the rate of excretion in the feces [19]. The enzymatic effect of selenium can be counteracted in some cases by arsenic, and the metal's deleterious action on the liver may be reduced by the "sparing action" of high-protein diets and by the administration of methionine in the presence of alpha-tocopherol [31].

THALLIUM

Absorption

Thallium is rapidly and almost completely absorbed through the skin and via the mucous membranes of the mouth and gastrointestinal tract. Peak blood levels are reached 2 hr after oral administration, and thallium is very quickly demonstrated in the urine [46].

Distribution

Using radioactive thallium, the concentration of the element in the blood 1 hr after absorption was found to be low. Heavy accumulations occur in the bones and renal medulla; lower, but still considerable, concentrations are in the gastric and intestinal mucosa, pancreas, salivary glands, hair, eye lens, and epithelium of the tongue. No thallium is found in the central nervous system 1 hr post-absorption. Ten days after administration, changes in distribution are seen. The thallium concentration increases in the renal medulla, and the level in the central nervous system is higher than in the liver. A high concentration is still present in the hair, but the level in the bones has decreased much more rapidly than in the muscle. No thallium is present in the blood [4].

Thallium passes through the placenta into the fetus. It also is present in the milk of poisoned individuals in high enough concentration to be toxic to the suckling [44].

Observations from humans poisoned with thallium give varying tissue distributions, depending primarily on the duration of illness, but in general the highest thallium concentrations are found in the kidney, followed in decreasing order by the bones, stomach, and small and large intestines, spleen, liver, muscle, lung, and brain [38].

Biotransformation

The exact explanation of the metabolic action and fate of thallium is not known. Studies of sulfhydryl enzyme systems reveal no change due to thallium, although some inhibition of succinic dehydrogenase activity occurs with high concentrations of the metal [14]. At very low concentrations the muscle membrane cannot distinguish between potassium and thallium ions. If such concentrations are left to act long enough, irreversible damage results to the muscle. The alopecia of thallium intoxication is thought to result from the action of the metal on the endocrine and vegetative nervous systems. Thallium also acts directly upon the hair follicles [33].

The effects of thallium are also thought due to its interference with the metabolism of compounds containing sulfur. Diets high in cystine protect rats against chronic thallium poisoning. Methionine and betaine dietary supplements also offer protection. Thallium has not been found to interfere with enzymes that contain sulfur. The metal does not block free sulfhydryl groups in the skin, and it does not react with serum albumin. Dimercaprol has no protective action. Monothioglycerol and mercaptopropane, both of which contain sulfur, have moderate protective activity. Aminoethylisothiuronium and aminopropylisothiuronium, which contain very reactive mercapto groups, and diphenylthiocarbazone, which contains sulfur, are relatively active in protecting against thallium 24 hr after administration. These compounds may protect against the effect of thallium by shielding the active

(sulfhydryl) groups of proteins by the formation of reversible disulfide
binding with thallium at the site of action [45].

Excretion

Thallium is very slowly excreted from the body. The biological half-
life is 3-8 days. Since the rate of loss is proportional to the amount
residing in the body, progressively smaller quantities are excreted with
time. The body clearance of thallium proceeds exponentially for at least
the first 3 weeks or until 1% of the administered dose remains [26]. Seventy to 90% of an administered dose is excreted within 4 weeks.

Excretion of thallium begins within hours after absorption and occurs
chiefly through the gastrointestinal tract and in the urine. The feces do
not contain thallium until the fourth day, and the metal is still present
in feces after 35 days [39]. While bile contains thallium, its secretion
by that route is not selective [27]. Digestive secretions probably provide
most of the thallium found in the feces.

Excretion via the urine is about half that occurring into the digestive
tract [46]. While it is one of the first routes of elimination, excretion
in the urine persists for up to 3 months after thallium administration. To
a lesser degree excretion takes place in the tears, sweat, and milk [39].

Retention of thallium in the organs for long periods may be due to the
fact that thallium is sparingly soluble. Forty-five percent of the administered dose is still present in the human body after 24 days. Thallium is
detected in cerebrospinal fluid after 25 days [39]. In dogs, significant
plasma concentrations are still detected after 26 days [11]. Approximately
half the oral dose of thallium salt is present in geese after 15 days [46].

The only exogenous compound capable of increasing the excretion of
thallium, and lessening the signs and lesions of poisoning, is diphenylthiocarbazone. This chemical forms a biological complex with thallium ions
and facilitates their elimination in the urine [11].

SUMMARY

Heavy metals are potent toxins. Their metabolic fate must be recognized to understand their actions and therapeutics. Solubility and binding
to tissue ligands are important factors in determining the metals' effect
on a biological system. The digestive tract, liver, and kidney are common
sites of attack and excretion.

Arsenic is relatively easily absorbed by a number of routes. It is distributed initially to the liver, kidney, lung, and digestive tract wall. After several days, accumulation in skin, hair, and bone occurs. The sulfhydryl groups of metabolic enzymes are attacked by the metallic ion and interference with oxidative phosphorylation results. Arsenic is slowly excreted in the urine. This excretion is increased by the administration of dimercaprol.

Inorganic lead is slowly absorbed by most routes except the skin. Solubility is a major factor. After absorption, quantities accumulate in bone, liver, and kidney, the latter being the best tissue for diagnostic work. Little is known of the true biotransformation of lead. Enzyme inhibition through sulfhydryl-group binding is postulated. Excretion occurs via the feces and urine. Bone is the last tissue to completely release its lead. A variety of metabolic and chelating agents are useful to increase excretion.

Ionic mercury is readily absorbed from any surface if in union with oxygen or sulfur. It is then found in highest concentration in the kidney. Specific biochemical action is with the sulfhydryl groups of various proteins, as well as with other binding sites. Inhibition of cellular respiration and protein precipitation are mercury's major effects. Excretion is primarily through the kidney and gastrointestinal tract. Retention by the kidney tissue is long term, but excretion may be hastened by the administration of dithiols and other chelating agents.

Selenium is rapidly absorbed from the gastrointestinal tract and lungs. It is distributed to all soft tissues but is highest in the liver and kidneys. During chronic exposure, accumulations occur in keratinized integumentary structures. Selenium is converted to selenate, selenite, and volatile methyl selenide. It may become bound in protein complexes. Its mechanism in the biological system is unclear; substitution for sulfur and enzyme inhibition through interference with oxidation have been suggested. Elimination of selenium is slow and occurs mainly via the urine. Specific antidotes are not available.

Elemental thallium is extremely absorbed via any route of application. General body distribution is diffuse, but slightly greater levels are found in the kidneys. With chronicity, the centers of concentration shift to include the central nervous system and hair. The metabolic action of thallium is not known. Interference with the metabolism of sulfur-containing compounds is under investigation. The protective ability of reagents

containing sulfur and active mercapto groups suggests that protein sulfhydryl groups may be the active site. The metal is very slowly excreted
from the gastrointestinal tract and kidneys. Considerable quantities remain in various body tissues. Diphenylthiocarbazone increases the urinary
elimination of thallium.

REFERENCES

1. K. R. Adam and M. Weatherall, J. Pharm. Pharmacol., 6, 403 (1954).

2. A. Albert, Selective Toxicity, Wiley, New York, 1965.

3. T. L. Althausen and L. J. Gunther, J. A. M. A., 92, 2002 (1929).

4. T. Andre, S. Ullberg, and G. Winquist, Acta Pharmacol. Toxicol., 16,
   229 (1960).

5. J. P. Arbuthnott, Nature, 196, 277 (1962).

6. M. Berlin and S. Gibson, Arch. Environ. Health, 6, 617 (1963).

7. L. I. Blaxter, J. Comp. Pathol. Ther., 60, 140 (1950).

8. L. I. Blaxter, J. Comp. Pathol. Ther., 60, 177 (1950).

9. H. O. Calvery, E. P. Lang, and H. J. Morris, J. Pharmacol. Exp. Ther.,
   64, 364 (1938).

10. D. J. Demis and A. Rothstein, Amer. J. Physiol., 180, 566 (1955).

11. R. L. Doak, R. P. Schmidtke, J. D. Wallach, L. E. Davis, and K. H.
    Niemeyer, Vet. Med. Small Animal Clin., 60, 1277 (1965).

12. H. S. Ducoff, W. B. Neal, R. L. Straube, L. O. Jacobson, and A. M.
    Brues, Proc. Soc. Exp. Biol. Med., 69, 548 (1948).

13. A. P. Dustin, Bull. Acad. Med. Belg., 13, 585 (1933).

14. P. J. Flesch, Investigative Dermat., 15, 345 (1950).

15. R. R. Forney and R. N. Harger, Fed. Proc., 8, 292 (1950).

16. H. L. Friedman, Ann. N. Y. Acad. Sci., 65, 461 (1957).

17. L. S. Goodman and A. Gilman, Pharmacological Basis of Therapeutics,
    Macmillan, New York, 1965.

18. M. Grinstein, H. M. Wikoff, R. P. de Mello, and C. J. Watson, J. Biol.
    Chem., 182, 723 (1950).

19. A. W. Halverson, P. L. Guss, and O. E. Olson, J. Nutr., 77, 459 (1961).

20. M. Heinreich and F. E. Kelsey, J. Pharmacol. Exp. Ther., 114, 28 (1955).

21. W. L. Hughes, Ann. N. Y. Acad. Sci., 65, 454 (1957).

22. R. A. Kehoe, J. Cholah, D. M. Hubbard, K. Bambach, R. R. McNary, and
    R. V. Story, J. Ind. Hyg. Toxicol., 22, 381 (1940).

23. R. A. Kehoe, F. Thamann, and J. Cholah, J. Ind. Hyg. Toxicol., 15,
    257 (1933).

24. R. A. Kehoe, F. Thamann, and J. Cholah, J. A. M. A., 104, 90 (1935).

25. G. R. Kingsley and R. R. Schaffert, Anal. Chem., 23, 914 (1951).

26. R. Lie, R. G. Thomas, and J. K. Scott, Health Physiol., 2, 334 (1960).

27. A. Lund, Acta Pharmacol. Toxicol., 12, 251 (1956).

28. K. P. McConnell, J. Biol. Chem., 173, 653 (1948).

29. K. P. McConnell and O. W. Portman, J. Biol. Chem., 195, 277 (1952).

30. K. P. McConnell and C. H. Wabnitz, J. Biol. Chem., 226, 765 (1957).

31. A. L. Moxon, A. E. Schaefer, H. A. Lardy, K. P. DuBois, and O. E. Olson, J. Biol. Chem., 132, 785 (1940).

32. O. H. Mueller and I. M. Weiner, J. Pharmacol., 118, 461 (1956).

33. L. Mulline and R. D. Moore, J. Gen. Physiol., 43, 759 (1960).

34. H. Passow, A. Rothstein, and T. W. Clarkson, Pharmacol. Rev., 13, 186 (1961).

35. S. A. Peoples, Fed. Proc., 20, 174 (1961).

36. R. A. Peters, H. M. Sinclair, and R. H. S. Thompson, Biochem. J., 40, 516 (1946).

37. V. R. Potter and C. A. Elvehjem, Biochem. J., 30, 189 (1936).

38. L. Reines and C. S. Leonard, Proc. Soc. Exp. Biol. Med., 29, 946 (1932).

39. J. F. Reith, in Thallium Poisoning (J. J. G. Prick, W. G. S. Smitt, and L. Muller, eds.), Elsevier, Amsterdam, 1955.

40. A. E. Rodin and C. N. Crowson, Amer. J. Pathol., 41, 297 (1962).

41. I. Rosenfeld and O. E. Beath, J. Biol. Chem., 172, 333 (1948).

42. A. Rothstein and A. D. Hayes, J. Pharmacol. Exp. Ther., 130, 166 (1960).

43. M. L. Scott, in Mineral Metabolism (C. L. Comar and F. Bronner, eds.), Vol. 2, Part B, Academic, New York, 1962.

44. T. Sollmann, Manual of Pharmacology, 8th Ed., W. B. Saunders, Philadelphia, 1957.

45. W. B. Stavinoha, G. A. Emerson, and J. B. Nash, Toxicol. Appl. Pharmacol., 1, 638 (1959).

46. C. P. Stewart and A. Stolman, Toxicology, Mechanisms and Analytical Methods, Vol. I, Academic, New York, 1960.

47. C. P. Stewart and A. Stolman, Toxicology, Mechanisms and Analytical Methods, Vol. II, Academic, New York, 1961.

48. L. A. Stocken and R. H. S. Thompson, Biochem. J., 40, 529 (1946).

49. A. Stolman and C. P. Stewart, in Progress in Chemical Toxicology (A. Stolman, ed.), Vol. 2, Academic, New York, 1965.

50. C. C. Tsen and B. H. Collier, Nature, 183, 1327 (1959).

51. C. Voegtlin, Physiol. Rev., 5, 63 (1925).

52. C. L. Wadkins, J. Biol. Chem., 235, 3300 (1960).

53. M. Weatherall, Pharmacol. Rev., 6, 133 (1954).

54. B. A. Westfall, Ph.D. dissertation, University of Missouri, Columbia, 1938.

55. B. B. Westfall and M. I. Smith, J. Pharmacol., 72, 245 (1941).

5

METABOLISM AND METABOLIC ACTION OF LEAD
AND OTHER HEAVY METALS

Paul B. Hammond*
University of Minnesota College of Veterinary Medicine
St. Paul, Minnesota

The heavy metals of principal toxicologic concern today are lead and mercury. At one time they posed major health problems in the industries
which mined and processed them. Today the concern has shifted to their
possible health effects in the general population of both man and animals.
Similar concerns also are emerging in regard to other heavy metals such as
cadmium. An increased emphasis on research concerning the environmental
impact of the heavy metals is clearly in evidence. It therefore seems
useful to focus attention on at least some of the toxicologic phenomena
which must be considered in planning future research on the toxicology of
metals and metallic compounds, particularly as to long-term, low-level
effects. A phenomenon assignable to the actions of one metal will not
necessarily be applicable in the case of another. But great similarities
do exist within certain groups.

In speculating as to what toxic properties a metal might have, it is
useful to examine available information on other metals having similar
physicochemical properties, such as ionic radius and electronic configuration. In this last regard, metals of the same group of the periodic table
of the elements are most similar as to the characteristics of the outer
electron shells. Arsenic and antimony, adjacent members of Group Vb of
the periodic table of the elements, are quite similar as to their chemical
and toxicologic properties, as are tin and lead, which are adjacent members
of Group IVb.

---

*Present Affiliation: University of Cincinnati Medical Center, Cincinnati,
Ohio.

METABOLISM OF METAL COMPOUNDS

Although metals are immutable and therefore not subject to degradation, they are frequently assimilated as discrete compounds, the fate and toxicity of which probably are due to the molecular species rather than simply to the metallic constituent alone. Thus, tetraethyllead has toxicologic properties quite different from those of inorganic lead salts, both as to the toxic dose of the lead and as to the nature of the biological effects. For inorganic salts of lead, the intravenous LD_{50} in the rat is approximately 70 mg/kg [1], whereas for lead in the form of tetraethyllead, the intravenous LD_{50} is approximately 10 mg/kg (15.4 mg/kg as tetraethyllead) [2]. Signs of poisoning also are quite different. Rats given tetraethyllead exhibit a high state of excitability, particularly to external stimuli, and generalized muscular tremors, whereas rats given inorganic salts of lead experience only a transient loss of appetite and die without manifestations of encephalopathy.

The conclusion that may be drawn from the above considerations is that tetraethyllead is an agent with toxic properties ascribable to the molecular entity rather than to the lead component. Actually, the conclusion is not correct. Tetraethyllead is susceptible to dealkylation in the body and, in fact, all evidence points to the dealkylated metabolite triethyllead as being responsible for the toxic effects of tetraethyllead. Dealkylation occurs in the liver. Evidence that triethyllead is the true toxicant is provided by data relating to the in vitro effects of the tri- and tetraethyl forms. Tetraethyllead has no effect on the respiration of brain slices when added in vitro except at very high concentrations but is quite active when given in vivo. The effect following in vivo administration occurs with concentrations of the metabolite, triethyllead, which produces similar effects whether the compound is administered in vitro or in vivo (Table 1).

A further dealkylation of tetraethyllead would yield diethyllead. If such a reaction does occur, the resultant metabolite probably plays little or no role in the manifestations of tetraethyllead toxicity [2]. As a matter of fact, there is evidence to suggest that little if any diethyllead is even formed following administration of tetraethyllead [3].

Tetramethyllead, which is also used as a fuel additive, bears the same relation to its metabolite trimethyllead as tetraethyllead bears to triethyllead. Similar evidence has been presented for the in vivo conversion of tetraethyltin to triethyltin and, again, it appears that the process is one converting a relatively nontoxic compound to a highly toxic one [5].

TABLE 1. Comparison of Activity of Brain Slices with Triethyllead Added in Vitro and Slices from Rats Given Either Tri- or Tetraethyllead in Vivo[a]

Route of Administration	Concentration of Triethyllead in Brain Fluid [17] $(x 10^{-5})$	Brain Cortex Slices (% of Control)		
		$Q O_2$	Lactate	Pyruvate
Triethyllead, in vitro	1.6[b]	86	285	66
	4.6[b]	45.5	355	38
Triethyllead, in vivo	1.8	54.5	330	58.5
	3.4	53	300	52
Tetraethyllead, in vivo	1.02	50	295	46
	8.15	33	280	36

[a]Reprinted from Ref. 2 by permission.

[b]Concentration in reaction vessel.

The demethylation of monomethylmercury has been under intensive study in recent years as a result of the widely publicized contamination of fish with methyl mercury. To the extent that it occurs in animals, demethylation of mercury is considered to be a detoxification mechanism (in contrast to the consequences of demethylation of tetramethyllead). This is so in spite of the fact that the acute LD_{50} for inorganic salts of mercury is somewhat lower than for methyl mercury salts [6]. Under conditions of acute poisoning, demethylation could hardly qualify as a detoxication mechanism. But with chronic exposure, the dealkylation process is a detoxifying mechanism since methyl mercury is considerably more toxic than inorganic mercury under these conditions. This is because methyl mercury is more cumulative than inorganic mercury [7].

The magnitude of demethylation is very substantial. With the first 10 days after administration of radiolabeled methyl mercury, approximately 50% of the radiolabeled mercury is excreted in the urine and feces as inorganic mercury [8]. But, even so, for many organs the fraction of the mercury retained as the inorganic form increases rapidly with time and may eventually equal or exceed the fraction retained in the original methylated form (Figs. 1 and 2).

In some organs, notably the brain, the fraction of mercury retained in the inorganic form is extremely small ($< 5\%$). This is not due to a more rapid clearance of the inorganic form, at least not in the case of the brain, since the fraction retained as inorganic mercury does not decrease

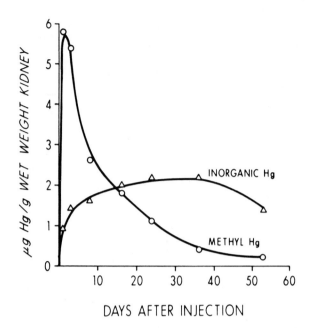

FIG. 1. Organic and inorganic mercury in kidneys of rats after intravenous methyl mercury. Reprinted from Ref. 8 by permission (copyright 1970, American Medical Association).

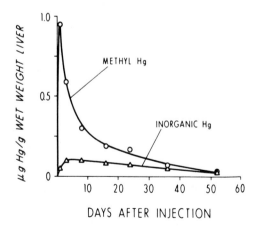

FIG. 2. Organic and inorganic mercury in the liver of rats after intravenous methyl mercury. Reprinted from Ref. 8 by permission (copyright 1970, American Medical Association).

perceptibly over 28 days [8]. The metabolism of ethyl mercury has also been studied to some extent. Here, too, dealkylation occurs to a substantial degree [9].

While dealkylation of mercury is a detoxifying metabolic pathway, the reverse process of alkylation would serve even better as a means of accelerating the rate of loss of methyl mercury from the body. Methylation of methyl mercury yields dimethyl mercury, which is rapidly excreted via the lungs. In mice, about 60% of a single intravenous dose of dimethyl mercury is excreted via the lungs within 3 hr (Fig. 3) [10]. The termination of this excretory process after approximately 6 hr is probably due to the simultaneous conversion of dimethyl mercury to monoethyl mercury, which is not readily exhaled. Interest in the metabolism of dimethyl mercury results largely from the knowledge that methanogenic bacteria which form monomethyl mercury in the aquatic environment also form dimethyl mercury

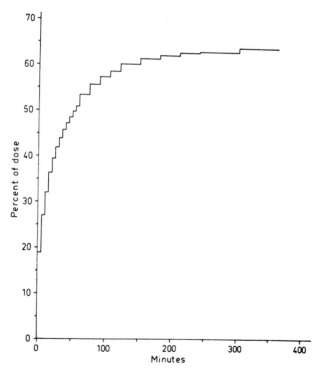

FIG. 3. Accumulated excretion of radiomercury via the lungs after intravenous injection of dimethylradiomercury in mice. Reprinted from Ref. 10 by permission.

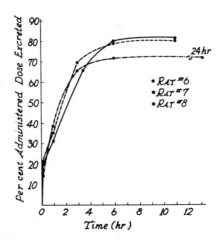

FIG. 4. Respiratory excretion of dimethylselenide by the rat after subcutaneous injection. Reprinted from Ref. 13 by permission.

[12]. It does not of itself accumulate in the aquatic environment since it
is highly volatile and is readily released into the air.

While methylation of mercury does not appear to occur in animals,
methylation of selenium does, and, as in the case of dimethyl mercury, the
methylated form is rapidly cleared from the body via the lungs (Fig. 4)
[13].

Dimethylselenide is probably a significant metabolite of inorganic
selenium only at toxic or near-toxic levels, while at normal or near-normal
dietary levels, formation of trimethylselenide seems to be the major route
of detoxification [14]. The metabolite, dimethylselenide, is of a very low
order of toxicity. The LD_{50} by the intraperitoneal route is 1600 and 1300
mg/kg in the rat and the mouse, respectively [15]. By contrast, inorganic
forms of selenium are extremely toxic. The I.P. LD_{75} of selenium as sodium
selenite the the rat, for example, is only 3.25-3.50 mg/kg [16].

The reverse process of demethylation of dimethylselenide must be occurring only at a very slow rate if it occurs at all, since even very large
doses of dimethylselenide are nontoxic and hence must not be giving rise to
any appreciable amount of demethylated or inorganic form.

DISTRIBUTION AND TRANSLOCATION

Metals have characteristic affinities for certain organs and systems.
The consequences of preferential localization in certain organs of the body
are not necessarily bad. As an example, lead has a great affinity for bone.
More than 90% of the lead in the body is localized in bone. Yet, the toxic

effects of lead are the result of its presence in much lower concentrations elsewhere in the body, notably in the hematopoietic system and in the central nervous system. In the case of methyl mercury, the highest concentration is in the kidney, but the manifestations of toxicity result from its simultaneous presence in the brain in a much lower concentration. The presence of toxic metals in what amounts to storage sites is nonetheless undesirable since there always exists the possibility that the metal might suddenly be mobilized as a result of some alteration in body function. There is little known concerning circumstances under which significant mobilization might occur. It has been stated that acidosis induced with ammonium chloride enhances the excretion of lead by mobilization from the bone [17]. It is conceivable that under some circumstances acidosis could convert a latent toxic burden to an active one. Translocation of metals from one point in the body to another has been shown to result from the administration of chelating agents. In the case of lead, administration of the chelating agent EDTA (ethylenediamine tetraacetate) causes translocation of lead from bone into muscle [18]. The effect is transient and of doubtful toxicologic significance. The translocation of methyl mercury has been demonstrated using dimercaprol (BAL) as the chelating agent [19]. In this study translocation occurred from the blood into the brain as well as into other organs. The implications remain to be explored.

Metals translocated under the influence of chelating agents are in the form of the metal-chelate complex. Once translocation is achieved, the metal may be released and exert a toxic effect. Alternatively, it may remain as the complex which could be either toxic or nontoxic. Evidence for a toxic translocation is seen in the case of the treatment of experimental cadmium poisoning with dimercaprol [20]. With cadmium alone, animals die within 24 hr with characteristic signs of muscular flaccidity. With cadmium plus dimercaprol, death is delayed beyond 24 hr, and the signs characteristic of acute cadmium poisoning are not observed. The animals ultimately die in even greater numbers than with cadmium alone, but with morphologic and biochemical evidence of severe renal damage. In this case it seems that the chelating agent serves to translocate the metal to the kidneys in amounts sufficient to cause fatal nephropathy.

Translocation and accelerated excretion of some nutritionally essential elements (copper and zinc) within the body has been shown to occur in experimental carbon disulfide poisoning [21]. Dietary supplementation with the translocated metals decreases the toxicity of the carbon disulfide.

The data suggest that metal ion shifts bear a causal relation to carbon disulfide poisoning. It is postulated that carbon disulfide reacts with free
amino groups in body structures to form dithiocarbamate, a chelating structure with a strong affinity for zinc and copper, the two metals shown to be
translocated.

DOSE-RESPONSE RELATIONSHIPS

The toxic actions of lead, mercury, and other heavy metals have been
the subject of extensive investigation ranging from purely clinical descriptions to studies of interactions with discrete enzymes. At one time
the major concerns were diagnosis of intoxication and therapy. Later the
emphasis shifted to the prevention of intoxification, mainly in the industries where exposure was particularly hazardous. The concept of tolerance
limit values (TLVs) was introduced. What, in effect, were the highest
levels of exposure which could be tolerated without compromising human
health? The highest tolerable levels could not be established without
some knowledge of the minimal toxic levels. The estimation of minimal
toxic levels of heavy metal intake is no longer solely a matter of concern
in industry. The widespread pollution of the general environment has resulted in efforts to assess minimal toxic levels of intake for the general
human population and for populations of animals thought to be at risk.

The search for endpoints of toxicity has mainly utilized the biochemical approach because of the sensitivity of the methodology and because the
effects are readily expressed in quantitative terms. Further, metabolic
alterations as reflected in abnormal concentrations of metabolites and altered enzyme activity often appear before functional damage is apparent.

A good example of the manner in which this approach has been utilized
is provided by lead.

Although lead inhibits a number of enzymes in vitro, its effects on
the enzymes which enter into the synthesis of heme are those best understood. Lead is a strong inhibitor of aminolevulinic acid dehydratase (ALAD)
which converts aminolevulinic acid to porphobilinogen (Fig. 5). It also
inhibits heme synthetase, which is responsible for the introduction of iron
into the tetrapyrrole ring (Fig. 5). The effects of the inhibition of heme
synthetase are manifested in the intact animals as protoporphyrinemia and
as elevated coproporphyrin excretion in the urine. Similarly, the inhibition of ALAD is manifested as elevated ALA excretion in the urine. In man

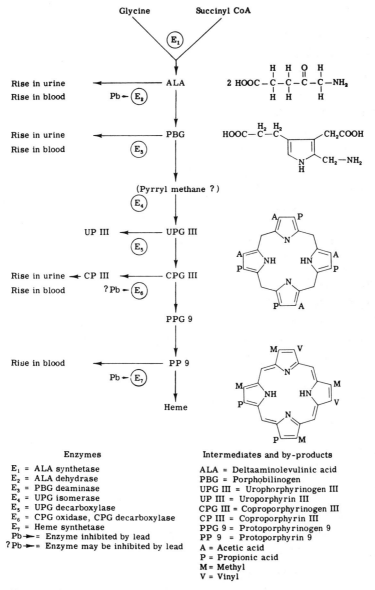

FIG. 5. The synthesis of heme. Modified from Haeger-Aronsen (1960). Reprinted from Ref. 28 by permission.

at least, elevated ALA excretion is a more sensitive and specific index of lead exposure than is elevated coproporphyrin excretion [22]. Abnormal ALA excretion occurs only when the blood level of lead exceeds 40 µg/100 g, a

value encountered in only a small fraction of the general population [23].
This biological effect is considered undesirable since it represents an
interference with the utilization of precursors for the synthesis of a
normal and essential body constituent (heme). One would presume that a deficiency of hemoglobin would result. Actually such is not the case. The
inhibition is compensated for, probably by increased production of ALA.
The capacity of animals to compensate for this inhibition is considerable.
Dogs fed lead in quantities sufficient to almost totally inhibit the ALAD
circulating in the peripheral blood remain apparently healthy. They respond as well to acute depletion of 50% of their blood volume as do control
animals in terms of regeneration of lost hemoglobin [24].

In contrast to the threshold nature of elevated ALA excretion, inhibition by lead of the activity of the enzyme ALAD in peripheral blood appears
to be of a nonthreshold character [25]. The enzyme activity decreases exponentially in proportion to the concentration of lead in the blood (Fig. 6).

It appears that this relationship holds for the full range of blood
lead concentrations encountered in the general population and in industrially exposed populations. Lead therefore seems to have a biological effect
even at exposure levels below those associated with abnormal excretion.
ALAD inhibition in this lower range can hardly be considered to be a toxic
effect, since no compromise in body function appears to occur. The inhibition probably represents functional loss of reserve enzyme.

Mercury binds strongly to sulfhydryl groups and would therefore be
expected to inactivate sulfhydryl enzymes, such as ALAD and heme synthetase. The effect of exposure to mercury on ALAD activity in the blood and
on excretion of coproporphyrin and ALA in the urine has been examined in
man. Two separate groups of investigators have reported a correlation between mercury exposure and coproporphyrin excretion in the urine [26,27].
The correlation was weak. There also was a correlation between urinary
excretion of mercury and inhibition of ALAD in the blood [27]. Again, the
correlation was weak and there was no corresponding increased excretion of
ALA in the urine. Other investigations of the relation between mercury exposure and metabolic abnormalities have not been particularly fruitful in
the sense that the dose-response relationship has proved to be very weak
and to lack sufficient sensitivity to serve as subclinical indices of toxicity.

The examples given for the study of metabolic effects of lead and mercury have been for man. For domesticated animals and wildlife, the study

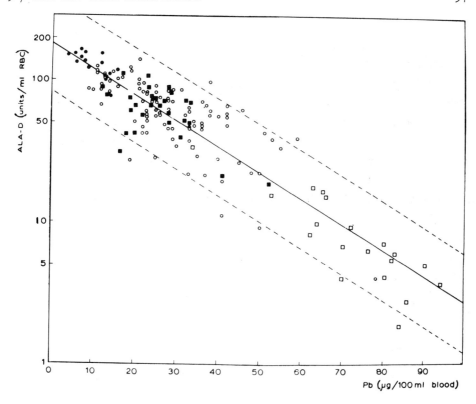

FIG. 6. Correlation between ALAD in blood and lead in blood. Solid circle
= medical students. Open circle = workers in printing shops. Solid square
= auto repair workers. Open square = lead smelters and ship scrappers.
Reprinted from Ref. 25 by permission (copyright 1970, American Medical
Association).

of subtle metabolic alterations at exposure levels of metals below those
causing overt signs of illness has not been pursued, except insofar as
these animals are used to study human health problems. Even our knowledge
of minimal lethal doses for the metals is inadequate. The likelihood that
such deficiencies will be rectified in the foreseeable future improves as
an increasing number of federal agencies and investigators make a commit-
ment to toxicology.

    SUMMARY

    Short-chain alkyl compounds of lead and mercury are susceptible to
metabolic degradation in animals. Dealkylation of tetraethyllead yields
triethyllead, to which can be ascribed the toxic properties of tetraethyllead. By contrast, dealkylation of methyl mercury is a detoxification

mechanism. The reverse process of methylation is a detoxification mechanism
in the case of selenium. Preferential localization in certain organs is
characteristic of many toxic metals. There is no correlation between affinity of metals for tissues and toxicity. Thus, methyl mercury occurs in
highest concentration in the kidney, and lead occurs in highest concentration in bone. Yet for both of these metallic species, the critical sites
of organ damage do not involve these organs. The translocation of metals
within the body under the influence of chelating agents, disease, or nutritional imbalances can be of toxicological significance either by depleting
the body of essential minerals or by transferring toxic metals to metal-
sensitive sites.

The elucidation of dose-response relationships is a major need today.
Clarifications are needed in this regard if the no-effect levels of environmental exposure are to be defined. Some success has been achieved in
developing dose-response relationships in the case of man's exposure to
lead.

REFERENCES

1. J. W. Fried, M. W. Rosenthal, and J. Schubert, Proc. Soc. Exp. Biol.
    Med. N. Y., 92, 331 (1956).

2. J. E. Cremer, Brit. J. Ind. Med., 16 (1959).

3. W. Bolanowska, Brit. J. Ind. Med., 25, 203 (1968).

4. J. E. Cremer and S. Callaway, Brit. J. Ind. Med., 18, 277 (1961).

5. J. E. Cremer, Biochem. J., 68, 685 (1958).

6. A. Swensson and V. Ulfvarson, Occup. Health Rev., 15, 5 (1963).

7. A. Swensson and V. Ulfvarson, Acta Pharmacol. Toxicol., 26, 273 (1968).

8. T. Norseth and T. W. Clarkson, Arch. Environ. Health, 21, 717 (1970).

9. Y. Takeda and T. Ukita, Toxicol. Appl. Pharmacol., 17, 181 (1970).

10. K. Ostlund, Acta Pharmacol. Toxicol., 27 (Suppl. 1), 112-115 (1969).

11. K. Ostlund, Ref. 10, pp. 116-119.

12. J. M. Wood, F. Scott Kennedy, and C. G. Rosen, Nature, 220, 173 (1968).

13. K. P. McConnell and O. W. Portman, J. Biol. Chem., 195, 277 (1952).

14. J. L. Byard, Arch. Biochem. Biophys., 130, 556 (1969).

15. K. P. McConnell and O. W. Portman, Proc. Soc. Exp. Biol. Med. N. Y.,
    79, 230 (1952).

16. K. W. Franke and A. L. Moxon, J. Pharmacol. Exp. Ther., 58, 454 (1936).

17. J. C. Aub, L. T. Fairhall, A. L. Minod, and P. Reznikoff, Medicine, 4,
    1 (1925).

18. P. B. Hammond, Toxicol. Appl. Pharmacol., 18, 296 (1971).

19. M. Berlin, L. G. Jerksell, and C. Nordberg, Acta Pharmacol. Toxicol.,
     23, 312 (1965).

20. A. Gilman, F. S. Phillips, R. A. Allen, and E. S. Koelle, J. Pharmacol.
     Exp. Ther., 87 (Supp.), 85 (1946).

21. L. D. Scheel, Amer. Ind. Hyg. Ass. J., 26, 585 (1965).

22. A. de Bruin and H. Hoolboom, Brit. J. Ind. Med., 24, 203 (1967).

23. Airborne Lead in Perspective, National Academy of Science, Washington,
     D.C., 1972.

24. L. B. Tepper and E. A. Pfitzer, Report of a Symposium, Department of
     Environmental Health, College of Medicine, University of Cincinnati,
     Cincinnati, 1970.

25. S. Hernberg, J. Nikkanen, G. Melin, and H. Lilius, Arch. Environ.
     Health, 21, 140 (1970).

26. L. J. Goldwater and M. M. Joselow, Arch. Environ. Health, 15, 327
     (1967).

27. O. Wada, K. Toyokawa, T. Suzuki, S. Suzuki, Y. Yano, and K. Nakao,
     Arch. Environ. Health, 19, 485 (1969).

28. P. B. Hammond, Lead Poisoning, An Old Problem with a New Dimension,
     Essays in Toxicology, Vol. 1, Academic, New York, 1969, pp. 115-155.

LEAD AND THE NERVOUS SYSTEM

Gary A. Van Gelder

University of Missouri College of Veterinary Medicine
Columbia, Missouri

The presence of lead residues in many components of the biota is generally recognized. This is not unexpected since lead is a naturally occurring element. Of concern is the relationship of these residues to human and animal health. Medical research thrusts have been in the areas of nutritional interactions, hematologic effects, reproductive effects, and neurologic effects. The first three areas are of interest to both human and veterinary medicine. The neurologic effects manifested as acute clinical neurotoxicosis are primarily a veterinary medical problem in cattle and dogs. The real or hypothesized subclinical neurologic problems and mental retardation are primarily of human pediatric significance.

Gilfillan [1] drew on the sciences of toxicology, vital statistics, and archeology and on evidence such as old bones and old Roman recipes and lead-lined pots for brewing poisons esteemed as delicious by the ancient. well to do in developing his theory on "Lead Poisoning and the Fall of Rome." His argument centers on the evidence of extensive oral lead exposure in the upper classes and the hypothesized relationship to the decay of Roman culture, debasement of her arts, ending of hopefulness and progress, and the connection of these declines of outward, evidenced intellectuality with a loss of innate, inherited brains. This is a very interesting paper that should be read by students of toxicology.

LEAD IN THE ENVIRONMENT

Waldron [2] has recently written an excellent and extensive review titled Subclinical Lead Poisoning covering human lead-associated problems as well as many aspects of experimental lead research. An extensive review of the biochemical effects of mercury, cadmium, and lead was published by Vallee and Ulmer in 1972 [3]. A review titled Lead in the Canadian Environment was published in 1973 [4]. This report reviewed lead contamination of air, vegetation, wildlife, and foodstuffs with emphasis on the Canadian environment.

Shukla and Leland [5] reviewed the history of lead use, the changing patterns of lead sources, atmospheric contamination, and presence of lead in rain and surface waters. They cite 118 references related to the presence of lead in air, water, and soil. They point out that modern man contains 100 times more lead than prehistoric man. Examples of increased environmental lead levels today compared to yesteryear include the 200-fold increase in lead content of North Pole and Greenland ice from 0.001 μg/kg ice in 800 B.C. to 0.200 μg/kg found in ice today, and the fourfold increase from 20 mg/kg to 80-90 mg/kg in mosses from 1860-1875 samples to present-day samples, respectively.

Background level in air for unpolluted northern Greenland is 0.005 μg per m^3. Chow et al. [6] report the base-line air concentration of lead in the continental United States to be 0.008 μg/m^3. Air levels in U.S. cities have been reported in the 0.1 to 18 μg/m^3 range. At least as significant as the absolute concentration is the fact that the average mass median diameter of the particles was 0.25 μm, which is in the respirable range for deep penetration into the lungs. The World Health Organization has set a limit of 2 μg lead/m^3 air [7].

The content of lead in rainwater ranged from 3 to 300 μg/liter in one study with an average of 40 μg/liter [8]. Another study reported an average concentration of 34 μg/liter for 32 sampling areas throughout the United States [9]. Brown algae and phytoplankton have been reported to show an affinity for lead. The average concentration of lead in soils is reported to be 16 mg/kg. These few examples point out the general presence of lead in the environment and evidence that the levels have increased over a period of years. Man is certainly exposed, but exposure does not necessarily mean that adverse health effects follow in a linear fashion.

Hicks [10] reviewed airborne lead as an environmental toxin in 1972 and cited 206 references. In the section on neurological effects it was pointed out that while lead encephalopathy was generally regarded as a disease of the infant there were very few experimental studies in young animals.

Recent research efforts have been directed at assessing the presence
or absence of functional changes following prenatal and early postnatal exposure. Some of the interesting literature in this area is reviewed herein.

Goyer [11] listed three problems relating to lead of concern to environmental health scientists: (1) actual level of lead in people, particularly young children, at which adverse health effects occur, (2) recognition
of this level of lead, and (3) identification of the sources of lead exposure.

NEUROLOGIC MANIFESTATIONS OF LEAD IN HUMANS

There are many reports [12-18] of the detrimental effects of lead
poisoning in children. Large exposures can result in convulsive seizures,
severe CNS depression, and death. It is generally accepted that some degree
of residual mental retardation results from severe lead encephalopathy.
However, it is not known if mental deficiency is associated with asymptomatic lead exposure [13].

Lead residues are present at birth and increase through the second
decade of life and then remain constant unless significant accidental or
occupational exposure occurs [19,20].

Lead crosses the placental barrier and is present in cord blood at
birth. Average values reported include 13.8 μg/100 ml [20] and 10.8 μg
per 100 ml [21]. Carpenter [22] found an average human cord blood of 13
μg/100 ml with a range of 10-90 μg/100 ml. Barltrop [21] found a strong
positive correlation between maternal and fetal blood lead values.

Levels in infants and young children range from 0.01 ppm in muscle to
0.46 ppm in liver. By 20 years the values have increased to 0.06 ppm in
muscle and 1.35 ppm in liver. Lead levels in bone increased from about
1 ppm in infants to 40 ppm in persons 50 years of age. About 95% of the
total lead body burden is present in bone. Levels in the brain cortex
averaged about 0.1-0.2 ppm varying somewhat with age [19] in one study from
England. A second study [23] using 191 autopsy samples from Chicago area
hospitals found an average brain lead residue of 0.5 ± 0.68 ppm. The findings of Zaworski and Oyasu [23] were in agreement with those of Barry and
Mossman [19] in that the significant increase in soft tissue lead residues
occurred during the first two decades of life.

Kolbye et al. [24] using lead residue data from a 1972 FDA market bas-
ket survey involving 12 food categories calculated an estimated daily lead

intake of 57 to 233 μg for an average 18-year-old male. The range results
from various assumptions of what a "trace" residue really means. The exposure for a 2-year-old child was estimated to be 75 μg/day. The exposure
for a 6-month-old infant was calculated to be 100-140 μg/day, depending on
the foods eaten.

Pueschel [25] studied 58 children with history of elevated lead exposure based on analysis of hair (> 100 ppm), blood lead (> 0.5 ppm), or
urinary output exceeding 500 μg/24 hr. The children were reevaluated 1.5
to 3 yr later. About one-third of the children were presented with a history of irritability and clumsiness. Under testing, 23% of the children
displayed problems related to fine motor control. Some had gross motor
impairment including abnormal gait, poor balance, and muscle weakness. No
differences were found in IQ scores or performance on the Sequin Formboard,
Visual-motor Integration Test, and Behavior Ratings between the lead burden
group and a matched control group.

It is noteworthy that this study population had not reached the stage
of lead encephalopathy but was basically subclinical and was identified as
lead exposed as the result of a screening program. When retested, some
children still had balance and coordination problems. Others [26,27] have
also reported problems of fine motor control.

Perlstein and Attala [15] reported that 39% of 425 children with lead
poisoning in follow-up examinations showed neurologic impairment. The patients with lead encephalopathy later displayed more serious handicaps, with
half of these classified as mentally retarded.

Byers [14] reported one-third of 45 lead-poisoned children admitted to
the hospital to be mentally deficient.

Some investigators [27,28] have not found evidence of mental deficiency
in asymptomatic lead poisoning. However, differences in sample size and
most importantly the use of different behavioral measures could account for
the failure to detect evidence of neurologic damage in subclinical exposures.

David [29] reported a study involving hyperactive children. Hyperactive children had blood lead levels significantly greater than the controls.
Also eight children with known lead exposure scored high on the hyperactivity
rating scale. Hyperactivity is characterized by a high level of motor activity and is usually coupled with a short attention span, low frustration
tolerance, and hyperexcitability. The hyperactive children also had higher
postchelation urine lead levels. While this study does not prove a cause-
effect relationship, it has opened up another approach for study.

Goldberg [30] studied a human population in the United Kingdom that was exposed to elevated lead levels in the domestic water supply resulting from storage in lead-lined tanks and the use of lead pipes. The average level in the contaminated water was 943 μg/liter (about 0.9 ppm) with some samples as high as 2-3 ppm. The average blood lead was 28 μg/100 ml or 0.28 ppm. This level was sufficient to depress aminolevulinic acid (ALA) dehydrase activity. Several were thought to be suffering from clinical lead toxicosis with symptoms including acute abdominal pain, tremor, hematological abnormalities, and hyperuremia. In two families there was clinical improvement following repair of the water system.

Assuming a body weight of 70 kg and a water consumption consisting of 1200 ml of contaminated water and 1200 ml of water from clean sources such as foods and beverages, this results in an exposure from the water of 0.017 mg/kg. There undoubtedly were other sources of lead exposure since families in the area with water-lead levels of 0.1 ppm had blood leads of about 20 μg/100 ml. If all other factors, such as air, diet, condition, and type of paint in residence and traffic, were the same, then the additional exposure of 0.017 ml/kg was sufficient to raise the blood lead by 0.08 ppm.

Kostial et al. [31] showed that the uptake and retention of lead from the intestine of newborn rats was 55% of the dose compared to only 1% for adult rats. Alexander [32] reported an average of 53% absorption and 18% retention of dietary lead in children. The daily intake was 10-28 μg/kg. This is considerably higher than the 5% absorption for adults which is often assumed to be representative. In the Alexander study there was no relationship between blood lead levels and IQ scores.

EXPERIMENTAL STUDIES OF THE EFFECTS OF LEAD
ON THE CENTRAL NERVOUS SYSTEM

One is impressed after reviewing a considerable amount of the lead toxicology literature by the wide variety of methods and occasional carelessness of reporting lead exposure data. For example, one may find reports very precisely stating the concentration of lead in the drinking water and further that distilled, ion-free water was used. However, when the investigator fails to provide water consumption data, it is impossible to interpret with certainty the exposure and, ultimately, the toxicity. It is well to recall that in most instances in mammals toxicity is expressed as units of toxicant per unit of body weight. With the water consumption problem, simply consider for a moment the laboratory animal facilities. How can one know if the room was hot, thus increasing water consumption by 50%, or cold, resulting in a decreased water consumption? Consider also the impact of relative humidity in the summer versus winter. The toxicologist may be interested in the amount of toxicant in the water or feed but is certainly more concerned about the amount consumed by the animal.

A companion problem is often encountered when the weights of the animals are not specified. Consequently, it has been necessary to make some approximate calculations in order to provide a basis for comparing effects in studies reviewed.

Two animal models involving feeding high levels of lead in the diet of early postparturient mice [34] or rats [35,36] have been used for studying lead encephalopathy.

Effects in Mice

Silbergeld and Goldberg [37] used lead acetate in drinking water to expose female mice. The weanlings also drank lead acetate in water. The three levels used were 2, 5, and 10 mg/ml. Controls received Na acetate. Litters were normalized to six at 48 hr. The lead exposure continued for 60 days. Growth depression occurred in all lead-exposed young. There was no weight loss in the dams indicating that food consumption was not impaired. In addition, eye opening, full incidence of body hair, coordinated walking, and weaning were delayed by as much as 8 days in the lead-treated offspring.

At 30 days of age, offspring in the 5- and 10-mg groups began showing motor deficits including ataxia and characteristic splayed gait. Deaths occurred in some of the 10-mg animals after 90 days. Of considerable interest was the fact that grid-crossing activity was elevated in the lead-treated animals when tested at 40-60 days of age. The increase in activity was the same for all three treatment groups.

Paradoxical responses to certain drugs were observed in the lead-treated offspring. Amphetamine caused a decrease in motor activity instead of an increase. Phenobarbital increased motor activity instead of causing a decrease as in the control group. Chloralhydrate depressed activity in the same way in both the control and lead-treated animals.

Assuming a water consumption of 6 ml/day, which is probably low for a lactating mouse, the lead exposure for a 35-g animal would be 342 mg/kg, 857 mg/kg, or 1714 mg/kg for the 2, 5, and 10 mg/ml lead/water concentrations, respectively.

Effects in Rats

The rat model developed by Pentschew and Garro [36] involves feeding 4% lead carbonate to the female rat following delivery. Enough lead, reported to be 4.6 mg% or approximately 46 ppm, is excreted in the milk to cause growth retardation and paraplegia to develop 26 to 30 days after birth. The dams tolerate the diet well. The cerebellum is damaged and the pathology involves capillary endothelial proliferation. About 85-90% of the young die in 2 weeks. The remaining animals recover in 2 to 3 weeks even though they continue consuming the 4% lead carbonate diet. It has been shown [38] that milk or powdered milk in the diet of 6-week-old rats significantly increases lead absorption 33- to 57-fold over rats receiving regular rat food.

The maximum amount of lead in drinking water that is tolerated for 10 weeks by rats and does not produce significant alteration in hematopoiesis, renal size, histology, or function is 200 μg Pb/ml of water [33]. Using a water consumption of 7.8 ml/100 g body weight/day for adult rats, the 200 μg Pb/ml is equivalent to 15.6 mg/kg body weight.

Krigman and Hogan [39] used the Pentschew and Garro model to study biochemic effects of lead on neuronal growth and maturation. The rats were weaned at 25 days and fed the 4% lead carbonate diet for an additional 5 days before tissues were harvested at 30 days of age. Urinary incontinence occurred at 22-24 days and caudal paraplegia occurred 2 to 4 days later. Focal hemorrhages in the cerebellar folia with endothelial proliferation were observed. The brain wet weight and total brain protein were diminished in the lead-treated animals. The brain lead residue was 9.2 ppm in the treated versus 0.3 ppm in the controls.

The number of brain cells was not reduced, thus leading to the deduction that the effect of early postnatal lead exposure in the rat is to restrict cellular growth and to delay maturation. There was a generalized reduction in mass of gray matter and a thinner cortical mantle. The postnatal cortical ontogenesis of the brain involves growth and differentiation of neurons, which is grossly manifested as enlargement of the cortical mantle. The growth process involves the soma, dendrites, and axons. There is a progressive subdivision of axon terminals and of dendrites. This increased branching results in a decrease in the average size of cell processes constituting the neuropil. The cellular processes of the cortical neurons in the lead-treated rats had a larger average size in the Krigman and Hogan study. This indicates a reduced or delayed subdivision of the dendrites and axons.

Biochemical measurements showed significant reductions in brain content of phospholipids, galactolipids, plasmalogens, and cholesterol. A study of the molecular subspecies of the brain phospholipids and gangliosides did not reveal any apparent changes in composition.

The work of Snowdon [40] clearly demonstrates that the behavioral and neurophysiologic effects of lead in the rat are greatest during the earliest stages of development. The behavioral task was a Hebb-Williams 12-problem maze series. Adult rats and weanling rats were given intraperitoneal injections of lead acetate at dosages of 0, 0.5, 0.8, or 1.2 mg/100 g. Exposure lasted for 21 days pretest and 16 days during testing. Lead exposure started at approximately 100 days of age in the adult and at 22 days of age in the weanlings. Urinary excretion of ALA was significantly increased in all lead-exposed rats at the end of the test indicating that the lead exposure was having a biologic effect. The rats receiving the 1.2-mg exposure developed signs of clinical lead poisoning and there were some deaths. There were no changes in behavior in any of the lead groups as measured by trials to criterion, number of errors, or time per trial on the closed-field maze problems.

Effects of previous lead exposure on retention (memory) of the task were determined in the weanling rats by retesting 6 weeks later. There was no additional lead exposure. The savings scores were not significantly different indicating that the lead exposure did not interfere with neurologic mechanisms involved in the retention of information about the maze problem.

These results are similar to those of Brown et al. [41] and Bullock et al. [42], where lead injected into adult or weanling rats failed to produce behavioral changes even though clinical toxicosis was produced.

Snowdon completed a second series of tests in which 0.8 mg lead/100 g was given to pregnant rats for 21 days and to a second group 21 days postpartum. The offspring were tested on the Hebb-Williams maze problems beginning at 43 days of age.

Although no overt clinical toxicosis was observed in the rats injected during pregnancy, no offspring were delivered. Thus, a dose of lead of 8.0 mg/kg (injected intraperitoneally) in the rat, while not causing behavioral changes as shown in the previous study, results in 100% reproductive failure.

The postnatal exposure group showed signs of developmental retardation. Eye opening was delayed and body weights were depressed. On behavioral testing the lead-exposed rats made more errors. The running times were not different, suggesting that there was not a motor deficit.

The exposure level of 0.8 mg/100 g is probably 5 to 10 times lower than the level achieved by feeding 4% lead acetate in the maternal diet. However, blood and tissue residue studies would be needed to verify this point. This approach should be continued to determine whether behavioral changes occur in offspring at lower exposure levels where no gross developmental or body weight changes occur.

Another study [43] has shown that the critical postpartum period for lead exposure in the newborn rat is during the first 10 days. The exposure level was 35 mg/kg and was given by gavage to the lactating dam. T-maze learning was impaired at 8 to 10 weeks of age in weanlings nursing dams dosed on days 1 to 10 postpartum but not in rats that had nursed dams dosed on days 11 to 20 postpartum. Growth rates were not altered.

In the rat, the most active myelination period is between 21 and 31 days [44]. This study demonstrates the critical exposure period and shows that the changes responsible for the learning deficit were present at 8 to 10 weeks of age even though the lead exposure occurred only during the early postnatal period.

Golter and Michaelson [45] used early postnatal oral administration of lead acetate in suckling rats to study effects on brain neurochemicals. The exposure level was a constant 1 mg lead/day apparently from day 1 to day 16. Using the published body weight curves, this would yield exposure levels of approximately 160 mg/kg on day 1 down to 28 mg/kg on day 16. From day 16 to day 35, the weanlings were continued on a diet containing 40 ppm lead. At an assumed average food consumption equivalent to 5% of the body weight this yields an estimated daily exposure of 2 mg/kg. This exposure resulted in an apparent increase in motor activity at 5 to 6 weeks of age. Dopamine levels were not affected. There was no effect on growth rate with this exposure method which significantly distinguishes this method from many others.

The authors advance the concept that central regulation of motor activity and behavior may be closely linked to a balance between interdependent elements, for example neurotransmitters. Therefore, it would be the matter of imbalance that would result in behavioral or activity changes. Such imbalances may not result in large absolute changes in neurochemical concentrations. It would seem feasible that low-level exposure to neurotoxicants might cause this type of alteration and result in subtle behavioral changes.

Clasen et al. [46] studied the ultrastructural features of lead encephalopathy by using the experimental model of feeding 4% lead acetate to lactating rats and studying the nursing pups at 24 to 40 days of age. Brain and kidney tissue was also obtained at autopsy from 11 humans with clinical diagnoses of lead toxicosis. In both the rat and human tissue the lead encephalopathy was characterized by exudative extracellular edema and perivascular PAS-positive globules. The globules occurred primarily in perivascular astrocytes. They were unable to produce the disease in rats by giving the lead after the age of 20 days or by using a 2% lead-containing diet. Of interest is the fact that the edema and vascular changes occur in the adult human and young rat but not in the adult rat. The primary site of injury is the blood vessel. In the rat this is thought to involve arrested development of growing blood vessels with the formation of intravascular strands.

In 6-week-old rats, 10 daily subcutaneous injections of 10 or 20 mg $Pb(NO_3)_2$/100 g body weight cause moderate proliferation of microglia and capillary endothelial cells in the cortex [47]. There is also a slight decrease in alkaline phosphatase in the capillaries and simultaneously an increase in acid phosphatase in neurons. Enzyme activity did not differ from controls if the rats were killed 9 or more days after the last exposure.

Mahaffey [48] summarized work done on lead-calcium-iron dietary interactions. Low-calcium diets (one-fifth of control) dramatically increase the soft tissue storage of lead such that rats on the low-calcium diet and given only 12 µg Pb/ml water (approximately equal to 0.9 mg/kg body weight) have renal lead residues comparable to rats on normal calcium diet drinking water containing 16 times as much lead. This same relationship holds for presence of renal inclusions, urinary delta-ALA, and kidney weight.

Low dietary calcium also increases Pb toxicity in the dog, horse, and pig [49-51].

Moore and Goyer [52] studied the composition of lead-induced inclusion bodies in renal tubular cells of rats. The inclusion bodies were separated by differential centrifugation and were found to be insoluble in physiological media. They were soluble in 6 M urea and sodium deoxycholate. The inclusion bodies contained 40-50 µg lead/mg protein. Other significant elements present were 13.4 µg calcium, 5 µg iron, and 1.3 µg zinc per mg protein.

The protein had a molecular weight of 27,500 and was found to be high in glutamic and aspartic acids, glycine and cystine. A second lead-binding protein soluble in saline was isolated from the nucleus.

Effects in Dogs

Zook [53] studied the effects of accidental lead exposure in 32 dogs diagnosed as having lead toxicosis. Lesions in the brain involved vascular damage consisting of swelling or necrosis of vascular endothelium in arterioles and capillaries and led to laminar necrosis in the cerebral cortex. In chronic cases there was proliferation of new capillaries and gliosis.

Neural lesions reported in canine lead toxicosis include degenerative changes in Purkinji cells, small hemorrhages, swelling and proliferation of capillary endothelium, edema, gliosis, cerebral neuronal degeneration, spongy necrosis of cerebral cortex, polyneuritis, and necrotizing myelopathy.

Lesions recognized included lead lines in radiographs of bones of immature dogs, hyperplasia of bone marrow, necrosis of occasional striated muscle fibers, decreased numbers of sperm and ovarian follicles, and peripheral neuropathy. Other lesions of lead toxicosis reported as occurring in the dog include nephrosis, nephritis, eosinophilic intranuclear inclusions in renal and occasional hepatic cells, degenerative changes or necrosis in liver cells, and periportal fibroplasia.

In Zook's experience, altered renal proximal tubular epithelium was to be expected in lead-poisoned dogs. Intranuclear inclusions were present in 27 of 32 dogs. The dogs without inclusions had lower liver lead residues. Dogs that had generalized convulsions were found to have brain lesions. Clinical neurologic signs included tremors, ataxia, champing of the jaws, ties, and hypersensibility.

Dogs with neurologic involvement persisting for 8 days or longer were found to have endothelial proliferation and budding of new capillaries with these changes most pronounced in the gray cerebral cortex.

Degeneration and necrosis of neurons in the gray cortex were common in dogs with neurologic involvement. The occipital and parietal cortical regions were more intensely involved. Similar changes were noted in the hippocampus and less commonly in cerebellar Purkinji cells.

Testicles were examined on five adult dogs. In four of these, fewer sperm were observed to be present. In one dog with a long clinical course no mature sperm were found.

The lesions of lead encephalopathy in children [35,54-57], cattle, monkeys, and dogs are similar in many respects. All are characterized by vascular damage. It is postulated that the neuronal damage is brought about by metabolic products passing the lead-damaged blood-brain barrier. One important difference between lead encephalopathy in dogs and children is the elevated cerebrospinal fluid (CSF) pressure in children which results in a swollen brain. The CSF pressure remains normal in dogs. Zook found evidence of peripheral neuropathy in two adult dogs.

Stowe et al. [58] fed dogs 100 ppm lead acetate from 6 to 18 weeks of age. This gave an equivalent exposure of 3.4 to 1.7 mg/kg at 3 and 11 weeks, respectively. A variety of biochemical and hematologic parameters were measured as well as tissue residues. Approximately 97% of the tissue lead was skeletal. The brain lead level was 1.24 ppm compared to 0.09 ppm for controls.

For comparison, the liver, kidney, and blood levels were 23, 32, and 1.7 ppm, respectively, in the lead-fed dogs.

Localized brain residues ranged from 0.59 ppm in the cerebellum to 2.36 ppm in the occipital gray cortex.

Effects in Sheep

Another experimental animal used to study the behavioral toxicologic effects of lead is the sheep [59-63].

In adult sheep 4 weeks of exposure to 100 mg lead/kg/day as the acetate is sufficient to significantly alter behavior on an auditory signal detection task [59]. The lead exposure markedly changed the day-to-day pattern of responding. While the usual pattern for individual subjects was to maintain about the same percentage of correct responses (±7.9%) on successive days, the lead-exposed subjects showed increased variation (±14.78%). It can be speculated that this type of erratic behavior, where on one day performance is high and the next day low, is experimental evidence of neurologic disease that may account for the increased number and more severe accidents reported for lead-exposed industrial workers [64].

Carson et al. [61] demonstrated that healthy adult sheep would tolerate daily exposures of 4.5 mg/kg/day for 6 months and would deliver healthy lambs. There was a small increase in packed cell volume in the high-dosed (4.5 mg/kg) group as compared to the low-dose (2.3 mg/kg) and control groups. Red blood cell and white blood cell counts, blood sugar, BUN, inorganic phosphorus, creatine phosphokinase, lactic dehydrogenase, and SGOT were not changed by lead exposure.

Following prenatal lead exposure in lambs, slowed learning was demonstrated in offspring at 10-15 months of age. The ewes were fed finely powdered elemental lead for 5 weeks before breeding and throughout gestation
which is approximately 5 months' duration in domestic sheep. The lead was
incorporated in the basal ration at a level sufficient to maintain blood
lead levels of 34 µg/100 ml in the high-dose group and 18 µg/100 ml in the
low-dose group. The control group had a background blood lead level of 4.7
µg/100 ml. The daily oral exposure levels averaged 4.5 and 2.3 mg lead/kg
body weight for the high and low groups.

At 2 weeks of age the lambs had blood lead values of 6, 17, and 24 µg
per 100 ml for the control, low, and high groups, respectively. These
levels are not excessively high and are as low as, or lower than, blood
levels reported for newborn human infants in some "at-risk" populations.

The lambs were tested from 7 days to about 3 months of age on a series
of closed-field maze problems [63]. There were no lead exposure related
differences in performance based on analysis of errors, slope of learning
curves, or time. Throughout the study, the offspring remained asymptomatic.
There was no intentional postnatal exposure to lead except for that in the
ewes' milk and the unavoidable background exposure.

Between the time the lambs were 5 months old and 15 months of age,
they were trained and tested on a simultaneous two-choice nonspatial visual
discrimination task [62]. There was a significant increase in the number
of days required by the high-group lambs to master a visual discrimination
involving two circles of unequal size.

The high-lead-exposed lambs did not have apparent difficulty in master-
ing simpler visual discriminations, that is, two-choice situations mastered
by all lambs in a shorter number of days. But the two problems requiring
more time to master by the control group required more than twice as long
for the high-lead group. The task used has some analogy to a visual-cogni-
tive task. The lambs were given adequate time to make the discrimination.
The learning criterion was based solely on consideration of correct versus
incorrect responding independent of response latency. There was no evidence
of physical impairment. The results of this study using prenatal exposure
strongly support the contention that subclinical lead exposure may result
in neurological damage.

In the sheep fetus, vascularization of the brain begins at 40 to 70
days and varies with the different areas of the brain. The importation of
lipid follows the vascularization. Myelination of the brain begins as
early as 60 days with most areas beginning between 70 to 100 days [65].

Effects in Primates

Daily oral dosages of lead acetate at 0.05, 0.5, and 5 mg/kg for 30 months in rhesus monkeys had no effect on performance of a delayed response task or on conditioned response behavior [66].

Infant rhesus monkeys raised with surrogate mothers with a lead base were found to have recurrent convulsions after 3 months [67]. Blood lead values were 160-400 μg/100 ml compared to background levels of 1-20 μg. In a subsequent study lead acetate was mixed in the formula of infant rhesus monkeys sufficient to provide exposures ranging from 0.5 to 9 mg lead acetate per kg. Seizures occurred when blood lead reached 300 μg/100 ml. The convulsions disappeared when lead exposure was stopped. The exposure of 0.5 mg lead acetate/kg was sufficient to cause the blood lead level to increase to 60-100 μg/100 ml after 4 weeks. These animals did not reach the convulsive stage but did display hyperactivity, insomnia, and a gradual decline in hemoglobin and packed cell volume.

Juvenile monkeys given lead acetate in drinking water at an equivalent exposure level of 20 mg/kg did not show the behavioral changes although the blood lead levels averaged 135 μg/100 ml. The infant rhesus model develops the anemia pattern as do young children.

Clasen et al. [68] used 5- to 6-month-old (newly weaned) rhesus monkeys as experimental models. The lead dose was 500 mg lead acetate with the subjects weighing 1 to 2 kg. This resulted in exposures in the four monkeys of 500-250 mg/kg/dose. The lead was administered three times a week. Blood leads ranged from 0.3 to 10 ppm. Overt toxicosis developed after 7 to 16 weeks. Clinical signs included anorexia, lethargy, muscular weakness, ataxia, and convulsions. At necropsy, one of the four monkeys showed intranuclear renal inclusions. Perivascular glial nodules were observed in the spinal cord and brain stem. PAS-positive cytoplasmic globules similar to those reported for man and rat [46] were observed in astrocytes in the gray and white matter of the brain stem, cerebellum, and cerebrum. Unlike the rat, but similar to the human, these globules were also seen within endothelial and perithelial cytoplasm.

There was pronounced brain edema which was observed grossly. Microscopically, the edema separated nerve fibers in the cerebellar and cerebral white matter. The authors considered the perivascular PAS-positive globules and the exudative edema to be characteristic of acute lead encephalopathy. They also reported finding these changes in a baboon. Unlike the young rat, the rhesus monkeys did not have hemorrhages in the cerebellum.

A question that needs to be explored in considering the use of the 6-month-old rhesus monkey as a model for childhood lead poisoning is the relative degree of CNS maturation at the time of lead exposure.

Cohen et al. [69] administered lead chloride **intravenously** to baboons at a dosage of 1.61 mg/kg/day. Death occurred in two subjects after 68 and 85 days with blood leads of 1500 μg/100 ml and 950 μg/100 ml. Based on the limited number of animals it was calculated that a dosage of 100 μg/kg for 2 months increased blood lead by 2 μg/100 ml, while a dosage of 500 μg/kg for 2 months increased blood lead by 9 μg/100 ml. Caution should be expressed, however, in that inspection of the preliminary data presented in their figures would **indicate** a tendency toward a sharp elevation in blood lead after 60-90 exposure days.

EFFECTS OF LEAD ON PERIPHERAL NERVES

Gombault [70] in 1880 described the degeneration and regeneration of myelin sheaths in peripheral nerves of chronic lead-poisoned guinea pigs. This process involves degeneration of the myelin sheaths between the nodes of Ranvier (the internode) and may result in some segments remaining intact while adjacent segments are lost. Remyelination occurs with the original myelin segment being replaced by multiple shorter segments. This phenomenon is known as segmental demyelination. Gombault reported that the axon was not involved. Some 86 years later Fullerton [71] reported evidence of axonal degeneration as well as demyelination in the chronic lead-poisoned guinea pig. Axonal degeneration and segmental demyelination also occur in lead-poisoned rats [72,73]. There are interesting species differences in the proportion of segmental demyelination to axonal degeneration [74]. Lead poisoning in the cat primarily affects the central nervous system [75,76] with only a small amount of peripheral nerve involvement [77]. In rabbits axonal degeneration is more common. In man only axonal degeneration occurs, and apparently segmental demyelination has never been described although specifically looked for in lead toxicosis [2].

The ultrastructural aspects of segmental demyelination have been described [73]. It should be pointed out that these changes are not peculiar to lead neuropathy but are also associated with diphtheria toxin, x-irradiation, and spinal barbotage [2].

Several investigators have examined the effect of lead on peripheral nerves using electrophysiologic methods. Fullerton [71] was able to demonstrate slowed mean maximum conduction velocity in some guinea pigs. In human cases there usually is no significant decrease [78]. This is explained by the fact that not all fibers in a nerve are affected at any one time, and consequently maximum conduction velocity remains near normal. There is some evidence [79] that changes may be detectable if conduction is measured in slower fibers. There was also evidence of diminished number of motor units and muscle fibrillations were recorded in 15 of 39 male lead workers.

Another syndrome called lower motor neuron disease has been associated with chronic lead toxicosis in man and was first described in 1907 [80]. This involves a distal symmetrical wasting and weakness of muscles. There is evidence of pyramidal tract disturbance in some and fasciculation has been recorded. Campbell et al. [81] more recently provided some evidence for a causal relationship of lead exposure in some patients with motor neuron disease. However, it must be understood that motor neuron disease has other causes since it also occurs in the absence of exposure to lead. Another interesting aspect of the Campbell study is the suggested relationship between skeletal demineralization and motor neuron disease.

Lead has an effect on synaptic transmission. Kostial and Vouk [82] perfused the superior cervical ganglion in cats with lead solutions containing 5 to 40 $\mu\underline{M}$ lead/liter (equivalent to approximately 1 to 8 ppm). With this preparation it is possible to separately electrically stimulate both the pre- and postganglion nerves. The observable response is contraction of the nictitating membrane. Lead at 5 $\mu\underline{M}$/liter for 12 min reduced acetycholine output to 35% of the initial value. The addition of calcium at 10.5 m$\underline{M}$/liter restored both the contraction and output of acetycholine. It was also shown that the lead-blocked ganglion was still sensitive to injected acetycholine. The results of this study support the concept of lead interfering with the release of acetycholine from the presynaptic nerve fibers.

Lead is known to have an affinity for SH groups (mercaptide reaction). The SH group is found as a functional group of coenzyme A which is essential for the acetylation of choline [83]. Sávay and Csillik [84] used this mercaptide reaction in their study of synaptic structures. A solution of lead nitrate, urea, and formalin was injected percutaneously into the short flexor muscle of the hind pad of anesthetized rats. The muscle tissue was taken after 15 min and frozen sections cut and immersed for a few seconds in a sodium sulfide solution. This treatment rendered the motor end plates clearly visible.

Nakamura et al. [85] expanded on the above study and demonstrated that lead, tin, cadmium, zinc, and copper ions would also react with motor end plates. Denervation did not affect the metal binding, indicating that the reaction is postsynaptic (muscle) rather than presynaptic (motor nerve terminal). Prior freezing of the tissues or formalin fixation inhibits the metal binding activity but does not greatly affect cholinesterase activity. They suggested a close relationship of the site of binding of divalent metal ions in the motor end plate to the site of calcium release, and a close but not identical relationship to the site of cholinesterase activity and the acetycholine receptor.

FUTURE DIRECTIONS

Some of the most interesting work being done on the public health aspects of lead toxicosis are the studies on the effects of lead on the developing brain. Over the past few decades a number of investigators have been postulating subclinical effects particularly in children involving mental retardation, hyperactivity, or other forms of neurologic impairment manifested as behavioral changes. It has only been in the last several years that appropriate experimental studies in animals have been conducted to prove or disprove the idea of subclinical neurologic impairment. It is readily apparent that studies in adult animals are not adding to the definition and understanding of pediatric lead toxicosis.

Animal studies must involve either prenatal or very early postnatal exposure depending on the species. Future studies should involve a larger number of species. Studies that involve rats need to concentrate on obtaining lead-exposed rats that do not show changes in growth or gross development in order to better define the type and extent of subclinical neurologic impairment.

There are a large number of behavioral measures that can be used to help define the nature of the effect of lead on the central nervous system. The maze studies in rats have been productive, but the question is whether other behavioral measures might be more sensitive in detecting lead-induced alterations. In the sheep studies the maze was not as sensitive a measure as the visual discrimination task. The surface has hardly been scratched in this area of research in behavioral toxicology. It would be naive to think that lead is the only substance affecting the developing nervous system. It can be expected that in the future an area of research known as behavioral teratology will develop in which the effects of prenatal exposure to not only metals and environmental contaminants but also drugs and perhaps varying levels of endogenous hormones and metabolites on the later development and function of the nervous system will be determined.

The mechanisms of these neurologic effects will need to be determined. Changes, as has been suggested, may involve delayed or interrupted neuronal growth. Perhaps not all of the neuronal interconnections and synapses are formed. Or perhaps the development or function of the supporting cells is impaired. The entire aspect of the nature of neurotransmitters, their synthesis, control mechanisms, storage, release, and ultimate destruction, awaits a better understanding.

Finally, it is apparent that more sensitive methods of evaluating peripheral nerve function are needed, to detect not only subclinical chronic lead effects but a wide variety of other neurotoxicants as well.

REFERENCES

1. S. C. Gilfillan, J. Occup. Med., 7, 53 (1965).
2. H. A. Waldron, Subclinical Lead Poisoning, Academic Press, New York, 1974.
3. B. L. Vallee and D. D. Ulmer, Ann. Rev. Biochem., 41, 91 (1972).
4. Associate Committee on Scientific Criteria for Environmental Quality, National Research Council of Canada, Lead in the Canadian Environment, Publ. No. BY73-7 (ES) of the Environmental Secretariate, Publications, NRCC/CNRC, Ottawa, Canada, 1973.
5. S. S. Shukla and H. V. Leland, J. Water Pollut. Control Fed., 45, 1319 (1973).
6. T. J. Chow, J. L. Earl, and C. B. Snyder, Science, 178, 401 (1972).
7. H. G. Goldsmith, J. Air Pollut. Control Ass., 19, 714 (1972).
8. T. J. Chow and J. L. Earl, Science, 169, 577 (1970).
9. A. L. Lazrus, E. Lorange, and J. P. Lodge, Environ. Sci. Technol., 4, 55 (1970).
10. R. M. Hicks, Chem.-Biol. Interactions, 5, 361 (1972).
11. R. A. Goyer, Environ. Health Perspect., 7, 1 (1974).
12. J. J. Chisolm and E. Kaplan, Dev. Med. Child Neurol., 7, 529 (1965).
13. G. Wiener, Public Health Rep., 85, 19 (1970).
14. R. K. Byers, Pediatrics, 23, 585 (1959).
15. M. A. Perlstein and R. Attala, Clin. Pediatr., 5, 292 (1966).
16. D. Barltrop, Postgrad. Med. J., 45, 129 (1969).
17. J. S. Lin-Fu, N. Engl. J. Med., 286, 702 (1972).

18. J. S. Lin-Fu, N. Engl. J. Med., 289, 1289 (1973).

19. P. S. I. Barry and D. B. Mossman, Brit. J. Ind. Med., 27, 339 (1970).

20. N. P. Kubasik and M. T. Volosin, Clin. Chem., 18, 1415 (1972).

21. D. Barltrop, in Mineral Metabolism in Pediatrics (D. Barltrop and W. L. Burland, eds.), F. A. Davis, Co., Philadelphia, 1969, pp. 135-151.

22. S. J. Carpenter, Environ. Health Perspect., 7, 129 (1974).

23. R. E. Zaworski and R. Oyasu, Arch. Environ. Health, 27, 383 (1973).

24. A. C. Kolbye, K. R. Mahaffey, J. A. Fiorino, P. C. Corneliussen, and C. F. Jelinek, Environ. Health Perspect., 7, 65 (1974).

25. S. M. Pueschel, Environ. Health Perspect., 7, 13 (1974).

26. B. del a Burde and M. S. Choate, J. Pediatr., 81, 1088 (1972).

27. D. Kotok, J. Pediatr., 80, 57 (1972).

28. H. D. Smith, Arch. Environ. Health, 8, 68 (1964).

29. O. J. David, Environ. Health Perspect., 7, 17 (1974).

30. A. Goldberg, Environ. Health Perspect., 7, 103 (1974).

31. K. Kostial, I. Simonovic, and M. Pisonic, Nature (London), 233, 564 (1971).

32. F. W. Alexander, Environ. Health Perspect., 7, 155 (1974).

33. R. A. Goyer, D. L. Leonard, J. F. Moore, B. Rhyne, and M. R. Krigman, Arch. Environ. Health, 20, 705 (1970).

34. W. I. Rosenblum and M. G. Johnson, Arch. Pathol., 85, 640 (1968).

35. A. Pentschew, Acta Neuropathol., 5, 133 (1965).

36. A. Pentschew and F. Garro, Acta Neuropathol., 6, 266 (1966).

37. E. K. Silbergeld and A. M. Goldberg, Environ. Health Perspect., 7, 227 (1974).

38. D. Kello and K. Kostial, Environ. Res., 6, 355 (1973).

39. M. R. Krigman and E. L. Hogan, Environ. Health Perspect., 7, 187 (1974).

40. C. T. Snowdon, Pharmacol. Biochem. Behav., 1, 599 (1973).

41. S. Brown, N. Gragann, and W. H. Vogel, Arch. Environ. Health, 22, 370 (1971).

42. J. D. Bullock, R. J. Whey, J. A. Zaia, T. Zarembok, and H. A. Schroeder, Arch. Environ. Health, 13, 21 (1966).

43. D. R. Brown, Toxicol. Appl. Pharmacol., 25, 466 (1973).

44. I. M. R. Krigman, R. C. Reitz, M. H. Wilson, L. R. Newell, and E. L. Hogan, Amer. J. Pathol., 66, 2 (1972).

45. M. Golter and I. A. Michaelson, Science, 187, 359 (1975).

46. R. A. Clasen, J. F. Hartmann, A. J. Starr, P. S. Coogan, S. Pandalfi, I. Laing, R. Becker, and G. M. Hass, Amer. J. Pathol., 74, 215 (1974).

47. A. Brun and U. Brunk, Acta Path. Microbiol. Scand., 70, 531 (1967).

48. K. R. Mahaffey, Environ. Health Perspect., 7, 107 (1974).

49. H. O. Calvery, E. P. Laug, and H. J. Morris, J. Pharmacol. Exp. Ther., 64, 364 (1938).

50. R. A. Willoughby, R. Thirapatsakun, and B. J. McSherry, Amer. J. Vet. Res., 33, 1165 (1972).

51. F. S. Hsu, L. Krook, J. N. Shively, J. R. Duncan, and W. G. Pond, Science, 181, 447 (1973).

52. J. F. Moore and R. A. Goyer, Environ. Health Perspect., 7, 121 (1974).

53. B. C. Zook, Vet. Pathol., 9, 310 (1972).

54. S. S. Blackman, Bull. Johns Hopkins Hosp., 61, 1 (1973).

55. N. Popoff, S. Weinberg, and I. Feigin, Neurology, 13, 101 (1963).

56. R. G. Christian and L. Tryphonas, Amer. J. Vet. Res., 32, 203 (1971).

57. B. C. Zook, Comp. Pathol. Bull., 3, 3 (1971).

58. H. D. Stowe, R. A. Goyer, M. M. Krigman, M. Wilson, and M. Cates, Arch. Pathol., 95, 106 (1973).

59. G. A. Van Gelder, T. L. Carson, R. M. Smith, W. B. Buck, and G. G. Karas, J. Amer. Vet. Med. Ass., 163, 1033 (1973).

60. G. A. Van Gelder, T. Carson, R. M. Smith, and W. B. Buck, Clin. Toxicol., 6, 405 (1973).

61. T. L. Carson, G. A. Van Gelder, W. B. Buck, and L. J. Hoffman, Clin. Toxicol., 6, 389 (1973).

62. T. L. Carson, G. A. Van Gelder, G. G. Karas, and W. B. Buck, Arch. Environ. Health, 29, 154 (1974).

63. T. L. Carson, G. A. Van Gelder, G. G. Karas, and W. B. Buck, Environ. Health Perspect., 7, 233 (1974).

64. F. E. Rieke, Arch. Environ. Health, 19, 521 (1969).

65. R. M. Barlow, J. Comp. Neurol., 135, 249 (1969).

66. J. W. Goode, S. Johnson, and J. C. Calandra, Toxicol. Appl. Pharmacol., 25, 465 (1973).

67. J. R. Allen, P. J. McWey, and S. J. Suomi, Environ. Health Perspect., 7, 239 (1974).

68. R. A. Clasen, J. F. Hartmann, P. S. Coogan, S. Pandolfi, I. Laing, and R. A. Becker, Environ. Health Perspect., 7, 175 (1974).

69. N. Cohen, T. J. Kneip, V. Rulon, and D. H. Goldstein, Environ. Health Perspect., 7, 161 (1974).

70. M. Gombault, Archives de Neurologie (Paris), 1, 11 (1880).

71. P. M. Fullerton, J. Neuropathol. Exp. Neurol., 25, 214 (1966).

72. W. W. Schlaepfer, J. Neuropathol. Exp. Neurol., 28, 401 (1968).

73. P. W. Lampert and S. S. Schochet, J. Neuropathol. Exp. Neurol., 27, 401 (1968).

74. A. Hopkins, Brit. J. Ind. Med., 27, 130 (1970).

75. A. Ferraro and R. Hernandez, Psychiatr. Q., 6, 121 (1932).

76. A. Ferraro and R. Hernandez, Psychiatr. Q., 6, 319 (1932).

77. T. M. Legge and K. W. Goadby, Lead Poisoning and Lead Absorption, Arnold, London, 1912.

78. M. J. Catton, M. J. G. Harrison, P. M. Fullerton, and G. Kazantzis, Brit. Med. J., 2, 80 (1970).

79. A. M. Seppäläinen and S. Hernberg, Brit. J. Ind. Med., 29, 443 (1972).

80. K. Wilson, Rev. Neurol. Psychiatr., 5, 441 (1907).

81. A. M. G. Campbell, E. R. Williams, and D. Barltrop, J. Neurol. Neurosurg. Psychiatr., 33, 877 (1970).

82. K. Kostial and V. B. Vouk, Brit. J. Pharmacol., 12, 219 (1957).

83. R. Reisberg, Biochem. Biophys. Acta, 14, 442 (1954).

84. Gy. Sávay and B. Csillik, Experientia, 15, 396 (1959).

85. T. Nakamura, T. Nambe, and D. Grob, J. Histochem. Cytochem., 15, 276 (1967).

7

LEAD POISONING

Vernon A. Green, George W. Wise, and John Corrie Callenbach
The Children's Mercy Hospital
Kansas City, Missouri

Certainly more has been written concerning poisoning by lead than any other metallic material and perhaps more than any other single toxicant. This is not surprising considering that the toxic properties of lead were known as early as the second century B.C. The ease of refining lead from galena and its malleable properties contributed much to the early widespread use of this metal. During the height of the Roman Empire, lead was used for plumbing, wine vessels, and cooking utensils by the wealthy. This widespread use, with reports of now recognized signs of plumbism, has led to speculation that chronic lead poisoning may have contributed to the fall of the Roman Empire [14].

Lead poisoning is a vast, many-faceted monster that has been around since antiquity. It is not the intent of this writing to discuss the historical, economic, political, and social aspects of the problem; therefore, discussion is limited to the medical aspects of the lead poisoning problem. The limitation is not an indication that the authors feel that the other aspects of the problem are of lesser importance than the medical. In fact, all of the many facets of the problem must be dealt with and actually are greatly overlapping and interrelated. To complicate the discussion of lead intoxication further, the medical aspects of the intoxication are neither as simple nor as straightforward as that of most toxicants. The difference between acute and chronic poisoning and the even greater disparity between the chronic poisoning in adults and that of young children add complexities to the problem. Further, the toxic syndrome produced by lead, especially chronic lead intoxication, is made up of diverse signs resembling numerous diseases and physiologic abnormalities, making the early diagnosis very difficult.

CHRONIC LEAD POISONING

With the institution of lead screening programs in many of the larger
cities, chronic lead poisoning, especially in children, has been confirmed
to be a major medical problem, and the treatment of acute toxic episodes a
medical emergency.   Intoxication is identified most often in the 1- to 5-
year-old age group, with the highest incidence of disease in the 12- to 36-
month-old children, most of whom have a history of pica and who live in
housing or regularly visit buildings built prior to 1955. The normal average daily intake of lead in children is probably less than 0.3 µg. Most
children exhibiting symptoms have been regularly ingesting lead, contained
in paint chips, for longer than 3 months with an intake in excess of 1.5 µg
per day, resulting in a progressive increase in the body burden. One small
paint chip may contain as much as 100 µg of lead. Children usually give up
their pica by age 5, but frequently teach the habit to younger siblings who
are likely to continually return to the paint chips because of their sweet
taste; therefore, the importance of screening siblings of patients with
plumbism should not be underestimated.

Many children present during an acute episode, usually during the summer
months, because the minor symptoms (anorexia, recurrent sporadic vomiting,
colicky abdominal pain, anemia with a hemoglobin less than 10 g, irritability, and constipation) are ignored until severe encephalopathic symptoms
appear. These include hyperirritability, agitation, ataxia, weakness, paralysis of the upper motor neurons, stupor, convulsions, and coma associated
with a high mortality rate and an even higher rate of morbidity. The reason
for the summer toxicity has not been clearly determined. Kehoe [23] indicates that toxicity is favored by dehydration and acidosis. Toxicity induced by high ambient temperatures is accompanied by a decrease in the
excretion of lead in the feces and urine [21]. Solar irradiation and
vitamin D enhance absorption of lead from the intestine and cause an increase in both accumulation and excretion of porphyrin intermediates [4].
It is doubtful that either or both of these explanations fully account for
the summer epidemics.

Lead poisoning should not be considered a summertime disease. At
present, a large number of cases are being reported in the winter months
due to health workers becoming aware of this problem. Some cases have
occurred during the winter when leaded battery casings were burned for
fuel and the fumes inhaled or there was prolonged contact with the ashes

[2,30]. Epidemiologic studies indicate that lead encephalopathy is more
frequent during the summer; however, asymptomatic lead poisoning is a year-
round disease [18].

The absorption, metabolism, and excretion of lead has been worked out
by Kehoe and his associates. The major routes of entry are the gastrointestinal tract and the lungs, with dermal absorption being relatively insignificant in most cases. Approximately 10% of the intestinal lead is absorbed
and vitamin D appears to be involved, as is competition with calcium. Lead
appears as a trace metal in virtually all foods and beverages, and the average adult consumes approximately 0.3 μg of elemental lead per day, of which
10% is absorbed. Another 0.3 μg per day is extracted from the atmosphere
by the lungs. Atmospheric lead is actively absorbed by the lungs in relation to its particle size. Of the finely divided lead in the air, 70 to
75% is discharged in expired air [3]. Of the 25 to 30% not returned, those
particles less than 0.1 μm in diameter are almost totally absorbed, whereas
40% of the retained lead of particle size 0.9 μm or greater in diameter is
trapped in the upper airways and swallowed where it follows the same pattern
as ingested lead [21]. Intermittent exposure and absorption are usually
associated with an occupational schedule, and there is a balance between
absorption and excretion. During time of exposure, intake is high and
output is relatively low. Once exposure is terminated, intake is low and
excretion of lead rises. When a significant portion of the inspired lead
is diverted to the intestinal tract, freedom from exposure is not eliminated as the gastrointestinal lead constitutes a source of continual absorption of lead, and the contribution of pulmonary absorbed lead equals
that of actual intestinal absorption. It now appears that inspired lead,
in some situations involving larger particle size, may even contribute additional lead to the intestine resulting in a prolonged exposure.

Once absorbed, lead is distributed initially to the body tissues in
accordance with vascularity and tissue affinity (see Table 1).

During periods of recent absorption, lead is initially deposited in
the soft tissues and flat bones. Over a period of time, with proper freedom from continuing exposure, the lead is gradually incorporated into the
long bones where a considerable pool may be built up. This appears as
"lead lines" in the metaphyseal arms of the long bones, and the width is
related more to the duration of exposure than to the severity of symptoms.
There is no known toxic significance of lead as a component of bone, but

TABLE 1. Tissue Lead Concentration[a]

	Individuals with No Exposure (mg/100 g tissue)	Individuals with Severe Exposure (mg/100 g tissue)
Kidney	0.05	0.22
Liver	0.12	0.71
Spleen	0.03	0.86
Muscle	0.03	0.10
Lungs	0.03	0.08
Brain	0.04	0.35
Flat bones	0.65	13.00
Long bones	1.78	8.00

[a]Reference 13.

it can serve as a source that can be mobilized during acidosis, alcoholism, and fractures.

Of the lead present in the circulation, over 90% is associated with the red cell membrane as lead phosphate. Kehoe has shown that during times of consistent excessive daily intake there is a relatively rapid attainment of a plateau in the blood lead level. In contrast, the total body burden of lead continues to increase until exposure is terminated [16,21].

It therefore appears that over a long period of time the concentration of lead in the body is not directly proportional to the concentration of lead in the blood. Neither, however, is the total body burden directly proportional to the clinical severity of intoxication. Much of this lead is eventually stored in the bone. Severity of symptoms is related to the general level of concentration of lead in the soft tissues, which is related primarily to the immediate exposure dose and the speed of absorption [21], and secondarily to bone reabsorption.

The excretion of lead is primarily through the gastrointestinal tract, and bile is the vehicle of excretion. Of the total average daily intake of 0.33 μg of lead, 0.30 μg is excreted in the feces [22]. A smaller amount is voided via the kidneys with the amount being proportional to glomerular filtration rate, except at higher concentrations where tubular reabsorption may play a role [31]. Approximately 35% of the average daily dose is eliminated in this way [22]. Because most of the body burden is stored in the bone, the excretion of this burden takes approximately twice as long as it did to accumulate [21]. Chelating agents are relatively ineffective in removing a significant portion of this bone lead.

The diagnosis is usually made on the basis of blood lead levels, although, as previously stated, this does not give an accurate estimate of the total body burden. Since much of the lead is attached to the red cell membrane, one must also consider the hematocrit prior to comparison with the levels in Table 2.

Early diagnosis is certainly preferable, and follow-up services should include those of a doctor, social worker, psychiatrist, child guidance counselor, health department, and visiting nurse association [10]. The child should be immediately removed from the source of exposure and further pica prevented. Treatment is indicated in Table 3. All levels greater than 80 μg% with symptoms should be considered medical emergencies since the onset of acute encephalitis can be both unpredictable and fulminant.

Occasionally blood lead levels are not readily attainable and other emergency tests can be employed. Many of these tests are based on the interference with synthesis by lead (Fig. 1).

Although it is recognized that several sulfhydryl enzymes are inhibited in heme synthesis, the rate-limiting step appears to be the inhibition of the incorporation of iron into protoporphyrin. Recent investigation of this one enzymatic reaction has led to a microphotofluorometric assay for protoporphyrin which some individuals believe to be the most sensitive and practical indicator of lead toxicity as yet devised [6]. The relative newness of this test means that most laboratories are not yet equipped to

TABLE 2. Normal and Elevated Blood Lead Levels

Blood Lead Level (μg%)	
0-20	Normal for most areas
20-40	Abnormal: suspect pica, water, vapors
> 60[a]	Follow blood levels; do provocative test
> 80 without symptoms	Provocative test indicated
> 80 with symptoms	Treatment indicated
> 100	Treatment indicated

[a]Surgeon General's Office has recommended 50 μg% as the intoxication level.

TABLE 3. Therapeutic Recommendations[a]

	Children	
1. Symptomatic		
a. With encephalopathy	BAL in oil, 4 mg/kg/dose every 4 hr	IM. First dose BAL only, then BAL and EDTA at separate sites. If quick response is noted, discontinue BAL after 3 days and decrease EDTA to 50 mg/kg/day b.i.d. or t.i.d. 5-7 days' total course.
	EDTA in 0.5% procaine, 12.5/kg per dose. Maximum dose 3/day	
	D-Penicillamine, 30-40 mg/day	PO. 3-6 months' course until blood level <60.
b. Without encephalopathy	BAL in oil every 4 hr	IM. See part (a) above.
	EDTA in 0.5% procaine <50 mg/kg per day	IM. Divided into 2-4 doses/day.
	<550 mg/kg/day	5-day treatment course.
2. Asymptomatic		
a. Blood level >100	BAL/EDTA, D-Penicillamine, 30-40 mg/kg/day b.i.d.	IM. See part (a) above. PO. 3-6 months' course.
b. Blood level <100	EDTA only, 50 mg/kg/day b.i.d.	IM. 3-5 days.

Adults

1. Symptomatic

   a. With encephalopathy — BAL 2.5 mg/kg/day, EDTA 8.0 mg/kg/day — IM. 5 days, 30 injections.

   b. Abnormal symptoms — BAL/EDTA, D-Penicillamine, 500–750 mg/day t.i.d. — IM. 3–5 days or until urinary level <500 µg/24 hr.

   c. Neuropathy — D-Penicillamine, 500–750 mg/day t.i.d. — PO. 1–2 months. Use BAL/EDTA if level >100 µg% in blood.

2. Asymptomatic

   a. >100 µg% blood — BAL/EDTA — IM. 3–5-day course. See part (a) above.

   D-Penicillamine — PO. See part (c) above.

   b. 80–100 µg% — D-Penicillamine, 500–750 mg/day t.i.d — PO. 1–2-months' course.

^aAdapted from Chisolm [8].

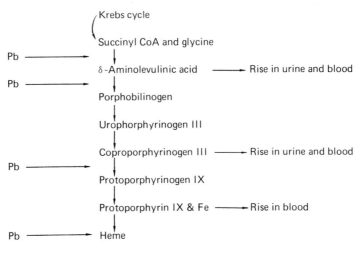

FIG. 1. Modified from Chisolm [4].

perform it. Qualitative urinary coproporphyrin levels are perhaps the
fastest and simplest test. A 3+ to 4+ test may roughly correspond to a
blood level of lead in excess of 100 μg% [1]. There appears to be a direct
linear relationship between the pretreatment UCP level and the quantitative
lead output in the first 24 hr of treatment. The test has a good correlation with the metabolically active (delta aminolevulinic acid) fraction of
lead in the soft tissues [4]. Urinary ALA levels are perhaps somewhat more
sensitive, but they are more difficult to perform. Lead determination on
whole blood is probably the most effective means of showing recent increased
lead intake.

Less specific tests include an abdominal film, which may reveal opaci-
ties in the bowel, and x-rays of the wrists and knees demonstrating "lead
lines," especially in the 2- to 5-year age range. Again, width is related
to duration of exposure rather than severity of symptoms. "Lead lines"
will continue to increase in width even after termination of exposure as
soft tissue lead is transported to the cancellous matrix. Presence of
these lines in children less than 2 years of age is unpredictable and variable. Lines in older bones (74 yr) may not be prominent. The blood smear
may show a microcytic (rarely normocytic) hypochromic anemia with hemoglobin less than 10 g. The reticulocyte count is usually elevated, which distinguishes it from iron deficiency anemia. Basophilic stippling in the pe-

ripheral smear is extremely variable. It was once thought that the anemia
was due to removal of stippled cells by the spleen. It now appears that
the spleen serves to mature the cells resulting in disappearance of the
stippling rather than RBC destruction [32]. Examination of the bone marrow
shows stippling in excess of 50% of the RBCs [4]. Urinalysis reveals glycosuria, proteinuria, cellular casts, elevated ALA greater than 6 mg/liter,
and aminoaciduria. A confirmatory provocative lead excretion test can be
done by measuring the 24-hr urine lead excretion provoked by BAL/EDTA chelation therapy. Levels greater than 1.5 mg Pb/24 hr in the adult and 1.0
mg Pb/24 hr in the child are indicative of lead intoxication [3].

The basic principles of treatment have been set down by Chisolm [5].
Treatment should be begun immediately. Adequate urine output should be
established over a 1- to 2-hr period. Should the patient fail to adequately
initiate urination a judicious use of mannitol, 1 g/kg in a 20% solution
given at 1 ml/min, may be employed. Since these patients occasionally exhibit an inappropriate antidiuretic hormone-type syndrome, once urine output
has been established, the patient should be placed on maintenance fluids to
maintain a urine flow of 350-500 ml/m^2/24 hr or 0.35 to 0.50 ml urine se-
creted/calorie metabolized in 24 hr. The use of diazepam and phenobarbital
certainly is probably not advisable in the face of disturbed porphyrin
metabolism. Paraldehyde may be the mainstay of seizure control during the
first several days, being implemented at the first suggestion of increased
intracranial pressure, which should be avoided. Osmotic diuresis may be
helpful. Steroid therapy has not proven to be of any benefit and may actually increase the renal toxicity of EDTA (Table 3).

Treatment usually depends upon the severity of symptoms and available
laboratory data. Exposed individuals should be followed closely with repeat blood levels and removed from continued exposure. Those persons with
increased blood levels, either symptomatic or asymptomatic, should be treated
according to the recommended treatment schedule (see Table 3). Such persons
should be removed from the source of exposure and followed on a regular
basis.

The BAL/EDTA combination has been in use since 1964. Prior to that
date, children with encephalopathy had a 25 to 30% mortality rate when
treated with EDTA alone [5]. In fact, there was occasionally a prompt de-
terioration of their clinical condition associated with an increased level
of ALA in the urine and plasma 48 to 72 hr after initiation of EDTA therapy
[4]. Although increased porphyrin levels have not been shown to cause the

toxicity seen in lead poisoning, there appears to be a strong direct connection. In order to establish proper chelation therapy, it is necessary to
achieve a sufficient molar excess of chelate over metal in order to minimize
dissociation of the metal-chelate complex during mobilization and excretion,
which is apparently not achieved with EDTA alone in its maximum dose of 75
mg/kg/day [29]. BAL (Dimercaprol) was chosen to be added to the regimen
because it removes lead from the red cell membrane, and it is excreted in
the bile and can therefore chelate intestinal lead as well. The rate of
decrease in the blood lead level is exponential. There is a 50% decrease
in 15 hr with BAL/EDTA compared with a 50% decrease in 68 hr with EDTA alone,
and blood levels approach normal in 5 days with the present therapy [5].

Chelating agents are not without their own toxicity. BAL, aside from
its odor, can cause nausea, vomiting, headache, abdominal pain, burning of
the mouth, increased systolic and diastolic pressure with tachycardia and
tremors. Symptoms usually disappear over a 2-hr period following termination of exposure. It is vitally important that all patients be taken off
their iron medications during treatment since BAL forms a toxic complex
with iron [12]. EDTA is usually excreted unchanged by the kidney and can
cause proteinuria, anemia, latent polyuria, urinary urgency, hematuria,
hypoglycemia with Trousseau's sign, hyperreflexia, hypertension, headache,
sneezing, nasal congestion, lacrimation, GI distress and hypermotility,
fever, chills, myalgia, nausea, and loss of appetite. Chronic poisoning
produces lesions that respond to B-complex vitamin therapy, especially to
B6. If given IV, the solution should be less than 0.5% with a maximum IV
rate of 38 mg EDTA/kg/hr in order to decrease the incidence of thrombophle-
bitis and renal damage. Steroids should be avoided during EDTA therapy
since they intensify the renal toxicity of EDTA. Although there is some
debate over the issue, it is probably best to rule out the possibility of
recurrent ingestion of lead because oral EDTA in an already lead-burdened
patient may result in increased absorption of lead from the gastrointestinal tract. Careful monitoring of the electrolytes, BUN, calcium, phosphorus,
alkaline phosphatase, and urinalyses is necessary during EDTA therapy.
Penicillamine is related to the penicillins and frequently displays its
toxicity as allergic reactions; eosinophilia, rash, ecchymosis, fever,
leukopenia, agranulocytosis, thrombocytopenia, and a nephrotic-like syndrome; D-penicillamine should be given on an empty stomach approximately
1-1/2 hr before a meal.

Follow-up is mandatory following the initial course of therapy, which
is usually 5 to 7 days. In symptomatic children a short rest period should
be given following the initial course. A minimum of 2 days is required and
14 to 21 days is preferred. A second course of therapy with BAL/EDTA is
given if blood levels remain greater than 80 $\mu g\%$. For levels less than 80
$\mu g\%$ a 3- to 6-month course of D-penicillamine (500-750 mg/day divided into
three oral doses) should be initiated and continued until blood levels are
less than 60 $\mu g\%$. Again, it is important to rule out recurrent ingestion
since oral chelates may result in increased absorption of intestinal lead.

It takes approximately 6 to 12 months for soft tissue lead to be transported to and safely stored in bone. An acute illness may mobilize lead
from these tissues causing an exacerbation with increased UCP and/or urinary
ALA. A 3-day course of EDTA at 50 mg/kg/day is then recommended. Repeat
ingestion is always a risk and should be treated appropriately. During the
first year of convalescence, blood lead levels may actually rise because of
the resolving anemia and increasing hematocrit. Radiographic studies will
show widening of the metaphyseal densities as lead is deposited in bone.
According to Chisolm, several ALA, UCP, and hematologic studies are perhaps
the most satisfactory means of detecting toxicity or recurrent exposures
during the convalescent period [4].

With our present therapy, the mortality rate is approximately 5% of
those cases with signs and symptoms of encephalopathy. The U.S. Public
Health Service estimates that each year 400,000 children have increased
blood lead levels and 16,000 of these require treatment. Two hundred
children die annually of acute encephalopathy and another 800 are severely
brain damaged and require permanent care. Another 3200 suffer moderate to
severe brain damage and require special medical and rehabilitation care
[11]. Chisolm reports that at least 25% of survivors of acute encephalopathy
sustain permanent brain damage [9]. This includes mental retardation, seizure disorders, cerebral palsy, optic atrophy, behavior disorders, and a
wide range of minimal brain dysfunction (MBD) [8,27]. Studies of MBD children with and without encephalopathy have been conflicting. An estimated
22 to 27% display minor neurologic dysfunction and various forms of motor
impairment [26]. Kotok recently reported a group of children with increased
lead levels with a mean of 81 $\mu g\%$. Some were symptomatic but none had symptoms of encephalopathy. Both the patients and carefully matched controls
demonstrated similar deficiencies in fine motor, adaptive, and language

functions, indicating inadequacies in the child's environment rather than
sequelae secondary to lead toxicity [24]. Further investigation of this
group without encephalopathy is certainly indicated.

Additional pathology is seen in the form of late onset renal insufficiency. Onset is usually during or after adolescence and is progressive.
Ninety-four of 352 patients in one study died of chronic nephritis 15 to
40 years after the initial intoxication [7]. This type of lead nephropathy
is probably restricted to those with a history of prolonged lead ingestion.

LEAD POISONING IN ADULTS

In contrast with poisoning in children, lead intoxication in adults
is usually in the form of acute ingestions and is usually related to industrial exposure. The occasional nonindustrial acute poisonings are related
to ingestion of highly soluble lead salts in illicit whiskey or "white
lightning," or from eating acidic foods which have leached the lead out of
improperly glazed ceramic containers. These salts are usually in the form
of acetates, citrates, carbamates, chromates, or salts of various acids.

Occupational poisoning may be chronic or acute and the distinction
between the two types is often vague. Industrial workers at risk are those
involved in lead mining, smelting, and refining, storage battery manufacturing, ship breaking, autobody painting, painting, and pottery glazing.
Those occupations occasionally at risk are those related to petroleum products, cable construction, ceramics, ammunitions, preventive shielding, and
noise and vibrational control [33].

After an acute ingestion, signs and symptoms are primarily related to
gastrointestinal irritation and include a metallic taste, dry mouth, nausea,
vomiting, and burning abdominal pain. Systemic symptomatology and signs
may include anemia, proteinuria, weakness, headache, paresthesias, pain and
cramps in the legs, depression, coma, and death.

Twenty to thirty grams of soluble lead salts normally constitute an
acute oral dose for an adult. Children eating lead paint chips, such as
gray porch paint, may develop an acute episode of lead poisoning. Frequently, a paint chip the size of a thumbnail may contain 100 mg of a lead
salt. Occasionally a hemolytic crisis and/or severe renal damage may follow
ingestion. Blood and urine lead levels are elevated and the provocative
lead excretion test is positive. Karpatkin reviewed the literature on
acute poisoning and found that his patient plus six to seven reported cases

all developed significant anemia, usually within 2 weeks. Recovery was
relatively rapid when compared with the protracted anemia of chronic poisoning. Several of Karpatkin's patients had hematuria, albuminuria, ne-
phrotic syndrome, headache, hypertension, and blood levels ranging from 60
to 150 µg% [26].

The ingestion of a metallic lead object does not cause acute poisoning.
Rather, it is retained as a source of chronic poisoning with gradual absorption. An acute symptomatic episode may occur days to weeks after ingestion
and is indistinguishable from chronic plumbism. Treatment is the same as
for chronic poisoning.

Tetraethyl lead (TEL) was introduced as an antiknock fuel additive in
1923 and has been the subject of much concern since that time. The problem
has threatened to affect whole populations as more and more of our lead
production is converted to TEL, of which 70 to 80% is eventually exhausted
into the atmosphere [17]. Automotive pollution control has tried to minimize this threat. At the present time, intoxication is limited almost entirely to the petroleum industry and the handling of the antiknock additive
up to the point where it is added to the petroleum storage tanks. Exposure
to tetraethyl or tetramethyl lead (TML) may result in toxicity either by
absorption through the skin or by inhalation. In contrast to lead salt,
TEL and TML can be absorbed in large quantities through the skin and can be
found in all tissues [4,25]. Absorbed tetraalkyl lead (TEL and TML) is distributed in the body as an oil-soluble material and is rapidly decomposed
by the tissues to the trialkyl forms. In 3 to 14 days, all this lead is
distributed through the body as if it were a water-soluble lead compound
[23]. Inhaled lead is handled in a similar manner.

Symptoms of TEL or TML intoxication include insomnia, wild terrifying
dreams, toxic psychosis, headaches, hyperactivity, ataxia, emotional instability, erratic behavior, delusions, convulsions, and mania. There are
rarely any hematologic abnormalities [3,28]. Urinary lead levels are usually quite elevated while blood levels may be slightly elevated or normal
[28]. There is a 20% mortality rate which accompanies this disease, but if
the patient recovers, sequelae are rare [3,28].

Industry is now required to monitor and control lead exposure. A safe
level reported for lead in the form of TEL is 75 µg Pb/m^3 air for a 40-hr
work week at 8 hr/day. Occasional exposures are evaluated by dose and
duration [28].

1000 μg Pb/m^3 for 1 hr

600 μg Pb/m^3 for 2 hr

400 μg Pb/m^3 for 3 hr

280 μg Pb/m^3 for 4 hr

These levels are well tolerated because the lead is quickly excreted in the urine during periods of nonexposure and no source of continued absorption is present. The recommended levels for soluble particulate inorganic compounds is 200 μg Pb/m^3 for a regular work week. This level will result in only mild intoxication or not at all [20]. However, this level is influenced by particle size. Large particles may be caught in the upper airways and swallowed, thereby constituting a potential source for chronic poisoning (Table 4).

In addition to monitoring atmospheric lead levels, the workers themselves are monitored, according to Table 4. Urinary coproporphyrin and δ-aminolevulinic acid levels probably are the most reliable indicators of body lead load and soft tissue concentrations of lead. ALA is somewhat more specific and sensitive [15]. Fleming recommends that if the UCP is 110, the worker be placed on a watch list. If it rises to 150, the worker should be transferred to a low-exposure area. At levels of 200, the patient should be examined by a physician and appropriate treatment be initiated [19].

Prevention is directed toward avoidance of exposure. Careful handwashing, proper ventilation, and provision of eating facilities that are separate from the actual working and exposure areas are mandatory.

TABLE 4. Laboratory Tests Used in Industrial Medicine to Monitor Occupational Exposure [3]

	Nonexposed Worker	Increased Absorption (Worker Healthy)	Dangerous Levels May Be Symptomatic
Blood lead, μg Pb/100 g	<40	55–80	>80
Urine level, μg Pb/liter	<80	<150	>200
Hemoglobin, g/100 ml	>13	>13	<13
UCP, μg/liter	<280	<500	>800
Qualt, UCP	0-++	+++	++++
Urinary ALA, mg/liter	<b	<13	>19

The primary principle of treatment is to remove the patient from exposure. If dermal absorption is suspected, the skin should be thoroughly
washed. Treatment of a single oral ingestion is to induce emesis or perform gastric lavage with a 1% solution of sodium or magnesium sulfate.
Thirty milliliters of magnesium sulfate in a 10 to 15% solution is left in
the stomach to promote catharsis. Demulcents should be given. The remainder
of therapy is supportive. Chelation therapy is rarely indicated. The patient should be observed for 10 days with initial and follow-up blood lead
and urine levels and/or UCP or ALA levels (see Tables 2 and 4). Treatment
of intoxications with lead vapors is basically supportive with prolonged
sedation with short-acting barbiturates. No specific therapy is recommended.
Careful attention should be paid to fluids and electrolytes.

After emergency treatment patients with acute lead ingestions or acute
poisoning should be treated similarly to the protocol for chronic lead
poisoning.

REPRESENTATIVE CASES OF LEAD INGESTION

Case No. 1

A 7-month-old infant was admitted because of vomiting, fever, and
respiratory distress. He was found to have microcytic, hypochromic red
cells with basophilic stippling. A blood lead of 40 µg% was considered
high for this age.

Lead source was from water concentrated by boiling in the family kettle
used in formula preparation. Tap water contained lead of 0.1 mg/liter.
Kettle scrapings contained 50 ppm of lead. Leached lead from pipe joints,
as a result of recent installation of a water softener, was incriminated.

Case No. 2

Vomiting, lethargy, albuminuria, glyosuria, hematocrit 23%, and basophilic stippling prompted the summer admission for this 2-1/2-year-old girl.
A blood lead level was 100 µg%, urinary ALA was 10 mg%, spinal fluid was
51 µg% protein, and radiopacities in G.I. tract confirmed lead intoxication
and possible encephalopathy even though pica was denied. Treatment with
BAL/EDTA was given for 5 days. Twenty-four hour daily urine samples showed
13, 15, 4, and 3.9 mg of lead excreted. A few days after treatment blood
lead was 54 µg%.

From 2-1/2 to 6-1/2 years she was followed vigorously. She was admitted five times for chelation with levels above 75 µg%. Penicillamine was used on an outpatient basis. Constant movement of this child from house to house and even out of the state made it difficult to identify the many sources. Continued efforts by physicians, nurses, housing inspectors, and a toxicologist were used. At 6-1/2 years her blood lead was 35 µg%.

Case No. 3

Asymptomatic toddler was seen because mother was concerned about him eating plaster at home and at the grandmother's. Mother was a clay eater. She had a friend whose child died of lead intoxication from eating plaster.

Blood lead level was 120 µg%, hct. 29%. He was treated with penicillamine and parents instructed. Three months later lead level was 91 µg% and radiopaque flecks were again seen in the intestine, supporting evidence of continued pica. One month later, after chelation, blood lead level was down to 56 µg%. The decrease was likely due to efforts of the social worker in identifying the source and reducing the pica.

Case No. 4

An 18-month-old boy was seen at an area hospital for irritability, vomiting, and refusal to play with peers. A "LEAD ALERT" intern found blood lead to be 191 µg%, hct. 27%, and basophilic stippling.

On admission urine ALA was 6.5 mg%, radiopaque objects were found in the colon, and a wrist drop was observed.

There was a history of pica for dirt, the bedposts of his crib, and the posts of the staircase. Paint chips from the home contained 40 ppm of lead. BAL/EDTA chelation was given for 6 days. Twenty-four hour urine for the first 3 days showed 7.95, 5.14, and 3.46 mg of lead excreted. Treatment was followed with penicillamine and correction of the household problem.

Case No. 5

This 2-year-old black male was admitted because of vomiting, intermittent abdominal pain, and constipation for 1 month. He was in no distress but was very withdrawn and shy. Blood smear showed hypochromic anemia with basophilic stippling. Urine for coproporphyrin was positive. Spinal fluid protein was 75 mg%. X-ray of the metaphasis revealed dense markings. Chelation with EDTA was started. Convulsions were not controlled by Dilantin and phenobarb; however, after diuresis with urea the convulsions ceased.

Child died the next day. (This death was due to lead encephalopathy, before stat blood leads were available.)

Case No. 6

This 20-month-old black male while walking along the street in midsummer suddenly developed rhythmic shaking of extremities. He presented in the emergency room in semicoma and moderate rigidity. Laboratory: hct. 33%, WBC 25,650, basophilic stippling, albuminuria, glyosuria, spinal fluid protein 72 mg%, calcium and phosphate normal, urine coproporphyrins positive, blood lead 131 μg%, EEG diffusely abnormal. BAL/EDTA chelation plus urea-induced diuresis brought improvement in clinical status in 48 hr. By the seventh day he was mobile enough to crawl over the elevated crib side, and after a second course of chelation, he was dismissed exhibiting a mild ataxia.

Home inspection did not reveal source of lead. Four subsequent episodes of chelation were performed. Sequelae were seizures and speech problems. One and one-half years later blood lead was 25 μg% and the EEG was markedly improved.

REFERENCES

1. D. E. Benson and J. J. Chisolm, A Reliable Qualitative Urine Coproporphyrin Test for Lead Intoxication in Young Children, J. Pediatr., 56, 759 (1960).

2. R. K. Byers, Lead Poisoning. Review of the Literature and Report on 45 Cases, Pediatrics, 23, 585 (1959).

3. J. J. Chisolm, Treatment of Lead Poisoning, Modern Treatment, 8, 593 (1971).

4. J. J. Chisolm, Disturbances in the Biosynthesis of Heme in Lead Intoxication, J. Pediatr., 64, 174 (1964).

5. J. J. Chisolm, The Use of Chemistry Agents in the Treatment of Acute and Chronic Lead Intoxication in Children, J. Pediatr., 73, 1 (1968).

6. J. J. Chisolm, Screening for Lead Poisoning in Children, J. Pediatr., 51, 280 (1973).

7. J. J. Chisolm, Chronic Lead Intoxication in Children, Develop. Med. Child Neurol., 7, 529 (1965).

8. J. J. Chisolm and H. E. Harrison, Treatment of Acute Lead Encephalopathy, J. Pediatr., 19, 2 (1957).

9. J. J. Chisolm, The Exposure of Children to Lead, J. Pediatr., 18, 943 (1956).

10. J. J. Chisolm and E. Kaplan, Lead Poisoning in Childhood — Comprehensive Management and Prevention, J. Pediatr., 73, 942 (1968).

11. Control of Lead Poisoning in Children, Bureau of Community Environ-
     mental Management, Public Health Service, 1970.

12. N. D. Edoe and G. E. Somers, The Effect of BAL in Acute Iron Poisoning,
     Quart J. Pharmacol., 21, 364 (1948).

13. A. J. Fleming, Industrial Hygiene and Medical Control Procedures, Arch.
     Environ. Health, 8, 266 (1964).

14. S. C. Gilfillan, Lead Poisoning and the Fall of Rome, J. Occup. Med.,
     7, 53, 60 (1965).

15. B. Haeger-Aronsen, An Assessment of the Laboratory Tests Used to Moni-
     tor the Exposure of Lead Workers, Brit. J. Ind. Med., 28, 57 (1971).

16. P. B. Hammond, Lead Poisoning, An Old Problem with a New Dimension,
     Essays in Toxicology, Vol. 1, Academic, New York, 1962, Chap. 4.

17. D. A. Hirschler, L. F. Gilbert, F. W. Lamb, and L. M. Miebylski, Par-
     ticulate Lead Compounds in Automobile Exhaust Gas, Ind. Eng. Chem.,
     49, 1131 (1957).

18. H. Jacobziner, Lead Poisoning in Childhood: Epidemiology, Manifesta-
     tions and Prevention, Clin. Pediatr., 5, 277 (1966).

19. S. Karpatkin, Lead Poisoning after Taking Lead Acetate with Suicidal
     Intent. Report of a Case with a Discussion of the Mechanism of Anemia,
     A. M. A. Arch. Environ. Health, 2, 679 (1961).

20. R. A. Kehoe, Industrial Lead Poisoning, Industrial Hygiene and Toxicity,
     2nd rev. Ed., Vol. II, Wiley (Interscience), New York, 1967.

21. R. A. Kehoe, Metabolism of Lead in Man in Health and Disease. The Hur-
     ben Lectures, J. Roy. Inst. Public Health, 24, 81-120, 124-143, 177,
     203 (1960).

22. R. A. Kehoe, Normal Metabolism of Lead, Arch. Environ. Health, 8, 232
     (1964).

23. R. A. Kehoe, J. Cholak, D. M. Hubbard, K. Burnbach, R. R. McNary, and
     D. V. Story, Experimental Studies on Ingestion of Lead Compounds, J.
     Ind. Hyg., 27, 381 (1940).

24. D. Kotok, Development of Children with Elevated Blood Lead Levels, J.
     Pediatr., 80, 57 (1972).

25. E. D. Lang and F. M. Kunze, Penetration of Lead Through the Skin, J.
     Ind. Hyg. Toxicol., 30, 256 (1948).

26. S. M. Perischel, L. Kupito, and H. Schwachman, Children with Increased
     Lead Burden, J. A. M. A., 222, 462 (1972).

27. M. A. Perlstein and R. A. Attala, Neurologic Sequence of Plumbism in
     Children, Clin. Pediatr., 5, 292 (1966).

28. L. W. Sanders, Tetraethyllead Intoxication, Arch. Environ. Health, 8,
     270 (1964).

29. A. Shulman and F. P. Dloyer, Chelating Agents and Metal Chelates,
     Academic, New York, 1964, Chap. 9, p. 388.

30. Statement of Diagnosis and Treatment of Lead Poisoning in Childhood
     by Subcommittee on Accidental Poisoning of American Academy of Pedi-
     atrics, Pediatrics, 27, 676 (1961).

31. J. Vostal, Study of the Renal Excretory Mechanism of Heavy Metals,
     Intern. Congr. Occupational Health, 15th, Vienna, 3, 61 (1966).

32. H. A. Waldron, The Anemia of Lead Poisonings: A Review, Brit. J. Ind.
     Med., 23, 53 (1966).

33. R. L. Ziegfeld, Importance and Uses of Lead, Arch. Environ. Health, 8,
     202 (1964).

8

EPIDEMIOLOGY OF LEAD POISONING IN ANIMALS[*]

Gary D. Osweiler and Gary A. Van Gelder
University of Missouri College of Veterinary Medicine
Columbia, Missouri

William B. Buck[†]
Iowa State University College of Veterinary Medicine
Ames, Iowa

Lead has been an enduring toxicologic problem in the recorded history of mankind [1]. Lead toxicosis continues to be a significant problem in both man and animals. A number of papers (a representative sample is cited) have reviewed selected aspects of the toxicology of lead in man [2-11], animals [12-21], wildlife [22,23], and general environment [24,25]. The clinical aspects of lead toxicosis in animals are discussed in current toxicology textbooks [26-28].

Lead has generally been considered to affect primarily the gastrointestinal, nervous, and hematopoietic systems. However, there is also broad interest in its general biologic effects [29]. As evidenced by recent reports, lead does have general cellular effects. Current research efforts have focused on the effects of lead on the kidney [30-40], interaction with nutritional factors [41], interaction with other contaminants [42], effect on renin-aldosterone [43], thyroid [44], chromosomes [45], immune

[*]Preparation of this manuscript was sponsored in part by the Toxicology Information Program/National Library of Medicine/National Institutes of Health. Consultant during the preparation of this review to the Toxicology Information Response Center/Oak Ridge National Laboratory, operated by Union Carbide Corporation Nuclear Division for the U.S. Energy Research and Development Administration.
[†]Present Affiliation: University of Illinois College of Veterinary Medicine, Urbana, Illinois.

mechanisms [46-51], hematologic dynamics [52-55], behavior [56-60], and electrical activity of nerves [61-63].

Domestic consumption of lead in the United States has increased from 1,110,000 tons in 1962 to 1,445,000 tons in 1972 [64]. Mine production was 618,361 tons in 1972 with an additional 610,000 tons recovered from secondary sources, of which reclaimed batteries accounted for 63%. The major uses of lead are shown in Table 1 and are based on the estimated 1972 consumption data [64].

TABLE 1. Uses of 1,445,600 Tons of Lead in the United States for 1972[a]

Use	Tons	Percent of Total[b]
Storage batteries	695,000	48.0
Antiknock additives	278,300	19.3
Ammunition	85,567	5.9
Red lead and litharge	69,000	4.7
Solder	71,400	4.9
Cable covering	48,700	3.4
Caulking	23,000	1.6
Sheet lead	23,000	1.6
Brass and bronze	18,600	1.3
Pipe	18,500	1.3
Weights	18,300	1.3
Type metal	18,000	1.2
Pigment colors	16,300	1.1
Bearing metals	15,000	1.0
Casting metals	6,500	0.4
Foil	4,500	0.3
Terne metal	4,500	0.3
Annealing	4,100	0.3
Collapsible tubes	3,800	0.3
White lead	2,900	0.2
Galvanizing	1,300	0.1
Other	18,900	1.3

[a]Adapted from Ref. 64.

[b]Rounded to nearest 0.1%.

OCCURRENCE OF LEAD POISONING IN ANIMALS

Lead poisoning is a major toxicologic problem in animals. Lead toxicosis has been reported in horses [65-70], cattle [12-18,65,68,69,71-81],
sheep [76,78,82-85], dogs [20-21,86-104], cats [105-108], primates [109-
117], fruit bats [118], wildfowl [119-129], raccoon [130], parrots [130],
and armadillos (personal communication, W. Wass, Iowa State University,
Ames, Iowa).

Among the domestic animals lead poisoning is most often diagnosed in
cattle. Some suggest that lead is one of the most common toxicants [13,17]
or the most common metal toxicant [14] of cattle.

Cattle

A definitive seasonal increase in lead poisoning appears to occur in
cattle [131]. In a 6-year period of study by the Iowa Veterinary Diagnostic Laboratory, June 1, 1966, through December 31, 1972, 121 cases of bovine
lead poisoning were recorded (Table 2). Of these cases, 67% occurred in
the first half of the calendar year and only 9% in the last quarter.

TABLE 2. Cumulative Seasonal Occurrence of Bovine Lead Poisoning from
Records of the Iowa Veterinary Diagnostic Laboratory

Month	No. of Cases	Percent of Total	Seasonal % Total
January	8	7	27
February	15	12	
March	10	8	
April	14	12	40
May	23	19	
June	11	9	
July	16	13	23
August	8	7	
September	4	3	
October	3	2	9
November	3	2	
December	6	5	
Total cases	121		

In cattle, the increased incidence has been associated with springtime access to areas of lead contamination (machinery, oil-soaked soil, trash piles in pastures). Another theory is that spring forages induce hypophosphatemia and may influence pica, which leads to ingestion of lead. Still another theory is that the increased sunlight during the summer months causes increased vitamin D levels and thus increased absorption of lead via the gastrointestinal tract [6].

In cases from the authors' files where sufficient information was known, mortality and morbidity statistics for bovine lead poisoning were determined (Table 3). Used oil (category 2) as a cause of clinical lead poisoning produced a lower morbidity rate than did category 1 (lead batteries, paint, grease), but a higher percentage (83%) of those affected from oil sources died than of those exposed to other sources (56-61%).

TABLE 3. Morbidity and Fatality Rates in Relation to Source of Clinical Lead Toxicoses in Cattle[a]

	Category 1[b]	Category 2[c]	Category 3[d]	All Sources
No. of episodes	15	9	12	37
Total no. animals in herd	864	374	1030	2268
Total no. animals clinically affected	130	46	110	286
Total no. animals died	73	38	62	173
Morbidity: $\frac{\text{No. affected}}{\text{Total no. in herd}} \times 100$	15%	12%	11%	13%
Case specific death rate: $\frac{\text{No. died}}{\text{Total no. in herd}} \times 100$	9%	10%	6%	8%
Case fatality rate: $\frac{\text{No. died}}{\text{No. affected in herd}} \times 100$	56%	83%	56%	61%

[a] Data from Iowa Veterinary Diagnostic Laboratory.

[b] Grease, batteries, and paint.

[c] Used crankcase oil.

[d] Junk piles and other unknown sources.

No definitive age relationship has been reported for lead poisoning in cattle, although in the authors' experience a majority of cases involve suckling calves or cattle less than one year of age.

Dogs

Lead poisoning in dogs has been reported as primarily an urban problem [132]. The reported incidence of lead poisoning in dogs appears to be increasing, due most probably to increased awareness by the veterinary profession and to improving diagnostic techniques.

Several reports of canine lead poisoning have suggested that the similarity in clinical syndromes between lead poisoning and viral canine distemper, which is a relatively common disease of young dogs, has resulted in a low rate of detection of clinical lead poisoning [21,90,92,97,102].

In one study [20] of canine lead poisoning it was found that 81% of the afflicted dogs were under 1 year of age. Of hospital admissions, the occurrence of lead poisoning was significantly greater in poodles (21%).

The authors have seen only a few cases of canine lead poisoning. According to Zook [21] the highest incidence of lead poisoning in dogs correlates with high population density in low-income areas with older housing. Since the authors' area of experience is mainly rural, small town, and of moderate income, this could explain an apparent low incidence of canine lead poisoning observed.

For dogs, Zook [20,21] reported 63% of canine lead poisoning occurring in the 5 months from June through October, with July being the peak month. The increased warm-season incidence of lead poisoning in cattle and dogs parallels the seasonal incidence in children [133].

Cats

Only a few cases of lead poisoning are reported in the cat [134]. The relatively more fastidious eating habits of cats may account for the low reported incidence of poisoning in that species. Although licking of the haircoat, commonly practiced by cats, could contribute to an increased body burden of lead deposited in areas of aerial contamination, this possibility has not been investigated.

Horses

Equine plumbism has been recorded in a number of instances [67-70,135]. Poisoning most often occurs in the vicinity of lead mines or smelters as a result of consumption of lead-contaminated forages. Both acute and chronic

syndromes are recognized in the horse. No age or seasonal relationships
are recorded.

Swine

Lead toxicosis is rarely reported in swine, and pigs were found experimentally to be relatively resistant to lead [136]. In addition, the
eating and foraging habits of swine do not suggest their consumption of
foreign objects.

Primates

During the past several years an old disease of caged Old World monkeys has been shown to be lead poisoning. The disease known as amaurotic
epilepsy or demyelinating encephalomyelopathy was described in the 1930s
[113]. Recent evidence points to the consumption of lead-based paints on
the cages as the cause, based on tissue residues averaging 52 ppm in liver
[115] and the presence of eosinophilic acid-fast inclusion bodies in renal
proximal tubular epithelium. The paint used on the cages contained 2.6-
67% of lead.

New World primates apparently do not chew on their cages as frequently
as the Old World primates and consequently do not develop acute amaurotic
epilepsy, now known to be lead poisoning.

Another demyelination disease of captive Old World primates, associated with but not proven to be synonymous with lead poisoning, is leucoencephalomyelosis, also known as cage paralysis, confluent leucoencephalosis,
and perivascular myelosis [116]. One theory suggests that vitamin B12 deficiency may be involved. Lead exposure may be either a confounding or
contributing factor as five of seven cases of leucoencephalomyelosis studied
also had elevated liver lead levels and acid-fast intranuclear inclusions
in renal proximal tubular cells.

Birds

Lead poisoning is often observed in surface-feeding birds such as
mallards and pintail ducks, with the number of lead shot found in gizzards
ranging from 1 to 22 [22,23,123,127]. Pellets are often worn down to
nothing more than a thin flake of lead. Soft-bottomed ponds allow lead
shot to sink, while hard-surfaced ponds hold shot available for long periods
of time.

SOURCES OF LEAD

The major source of lead for the reported cases of lead poisoning in animals continues to be lead-containing paints [16,20,75,131,134]. Many old paints contain in excess of 10% lead compounds. Their occurrence is most often associated with older houses, barns, fences, and discarded paint cans in junk piles. The tendency for older paints to flake or peel makes them special targets for ingestion by puppies, cattle, and children, all of whom may mouth or chew on painted objects at various times. Older buildings may also be plagued by crumbling putty, broken linoleum, exposed tar paper, and lead water pipes. Thus the high lead content and deteriorating nature of these products lend them to the hazard of easy ingestion.

In the United States, the contribution of gasoline to lead in the atmosphere, and subsequently forage, is estimated to be 20 times that from burning coal [137]. The lead content of air drops off rapidly with distance from the vehicular source or roadway.

Several reports have documented the presence of elevated lead residues in grass forages growing adjacent to heavily traveled roads. Others describe the nature of lead products in automotive exhausts resulting from burning leaded gasoline [138,139]. Motto [140] found levels as high as 255 mg lead/kg dry weight grass immediately adjacent to roadways. The levels decreased to 165 mg/kg at 7.6 m, 99 mg/kg at 22.8 m, 67 mg/kg at 38.1 m, 55 mg/kg at 53.3 m, and 46 mg/kg at 68.6 m from the road. Since a cow can be expected to eat 22.5 g dry matter/kg/day, the lead exposure would range from 5.7 to 1.0 mg/kg from eating the contaminated forage. Potential exposure in horses would be equivalent since a horse averages eating 21 g dry matter/kg/day. Consumption of only the most heavily contaminated forage would be likely to cause any overt adverse health effects. Grass growing near heavily trafficked roads is generally not available to most grazing animals so the total hazard is low. To date, no instances of poisoning from motor vehicle pollution of forage have been reported [141]. However, the lead intake from contaminated pasture or hay would be expected to contribute to the total body burden of lead.

Cattle

For cattle, in addition to paint, major sources include used motor oil, discarded lead storage batteries, machinery grease, lead caulking, and isolated instances of chewing or eating other lead-containing products [13].

Many times these items are available in rural trash piles within the confines of a pasture. Such articles may also be found around barns where calves may be able to gnaw at them.

In a survey of lead poisoning in livestock conducted at Iowa State University, lead poisoning was diagnosed in a total of 63 separate cases from 1965 through 1970 [16]. The results of that study are summarized in Table 4. All livestock lead poisoning was confined to cattle.

Paint and petroleum products accounted for 60% of all lead poisoning diagnosed in the 6-yr period of study. Unknown or general sources such as trash piles involved 35% of the cases.

Recent work (1971 and 1972, inclusive) at Iowa State University involved 57 additional cases of lead poisoning. Results again were similar to the 6-yr study in that paint and used motor oil were the leading sources. The higher yearly incidence of poisoning in 1971 and 1972 most likely represents increased awareness of potential toxicosis and improved and refined diagnostic methods.

None of the unknown-source episodes from the authors' files appeared related to air pollution as a source of lead. Heavy industry and lead smelting or reclamation are relatively uncommon in the authors' area (Iowa), and this would account for the lack of industrially related lead toxicosis.

In horses most episodes have involved contaminated pastures near smelters [66-69]. Airborne particulate lead settles upon forage and vegetation around smelters and is ingested with the natural herbage intake of the animals [66,68,135]. In a relatively old report, Haring [66] investigated 32 farms located near a smelter. Twelve cases of laryngeal paralysis were

TABLE 4. Sources of Clinical Lead Poisoning in Cattle

Source	No. of Episodes	Percent of Total
Paint	18	29
Oil	16	25
Unknown	15	24
Trash piles	7	11
Grease	4	6
Batteries	3	5
Total	63	100

identified in horses grazing contaminated pastures. The estimated lead
dose was 250 mg/horse/day. Cattle and hogs raised in the area were not
affected.

Aronson [69] reviewed epizootics of lead poisoning in horses and cattle.
Based on the level of lead in contaminated forages, exposure for 2 months
of an estimated dosage of 5-6 mg/kg proved lethal in some cattle, while only
1.7 mg/kg fed for several months was fatal to horses. In some cases involving contaminated forages, cattle remained healthy while horses became ill
grazing the same pasture.

The impact of lead mining and ore processing activities on lead contamination in an area in southeast Missouri that produces 75% of the current
U.S. supply has been reported by Dorn [135]. Test cows placed in the area
were estimated to have their lead exposures increased from a background of
0.8 mg/kg up to 8.6 mg/kg in the test site. As in other similar situations,
the horse seemed either to be more sensitive or somehow to have achieved a
higher total exposure.

Reports of horses affected while cattle grazing in the same area remain unaffected suggest that horses are more susceptible to lead toxicosis.
Aronson cautions, however, that since horses tend to eat grass closer to the
ground and occasionally pull plants up by the roots they may receive additional exposure from contaminated soil.

Dogs

The major identified lead sources were paint and linoleum for a population of city-dwelling dogs [20,132]. Other sources for dogs have included
lead drapery weights, fish sinkers, and other miscellaneous objects. In
addition to the above sources for dogs, Clarke [134] reviewed reports of
poisoning from contaminated soil and from dust subsequent to high-speed
mechanical removal of paint. Of several confirmed cases of lead poisoning
in dogs observed by the authors three had an identified lead source. One
of these was related to gnawing on boards painted with lead-containing
paint; another involved ingestion of Christmas tinsel with high lead content. The third involved a mechanic's dog kept constantly in an area contaminated with used oil and gasoline.

Primates

In captive primates the major sources have been the paint used on the
cages. Houser and Frank [112] reported a case of accidental lead poisoning

in a 4-month-old rhesus monkey being raised on a surrogate mother mounted
on a solid lead base.

Birds

Wild waterfowl ingest and retain in the gizzard spent lead shot which
results in lead poisoning [119,123,127]. More recently some upland game
birds have been shown to have elevated lead exposures [124].

The approximate toxic dose for wildfowl such as mallard ducks is 1 g,
based on feeding trials where birds were fed eight no. 6 shotgun pellets
[126]. The nonlethal dose is approximately 380 mg (three no. 8 pellets)
[119]. Birds have a remarkable capacity for retaining lead shot in the
gizzard; frequently six or eight shot are retained [125,126]. In some
cases birds die while others develop nonlethal poisoning. Ducks have been
shown experimentally to be susceptible to poisoning from consumption of
marsh soil containing disintegrated lead shot [142].

CLINICAL SIGNS

In most animals, the overt clinical signs of lead toxicosis relate to
the nervous and gastrointestinal systems [28]. Hematopoietic changes are
usually discovered secondarily.

Cattle

In a survey conducted by the authors [131], 90% of affected cattle
exhibited signs of central nervous system (CNS) involvement, and 60% exhibited signs of gastrointestinal involvement. This information is summarized
in Table 5. Body temperatures were recorded in only 15 cases and were normal in eight but increased (up to 111° F) in the other seven. Acute death
(less than 24 hr after onset of clinical signs) was reported in about one-
third of the episodes. Many animals, however, survived for a longer period
of time, ranging from a few hours to about 10 days.

Horse

The horse appears to show more peripheral nerve involvement with muscular weakness and "roaring" which results from paralysis of the recurrent
laryngeal nerve [66-68]. There is also loss of weight, stiffness of joints,
progressive arching of the back, and cachexia [67]. Young foals are more
severely affected than older horses [67].

TABLE 5. Clinical Signs Observed in Lead-Poisoned Cattle[a]

System	Clinical Sign	No. of Episodes	Percent of Total Episodes
CNS	Blindness	32	51
	Muscle twitching	25	40
	Hyperirritability	21	33
	Depression	20	32
	Convulsions	20	32
	Grinding teeth	15	24
	Ataxia	11	18
	Circling	10	16
	Pushing against objects	7	11
	(One or more of these signs were reported in 90% of the episodes)		
Gastrointestinal	Excessive salivation	26	45
	Anorexia	16	21
	Tucked abdomen	6	10
	Diarrhea	6	10
	(One or more of these signs were reported in 60% of the episodes)		
Other	Acute death	22	35
	Bellowing	8	13

[a]From records of the Iowa Veterinary Diagnostic Laboratory.

[b]Other signs reported (in $< 8\%$ of the cases) were abnormal posture, decreased milk production, down and unable to rise, rhythmic head jerking, paddling, emaciation, abortion, nasal and eye discharge, gray gum line, stiffness, dehydration, pharyngeal paralysis, rumen atony, aimless walking, constant dribbling of urine, dyspnea, vomiting, blistered teats, hyperkeratinized skin areas, protruding tongue, pica, opisthotonus, nystagmus, and coma.

Sheep

Calves but not sheep tend to show episodes of excitation a few hours before death [76]. In sheep, a transient constipation occurs 24 hr before anorexia.

Steward and Allcroft [143] reported specific motor disabilities in lame and poor-thriving lambs raised in a lead mining area. Tissue lead levels were elevated, but typical signs of lead poisoning did not occur. Others [85,144] have reported osteoporosis in young lambs raised in an old lead-mining region. Although unusually high tissue lead values were present, the authors did not consider that the condition presented the classical pathological picture of lead poisoning. Pregnant ewes fed powdered lead throughout gestation with their grain ration at a level to maintain blood levels at 50-60 µg/100 ml developed clinical signs of anorexia, CNS depression, weakness, and abortion [84]. No convulsions or hyperactivity were observed.

A relatively low exposure of 1 mg/kg to ewes in poor condition resulted in fetal deaths and abortions. Nonpregnant ewes tolerated 2-6 mg/kg for 1 yr [76].

Dogs

The clinical picture of lead toxicosis in dogs usually begins with anorexia, vomiting, colic, diarrhea, or constipation. The pet owner's attention is usually not attracted until signs of neurologic disorders occur, such as CNS depression, hyperexcitability, hysterical barking, champing fits, convulsions, opisthotonos, paraplegia, muscular spasms, hyperesthesia, or blindness. Clarke [134] cites evidence that gastrointestinal signs are more common in older dogs.

Gastrointestinal symptoms occurred in 87% of 60 clinical cases reviewed, while nervous system disorders occurred in 76% of the same cases [21]. Although a lead gum line has been reported twice [145,146], it appears that this is a rare finding and may have resulted from confusion with natural gingival pigments.

Because of some similarities between the clinical signs of lead poisoning and infectious canine distemper disease of dogs, it has been suggested [20,21,88,90,92,93] that the incidence of lead poisoning in dogs may be even higher than presently recognized. It is difficult to establish a firm diagnosis of canine distemper and many clinicians may not suspect lead poisoning. The basis for differentiating the two diseases has been reported [20]. The main features include reticular stippling of red blood cells and the presence of 5-40 nucleated red blood cells (metarubricytes or rubricytes) per 100 white blood cells. Differential diagnoses to be considered in canine lead poisoning include intussusception, acute pancreatitis, hepatitis, heat stroke, encephalitis, rabies, distemper, chlorinated hydrocarbon pesticide toxicosis, and metaldehyde poisoning.

Primates

Clinical signs in young rhesus monkeys included short convulsive seizures, eye blinking, decreased activity, and visual impairment. There was
no apparent effect on hemoglobin or packed cell volume. Intranuclear inclusions were found in a kidney biopsy. After 6 months there was no gross
evidence of permanent neurologic damage [112].

Birds

The clinical signs in waterfowl include weakness, anorexia, weight
loss, anemia, and paralysis of wings and legs.

PHYSIOPATHOLOGY

Pathologic Changes

Diseases which must be differentiated from bovine lead poisoning include polioencephalomalacia, infectious thromboembolic meningoencephalitis,
nervous form of coccidiosis, rabies, listeriosis, and organic insecticide
poisoning [13,28]. In the authors' experience, most cases of bovine lead
poisoning result in acute onset of clinical symptoms and rapid deaths; consequently few cellular lesions have been observed.

Some sources [17,147] suggest that one or more of the following brain
changes, ranging from mild to severe, are observed in acute lead poisoning
in the bovine: moderate brain edema, swelling, severe congestion of cerebral cortical tissue, prominence of capillaries and endothelial swelling,
petechial hemorrhage, and laminar neuronal necrosis.

A recent paper has reviewed the brain histopathology of lead poisoning
in cattle [80]. The more severe changes are limited to protracted cases
where animals linger with marked signs for 1-2 weeks. The histopathology
of prolonged cases may include laminar cortical necrosis [15,147], endothelial and astrocytic proliferation, microglial accumulation, and eosinophilic infiltration of leptomeninges [17]. A study of 55 confirmed cases
of lead poisoning revealed that 63% of the animals had intranuclear acid-
fast inclusion bodies characteristic of lead poisoning [148].

Necropsies were done in 37 of the 63 episodes in the authors' survey.
Of these, no gross lesions were observed in 10. In the remaining cases,
the following lesions were observed: oil in the gastrointestinal tract
(30%), gastritis and/or enteritis (24%), petechiation of epicardium and/or
myocardium (21%), pulmonary congestion (16%), and kidney degeneration
(16%).

Other lesions observed (in less than 10% of the episodes) were fatty liver; pale and watery muscle; petechiation of subcutaneous tissues, thymus, and trachea; cystitis; cloudy cornea; ocular hemorrhage; brain edema and hyperemia; metal or paint materials in rumen and reticulum; and swollen mesenteric lymph nodes.

Background urinary ALA (delta-aminolevulinic acid) levels in normal cows averaged 139.2 ± 75 μg/100 ml and elevated ALA levels above 500 μg were associated with bovine lead toxicosis [149]. Sharma reported background urinary ALA levels in sheep to be less than 50 μg/100 ml [84]. Blood lead and urinary ALA levels were directly correlated. Clinical signs of lead poisoning were accompanied by ALA levels of 300-400 μg/100 ml of urine. Chronic low-level lead ingestion has resulted in definite hematological lesions prior to onset of overt clinical signs [150-153]. A normocytic, normochromic anemia developed accompanied by elevated blood lead, increased serum and urine delta-aminolevulinic acid, and depressed amino levulinic acid dehydrase in erythrocytes. In ruminants these parameters appear to be early signals of impending lead poisoning.

Lead crosses the placental barrier. Female sheep fed 50 mg lead during gestation produced lambs with 37 ppm lead in their livers [78]. Lambs born from ewes grazing lead-contaminated pastures had liver lead residues of 4-8 ppm, while ewes with lambs on noncontaminated pastures had liver lead levels less than 1 ppm [74]. Nine of the 12 ewes given lead-contaminated feed sufficient to maintain blood lead levels of 50-60 μg/100 ml either aborted or resorbed fetuses [84]. Fetal tissues ranged from less than 1 ppm to 43 ppm on a wet-weight basis.

Uptake and elimination studies done in sheep [82] showed that only 1-2% of oral doses of either lead acetate or lead carbonate were absorbed. The maximum urinary excretion was 0.8 mg lead/day.

Sharma [84] reported blood background levels of 20 μg/100 ml or less in sheep and levels of 60 μg lead/100 ml blood to be associated with clinical signs of poisoning.

Zook described gross and microscopic lesions in dogs affected with lead poisoning [155]. Gross lesions, while generally nondiagnostic included abnormally red bone marrow, brain congestion, petechial hemorrhages, esophageal dilatation, and reduction in thickness of adrenal cortices. All lead-poisoned dogs had histologic lesions of renal tubular damage and a majority of affected dogs had intranuclear inclusion bodies in proximal

renal tubules. In the bone, metaphyseal sclerosis was observed in young
dogs poisoned by lead. The most obvious and consistent lesions in the
brain included vascular endothelial swelling and degenerative to necrotic
vascular changes. The reader is referred to the specific paper cited for
a detailed account of histopathological lesions. Necrotic gastroenteritis
in lead-poisoned dogs has also been reported [99].

Electroencephalographic (EEG) changes consisting of high-amplitude
(200 µV) delta waves (1-2 Hz) have been observed in lead-poisoned dogs [20].
The EEG changes were reversed after EDTA treatment.

Analysis of cerebrospinal fluid did not reveal consistent changes in
pressure or protein and cellular content [20].

Background urinary ALA levels in two dogs were 120 and 190 µg/100 ml.
In two experimental dogs fed lead, the ALA levels increased in a somewhat
erratic pattern with values of 200-5960 µg/100 ml [149].

No definitive brain lesions have been reported for lead-poisoned waterfowl. Vacuolation of liver cord cells and acid-fast intranuclear inclusions
in proximal convoluted kidney tubules have been reported [125,154]. Evidence
has been reported for lead-induced anemia in wildlife [126].

The presence of intranuclear inclusion bodies is frequently observed.
Richter et al. [156] demonstrated that the inclusions were distinct from
the nucleoli and that some were surrounded by Feulgen positive material but
did not contain DNA. The core of the inclusion was compact and amorphous,
probably containing proteins other than histones. The inclusion was surrounded by a fringe of microfibrils.

Diagnosis

As with other toxicoses, consideration must be given to obtaining a
good history, noting clinical signs, determining the likelihood of exposure,
and interpreting the chemical analyses. The clinical signs, important histopathologic factors, and hematologic findings, especially in dogs, have
been discussed previously.

The major analyses presently used to assist in establishing a diagnosis
of lead toxicosis are blood, liver, kidney, and stomach content lead levels.

Because of the variability in urinary lead levels and the difficulty of
routinely obtaining 24-hr urine specimens in domestic animals, urine lead
levels are of little use in establishing a diagnosis. The best use of urine
specimens is in comparing pretreatment and posttreatment levels after EDTA
chelation therapy. A large increase in lead levels should be observed [103].

One investigator has found that tissues stored in formalin can still be used for lead analysis to obtain retrospective information on lead poisoning [115].

The analysis of hair specimens for lead and other metals is of use in human studies [145] but of more limited value in animals because of seasonal growth changes and shedding [103]. In addition, studies have shown that metallic content comparisons in humans must be limited to a narrow age range and to one sex [157].

Kidney tissue contains high lead levels in poisoned animals. However, the investigator must be aware that kidney cortex contains higher levels than medulla [14]. Consequently, the investigator who uses kidney tissue for diagnostic purposes should very carefully separate cortical and medullary tissue.

One author has developed a point system for evaluating clinical findings in establishing a diagnosis of lead poisoning in dogs [20]. Key factors included afebrile colic and/or afebrile neurologic signs, 20 or more nucleated red blood cells/100 white blood cells, 26 or more stippled RBCs per 10,000 RBCs, metaphyseal sclerosis, marked improvement within 72 hr after Ca EDTA treatment, blood lead above 50 μg/100 ml, liver lead residues above 10 ppm, and 820 or more μg lead/liter urine 24 hr after chelation therapy is initiated.

Currently, lead levels in liver, kidney, other organs, and gastrointestinal contents are best determined by atomic absorption spectrophotometry [159]. Determinations of levels of lead in whole, unclotted blood are made using the procedure described by Hessel [160]. The use of EDTA salts as an anticoagulant is avoided.

Lead Concentration in Tissues and Blood

Before the diagnostic toxicologist can interpret the significance of tissue lead levels, the normal background levels should be known. Background tissue lead levels (Table 6) were obtained from animals submitted to the Iowa Veterinary Diagnostic Laboratory for which a diagnosis other than lead poisoning was made. Representative background tissue lead levels for domestic animals previously reported are summarized in Table 7.

Hopkins [161] reported an average background blood lead level for 27 baboons of 11.7 μg/100 ml. Locke and Bagley [124] reported that 37/40 randomly shot doves had liver lead levels of 5 ppm or less with most values below 3 ppm. Cook and Trainer [121] found background blood lead values of 18-37 μg/100 ml in 10 Canadian geese.

TABLE 6. Background Tissue (Wet Weight) Lead Levels for Domestic Animals

Species and Tissue	No. of Samples	Lead (ppm Mean ± S.D.)
Equine		
Blood	2	0.18 ± 0.05
Liver	9	0.82 ± 0.58
Kidney	7	0.93 ± 0.59
Bovine		
Blood	92	0.103 ± 0.044
Liver	197	1.12 ± 1.36
Kidney	181	1.21 ± 1.69
Rumen contents	52	1.07 ± 1.44
Canine		
Blood	20	0.093 ± 0.031
Liver	31	1.00 ± 0.98
Kidney	27	0.81 ± 0.82
Stomach contents	7	0.91 ± 1.36
Porcine		
Blood	2	0.12 ± 0.09
Liver	27	0.93 ± 0.73
Kidney	24	0.95 ± 0.97
Stomach contents	8	0.95 ± 0.88
Ovine		
Blood	2	0.09 ± 0.05
Liver	13	0.72 ± 0.58
Kidney	13	0.72 ± 0.58
Rumen contents	4	0.55 ± 0.30

Martin [162] reported whole body lead residues averaging 3.18 ppm in starlings collected from 23 survey sites in 1970. Martin and Nickerson [163] in 1971 found whole body lead residues of 0.12-6.6 ppm in starlings collected from 50 sites throughout the United States. Levels were lowest in birds collected from rural sites and higher in urban-collected birds.

The tissue lead levels associated with clinical bovine lead poisoning based on cases submitted to the Iowa Veterinary Diagnostic Laboratory are shown in Table 8.

TABLE 7. Background Tissue Levels in Domestic Animals

Animal	Tissue	No. of Samples	Lead Level	Refs.
Horse	Blood	2	0.14 μg/100 ml	[77]
Cow	Liver	13	0.3-1.5 ppm	[77]
Cow	Kidney	13	0.3-1.5 ppm	[77]
Calves	Blood	30	12.9 μg/100 ml	[77]
Calves	Liver	10	0.4-1.2 ppm	[77]
Calves	Kidney	10	0.4-1.0 ppm	[77]
Sheep	Blood	12	13.9 μg/100 ml	[77]
Sheep	Liver	5	0.6-1.2 ppm	[77]
Sheep	Kidney	5	0.3-0.8 ppm	[77]
Lambs	Liver	3	< 1.0 ppm	[77]
Lambs	Kidney	3	< 1.0 ppm	[77]
Goat	Blood	4	13.0 μg/100 ml	[77]
Dog	Blood	40	19.0 μg/100 ml	[103]
Dog	Liver	23	1.8 ppm	[103]
Dog	Liver	12	0.7 ppm	[90]

Previously reported tissue lead levels associated with toxicosis are summarized for domestic animals (Table 9), primates (Table 10), and birds (Table 11).

At the Iowa Veterinary Diagnostic Laboratory, it is considered of diagnostic significance when liver or kidney tissues contain at least 10 ppm lead on a wet basis. Whole blood lead levels of at least 0.35 are considered significant in cattle and probably other ruminants. Furthermore, for a

TABLE 8. Levels of Lead in Tissues and Rumen Contents Associated with Clinical Lead Toxicosis in Cattle

Tissue	No. of Cases	Mean	Range
Liver	100	26.40	1.0-83.0 ppm
Kidney	105	50.30	3.0-200 ppm
Blood	50	0.81	0.19-3.80 ppm
Rumen contents	52	400.80	0.0-11,875 ppm

TABLE 9. Tissue Lead Levels in Domestic Animals Associated with Lead Toxicoses

Animal	Tissue	No. of Samples	Lead Level	Refs.
Horses	Blood	6	39 µg/100 ml	[67]
	Liver	6	18 ppm	[67]
	Kidney	6	16 ppm	[67]
Horses	Liver	2	50, 98 ppm	[173]
Calves	Liver	10	38 ppm	[77,78]
	Kidney cortex	10	173 ppm	[77,78]
	Kidney medulla	8	10.3 ppm	[77,78]
Calves	Blood	5	100 µg/100 ml	[79]
Cow	Liver	2	16.1, 11.1 ppm	[15]
	Kidney	2	39.0, 61.4 ppm	[15]
Cattle	Kidney	158	137 ppm	[14]
	Liver	170	43 ppm	[14]
	Rumen contents	133	3427 ppm	[14]
Dog	Blood	209	94 µg/100 ml	[103]
	Liver	30	24 ppm	[103]

TABLE 10. Tissue Lead Levels in Primates with Lead Poisoning

Animal	Tissue	No. of Samples	Level	Refs.
Barbary ape	Liver	1	110 ppm	[113]
	Kidney	1	120 ppm	[113]
Red-faced macaque	Liver	1	65 ppm	[113]
	Kidney	1	90 ppm	[113]
Spot-nosed guenon	Liver	1	10 ppm	[113]
Rhesus	Blood	9	156 µg/100 ml	[112]
Not specified	Liver	12	52 ppm	[115]
Baboons	Blood	2	400-4550 µg/100 ml	[148]

TABLE 11. Tissue Lead Levels in Lead-Poisoned Birds

Animal	Tissue	No. of Samples	Level	Refs.
Dove	Liver	1	72 ppm	[124]
Geese	Liver	7	5-32 ppm	[121]
	Blood	10	81-1680 µg/100 mg	[121]
Mallard duck	Liver	10	33 ppm	[126]
	Brain	10	5 ppm	[126]
Andean condor	Liver	1	34 ppm	[128]
23 species	Liver	1-37/species	0.5-3.7 ppm	[22]

clinician to render a positive diagnosis of lead poisoning, it is considered imperative that symptomatology, history, or circumstantial evidence be compatible with lead poisoning. In some cases diagnoses may be confirmed at less than 10 ppm liver lead if clinical signs, exposure, and other tissue levels are known.

Generally those animals with blood lead concentrations in excess of 1.0 ppm should be given a guarded prognosis. However, recovering animals may maintain blood levels well in excess of the clinically significant level of 0.35 ppm and still continue to improve. In the authors' experience, blood lead levels may range from 0.4 to 0.8 ppm for 1 to 3 weeks. Thus any one blood level at one point in time may not be sufficient as a prognostic tool. Knowledge of concomitant chelation therapy, peak blood lead concentration, time since clinical disease first occurred, and continued exposure to lead should all enter into evaluation of the animal.

The possibility exists that liver and kidney levels from 1 to 10 ppm and bovine blood levels from 0.10 to 0.35 ppm are significant either as a primary etiological agent or as a predisposing or contributory factor. Dodd and Staples [92] found that young dogs with lead poisoning sometimes had liver lead levels less than 10 ppm.

THERAPY OF LEAD POISONING

Chisolm has recently published a number of papers describing in considerable detail the treatment of childhood lead poisoning [164-166]. These papers are worthwhile reading for veterinary clinicians.

The major drug used in treating animal lead toxicosis has been calcium EDTA (ethylenediaminetetraacetate) [28]. Calcium EDTA was introduced by Bessman et al. [167] in 1952 for treating lead poisoning in man. Early application of this antidote in horses was reported by Holm et al. [70] in 1953, cattle by Lewis and Meikle [168] in 1956, and dogs by Pettit et al. [169] in 1956. The plasma half-life of EDTA in the bovine is 65 min [170]. EDTA is also toxic and a single intravenous dose of 440 mg/kg is lethal to cattle. Cattle will tolerate 220 mg/kg for 5 days if given slowly, although diarrhea may occur. In cattle, two doses of 110 mg/kg 6 hr apart are beneficial in increasing lead excretion [170].

Others subsequently reported the use of Ca EDTA in cattle [171-174] and dogs [21,93,97]. The successful treatment of one lead-poisoned bird with twice-daily doses of 0.5 ml of a 6.6% EDTA solution given for 3 days has been reported [120].

LEAD RESIDUES IN ANIMALS

Animal tissues which would constitute the highest exposures for man are liver and kidney. Lead concentration in milk, although elevated following lead exposure, is generally much lower than blood levels.

Subclinical effects of lead in cattle were reported earlier in this chapter to cause anemia and disturbances in the hematopoietic enzymes.

Dinius et al. [175] reported that dietary levels of 100 ppm lead as the chromate resulted in accumulation of lead in tissues, but no cellular ultrastructural changes were detected. Liver from cattle fed 100 ppm for 100 days contained 2.3 ppm lead vs. 0.6 ppm for controls. At the 100-ppm level, however, no lead was detected in longissimus muscle.

White et al. [74], in a case of heavy lead exposure that killed 18 of 20 cows in 7 days, reported a milk lead level of 2.26 ppm from one surviving cow. Dorn et al. [135] found milk lead levels of 0.034-0.25 ppm in cows placed in a lead-contaminated test site. Blood levels were 0.28-0.58 ppm. Background levels were 0.06-0.13 ppm in milk and 0.06-0.18 ppm in blood.

Blaxter [82] reported finding 0.2-0.5 ppm lead in milk of lead-poisoned sheep.

REPRODUCTIVE EFFECTS OF LEAD

Reproductive effects of lead exposure in livestock have been reviewed [141]. Transplacental transfer of lead has been documented. In addition, evidence of embryopathic effects and infertility has been established [84, 141]. Abortions, however, have been associated with overt lead poisoning or with maternal blood lead levels which would be considered in the toxic range. Neither teratogenic nor mutagenic effects have been described in domestic animals as a result of lead exposure. Genetic and teratogenic effects have been ascribed to lead at high levels in experimental animals [45,141] exposed at the time of gametogenesis or during gestation. Permanent reproductive effects during subsequent cycles do not appear to have been described.

FACTORS INFLUENCING SUSCEPTIBILITY
TO LEAD POISONING

Both environmental and animal factors may be involved in the expression of lead toxicosis in animals. The relationship of age and season of the year has been explained earlier in this chapter. Other factors variously implicated in susceptibility of man or animals to lead include calcium-phosphorus status, iron deficiency, dietary protein deficiency, vitamin D deficiency, ascorbic acid and nicotinic acid deficiency, and presence of excess zinc or cadmium in the diet.

Of the above, calcium deficiency has been shown experimentally to enhance lead toxicity in rats. Other experimental studies of deficiency states enhancing lead poisoning or storage in laboratory animals have included iron, vitamin D, and ascorbic acid deficiencies. These works have been reviewed recently [41,141].

In growing horses, toxic levels of dietary zinc have been found to alleviate or mask signs of lead poisoning [171].

Cadmium has been found to act in a synergistic way to increase some teratogenic effects of lead while reducing others [41,141].

Lead salts (phosphate, acetate) have been shown to induce tumors such as renal adenomas and adenocarcinomas in rodents [141].

SUMMARY

Acute or overt lead poisoning is a major toxicological problem among many domestic and wild species. Notably cattle, horses, dogs, wildfowl, and captive nonhuman primates have been affected. The cumulative nature of lead in the body predisposes to chronic intake which may culminate in severe clinical disease.

Lead is available to animals in the form of paints, plumbers materials, waste crankcase oil, batteries, lead shot, and contaminated forage. Much of the access is due to mismanagement and carelessness on the part of owners or attendants.

Clinical disease in all species invariably involves the nervous system and often the gastrointestinal tract. Animals with a combination of these characteristic signs should be considered as potential lead poisoning suspects along with other tentative diagnoses.

Confirmation of lead poisoning can be made from clinical evaluation, hematologic changes, and lead levels in blood and tissue. Both background and acutely toxic levels of lead in animals have been reviewed.

Factors in the epidemiology of lead poisoning to which particular attention should be given include source and form of lead, age of animal, season of the year, and potential nutritional influences.

Lead is an ubiquitous element with many overt and subtle biological effects. Unfortunately, many of the subtle effects no doubt remain to be discovered. Behavioral effects and immune suppression are but two of these. By close attention to patterns of problems occurring in animals, additional areas of research and study concerning lead may be delineated.

REFERENCES

1. S. C. Gilfillan, J. Occup. Med., 7 53 (1965).

2. J. J. Chisolm and E. Kaplan, Dev. Med. Child Neurol., 7, 529 (1965).

3. G. Wiener, Public Health Reports, 85, 19 (1970).

4. H. L. Hardy, R. I. Chamberlin, C. C. Maloof, G. W. Boylan, and M. C. Howell, Clin. Pharmacol. Ther., 12, 982 (1971).

5. American Petroleum Institute, Air Quality Monographs, No. 69-7, American Petroleum Institute, New York, 1969.

6. R. K. Byers, Pediatrics, 23, 585 (1959).

7. M. A. Perlstein and R. Attala, Clin. Pediatr., 5, 292 (1966).

8. D. Barltrop, Postgrad. Med. J., 45, 129 (1969).

9. Institutional Consortium on Edemic Lead Poisoning, Clin. Toxicol. Bull., 3 (1 and 2), 27 (1973).

10. Conference on Inorganic Lead, Arch. Environ. Health, 23, 245 (1971).

11. F. E. Rieke, Arch. Environ. Health, 19, 521 (1969).

12. R. Allcroft and K. L. Blaxter, J. Comp. Pathol., 60, 190 (1950).

13. W. B. Buck, J. Amer. Vet. Med. Ass., 156, 1468 (1970).

14. R. C. Hatch and H. S. Funnell, Can. Vet. J., 10, 258 (1969).

15. D. C. Kradel, W. M. Adams, and S. B. Guss, Vet. Med. Small Anim. Clin., 60, 1045 (1965).

16. S. L. Leary, W. B. Buck, W. E. Lloyd, and G. D. Osweiler, Iowa State Univ. Vet., 3, 112 (1970).

17. P. B. Little and D. K. Sorensen, J. Amer. Vet. Med. Ass., 155, 1892 (1969).

18. J. R. Todd, Vet. Rec., 74, 116 (1962).

19. M. R. Wilson and G. Lewis, Vet. Rec., 75, 787 (1963).

20. B. C. Zook, J. L. Carpenter, and R. M. Roberts, Amer. J. Vet. Res., 33, 891 (1972).

21. B. C. Zook, J. L. Carpenter, and E. B. Leeds, J. Amer. Vet. Med. Ass., 155, 1329 (1969).

22. G. E. Bagley and L. N. Locke, Bull. Environ. Contam. Toxicol., 2, 297 (1967).

23. F. C. Bellrose, Ill. Nat. Hist. Surv. Bull., 27, 235 (1959).

24. National Academy of Science, Committee on Biologic Effects of Atmospheric Pollutants, Washington, D. C., 1972.

25. C. C. Patterson, Arch. Environ. Health, 11, 344 (1965).

26. E. G. C. Clarke and M. L. Clarke, in Garners Veterinary Toxicology, 3rd Ed., Williams and Wilkins Co., Baltimore, 1967.

27. R. D. Radeleff, Veterinary Toxicology, 2nd Ed., Lea and Febiger, Philadelphia, 1970.

28. W. B. Buck, G. D. Osweiler, and G. A. Van Gelder, Clinical and Diagnostic Veterinary Toxicology, Kendall-Hunt, Dubuque, 1973.

29. A. de Bruin, Arch. Environ. Health, 23, 249 (1971).

30. R. A. Goyer, P. May, M. M. Cates, and M. R. Kringman, Lab. Invest., 22, 245 (1970).

31. R. A. Goyer, A. Krall, and J. P. Kimball, Lab. Invest., 19, 78 (1968).

32. R. A. Goyer, Lab. Invest., 19, 71 (1968).

33. R. A. Goyer, Amer. J. Pathol., 64, 167 (1971).

34. R. A. Goyer, D. L. Leonard, J. F. Moore, B. Rhyne, and M. R. Kringman, Arch. Environ. Health, 20, 705 (1970).

35. R. F. Macadam, Brit. J. Exp. Pathol., 50, 239 (1969).

36. G. W. Richert, Y. Kress, and C. C. Cornwall, Amer. J. Pathol., 53, 189 (1968).

37. R. A. Goyer, A. Krall, and J. P. Kimball, Lab. Invest., 19, 78 (1968).

38. R. A. Goyer, D. L. Leonard, P. R. Bream, and T. G. Irons, Proc. Soc. Exp. Biol. Med., 135, 767 (1970).

39. G. D. Secchi, L. Alessio, and A. Cirla, Clin. Chim. Acta, 27, 467 (1970).

40. W. Bielecka, Acta Pol. Pharm., 29, 103 (1972).

41. R. A. Goyer and K. R. Mahaffey, Environ. Health Perspect., 2, 73 (1972).

42. W. E. J. Phillips, D. C. Villeneuve, and G. C. Beching, Bull. Environ. Contam. Toxicol., 6, 570 (1971).

43. H. H. Sandstead, A. M. Michelakis, and T. E. Temple, Arch. Environ. Health, 20, 356 (1970).

44. H. H. Sandstead, E. G. Stant, A. B. Brill, L. I. Arias, and R. T. Terry, Arch. Intern. Med., 123, 632 (1969).

45. L. A. Muro and R. A. Goyer, Arch. Pathol., 87, 660 (1969).

46. F. E. Hemphill, M. L. Kaeberle, and W. B. Buck, Science, 172, 1031 (1971).

47. Z. Palmieniak and R. Smolik, Pol. Tyg. Lek., 17, 358 (1963).

48. M. T. Rakhimova, Gig. Tr. Prof. Zabol., 12, 39 (1968).

49. M. H. M. Soliman, Y. M. El-Sadik, K. M. El-Kashlan, and A. El-Waseef, Arch. Environ. Health, 21, 529 (1970).

50. J. Sroezynski, A. Kujawaska, and B. Perkarsk, Med. Pracy., 15, 77 (1964).

51. F. E. Hemphill, Doctoral Dissertation, Iowa State Univ. Library, Ames, 1973.

52. H. A. Waldron, Brit. J. Ind. Med., 23, 83 (1966).

53. S. L. M. Gibson and A. Goldberg, Clin. Sci., 38, 63 (1970).

54. L. E. Rogers, N. D. Battles, E. W. Reimold, and P. Sartain, Arch. Toxikol., 28, 202 (1971).

55. B. S. Morse, G. J. Germano, and D. G. Giuliani, Blood, 39, 713 (1972).

56. G. A. Van Gelder, T. L. Carson, R. M. Smith, and W. B. Buck, Clin. Toxicol., 6, 405 (1973).

57. G. A. Van Gelder, T. L. Carson, and W. B. Buck, Toxicol. Appl. Pharmacol., 25, 466 (1973).

58. J. D. Bullock, R. J. Wey, J. A. Zaia, I. Zarembok, and H. A. Schroeder, Arch. Environ. Health, 13, 21 (1966).

59. S. Brown, N. Gragann, and W. H. Vogel, Arch. Environ. Health, 22, 370 (1971).

60. D. R. Brown, Toxicol. Appl. Pharmacol., 25, 466 (1973).

61. P. M. Fullerton, J. Neuropathol. Exp. Neurol., 25, 214 (1966).

62. G. A. Van Gelder, T. L. Carson, R. M. Smith, W. B. Buck, and G. G. Karas, J. Amer. Vet. Med. Ass., 163, 1033 (1973).

63. M. J. Catton, M. J. G. Harrison, P. M. Fullerton, and G. Kazantzis, Brit. Med. J., 2, 80 (1970).

64. Annual Review 1972, U.S. Lead Industry, Lead Industries Assn. Inc., New York, 1972.

65. W. Hughes, Vet. J., 79, 270 (1923).

66. C. M. Haring and K. F. Meyer, U.S. Bureau of Mines Bull., 98, 474 (1915).

67. N. Schmitt, G. Brown, E. L. Devlin, A. A. Larsen, E. D. McCausland, and J. M. Savile, Arch. Environ. Health, 23, 185 (1971).

68. P. B. Hammond and A. L. Aronson, Ann. N. Y. Acad. Sci., 111, 595 (1964).

69. A. L. Aronson, Amer. J. Vet. Res., 33, 627 (1972).

70. L. W. Holm, J. D. Wheat, E. A. Rohde, and G. Firch, J. Amer. Vet. Med. Ass., 123, 383 (1953).

71. B. G. Steyn, Farming in South Africa, 15, 350 (1940).

72. C. L. Shrewsbury, F. G. King, E. Barrick, J. A. Hoeper, and L. P. Doyle, J. Animal Sci., 4, 20 (1945).

73. R. Fenstermacher, B. S. Pomeroy, M. H. Roepke, and W. L. Boyd, J. Amer. Vet. Med. Ass., 108, 1 (1946).

74. W. B. White, P. A. Clifford, and H. O. Calvery, J. Amer. Vet. Med. Ass., 102, 292 (1943).

75. E. G. White and E. Cotchin, Vet. J., 104, 75 (1948).

76. R. Allcroft and K. L. Blaxter, J. Comp. Pathol., 60, 209 (1950).

77. R. Allcroft, J. Comp. Pathol., 60, 190 (1950).

78. R. Allcroft, Vet. Rec., 63, 583 (1951).

79. J. F. Harbourne, C. T. McCrea, and J. Watkinson, Vet. Rec., 83, 515 (1968).

80. R. G. Christian and L. Tryphonas, Amer. J. Vet. Res., 32, 203 (1971).

81. D. A. Egan and T. O'Cuill, Vet. Rec., 84, 230 (1969).

82. K. L. Blaxter, J. Comp. Pathol., 60, 140 (1950).

83. T. L. Carson, G. A. Van Gelder, W. B. Buck, L. J. Hoffman, D. L. Mick, and K. R. Long, Clin. Toxicol., 6, 389 (1973).

84. R. M. Sharma, M.S. Thesis, Iowa State Univ., Ames, 1971.

85. F. G. Clegg and J. M. Rylands, J. Comp. Pathol., 76, 15 (1966).

86. M. K. Horwitt and G. R. Cowgill, J. Pharmacol. Exp. Ther., 66, 289 (1939).

87. H. O. Calvery, E. P. Laug, and H. J. Morris, J. Pharmacol. Exp. Ther., 64, 364 (1938).

88. L. L. Liberman, N. Amer. Vet., 29, 574 (1944).

89. E. Bond and R. Kubin, Vet. Med., 44, 118 (1949).

90. E. L. J. Staples, N. Z. Vet., 3, 39 (1955).

91. G. D. Pettit, L. W. Holm, and W. E. Ruchworth, J. Amer. Vet. Med. Ass., 128, 295 (1956).

92. D. C. Dodd and E. L. J. Staples, N. Z. Vet., 4 (1), 1 (1956).

93. W. J. Hartley, N. Z. Vet., 4, 147 (1956).

94. W. T. Oliver, L. W. Geib, and B. Sorrell, Can. J. Comp. Med., 23, 21 (1959).

95. H. M. Scott, Vet. Rec., 75, 830 (1963).

96. R. E. Lewis, W. B. Henry, G. W. Thornton, and C. E. Gilmore, in Scientific Proceedings, 100th Ann. Meeting Amer. Vet. Med. Ass., 1963, p. 140.

97. M. R. Wilson and G. Lewis, Vet. Rec., 75, 787 (1963).

98. A. P. Berry, Vet. Rec., _79_, 248 (1966).

99. G. D. Osweiler, J. Amer. Vet. Med. Ass., _155_, 2011 (1969).

100. C. M. Robertson, Vet. Rec., _86_, 195 (1970).

101. B. C. Zook and J. L. Carpenter, in Current Veterinary Therapy, 4th Ed., W. B. Saunders, Philadelphia, 1971.

102. B. C. Zook, G. McConnell, and C. E. Gilmore, J. Amer. Vet. Med. Ass., _157_, 2092 (1970).

103. B. C. Zook, L. Kopito, J. L. Carpenter, D. V. Cramer, and H. Shwachman, Amer. J. Vet. Res., _33_, 903 (1972).

104. B. Sass, J. Amer. Vet. Med. Ass., _157_, 76 (1970).

105. A. Ferraro and R. Hernandez, Psychiatr. Q., _6_, 121 (1932).

106. A. Ferraro and R. Hernandez, Psychiatr. Q., _6_, 319 (1932).

107. K. Goadby, J. Hyg. (Cambridge), _9_, 122 (1909).

108. A. S. Minot, J. Ind. Hyg., _6_, 137 (1924).

109. L. E. Fisher, J. Amer. Vet. Med. Ass., _125_, 478 (1954).

110. G. J. Vernande-Van Eck and J. W. Meigs, Fertil. Steril., _11_, 223 (1960).

111. R. Hausman, R. A. Sturtevant, and W. J. Wilson, J. Forensic Sci., _6_, 180 (1961).

112. W. D. Houser and N. Frank, J. Amer. Vet. Med. Ass., _157_, 1919 (1970).

113. R. M. Sauer, B. C. Zook, and F. M. Garner, Science, _169_, 1091 (1970).

114. B. C. Zook, Comp. Pathol. Bull., _3_, 3 (1971).

115. B. C. Zook, R. M. Sauer, and F. M. Garner, J. Amer. Vet. Med. Ass., _161_, 683 (1972).

116. B. C. Zook and R. M. Sauer, J. Wildl. Dis., _9_, 61 (1973).

117. A. Hopkins, Brit. J. Ind. Med., _27_, 130 (1970).

118. B. C. Zook, R. M. Sauer, and F. M. Garner, J. Amer. Vet. Med. Ass., _157_, 691 (1970).

119. J. S. Jordan and F. C. Bellrose, Trans. N. Amer. Wildl. Conf., _15_, 155 (1950).

120. G. Wobeser, Bull. Wildl. Dis. Ass., _5_, 120 (1969).

121. R. S. Cook and D. O. Trainer, J. Wildl. Mgt., _30_, 1 (1966).

122. D. O. Trainer and R. A. Hunt, Avian Dis., _9_, 252 (1965).

123. L. N. Locke, G. E. Bagley, and H. D. Irby, Bull. Wildl. Dis. Ass., _2_, 127 (1966).

124. L. N. Locke and G. E. Bagley, J. Wildl. Mgt., _31_, 515 (1967).

125. J. W. Grandy, L. N. Locke, and G. E. Bagley, J. Wildl. Mgt., _30_, 483 (1968).

126. F. Y. Bates, D. M. Barnes, and J. M. Higbee, Bull. Wildl. Dis. Ass., _4_, 116 (1968).

127. M. Kolacek, P. Satran, and L. Stastny, Veterinarstivi, _19_, 317 (1969).

128. L. Locke, G. E. Bagley, D. N. Frickie, and L. T. Young, J. Amer. Vet. Med. Ass., 155, 1052 (1969).

129. M. W. Barrett and L. H. Karstad, J. Wildl. Mgt., 35, 109 (1971).

130. B. C. Zook, Clin. Toxicol. Bull., 3, 91 (1973).

131. G. D. Osweiler, W. B. Buck, and W. E. Lloyd, Clin. Toxicol., 6, 367 (1973).

132. B. C. Zook, Clin. Toxicol., 6, 377 (1973).

133. V. F. Guinee, Amer. J. Med., 52, 283 (1972).

134. E. G. C. Clarke, J. Small Animal Pract., 14, 183 (1973).

135. R. C. Dorn, J. O. Pierce, G. R. Chase, and P. E. Phillips, Report for Environmental Protection Agency Contract 68-02-0092, University of Missouri, Columbia, 1972.

136. R. P. Link, Amer. J. Vet. Res., 27, 759 (1966).

137. R. M. Hicks, Chem.-Biol. Interactions, 5, 361 (1972).

138. D. A. Hirschler and L. F. Gilbert, Arch. Environ. Health, 8, 297 (1964).

139. J. Cholok, L. J. Schafer, and D. Yeager, Amer. Ind. Hyg. Ass. J., 29, 562 (1968).

140. H. L. Motto, R. H. Daines, D. M. Chilko, and C. K. Motto, Environ. Sci. Technol., 4, 231 (1970).

141. National Research Council (Canada), Environ. Secretariat, No. BY-73-7(ES), December, 1973.

142. J. C. Irwin and L. H. Karstad, J. Wildl. Dis., 8, 149 (1972).

143. W. F. Steward and R. Allcroft, Vet. Rec., 68, 723 (1965).

144. E. J. Butler, D. I. Nisbet, and J. M. Robertson, J. Comp. Pathol., 67, 378 (1957).

145. P. Fauts and J. Page, Amer. Heart J., 24, 329 (1942).

146. K. Saloman and G. R. Cowgill, J. Ind. Hyg., 30, 114 (1948).

147. K. V. Jubb and P. C. Kennedy, Pathology of Domestic Animals, Vol. 2, Academic Press, New York, 1963, p. 332.

148. R. G. Thomson, Can. Vet. J., 13, 88 (1972).

149. B. J. McSherry, R. A. Willoughby, and R. G. Thomson, Can. J. Comp. Med., 35, 136 (1971).

150. D. J. Kelliher, E. P. Hillard, D. B. R. Poole, and J. D. Collins, Irish J. Agr. Res., 12, 61 (1973).

151. E. P. Hillard, D. B. R. Poole, and J. D. Collins, Brit. Vet. J., 129, lxxxii (1973).

152. H. J. Hapke and E. Prigge, Berliner und Munchener Tierarztliche Wochenschrift, 86, 410 (1973).

153. R. A. Green, A. W. Monlux, and T. C. Randolph, Bovine Pract., 8, 30 (1973).

154. G. Del Bono and G. Braca, First Patologia Vet., 10, 77 (1972).

155. B. C. Zook, Vet. Pathol., 9, 310 (1972).

156. G. W. Richter, Y. Kress, and C. C. Cornwall, Amer. J. Pathol., 53, 189 (1968).

157. L. Kopito, R. K. Byers, and H. Shwachman, N. Engl. J. Med., 276, 949 (1967).

158. H. G. Petering, D. W. Yeager, and S. O. Winthrop, Arch. Environ. Health, 27, 327 (1973).

159. Analytical Methods for Atomic Absorption Spectrophotometry, Perkin Elmer Corporation, Norwalk, Connecticut, 1968.

160. D. W. Hessel, Atomic Absorption Newsletter, 7, 55 (1968).

161. A. Hopkins, Brit. J. Ind. Med., 27, 130 (1970).

162. W. E. Martin, Pestic. Monit. J., 6, 27 (1972).

163. W. E. Martin and P. R. Nickerson, Pestic. Monit. J., 7, 67 (1973).

164. J. J. Chisolm, J. Pediatr., 73, 1 (1968).

165. J. J. Chisolm, Mod. Treat., 8, 710 (1971).

166. J. J. Chisolm and E. Kaplan, J. Pediatr., 73, 942 (1968).

167. S. P. Bessman, H. Reid, and M. Rubin, Med. Ann. D. C., 21, 312 (1952).

168. E. F. Lewis and J. C. Meikle, Vet. Rec., 68, 98 (1956).

169. G. D. Pettit, L. W. Holm, and W. E. Rushworth, J. Amer. Vet. Med. Ass., 128, 295 (1956).

170. A. L. Aronson, P. B. Hammond, and A. C. Strafuss, Toxicol. Appl. Pharmacol., 12, 337 (1968).

171. J. R. Todd, Vet. Rec., 69, 31 (1957).

172. E. F. Lewis and J. C. Meikle, Brit. Vet. J., 114, 69 (1958).

173. P. B. Hammond and D. K. Sorenson, J. Amer. Vet. Med. Ass., 130, 23 (1957).

174. P. B. Hammond and A. L. Aronson, Ann. N. Y. Acad. Aci., 88, 498 (1960).

175. D. A. Dinius, T. H. Brinsfield, and E. E. Williams, J. Animal Sci., 37, 169 (1973).

176. R. A. Willoughby, E. MacDonald, B. J. McSherry, and G. Brown, Can. J. Comp. Med., 36 (4), 348 (1972).

9

OUTBREAKS OF PLUMBISM IN ANIMALS
ASSOCIATED WITH INDUSTRIAL LEAD OPERATIONS

Arthur L. Aronson

Cornell University New York State College of Veterinary Medicine
Ithaca, New York

Lead poisoning is considered to be the most common cause of accidental poisoning in domestic animals. The condition is diagnosed most frequently in cattle and dogs. Common histories of exposure in dogs include chewing on objects painted with lead-base paints (for example, when home remodeling entails scraping of plaster and old paint), eating linoleum, or ingesting lead materials such as shotgun slugs or curtain weights [23]. The most common sources of lead incriminated in poisonings of cattle include lead-base paint (either from discarded paint cans or paint peeling from walls), used motor oil, discarded oil filters, storage batteries, certain types of greases and putty, and linoleum [4,9,12,19]. These sources have been incriminated on the basis of (1) evidence of ingestion, (2) clinical signs, and (3) finding elevated concentrations of lead in the tissues. The sources common for cattle often can be found in the vicinity of farm buildings and in dumps located in pastures. It is interesting that these sources rarely are incriminated in lead poisoning in horses. Horses are more selective than cattle in their eating habits. They usually do not lick old paint cans, storage batteries, or peeling paint; nor do they seem to find the taste of used motor oil attractive.

Several outbreaks of lead poisoning in domestic animals have been recorded in North America and throughout the world where the source of the metal was contamination of pasture or crops by industrial lead operations [3,6-8,10-11,13-17,20,22]. These outbreaks differ from the common cases of lead poisoning previously described in that several animals were involved. Pastures and crops are contaminated by fumes and dusts emitted from lead

industries and settling out on the surrounding countryside. Animals eating
this vegetation can accumulate amounts of lead sufficient to produce clinical signs of lead poisoning. A number of studies have been made to determine if the lead found in vegetation is the result of direct airborne origin
or due to translocation from soil. These studies recently have been reviewed
by Mueller and Stanley [21]. They conclude, on the basis of their work and
the work of others, that translocation from soil does not contribute more
than 15 µg lead/g dry weight of forage even when plants are grown in soil
containing up to 700-3000 µg lead/g. Thus, amounts of lead in plants in
excess of 15 µg/g most likely are due to direct aerial fallout. The extent
to which contamination can occur is illustrated by finding concentrations
of 3200 µg lead/g dry weight in corn leaves located 75 yards from a lead
smelter in one outbreak [8].

It has been possible to estimate that a daily intake of 6-7 mg lead
per kg body weight constitutes a minimum cumulative fatal dosage of lead
for cattle [8]. This intake represents a concentration in excess of 200 ppm
lead in the total diet. These cattle were located approximately 2 miles
from the smelter, but were fed lead-contaminated hay and corn silage grown
in fields adjacent to the smelter. A fatal case of lead poisoning occurred
following 2 months of this diet. An intake of approximately half this dosage had no observable effect on cattle at another farm the previous winter.
It is of interest to note that daily dosages of 5-6 mg lead/kg body weight
have been fed to cattle for a period of 2 years with no observable clinical
effects [1], but that longer intake at this rate may be fatal [2]. Calves
were not adversely affected by the ingestion of 100 ppm lead in the total
diet for 100 days as judged by respiratory rate, electrocardiogram, or the
ultrastructural appearance of the cerebral cortex, liver, and kidney cortex
[5]. Although concentrations of lead in the cerebral cortex and muscle
remained similar to control animals, the feeding of 100 ppm lead did result
in accumulations of the element in liver and kidney [5].

There is some evidence suggesting that horses may be more susceptible
to the chronic ingestion of lead than cattle. Whereas in one outbreak
horses contracted lead poisoning on pastures adjacent to a lead smelter,
cattle grazing in the same area appeared healthy [18]. In another outbreak
at one farm adjacent to a smelter, horses succumbed to lead poisoning in
March following a winter intake in their hay of 2.4 mg lead/kg/day [8]. It
was not possible to determine lead intake from pasture grazing the previous

summer. However, since cattle and horses had similar pasture that summer,
and since the winter ration for the horses contained appreciably less lead
than did that for the cows, it would seem that cumulative toxicity occurred
more readily in horses. In still another outbreak, it is of interest that
pasture grass containing approximately 80 μg lead/g dry weight was toxic to
horses [21]. A horse eats approximately 21 g dry matter per kg body weight
per day. This represents a daily intake of 1.7 mg lead/kg/day, a figure
close to that estimated for the previous outbreak.

Although the evidence suggests that horses might be more susceptible
to lead than cattle, a consideration of the grazing habits of horses pre-
cludes any firm conclusions. Horses occasionally will pull forage out by
the roots and eat the roots and attendant soil along with the forage.
Cattle rarely, if ever, do this, probably because they lack the teeth and
jaw structure to make it possible. The soil near smelters usually contains
far greater amounts of lead than does the forage. It is apparent that a
horse showing a marked tendency toward grass pulling could ingest far greater
quantities of lead than would be estimated from the analysis of forage alone.

It is natural that humans residing near smelters around which animals
are dying of lead poisoning would be concerned about their own health. In
many cases these people are eating produce from home gardens. Analysis of
blood and urine from these people by local public health officials has not
revealed evidence of increased lead absorption. Keep in mind that horses
and cattle are vegetarians. If their hay or pasture is contaminated with
lead, their entire diet may consist of contaminated vegetation. Probably
only a small fraction of the total diet of human beings would consist of
food grown in the vicinity of a lead operation. Moreover, it is customary
for people to wash garden produce (or husk corn) before its consumption.
This practice would undoubtedly remove appreciable quantities of surface
lead. Since the animal and human population near the smelters breathe the
same air, and since residents in the area have not shown evidence of increased lead absorption, it may be justified to conclude that the animals
received virtually all their lead burden through oral ingestion.

SUMMARY

Lead poisoning in cattle most commonly occurs as a result of a single
ingestion of material containing a large quantity of lead. Poisoning also
can occur if cattle are required to ingest crops or pasture forage contami-

nated by lead from fumes and dusts emitted from industrial lead operations. The ingestion of lead-contaminated hay or forage appears the most common source of metal in the poisoning of horses. Whereas a daily intake of approximately 2 mg lead/kg body weight can produce clinical poisoning in horses, a daily intake of approximately 6-7 mg lead/kg body weight is required to produce poisoning in cattle. Although animals have died of lead poisoning in close proximity to humans residing in the same area, no evidence of human intoxication has been found. This probably is due to the entire diet of animals consisting of lead-contaminated feed.

REFERENCES

1. R. Allcroft, Lead as a Nutritional Hazard to Farm Livestock. IV. Distribution of Lead in Tissues of Bovines after Ingestion of Various Lead Compounds, J. Comp. Pathol., 60, 190-208 (1950).

2. R. Allcroft, Lead Poisoning in Cattle and Sheep, Vet. Rec., 63, 583-590 (1951).

3. J. A. Beijers, Loodvergiftging, Tijschr. v. Diergeneesk, 77, 587-605 (1952).

4. W. B. Buck, Lead and Organic Pesticide Poisonings in Cattle, J. Amer. Vet. Med. Ass., 156, 1468-1472 (1970).

5. D. A. Dinius, T. H. Brinsfield, and E. E. Williams, Effect of Subclinical Lead Intake on Calves, J. Animal Sci., 37, 169-173 (1973).

6. D. A. Egan and T. O'Cuill, Opencast Lead Mining Areas: A Toxic Hazard to Grazing Stock, Vet. Rec., 84, 230 (1969).

7. D. A. Egan and T. O'Cuill, Cumulative Lead Poisoning in Horses in a Mining Area Contaminated with Galena, Vet. Rec., 86, 736-738 (1970).

8. P. B. Hammond and A. L. Aronson, Lead Poisoning in Cattle and Horses in the Vicinity of a Smelter, Ann. N. Y. Acad. Sci., 111, 595-611 (1964).

9. P. B. Hammond, H. N. Wright, and M. H. Roepke, A Method for the Detection of Lead in Bovine Blood and Liver, Univ. Minn. Agr. Exp. Sta. Tech. Bull., 221 (1956).

10. J. F. Harbourne, C. T. McCrea, and J. Watkinson, An Unusual Outbreak of Lead Poisoning in Calves, Vet. Rec., 83, 515-517 (1968).

11. C. M. Haring and K. F. Meyer, Investigation of Livestock Conditions and Losses in the Selby Smoke Zone, U.S. Bur. Mines Bull., 98, 474-502 (1915).

12. R. C. Hatch and H. S. Funnell, Lead Levels in Tissues and Stomach Contents of Poisoned Cattle: A 15-Year Survey, Can. Vet. J., 10, 258-262 (1969).

13. L. W. Holm, J. D. Wheat, E. A. Rohde, and G. Firch, The Treatment of Chronic Lead Poisoning in Horses with Calcium Disodium Ethylenediaminetetraacetate, J. Amer. Vet. Med. Ass., 123, 383-388 (1953).

14. W. Hughes, Lead-Poisoning in Horses and Cattle, Vet. J., _79_, 270-271 (1923).

15. E. Hupka, Über Flugstaubvergiftungen in der Umgebunf von Matellhütten, Weiner Tierarztl. Monatsschr., _42_, 763-775 (1955).

16. H. D. Knight and R. G. Burau, Chronic Lead Poisoning in Horses, J. Amer. Vet. Med. Ass., _162_, 781-786 (1973).

17. D. C. Kradel, M. W. Adams, and S. B. Guss, Lead Poisoning and Eosinophilic Meningoencephalitis in Cattle: A Case Report, Vet. Med., _60_, 1045-1050 (1965).

18. A. A. Larsen, Report of the Interagency Committee on Environmental Study Arising out of a Debilitating Disease in Young Horses, Trail Area, 1969.

19. S. L. Leary, W. B. Buck, W. E. Lloyd, and G. D. Osweiler, Epidemiology of Lead Poisoning in Cattle, Iowa State Univ. Vet., _32_, 112-117 (1970).

20. H. Miessner, Shädigung der Tierweit durch Industrie and Technik, Deut. Tierarztl. Wchnschr., _39_, 340-345 (1931).

21. P. K. Mueller and R. L. Stanley, Origin of Lead in Surface Vegetation, AIHL Report No. 87, State of California Department of Public Health, Air and Industrial Hygiene Laboratory, Berkeley, 1970.

22. N. Schmitt, G. Brown, E. L. Devlin, A. A. Larsen, E. D. McCausland, and J. M. Savile, Lead Poisoning in Horses, Arch. Environ. Health, _23_, 185-195 (1971).

23. B. C. Zook, J. L. Carpenter, and E. B. Leeds, Lead Poisoning in Dogs, J. Amer. Vet. Med. Ass., _155_, 1329-1342 (1969).

10

LEAD INTOXICATION IN URBAN DOGS

Bernard C. Zook
The George Washington University Medical Center
Washington, D. C.

Pet animals share man's urban environment, dwelling even in his home, and are thus exposed to the same ecologic hazards. The study of lead poisoning and other toxicologic diseases in dogs and other pets may afford valuable insight into human environmental health problems.

The purpose of this report is to summarize data gathered from a large number of accidentally lead-poisoned dogs studied at a large urban animal hospital (Angell Memorial Animal Hospital, Boston, Massachusetts), to review the published data of others on lead intoxication in accidentally and experimentally poisoned dogs, and to compare features of lead poisoning in dogs with those in man. The information contained herein is offered to alert the medical profession to the occurrence and nature of lead poisoning in dogs, to establish the value of dogs as natural and experimental models for lead toxicity studies, and to indicate the usefulness of dogs and other pet animals as biologic barometers of environmental poisons dangerous to man.

OCCURRENCE

Cases of accidental lead poisoning in dogs reported by others include 22 in the United States (all but five occurred in urban areas of New England) [1-9], 48 in New Zealand [10-12], approximately 20 in Rhodesia [13], 19 in England [14-16], seven in Europe [17-19], and two in Canada [20,21]. A total of 502 canine lead poisonings were diagnosed at the Angell Memorial Hospital in Boston, Massachusetts, over a 10-year period, 1961 to 1971. The yearly occurrence of lead poisoning in dogs (Table 1) increased during the latter years owing to studies of this disease which brought about in-

179

TABLE 1.   Yearly Occurrence of Lead Poisoning in Dogs Admitted to the
Angell Memorial Animal Hospital, Boston, Massachusetts

Year	No. of Dogs
1961	11
1962	11
1963	6
1964	18
1965	19
1966	26
1967	49
1968	59
1969[a]	97
1970[a]	107
1971[a]	99
Total	502

[a]Rate of occurrence among all canine hospital admissions in 1969 through
1971 was 1.2%.

creased clinical awareness and ease of diagnosis. The average occurrence
rate among all canine hospital admissions for 1969 through 1971 was 1.2%.
The occurrence for dogs 1 year old or less was 1 in about every 25 admissions (4.2%). The relative occurrence of this disease in urban children
and dogs is difficult to estimate; however, it would seem from the above
data that overt lead intoxication may be even more common in young urban
dogs than in children.

The age occurrence of lead poisoning in dogs ranges from newborn puppies [14] to old age. The vast majority of dogs are less than 1 year old,
and over half are 5 months of age or less [22]. This corresponds well to
the incidence of this disease in man where 80-85% of cases occur in children 12 to 35 months old [23]. The paucity of reports of lead poisoning in
rural dogs, and the finding that this disease is more common in dogs dwelling in urban slums than in more economically advantaged areas [22], coincides with urban slum occurrence of this disease in children. The seasonal
pattern of lead poisoning among urban dogs and children is also similar
(Fig. 1), the majority of cases occurring in the summer and fall months
[22].

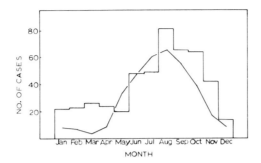

FIG. 1. Monthly occurrence of lead poisoning in 483 dogs (1961-1971) is given in bars. For comparison, average monthly reported cases of lead poisoning in Baltimore children (1931-1951) is given in curve.

SOURCES, ABSORPTION, AND TOXIC DOSES OF LEAD

The sources of lead responsible for poisoning dogs are often unknown. Of those cases where the source was determined, most were poisoned by ingesting lead-containing paint. Other sources include ingestion of linoleum, plumbers' materials, battery parts or water, grease, water from lead pipes, rug padding, and roofing material [22]. Additional sources reported by others include environmental contamination from nearby mining or smelting operations; lead objects such as weights, toys, and fishing sinkers; golf balls; putty; water from a lead-glazed drinking container; and lead dust from a linotype machine [7,13,16,19]. As in children, lead-containing paint is by far the most common source.

The smallest dose of ingested lead known to be toxic for dogs is 0.32 mg/kg body weight/day. That dose caused overt signs of toxicity when fed to a 5-month-old dog for a period of 5 months [24]. In other experiments, however, doses of approximately 1.0 mg/kg/day failed to cause signs in young dogs within 6 months [25]. Approximately 0.043 mg/kg/day is estimated to be toxic for adult men within 8 months [26]. The maximum permissible daily intake of lead for children is estimated to be about 0.025 mg/kg/day [27].

Absorption, body distribution, and excretion of ingested lead in dogs are comparable to those found in man. The amounts of lead excreted in the feces of lead-poisoned dogs are about 87-99% of the ingested dose [25,28, 29]. The amounts of lead excreted in the urine of dogs are approximately 1/100 to 1/500 the amount excreted in the feces [28,29]. Absorbed lead accumulates in teeth and bone (particularly the metaphyses of bone), where about 90-97% of the retained body lead is stored in dogs; liver, kidney, and spleen contain considerable amounts of lead, and brain and muscle the least [25,28-31]. The determination of lead in the liver is the best single postmortem test of lead poisoning in dogs; the upper limit of normal for healthy urban dogs is 3.6 ppm [32].

CLINICAL SIGNS

The physical signs of lead poisoning in dogs are referable to the gastrointestinal and nervous systems. Dogs are usually brought to clinicians with a history of colic, and often of vomiting, of approximately 7 to 10 days' duration. Nervous disorders have often occurred a few hours or days previously, prompting the owner to seek medical counsel. Clinical examination usually discloses an afebrile, nervous, or apprehensive dog in good nutritional condition with signs of abdominal pain. Signs or history of neurological disorders occur in approximately 75% of dogs and include clonic-tonic convulsions, psychomotor seizures (characterized by sudden apparent fright and blindly running about while continuously barking), tremors, ataxia, extreme nervousness, behavioral changes, Horner's syndrome, and hypersensibility. Burtonian lines are rarely, if ever, seen [1,3,16, 33].

Radiographic examination may reveal "lead lines" in long bones or ribs (Fig. 2) of immature dogs [33,34], radiopaque foreign material in the gastrointestinal tract, and rarely paralysis of the esophagus (Fig. 2), apparently due to vagal lead neuropathy [35]. Electroencephalographic recordings usually disclose irregular generalized slow wave activity of increased amplitude [22]. The physical changes of lead intoxication in dogs (with the exception of the type of psychomotor changes and paralysis of the esophagus) are similar to those in children.

Hematologic changes in dogs with lead poisoning are distinctive and may provide sufficient evidence for diagnosis. Nucleated erythrocytes (often 20-100 or more per 100 white blood cells) and basophilic stippling (15 or more per 10,000 red blood cells) occur in 97% and 95%, respectively, of lead-poisoned dogs. These changes, together with numerous immature (including reticulocytes) and abnormally sized and shaped erythrocytes in the absence of marked anemia, are essentially diagnostic [22,36]. Basophilic stippling, unlike that in man and certain animals, is seldom found in anemic dogs ill with diseases other than lead poisoning; as in man, however,

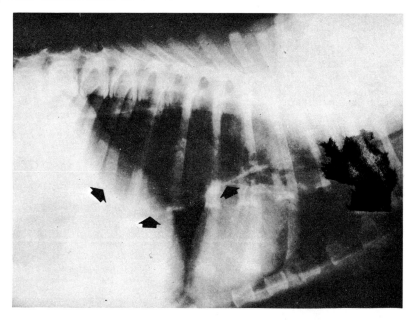

FIG. 2. Radiograph of thorax of immature lead-poisoned dog. Costochondral junctions in mid-thorax have radiodense bands. Paralyzed esophagus, apparently the result of vagal lead neuropathy, is outlined by arrows.

stippled erythrocytes are less commonly found after certain anticoagulants or fixatives are used on blood specimens [36,37]. Bone marrow aspirates are rich in erythroid elements, especially rubricytes and metarubricytes, and the M/E ratio is usually less than 1 [5,22].

Measurement of lead in blood is considered the best single test for the antemortem diagnosis of lead poisoning in dogs, as it is in children. The upper limit of normal for urban dogs is 35 µg/100 ml, which approximates that found in urban children. Blood lead values for lead-poisoned dogs range from 20 to 520 µg/100 ml, with a median of 80 µg/100 ml [32]. Analysis for lead in hair is seldom of diagnostic value in dogs because of irregularities in hair growth and shedding of hair [32].

Cerebrospinal fluid analyses in dogs with lead encephalopathy do not reveal consistent abnormalities [22]. This is in contrast to lead encephalopathy in children where pressure, protein, and cell counts are usually increased. Serum chemistry determinations in lead-intoxicated dogs are frequently normal.

Urinalysis usually discloses granular casts, sometimes mild proteinuria, and less commonly glycosuria [22]. The urine may contain excess

coproporphyrin [38] and delta amino levulinic acid [28]. Single urine
lead determinations are seldom of diagnostic significance; however, urine
lead values following 24 hr of treatment with calcium disodium ethylenediaminetetraacetate (Ca EDTA) are considerably increased in lead-poisoned
dogs and are diagnostically helpful [32].

TREATMENT AND OUTCOME

Treatment of lead intoxication in dogs with Ca EDTA is at least 95%
effective, and rapid clinical improvement is usually obvious in 24 to 48
hr [33]. Calcium EDTA is diluted and administered subcutaneously in divided doses at the rate of 50 mg/lb body weight for at least 5 days. Penicillamine is of somewhat less value, but is recommended in certain instances, especially when clinical signs are mild [39]. Sedatives, anticon-
vulsants, enemas, and dexamethazone are given as required. Recurrences of
lead intoxication following hospital discharge and return to the owner's
environment are not uncommon.

Death as a result of lead intoxication occurs in approximately 15% of
affected dogs. Some dogs are euthanatized at the owner's request primarily
to avoid the expense of treatment. Of dogs treated, between 2 and 5% die,
often because owners sought medical advice too late. The overall death
rate is likely to be higher than 15% for several reasons; e.g., some owners
may neglect to obtain proper diagnosis and treatment for their dogs, sick
unowned or lost dogs are not often taken to veterinary hospitals for treatment, the signs of lead poisoning may be readily confused with other serious diseases resulting in misdiagnoses and euthanasia, or lack of proper
treatment.

POSTMORTEM EXAMINATION

Thorough postmortem examination may reveal few gross changes in lead-
intoxicated dogs. Meningeal congestion is common, but obvious swelling or
herniation grooves (as may be seen in children) are rarely, if ever, found.
White bands (corresponding to radiographically visualized lead lines) may be
seen in the metaphyses of immature dogs. The bone marrow is abnormally red
and the gastrointestinal tract may contain foreign material, including paint
and other lead-containing substances [35].

Lead encephalopathy in dogs is marked by vascular lesions in the brain.
These lesions are characterized by dilatation of blood vessels, swelling and

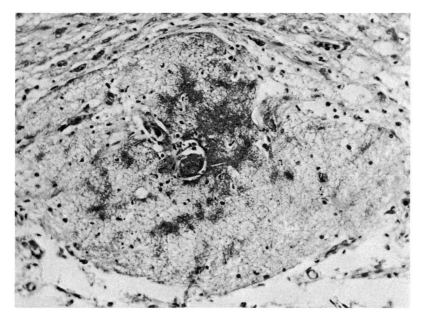

FIG. 3. Damaged cerebral vessel from dog with lead poisoning. Capillary
contains hyalin material (center) and is surrounded by fibrin-rich plasma
that has exuded from vessel and is compressing brain tissue.

necrosis of endothelial cells, hyalinization and necrosis of some arterioles,
and occasional thrombosis of capillaries [11,35]. Damaged vessels are often
surrounded by edema, fibrin, and hemorrhage (Fig. 3). Associated with the
vascular changes are lamina of vacuolation, gliosis, and necrosis of neurones
in the cerebral cortex [3,35,40]. In those dogs where neurologic signs persist for more than a week, proliferation of endothelial cells and new capillaries are seen in the cortical gray matter (Fig. 4) [35,40]. These lesions
are quite similar to those found in lead encephalopathy of children, except
that significant cerebral edema and lesions in the cerebellum are more common
in children [41,42].

Bone lesions in dogs with radiographic lead lines consisted of persistent heavily mineralized cartilaginous trabeculae covered with a thin layer
of bone extending from the epiphysis toward the diaphysis [34,35]. The bone
lesions may develop in as short a time as 10 days [34]. Similar metaphyseal
sclerosis is thought to require several months' ingestion of lead in children, probably because of their slower bone growth.

Lesions in other organs include: eosinophilic acid-fast intranuclear
inclusion bodies in renal and hepatic epithelia (Fig. 5) in most, but not

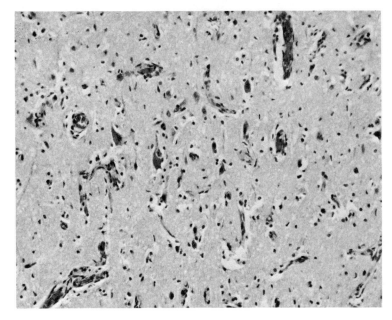

FIG. 4. Cerebral cortex from dog with prolonged nervous disorder caused by lead ingestion. Proliferation of new capillaries and gliosis is marked.

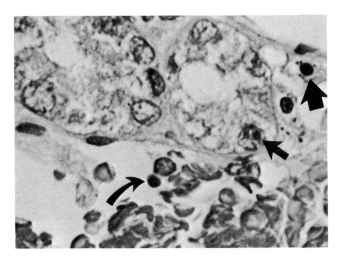

FIG. 5. Renal cortex from a lead-intoxicated dog. An acid-fast inclusion (thin straight arrow) occupies nucleus of proximal tubular epithelial cell. Some epithelial cell nuclei (broad straight arrow) are undergoing pyknosis. A metarubricyte (curved arrow) is present in a renal vein.

all, dogs; bone marrow hyperplasia, especially of erythroid elements; necrosis of random striated muscle fibers; peripheral neuropathy; paucity of developing follicles in ovaries and of sperm in testes; and hemosiderosis in liver and spleen [11,12,20,35].

Lesions or signs of peripheral neuropathy, similar to those occurring in plumbism of adult humans, are uncommon, but do occur. Three dogs with lead intoxication had paralyzed esophagi thought to be due to vagal neuropathy (the muscularis of the canine esophagus is primarily voluntary muscle). Similar vagal-esophageal changes occur in other heavy metal toxicities of dogs [43].

DISCUSSION

There are many similarities between lead poisoning in dogs and that in children. These similarities include: age, seasonal, and urban slum occurrence; source of lead; physical signs; hematologic and urinary findings; absorption, excretion, and distribution of lead in the body; response to chelating agents; and morphologic tissue changes. Because of these many similarities, it is logical to assume that many of the underlying mechanisms of lead intoxication are also similar. Thus, the dog would appear to be a useful model for the study of lead poisoning in man. Certain changes, such as the rapid development of typical hematologic and of metaphyseal bone changes offer advantages over other experimental animals. Study of the pathogenetic mechanisms of lead encephalopathy and of neurologic sequelae so common in children might be profitably studied in dogs, wherein the brain is of more appropriate size and complexity than many laboratory animals, and wherein brain changes are uncomplicated with marked cerebral edema.

The diagnosis of lead poisoning in a family dog should provide warning of lead-containing substances probably accessible to small children in the same dwelling. It is the experience of the author that young dogs develop signs of lead toxicity more rapidly than is generally believed to occur in children, and veterinary clinicians diagnosing lead poisoning in family pets may be the first to encounter evidence of a lead hazard possibly accessible for that family's children. It is hoped that such forewarned veterinarians would explain the dangers of lead poisoning to pet owners that have small children. A reporting system wherein family physicians are notified of lead poisoning and other germane toxicologic diseases diagnosed in their clients' pets might be a more effective warning procedure.

The potential usefulness of pet animals as indicators of environmental hazards has been effectively demonstrated in Minamata, Japan. Cats, dogs, and other animals sickened, convulsed, and died of organic mercury poisoning acquired from the same sources that months later poisoned over a hundred people. If the affected animals had been promptly studied, and if the cause and source of their violent neurologic disease had been quickly established, much suffering and many deaths might have been prevented among the people of Minamata.

It is noteworthy that domesticated cats, which also share our homes and urban environment, rarely are affected with lead poisoning. In the years 1961 to 1971, only two lead poisonings are recorded among approximately 43,600 feline admissions at the Angell Memorial Animal Hospital. Cats are specialized eaters and rarely mouth or ingest nonfood objects, thus eliminating the common sources of lead which poison children and dogs. Cats do, however, lick their fur repeatedly, and it must be considered that much of the atmospheric fallout of lead-containing particles would be ingested after landing upon a cat's fur. Urban cats must inhale airborne lead as well as ingest lead fallout, yet despite this intake of environmental lead, poisoning is apparently rare. Studies of lead intake and tissue accumulation of lead in city-dwelling cats might be of considerable value in determining the hazards of lead in our atmosphere for urbanized man.

SUMMARY

Lead intoxication is common among young urban dogs. The age, seasonal, and slum occurrence is similar to that in children. The clinical signs, radiographic, hematologic, and urinary changes, and the tissue lesions are also comparable to those in children. Because of these similarities, the study of accidental and experimental lead poisoning in dogs may be of considerable comparative interest. It is suggested that the study of known or suspected toxicologic diseases and the blood or tissue accumulation of various toxicants in urban pets, especially dogs and cats, might be of considerable value in determining the hazards of environmental lead and other toxins for urban man.

RECENT DEVELOPMENTS

Recently, contamination of commercial dog food with lead salts has resulted in the illness and death of hundreds of dogs. The ingredient that became adulterated with lead could have as well been used for human consumption. We must be acutely observant of our "biologic sentinels," and we must act upon the knowledge gained, if we are to save our children.

Several relevant works have been published since this chapter was prepared [44-48].

REFERENCES

1. E. Bond and R. Kubin, Vet. Med., 44, 118 (1949).

2. E. Estrada, Mod. Vet. Prac., 43, 72 (1962).

3. L. L. Lieberman, N. Amer. Vet., 29, 574 (1948).

4. W. L. Molpus, Auburn Vet., 14, 104 (1958).

5. G. D. Pettit, L. W. Holm, and W. E. Rushworth, J. Amer. Vet. Med. Ass., 128, 295 (1956).

6. C. M. Robertson, Vet. Rec., 86, 195 (1970).

7. T. W. Schrimsher, Vet. Med. Small Animal Clin., 66, 489 (1971).

8. B. Sass, J. Amer. Vet. Med. Ass., 157, 76 (1970).

9. D. E. Harling, J. Amer. Vet. Med. Ass., 156, 913 (1970).

10. D. C. Dodd and E. L. J. Staples, New Zealand Vet., 4, 1 (1956).

11. W. J. Hartley, New Zealand Vet., 4, 147 (1956).

12. E. L. J. Staples, New Zealand Vet., 3, 39 (1955).

13. H. M. Scott, Vet. Rec., 75, 830 (1963).

14. A. P. Berry, Vet. Rec., 79, 248 (1966).

15. M. A. Horrox, Vet. Rec., 75, 849 (1963).

16. M. R. Wilson and G. Lewis, Vet. Rec., 75, 787 (1963).

17. A. Agresti, S. Biondi, and G. Catellani, Acta Med. Vet., 4, (1958).

18. W. Kramer, Tijdschr. Diergeneesk., 82, 411 (1957).

19. M. J. Monot, Soc. Vet. Prat. Fr., 45, 217 (1961).

20. W. T. Oliver, L. W. Geib, and B. Sorrell, Can. J. Comp. Med., 23, 21 (1959).

21. C. A. Mitchell, Can. J. Comp. Med., 4, 170 (1940).

22. B. C. Zook, J. L. Carpenter, and R. M. Roberts, Amer. J. Vet. Res., 33, 891 (1972).

23. J. J. Chisolm, Sci. Amer., 224, 15 (1971).

24. L. L. Finner and H. O. Calvery, Arch. Pathol., 27, 433 (1939).

25. M. K. Horwitt and G. R. Cowgill, J. Pharmacol. Exp. Ther., 66, 289 (1939).

26. R. A. Kehoe, J. Roy. Inst. Public Health, 24, 81 (1961).

27. B. G. King, Amer. J. Diseases Child., 122, 337 (1971).

28. F. Takada, Osaka Shiritsu Daigaku Igaku Zasshi, 17, 425 (1968).

29. N. Wada, Osaka City Med. J., 4, 113 (1957).

30. S. C. Black, Arch. Environ. Health, 5, 423 (1962).

31. H. O. Calvery, E. P. Laug, and H. J. Morris, J. Pharmacol. Exp. Ther., 64, 364 (1938).

32. B. C. Zook, L. Kopito, J. L. Carpenter, D. V. Cramer, and H. Shwachman, Amer. J. Vet. Res., 33, 903 (1972).

33. B. C. Zook, J. L. Carpenter, and E. B. Leeds, J. Amer. Vet. Med. Ass., 155, 1329 (1969).

34. J. Caffey, Radiology, 17, 957 (1931).

35. B. C. Zook, Vet. Pathol., 9, 310 (1972).

36. B. C. Zook, G. McConnell, and C. E. Gilmore, J. Amer. Vet. Med. Ass., 157, 2092 (1971).

37. R. L. Clark, J. H. Jones, and J. F. Jones, J. Clin. Pathol., 20, 166 (1967).

38. B. J. McSherry, R. A. Willoughby, and R. G. Thomson, Can. J. Comp. Med., 35, 136 (1971).

39. B. C. Zook and J. L. Carpenter, Current Veterinary Therapy, Saunders, Philadelphia, 1977, pp. 128-133.

40. D. J. McCarthy, Univ. Penn. Med. Bull., 14, 398 (1902).

41. A. Penschew, Acta Neuropathol., 5, 133 (1965).

42. S. S. Blackman, Bull. Johns Hopkins Hosp., 61, 1 (1937).

43. B. C. Zook and C. E. Gilmore, J. Amer. Vet. Med. Ass., 151, 1, 206 (1967).

44. D. F. Kowalczyk, J. Amer. Vet. Med. Ass., 168, 428 (1976).

45. T. W. Craig, J. L. Rising, and J. K. Moore, J. Amer. Vet. Med. Ass., 167, 995 (1975); 169, 1237 (1976).

46. A. Azar, H. J. Trochimowicz, and M. E. Maxfield, Proc. Int. Symp. Environ. Health Aspects of Lead, 1973, p. 199.

47. H. D. Stowe, R. A. Goyer, M. M. Krigman, M. Wilson, and M. Cates, Arch. Pathol., 95, 106 (1973).

48. E. G. C. Clarke, J. Small Anim. Pract., 14, 183 (1973).

11

THE USE OF ANIMAL MODELS FOR
COMPARATIVE STUDIES OF LEAD POISONING

Nancy N. Scharding* and Frederick W. Oehme
Kansas State University
Manhattan, Kansas

Despite the long-standing recognition of lead as a clinically important poison, our knowledge concerning the basic metabolic alterations induced by toxic concentrations of lead in the body is surprisingly limited [11]. Much work remains to be done to explain lead's diverse toxic effects.

The use of man as an experimental animal in lead poisoning is a hazardous occupation, especially when it concerns the acute form of toxicity. The use of animals is, of course, much more practical and realistic. Animals permit the use of experimental procedures and intervention studies not possible in man. The early stages of the pathological processes ean be studied in detail. Furthermore, due to their short life spans, changes may be observed in a few years in animals which might be spread over half a century in the life of man [3].

In the past, the choice of an experimental animal for medical research has usually been confined to rodents, rabbits, cats, and dogs. Convenience has played a major role in the final choice of a species [7]. Today, however, there is a new awareness concerning the importance of selecting animals on the basis of their similarity to man, anatomically, physiologically, or biochemically [7,19]. Many heretofore unknown or ignored species are coming into common use [8]. The use of some of these species, such as the larger farm animals, may, at first, seem impractical, but upon reconsideration, their use is seen to be very reasonable [7]. Because of the fact

---

*Present Address: Box 840, Memphis, Tennessee.

that they live largely out of doors and graze forage which is possibly
contaminated, domestic animals have more exposure to environmental toxins
than does man. Domestic pets share man's very home, diet, and air. In
short, they live in the same environment as man and are exposed to similar
hazards.

As a result of the wide use and distribution of lead in our environment and the close association of our domestic animals to it, lead poisoning
is the most common form of metallic poisoning in livestock throughout the
world [6,11]. However, the response of different species of animals to
lead poisoning varies considerably. It is not within the scope of this
chapter to discuss all the ways in which lead poisoning of animals is similar to or different from lead poisoning in man. However, some of the more
interesting and important areas should be mentioned. Although the similarities are more useful in setting up actual "animal models," it should be
remembered that when we know why an animal has a different response from
man, we also gain knowledge concerning the mechanisms of action of lead.

Different species of animals vary considerably in the amount of lead
they can tolerate. Daily oral doses of 33 to 66 mg of lead acetate/kg of
body weight are required for periods of up to 14 weeks to have fatal effects
on pigs [17]. Rats can tolerate lead at 60 mg per day or more in their diet
for periods of greater than 1 year. This dose in man is roughly equivalent
to 8 g of lead per day [5]. Toxic effects in man have been produced by 2
to 3 g of lead acetate [22]. It would be very valuable to know how rats
and pigs can tolerate such high amounts of lead. Surely, it must be related
to their absorption, detoxification, and excretion mechanisms.

It is known that absorption of lead varies widely among species. For
example, man absorbs approximately 10% of the lead he ingests [14]. Cattle
and sheep absorb only 1 to 2% [4]. As a consequence, these latter two
species show great tolerance to the continual ingestion of relatively large
quantities of lead and rarely develop chronic lead poisoning [1,2].

Anemia is a constant finding in chronic lead poisoning in man. It is
also observed in dogs, cattle, rats, and rabbits. The hematologic disturbances produced experimentally in rabbits are similar to those found in human
cases of lead poisoning. In fact, rabbits and man are among the few species
in which lead poisoning interferes with porphyrin metabolism. Furthermore,
these two species, more so than any others, readily develop basophilic
stippling of erythrocytes [12].

Rabbits are also good models for the study of lead-induced kidney damage. The location and sequence of development of renal pathology in the rabbit resemble those of man more closely than do the renal changes in the rat (even though the rat has often been used as the experimental animal in studies of lead nephropathy [12]). Expanded use of rabbits may help to define the relationship between lead toxicity and chronic nephritis in man.

Utilizing the Mongolian gerbil, long-term administration of lead acetate has produced chronic progressive nephropathy with tubular degeneration, intranuclear inclusions, and interstitial fibrosis, as well as hypochromic anemia, indicating that the gerbil may be a useful model to study chronic lead poisoning in man [21].

In rats, intravenous injection of 20 to 40 mg of lead phosphate produced an increase in blood pressure, followed by hypertrophy of the media of the arterioles in the kidneys, intestines, pancreas, and brain [22]. In rabbits poisoned with lead, medial calcification in the proximal aorta was frequently found [12]. Thus, rats and rabbits may be of value in evaluating the contribution of lead poisoning to atherosclerosis in man.

The results of lead administration to sheep suggest that the pregnant ewe is more susceptible to lead poisoning. Abortions may follow continuous ingestion of 0.5 mg/lb of body weight, a quantity which is quite safe for nonpregnant sheep [2]. Congenital skeletal malformations consisting of varying degrees of stunting of the tail associated with abnormalities of the sacral vertebrae have been produced in hamster embryos by treating pregnant hamsters with various salts of lead [9]. More recently, the Siberian rodent, Microtus ochrogaster, has been shown to respond similarly to the administration of lead acetate and to be well suited physiologically and by reproductive traits for teratogenic studies [16]. The implications of these studies point out the need for more research to be conducted, if possible, in primates to determine what effect low levels of lead might have on the developing human embryo.

It is observed in man that the peripheral neuropathy of chronic lead poisoning is purely motor [13]. Paralysis affects those muscles which are used most constantly and is thought to be due to segmental degeneration of axons and myelin in the distal parts of motor fibers. Peripheral nerve degeneration also occurs frequently in chronic cases of lead poisoning in horses. This species develops swollen knees, gradual paralysis of the hind legs, and a characteristic shortness of breath (or "roaring") brought about by paralysis of the laryngeal muscles. The basis of the paralytic changes in horses appears not to have been described. However, it is probable that this pathology has the same histological basis as human peripheral neuropathy [13]. The horse would be a good animal in which to study the nerve-muscle changes associated with lead poisoning.

Growing children with lead poisoning have altered bone formation with transverse lines of greater density behind the diaphyses. Since similar changes have been produced in growing dogs [22,25], they could be used as excellent models for bone studies.

The effect of lead upon the central nervous system is probably the most important and tragic aspect of lead poisoning in man. In this area, more so than any other, animal models could be of value in helping to understand the effects of lead on the central nervous system, and thus aid in the treatment and prevention of the degenerative aftereffects. The nervous signs shown by calves poisoned by lead resemble those seen in children [24]. The syndrome is acute and involves various derangements of the central nervous system, including hyperexcitability and convulsions.

Young dogs with lead poisoning frequently show nervous disorders. The nervous involvement varies from abnormal attitude through hyperexcitability and hysteria to frank convulsions [25]. A study of tissues from dogs tentatively diagnosed as having rabies, but producing negative animal inoculation results, revealed that a significant number of specimens contained sufficient concentrations of lead to suggest that the animals were actually suffering from lead intoxication [23]. Since there is a high incidence of lead poisoning in young dogs and since cerebral disorders of unknown origin are relatively common in old dogs, studies to establish a cause-effect relationship between lead poisoning and cerebral disorders would benefit both veterinary and human medicine [25].

The discovery has been made that infant rats are excellent models for the study of lead-induced encephalopathy, despite the fact that adult rats are highly resistant to lead intoxication. Paresis can be produced in the 23- to 29-day-old rat by the transference of lead through its mother's milk. Constant, spectacular, and characteristic brain changes can be generated by this simple procedure. The lesions produced in the infant rat are quite similar to those seen in the human infant [15,18,20].

Another interesting action of lead on the nervous system is its effect on conditioned reflexes of the rat. Three conditioned reflex patterns were

established in young rats; two were positive responses to light and to a
bell, while the third was a negative response to a buzzer. The conditioned
rats were then exposed to an atmospheric lead concentration of 11 $\mu g/m^3$ for
6 hr daily. In 3 months there was an increase in the latent period of response to the stimuli and a decreased force of response [10]. This study
raises questions as to the effect of lead on man's reflexes. It would be
most interesting to repeat this work using squirrel monkeys or other primates with brains more similar to man's.

It may be seen that no one species of animal is an exact model for
man, and no one species responds to lead poisoning exactly as man does.
However, many species have one or more body organs or systems which do respond to lead in a manner similar to the corresponding organs or systems of
man. These species can be studied as models for detailed investigations
into the mechanisms of action of lead, and they are capable of adding valuable facts to our understanding of lead intoxication.

SUMMARY

A variety of common laboratory and domestic animals, as well as household pets, possess unique and relatively unknown physiological responses to
lead that make them well suited as models for investigations into the mechanisms of action of lead intoxication. This is especially true of studies
dealing with the effects of long-term ingestion of lead. These specific
characteristics should be utilized by researchers and clinical investigators
in detecting and studying the subtle biological effects of environmental
lead on man and animals.

REFERENCES

1. R. Allcroft, Lead as a Nutritional Hazard to Farm Livestock. IV.
   Distribution of Lead in the Tissues of Bovines after Ingestion of
   Various Lead Compounds, J. Comp. Pathol., 60, 190-208 (1950).

2. R. Allcroft and K. L. Blaxter, Lead as a Nutritional Hazard to Farm
   Livestock. V. The Toxicity of Lead to Cattle and Sheep and an Evaluation of the Lead Hazard Under Farm Conditions, J. Comp. Pathol., 60,
   209-218 (1950).

3. W. I. B. Beveridge, The Future of Comparative Medicine, Missouri Med.,
   February, 98-103 (1968).

4. K. L. Blaxter, Lead as a Nutritional Hazard to Farm Livestock. II.
   The Absorption and Excretion of Lead by Sheep and Rabbits, J. Comp.
   Pathol., 60, 140-159 (1950).

5. A. Cantarow and M. Trumper, Lead Poisoning, Williams and Wilkins,
    Baltimore, 1944.

6. E. G. C. Clarke and M. L. C. Clarke, Garner's Veterinary Toxicology,
    Williams and Wilkins, Baltimore, 1967, pp. 91-99.

7. R. E. Doyle, S. Garb, L. E. Davis, D. K. Meyer, and F. W. Clayton,
    Domesticated Farm Animals in Medical Research, Ann. N. Y. Acad. Sci.,
    147, 129-204 (1968).

8. C. G. Durbin and J. F. Robens, The Use of Laboratory Animals for Drug
    Testing, Ann. N. Y. Acad. Sci., 111, 696-711 (1964).

9. J. H. Ferm and S. J. Carpenter, Developmental Malformations Resulting
    from the Administration of Lead Salts, Exp. Mol. Pathol., 7, 208-213
    (1967).

10. M. I. Gusev, Limits of Allowable Lead Concentration in the Air of
    Inhabited Localities, in Limits of Allowable Concentrations of At-
    mospheric Pollutants, Book 4, English translation available from office
    of Tech. Serv., U.S. Dept. of Commerce, Washington, D. C.

11. P. B. Hammond, Toxic Minerals, in Veterinary Pharmacology and Thera-
    peutics, Iowa State Univ. Press, Ames, 1965, pp. 972-983.

12. G. M. Hass, Relations Between Lead Poisoning in Rabbit and Man, Amer.
    J. Pathol., 45, 691-727 (1964).

13. K. V. F. Jubb and P. C. Kennedy, Pathology of Domestic Animals, Vol.
    2, Academic Press, New York, 1970, pp. 388-391.

14. R. A. Kehoe, The Harben Lectures, 1960, The Metabolism of Lead in Man
    in Health and Disease. II. The Metabolism of Lead Under Abnormal
    Conditions, J. Roy. Inst. Public Health Hyg., 19-53 (1961).

15. M. R. Krigman and E. L. Hogan, Effect of Lead Intoxication on the
    Postnatal Growth of the Rat Nervous System, Environ. Health Perspect.,
    7, 187-199 (1974).

16. S. M. Kruckenberg, Ph.D. dissertation, Kansas State University, Man-
    hattan, 1972.

17. R. P. Link, Lead Toxicosis in Swine, Amer. J. Vet. Res., 27, 759-763
    (1966).

18. I. A. Michaelson and M. W. Sauerhoff, Animal Models of Human Disease:
    Severe and Mild Lead Encephalopathy in the Neontal Rat, Environ. Health
    Perspect., 7, 201-225 (1974).

19. F. W. Oehme, Species Differences: The Basis for and Importance of
    Comparative Toxicology, Clin. Toxicol., 3, 5-10 (1970).

20. A. Pentschew and F. Garro, Lead Encephalo-myelopathy of the Suckling
    Rat and Its Implications on the Porphyrinopathic Nervous Diseases,
    Acta Neuropathol., 6, 266-278 (1966).

21. C. D. Port and D. W. Baxter, The Mongolian Gerbil as a Model for
    Chronic Lead Toxicity, J. Comp. Pathol., 85, 119-131 (1975).

22. T. Sollmann, A Manual of Pharmacology, W. B. Saunders, Philadelphia,
    1957, pp. 1137-1351.

23. R. M. Thomas, Lead Toxicity in the Canine, Morris Animal Foundation
    Report, Denver, 1961.

24. E. G. White and E. Cotchin, Natural and Experimental Cases of Poison-
     ing of Calves by Flaking Lead Paint, Vet. J., 104, 75-91 (1948).

25. B. C. Zook, J. L. Carpenter, and E. B. Leeds, Lead Poisoning in Dogs,
     J. Amer. Vet. Med. Ass., 155, 1329-1342 (1969).

POLLUTION BY CADMIUM AND THE
ITAI-ITAI DISEASE IN JAPAN

Jun Kobayashi
Okayama University Institute for Agricultural and Biological Sciences
Kurashiki, Japan

After the Second World War, it was found that the inhabitants along the
lower stream of the Jinzu River, which pours into the Japan Sea at the
central part of the Japanese mainland, had been suffering from a very miserable and heretofore unrecorded osseous disease which was called the itai-
itai (ouch-ouch) disease. These people suffered from intense pains in their
bones, and the disease was so named because the patients continually moaned
"Itai-itai." This disease mostly attacked farm women above middle age who
had lived for more than 30 years in this area. The disease starts with
symptoms similar to rheumatism, neuralgia, or neuritis. However, as the
condition aggravates gradually, a peculiar form of gait appears, viz.,
wiggling the hips like a duck, due to pain in the pelvic region. It is
then that the patients realize that they are afflicted with the fatal itai-
itai disease. After a long progress extending over several years (in some
cases, more than 10 years), the patients become incapable of rising from
their beds, and even a slight outside force causes bone fractures in various
parts of the body. Among the subjects at autopsy, there was a patient who
had 28 fractures in the ribs and 72 in the whole body [1]. As the real
cause of the disease was unknown, no specific treatment had been found, so
it aroused considerable anxiety among the inhabitants.

Although the date of the first appearance of this disease is not clear,
it had already existed before World War II and markedly increased during
and after it. Because of the delay in recognizing itai-itai as a specific
illness, there is no exact number of cases available. But it is roughly
estimated that 200 had suffered from this disease and half of them had died
of it up to the end of 1965.

A review of this dramatic disease, which has been identified as chronic cadmium poisoning resulting from the contamination of river water, soil, and daily diet by cadmium released through mining activities, will be instructive both to illustrate the toxic nature of cadmium and to emphasize the danger of environmental pollution by cadmium.

BRIEF HISTORY OF THE DETERMINATION
OF THE CAUSE OF ITAI-ITAI

In 1955, Kohno started a study of the disease and drew attention to it, and medical circles began to take a special interest in this disease. After that, Kohno and co-workers [1-4], Murata [5-9], Nakagawa [10], Nakayama [11], Kajikawa et al. [12], and others clarified the characteristic symptoms of the disease, which were summarized as follows: (a) osteomalacia accompanied by osteoporosis due to a great loss of minerals from the bones, (b) renal dysfunction with increased proteinuria and glycosuria, (c) decreased inorganic phosphate and increased alkaline phosphatase in the blood serum. Kohno et al. [3] reported that administration of vitamin D in large doses is effective in reducing the pains and disorder of the bones.

However, as to the cause of the disease, they tried to attribute it to a combination of factors such as the excessive field labor of farm women, an unbalanced diet which depended too much on polished rice with deficient amounts of protein, vitamins, and calcium, and also to the climatic conditions, i.e., the lack of sunshine in wintertime which would cause the vitamin D deficiency. But these conditions could not make it clear why the appearance of the disease was concentrated in the areas along the two sides of the lower Jinzu River. Hagino, a doctor practicing at the center of the disease area since the end of the war, presented a hypothesis (1957) of mine poisoning due to zinc. However this hypothesis was disregarded or opposed by the mine and all the other doctors cited above because it had no supporting data.

On the other hand, Kobayashi [13] carried out (in 1960) a spectrographic analysis of the bones and other tissues of severely affected patients and happened, to his surprise, to find a markedly high content of cadmium, zinc, lead, and copper. These metals were the same as those that

he had found in the Jinzu River and the mining wastewater in 1959, with
the exception of arsenic. He had been investigating the quality of river
waters [13-16] and agricultural damage caused by mine poisoning since before
the war. His investigation of the poisons of the Kamioka Mine (Mitsui
Mining and Smelting Co., Ltd.), which is the biggest zinc mine in Japan,
located along the upper stream of the Jinzu River, goes back to 1943.
During World War II, with the rapid increase in the production of zinc and
lead together with the faulty treatment of the wastewater, the Jinzu River
became polluted and turbid. This caused great damage to rice cultivation
in Fuchu-machi and its neighboring farm villages along the lower reaches
of the river (the same areas as those of the itai-itai disease), as a result of which a severe dispute broke out between the farmers and the mine.
At that time, Kobayashi [17] investigated both the damaged fields irrigated
by the river water and the mine to observe the operating process and the
treatment of wastewater. As soon as he found poisonous metals in the patients, he intuitively suspected that the disease must have been caused by
the wastewater from the Kamioka Mine and encouraged Hagino, who was the
only member of medical circles to hold the hypothesis of mine poisoning.

In 1961, Kobayashi visited the Director of the Mine and the Toyama
Prefectural Governor to show his analytical data on metals found in the
patients' tissues and in various other samples such as rice, soil, and
water collected from the affected area and to inform them of his suspicion
that mine poisons were the cause of the disease. His information was reported by the local newspaper [18] attracting public attention. Hagino
and Yoshioka [19] also reported Kobayashi's earliest data and his methodology to medical circles, among whom neither cadmium toxicity nor the
spectrographic approach had been known. In that year, Kobayashi and Hagino
[20] reported on the cause of the disease, pointing out the osteomalacia
found by Nicaud et al. [21] at a cadmium factory in France.

Stimulated by these reports, two groups, in Toyama (1961) and Kanazawa
(1963), began to investigate the cause of the disease. The former investigation was supported by the Prefecture and the latter by the National
Government. However, in 1966 after several years of investigation, kept

at a distance from the great mine, the two groups announced the conclusion
[22] that the cause of the disease might be cadmium plus alpha (poor nutrition), but they made no mention of mine poisoning. In the meantime,
Kobayashi carried out three experiments, reported below [13,23], to test his
assumption that the disease had been induced by the poisonous metals found
in the patients.

Relation of the Itai-Itai Disease to the Agricultural Damage

When the damage to the rice crop along the downstream areas of the
Jinzu River caused by the poisonous wastewater from the Kamioka Mine had
reached its climax during the war, Kobayashi [17] investigated both the
mining operations and the damage to the paddy fields. In those days, the
mining company forced the production of zinc and lead to an extreme degree
with faulty treatment of the wastewater; 3000 tons of zinc-lead ore per
day were mined, pulverized, concentrated by flotation, and smelted. The
cadmium contained in the ore was abandoned as it had no utility value. The
wastewater including fine particles of the flotation tailings was discharged
into the river which, after running through steep mountains for about 30 km,
flowed into the paddy fields downstream (see Fig. 1). The particles were
deposited near the inlets for the irrigation of fields, and there the greatest injury to the growth of rice appeared. The only countermeasure for
farmers against the mine poisoning then was to dig a small sedimentation
pool (see Fig. 2) at each inlet into the fields and to remove the deposited
poisonous particles, in spite of the great expense of labor and little effectiveness. The farmers lived unsuspectingly on rice with excessive cadmium content and, moreover, habitually drank the poisonous river water introduced into their kitchens (see Fig. 3) through irrigation waterways. People
who had eaten polluted rice and drunk the river water over a period of 30
years, including the wartime, accumulated a large amount of the metals in
their bodies (see Table 1 and Fig. 8) and came to suffer from the disease.

Figure 4 shows the locality where the patients were found. It is all
within 3 km of the banks of the river and is restricted to the rice field
districts which were irrigated by the Jinzu River. The prevalence of the
disease was markedly concentrated in the triangular district (Fuchu-machi)
between the Jinzu River and the biggest irrigation canal (Ushigakubi Yosui)
on the left bank. Since no other water for irrigation and drinking could

be obtained except from the polluted Jinzu River, it is natural that both the heavy damage to rice plants and the prevalence of the disease occurred in this same district. In the neighboring farm areas where the water from the tributaries of the Jinzu River or other rivers was used, neither cases of the illness nor damage to rice crops appeared. Thus, the itai-itai disease and the damage to the growth of rice occurred in the same place, both increasing during the war and rapidly decreasing after the time when the wastewater came to be purified in a big settling basin built by the mining company in 1955-1956.

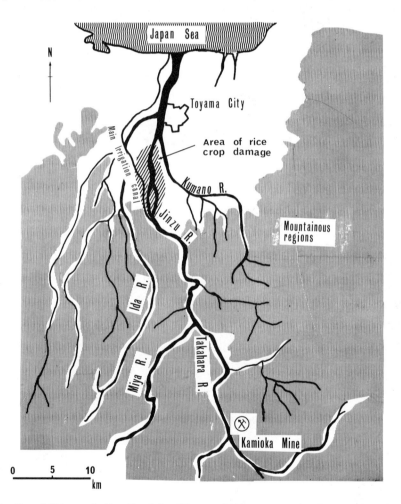

FIG. 1. Localities of the Kamioka Mine and the area of rice crop damaged.

FIG. 2. Sedimentation pool at the inlet to a damaged rice field in the polluted area.   No harvest was obtained near the inlet.

FIG. 3. Kitchen of a patient's house.   (The arrow indicates river water piped from irrigation waterway.)

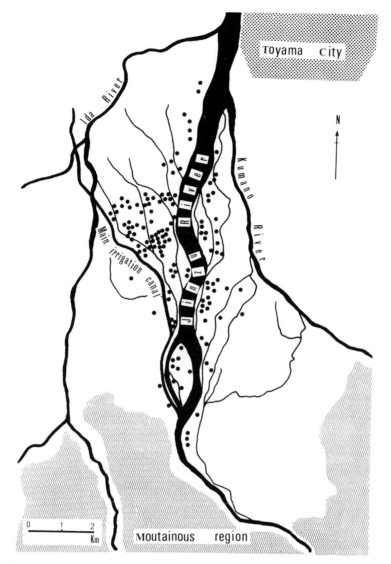

FIG. 4. The locality where the patients were found (dead patients included).

Animal Experiments

Animal experiments were undertaken over a period of 2 years, beginning in September 1962, in order to ascertain whether bone decalcification was caused by the poisonous metals found both in the wastewater and the patients. By continued determination of the amounts of calcium received from the diet and excreted in feces and urine, Kobayashi could quantitatively trace the changes in the balance of calcium metabolism in rats and found

that an oral administration of cadmium resulted in a significant loss of
calcium from the bones (see final section).

Existence of a Similar Disease
in Other Mining Districts

If a disease similar to the itai-itai disease could be found in other
mining districts polluted with cadmium, the proof for mine poisoning would
be well established. On Tsushima Island, Nagasaki Prefecture, located in
the channel between Japan and Korea, there was a mine (closed in 1973)
which had produced zinc, lead, and cadmium. In 1963, Kobayashi collected
samples of soil, rice, water, etc., and, on analyzing them, found that the
pollution by cadmium there was even greater than that of the Jinzu River
basin at that time. The highest value of cadmium content in well water,
which had been used in the past despite its nauseating taste, was 0.225 ppm,
and severe chlorosis of the leaves of sweet potatoes (see Fig. 5), which
turned bright yellow due to cadmium poisoning, was observed. Consequently,
the author went again to Tsushima in 1964, accompanied by Hagino, to ascertain if the itai-itai disease was present. As a result, one living patient
(who died in 1966) and two dead were discovered at Kashine, a small farm
hamlet which had been most severely contaminated by the mine [24].

FIG. 5. Sweet potato leaves which have turned yellow through loss of
chlorophyll.

By these three experiments, completed by the beginning of 1965, he could confirm his suspicion that the itai-itai disease was induced by cadmium in the wastewater from the mine. At the Annual Meeting of the Japanese Society for Hygiene in 1965, a professor of the Gifu Medical School, who had won a debate against Hagino on the NHK television in 1964, stood against the mine-poisoning theory presented by Kobayashi [25], citing Hirota's report [26] on an experiment in which animals suffered no harm from drinking the wastewater of the mine. However, Kobayashi pointed out that this experiment had been irrelevant because Hirota had used the wastewater which was purified in a big settling basin built by the mining company in 1955-1956, after which time both the agricultural damage and the occurrence of the itai-itai disease rapidly decreased. Thus, the mine-poisoning theory has established a foothold in medical circles. Later, Hirota was found to have been a doctor working at the hospital of the Kamioka Mine.

The study group set up by the Ministry of Health and Welfare in 1965 started research [27] in two mining districts, Tsushima Island in Nagasaki Prefecture and a district downstream from the Hosokura mine in Miyagi Prefecture as well as in the endemic area of the Jinzu River, but this group avoided tracing the source of the pollution of the Jinzu River up to the Kamioka Mine. However, 2 years later, at the House of Councilors 1967 [28], a member of an opposition party asked the Minister of Health and Welfare to hurry in clarifying the relation between the itai-itai disease and the pollution from the mine and to relieve the unfortunate patients. For the first time the study group [29,30] traced the origin of the cadmium pollution along the whole basin of the Jinzu River, from which the heavy pollution had practically disappeared by purifying the wastewater as said above. In that year, Kobayashi, illustrating the results of his study, advised the hesitating inhabitants to take legal proceedings against the mining company. These were begun before the Minister's courageous statement in 1968.

In May 1968, S. Sonoda [31], the Minister of Health and Welfare and a man of justice, officially declared that the itai-itai disease was a mining hazard induced by cadmium released through the activities of the Kamioka Mine and that national health should take precedence over industrial profits. It was an epoch-making event that the responsibility was laid on a big enterprise for the pollution it caused. The company, in league with some scientists, rejected the declaration and, furthermore, appealed against the judgment of a district court [32] which had been given in 1971 in favor of

the victims after several years of legal proceedings. In order to reverse
the judgment, Takeuchi [33] insisted at the higher court on his hypothesis
that the itai-itai disease might have been caused not by cadmium poisoning
but by an initial vitamin D deficiency and a subsequent poisoning by vitamin
D which had been administered in very large doses. The higher court [34,35],
however, rejected this hypothesis as it was not supported by any factual
data and established the final legal responsibility for the cause of the
disease. Thus, the theory of mine poisoning by cadmium was upheld by the
courts, and the mining company was ordered to compensate the unfortunate
victims. As this final decision was given on August 9, 1972, 17 years had
elapsed between the time that medical circles had begun the study and the
time when responsibility for the disease was legally established.

However, the causal mechanism of osteomalacia by cadmium has not yet
been established because the hypothesis, which has been predominant in
medical circles, that the decalcification of the bones is due to some renal
tubular dysfunction (Fanconi's syndrome) cannot be supported by animal
experiments (see final section).

NATURE OF THE DISEASE AND ITS IDENTIFICATION

As described in the preceding sections, this disease is a dramatic
and very miserable case of chronic poisoning resulting from drinking river
water and eating rice contaminated by cadmium released through mining activities. The deficiency of calcium, vitamins, and proteins in the farmers'
habitual unbalanced diet until quite recently in Japan must have played a
part in the increased absorption and toxication of cadmium. The main features of the fully developed syndrome of the itai-itai disease, which affected mostly farm women of older age who had given birth to 5-6 children
on the average, were intense pains in the bones due to severe decalcification (osteomalacia and osteoporosis), multiple pathological fractures, and
renal dysfunction causing proteinuria and glucosuria. Malnutrition and a
considerable decrease in stature due to compression fractures and deformation of the vertebrae were also noticed.

Hagino [36], who had treated the greatest number of patients, divided
the disease into the following five stages:

1. First stage (incubation period): Pain in the lower back and limbs during
   the farmers' busy season or after overwork, which decreases after a rest
   or bath. There are no specific symptoms excepting the increased urinary
   excretion of cadmium at intervals.

2. Second stage (warning period): Marked pain with or without proteinuria. Shortly after the war, when the pollution of the Jinzu River was still great, yellow rings were noticed by Hagino [37] at the lower part of the teeth in some of the patients.

3. Third stage (painful period): Pains in pelvic region and various other parts of the body due to decalcification, specific waddling gait, anemia, and osteoporosis detected by x-ray diagnosis. Reduced inorganic phosphorus and increased alkaline phosphatase in blood serum, and increased urinary excretion of cadmium and proteins with or without glucosuria were observed.

4. Fourth stage (deformation period): Intense pains in various bones of the body, difficulty in walking, shrinkage in stature due to compression fractures in the vertebrae. Symptoms of osteomalacia seen by x-ray, such as deformation of bones, Milkmann's syndrome, and pseudofractures. Continued proteinuria and glucosuria.

5. Fifth stage (multiple fractures period): Severe atrophy and decalcification leading to fractures in different parts of the body, even on slight application of external force. Figure 6 shows a serious patient, whose arms and legs are crooked from fractures, the right leg being tied with a towel to the wooden frame of a foot warmer. An x-ray photograph of fractured legs of another patient is illustrated in Fig. 7. Due to the extreme loss of minerals, the bones were as transparent as soft tissues.

Hagino's fourth or fifth stage can be diagnosed by x-ray examination and from other symptoms, but the third and earlier stages of the disease, from which many more people must have been suffering in cadmium-contaminated areas, are difficult to differentiate from other similar kinds of illness because the x-ray changes would not be evident until 30-40% of the bone minerals have been lost.

The recent standard procedures for finding suspected cases of itai-itai disease in polluted areas recommended by the Ministry of Health and Welfare [38] are as follows:

1. First screening: Interview and urine analysis for protein (trichloroacetic acid or sulfosalicylic acid method). Persons who were found to have proteinuria should be subjected to the second screening.

2. Second screening: Urine should be tested for cadmium, protein, glucose, and for pattern of proteins by disc electrophoresis. If a person was

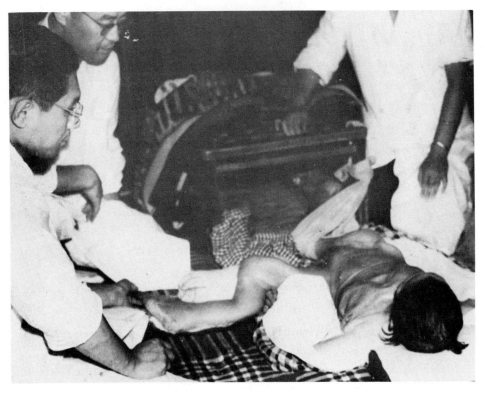

FIG. 6. A severe case of the itai-itai disease.

suffering at all from cadmium poisoning, urinary proteins of small molecular weight (20,000-30,000) indicating tubular damage should be found.
In cases where the disc electrophoretic pattern showed tubular damage at
the second screening, the third screening should be performed.

3. Third screening: A number of urine and blood tests are given, including
   the measurement of alkaline phosphatase and inorganic phosphorus in the
   serum. X-ray tests of the pelvis (lower part) and femur are given. The
   result of the third screening should be analyzed by the study group set
   up by the Ministry of Health and Welfare, and a final conclusion should
   be given. (Since 1971, this study group has been under the newly created
   Environment Agency of the Japanese Government.)

In Japan extensive investigations have been carried out in several areas
polluted by cadmium, as cited by Friberg et al. [39]. However, it remains
a mystery why in some of the polluted areas no responsibility has ever been
laid on industries for a number of possible patients in the earlier stages

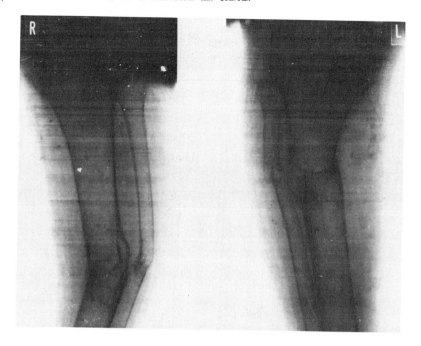

FIG. 7. X-ray photograph of the legs (crura) of an itai-itai disease patient (from Hagino).

whose health has been affected by cadmium. Nogawa et al. [40] reported the existence of a female itai-itai patient aged 73 and other deeply suspected cases in the Ichikawa River basin polluted by an old copper mine in Hyogo Prefecture. Through their examinations, the former patient was found to have the typical symptoms of the itai-itai disease which met all of the conditions of the standard method as stated above, i.e., severe osteomalacia, increased urinary excretion of cadmium, glucose, and proteins of tubular type on electrophoretic examination, and increased alkaline phosphatase in her blood serum. However, due to a lack of unanimity in the above-mentioned study group, the patient and the suspects could not be recognized as victims. Thus, up to date, the only patients and patient suspects of the itai-itai disease who have been compensated by a company have been those within the endemic area of the Jinzu River basin.

ACCUMULATION OF CADMIUM, ZINC, AND LEAD
IN PATIENTS' ORGANS

An unbelievably heavy accumulation of trace metals found by Kobayashi [13,23] in a patient who died (in 1956) of the itai-itai disease in inex-

TABLE 1. Accumulation of Trace Metals in Patient's Tissues (Patient Komatsu)

Tissue	Ash in Dry Matter (%)	Metal Content (ppm ash weight)			
		Cd	Zn	Pb	Cu
Sternum, upper part	4.2	9950	4270	260	180
Rib	10.2	11,500	3410	410	260
Costal cartilage	6.0	4760	2710	160	890
Femur, lower end	2.2	14,900	5720	280	140
Vertebra	22.4	6970	2910	260	--
Brain	2.1	290	3830	340	2100
Tongue	0.2	3100	1300	80	1200
Larynx	1.1	3600	5700	800	570
Lung	2.2	2550	3030	850	890
Heart	0.6	2200	6200	450	1320
Aorta	1.2	2700	2460	110	1380
Liver	1.4	7050	3550	230	2400
Stomach	3.1	3760	3170	660	5300
Small intestine	1.5	4080	2620	2220	560
Large intestine	1.4	3030	4400	1300	2300
Rectum	1.2	3600	4500	2300	1100
Kidney	2.0	4900	2700	1000	1680
Bladder	0.2	2300	2200	600	1200

pressible misery at the age of 48 is shown in Table 1 and Fig. 8. She had lived and died at her home without any treatment in the most polluted part of the endemic area, drinking the river water together with poisons which came from the mine and eating the damaged rice. The data were carefully checked with repeated analysis by spectrographic, highly sensitive square-wave polarographic and/or later by atomic absorption methods. Of course, as the content of the metals in ash weight must have been changed during several years' preservation in formalin solution at Hagino's hospital, the data cannot show the original content at the time of the autopsy. However, the reduced content of bone ash (obtained at $450°C$), indicating severe de-calcification, the extraordinary accumulation of those very poisonous metals found in the wastewater from the mine, and the high cadmium-to-zinc ratio in the victim's tissues were clearly noticed (see also Table 7).

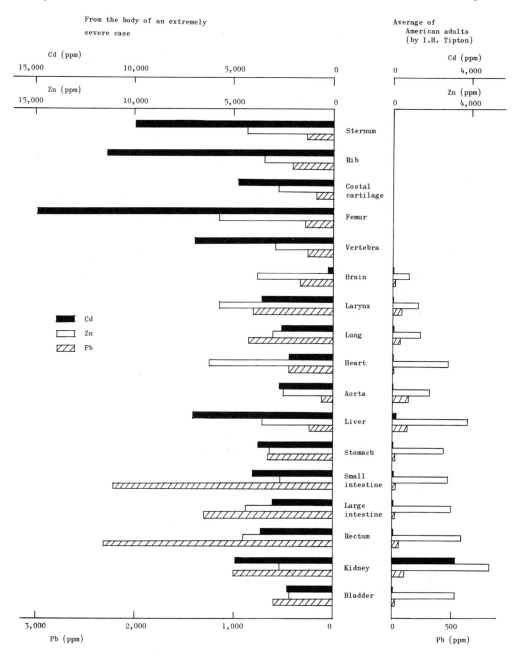

FIG. 8. Comparison of Cd, Zn, and Pb content in the tissues of an itai-itai disease victim and those of average Americans (ppm ash weight).

Furthermore, the data indicate that the patient had been affected by metals directly from the mine, by drinking river water containing the fine suspended particles which had flowed down from the flotation tailings, in addition to eating the heavily contaminated rice which tends to accumulate more of the cadmium to excessive content than of the other two poisonous metals, lead and zinc. Except for the data from Kobayashi, no other analysis was ever reported on such a severely affected victim of the itai-itai disease who had suffered very high and long exposure to the poisonous metals. Figure 8, in which the concentrations of cadmium, zinc, and lead in the patient's tissues are compared with those of Americans tested by Tipton [41], illustrates that the accumulation of cadmium among the three kinds of metal was greatest in the damaged bones and the liver. In the kidney and the brain, the cadmium content was not so high as compared with Tipton's data. In contrast to the mercury of the Minamata disease, neither an accumulation of cadmium in the brain nor excretion of it through the hair [20,42] has been observed. So hair analysis is useless for examining the effects on health of cadmium exposure. However, the phosphorus content in the patients' hair was found to be remarkably decreased [20], showing the same tendency as in the blood serum.

Later, Ishizaki et al. [43] reported the cadmium and zinc content in organs determined by the atomic absorption method from autopsies on five itai-itai patients, who died from other illnesses after treatment with large doses of vitamin D over a number of years, during which they took no more polluted rice or water. The results are given in Table 2. At the same time they [43] analyzed the liver and kidney from 41 controls who were autopsied at Kanazawa Medical School, after living in nonendemic areas in or near Kanazawa; the results are shown in Table 3. By comparing the two tables, no such heavy accumulation of metals as reported by Kobayashi was seen. However, the cadmium content of the liver was several times higher in itai-itai patients than in the controls because the average content of the former was 99 ppm wet weight, while that of the latter in the same age group for females was only 20 ppm. In contrast, the cadmium content in the kidney, where the highest values of cadmium deposit would be expected to appear, was found to be much lower in the itai-itai patients than in those of the controls. The authors stated that the low values in the patients' kidneys must be due to the release of cadmium through advanced renal damage.

TABLE 2. Cadmium and Zinc Content in Organs from Female Itai-itai Patients (ppm wet weight)[a]

| | Cd | | | | | Zn | | | | | Zn/Cd |
	Patient 1	Patient 2	Patient 3	Patient 4	Patient 5	Patient 1	Patient 2	Patient 3	Patient 4	Patient 5	Average
Age	79	72	61	67	73	79	72	61	67	73	
Cause of death	Stomach cancer	Bronchial pneumonia	Endocarditis verrucosa	Uremia	Stomach cancer	Stomach cancer	Bronchial pneumonia	Endocarditis verrucosa	Uremia	Stomach cancer	
Analysis:											
Liver	94	118	63	132	89	158	143	113	210	128	1.5
Renal cortex	41	32	20[b]	12	--	54	25	25[b]	28	--	1.4
Renal medulla	40	26	--	10	--	52	21	--	20	--	1.4
Lung	--	2.5	8.0	2.1	--	--	11	21	11	--	4.1
Spleen	--	6.8	6.2	6.0	--	--	19	21	33	--	3.9
Pancreas	45	65	--	52	--	62	57	--	72	--	1.2
Stomach	--	--	--	--	4.8	--	--	--	--	20	4.2
Small intestine	--	3.0	12	9.9	5.7	--	12	17	31	21	3.1
Large intestine	--	1.7	12	--	--	--	8	26	--	--	3.5
Ribs	--	--	2.8	2.6	--	--	--	51	53	--	19
Bone cortex	--	1.7	--	--	--	--	42	--	--	--	25
Bone marrow	--	1.1	--	--	--	--	6	--	--	--	5.5
Skin	--	4.6	5.1	--	3.9	--	10	11	--	6	2.0
Muscles	--	14	--	8.3	--	--	86	--	108	--	9.6
Brain	0.6	--	--	--	--	11	--	--	--	--	18

[a]From Ishizaki et al. [43].
[b]Cortex and medulla are not separated.

TABLE 3. Cadmium and Zinc Content in Liver and Kidney in Controls (ppm wet weight)[a]

| | | Cd | | | | | | Zn | | | | | | Zn/Cd | |
| | | Male | | | Female | | | Male | | | Female | | | Male | Female |
	Age at Death	No. of Samples	Average Content	S.D.[b]	No. of Samples	Average Content	S.D.[b]	No. of Samples	Average Content	S.D.[b]	No. of Samples	Average Content	S.D.[b]	Average	Average
Liver	20-39	3	6.6	4.4	2	13	--	3	128	76	2	77	--	19	5.9
	40-59	3	17	19	6	8.3	4.7	3	145	57	6	62	26	8.5	7.5
	60-79	14	9.9	4.9	6	20	7.7	14	106	64	6	75	21	11	3.8
	80-	2	11	--	5	19	9.9	2	59	--	5	124	62	5.4	6.5
Renal cortex	20-39	3	87	48	2	66	--	3	53	15	2	75	--	0.61	1.1
	40-59	3	87	37	6	99	20	3	99	26	6	75	30	1.1	0.76
	60-79	14	90	36	6	108	31	14	73	30	6	60	13	0.81	0.56
	80-	2	56	--	5	66	33	2	57	--	5	62	51	1.0	0.94
Renal medulla	20-39	3	39	31	2	38	--	3	42	20	2	58	--	1.1	1.5
	40-59	3	36	25	6	44	22	3	50	14	6	44	13	1.4	1.0
	60-79	14	49	29	6	55	28	14	49	25	6	37	10	1.0	0.67
	80-	2	27	--	5	33	16	2	29	--	5	50	35	1.1	1.5

[a]From Ishizaki et al. [43].
[b]S.D. = standard deviation.

CADMIUM CONCENTRATIONS IN RICE FROM
THE AFFECTED AND UNAFFECTED AREAS

It was stated in the second section that during the war in the itai-
itai disease area there had been damage to rice growth caused by the poisonous metals carried by the Jinzu River from the mining area upstream.
Some of the earlier data on the content of cadmium, zinc, and lead in the
paddy soil, rice roots, and polished rice in the affected area, as compared
with a nonaffected one obtained by Kobayashi [20] are reproduced in Table 4.
The latter area was selected along the basin of the Ida River, a tributary
of the Jinzu River, where the itai-itai disease was not found. The samples
were collected in 1961 and analyzed by a highly sensitive polarographic
method after dithizone extraction or spectrographic analysis, as the atomic
absorption method had not yet been developed in those days. From the results
in the table, it is clear that a large amount of poisonous metals still remained in the affected paddy fields even after the marked damage to rice
crops had disappeared. Taking 0.5% of the ash weight of polished rice into
account, the cadmium content in wet rice from the polluted area was 0.08-
1.25 (average 0.63) ppm, while that from the control area (the Ida River
basin) was 0.05-0.37 (average 0.15) ppm (also see Figs. 9 and 10). Furthermore, the table indicates that the cadmium content (6 ppm) in the polluted paddy soil was concentrated to as much as 200 or 20 times, respectively,
in the ash of rice roots and of polished rice grains. Thus the rice plant,
in absorbing minerals from the soil and water, tends to accumulate cadmium.
During the war, when the damage to rice growth had been so great due to the
faulty treatment of the wastewater of the mine, the soil and rice of the
affected area must have been much more contaminated.

In another earlier experiment to study the cadmium content in typical
Japanese rice, which is the staple food of the nation, Moritsugu and Kobayashi [44] collected more than 200 samples of polished rice from 41 prefectural agricultural experiment stations all over Japan. Rice samples collected from affected areas and those sent from the U.S. Rice Experiment
Station (Louisiana) and from other countries were added in this study.
These samples were collected from 1959 to 1961, before and after the analytical findings from the itai-itai disease, and were analyzed by the colorimetric method and checked by a highly sensitive square-wave polarographic
procedure after dithizone extraction as reported by Saino and Kobayashi
[45]. The results are shown in Figs. 9 and 10. By comparing Figs. 9 and

TABLE 4. Earlier Data Comparing Cd, Zn, and Pb Content (ppm ash weight) in Paddy Soil, Rice Roots, and Rice from Various Parts of the Affected Area with Those from Control Area in Toyama.

	Paddy Soil	Rice Root		Polished Rice	
	Affected Area	Affected Area	Control Area	Affected Area	Control Area
No. of samples	5	5	2	17	6
Mineral:					
Cd (range)		760–2200	4–320	16–250	10–74
(average)	6	1250	162	125	30
Zn (range)		1600–4000	1600–2100	2200–7500	2700–4100
(average)	1125	2600	1850	4700	3250
Pb (range)		100–1500	29–43	4–60	0.6–13
(average)	348	810	36	22	5

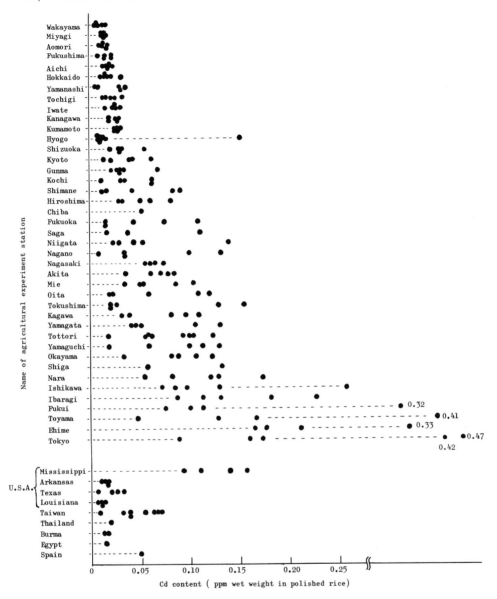

FIG. 9. Cd content in rice produced from various agricultural experiment stations all over Japan (1959) and from other countries (1960).

10 it is clear that the samples from polluted areas such as the Jinzu River basin are strikingly high in cadmium content as compared with the control areas. Further the results of this study indicated that (1) the content of cadmium in rice from the control areas differed more greatly with locality

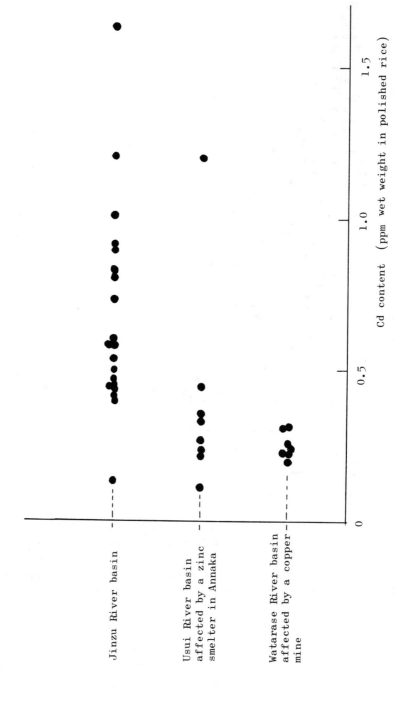

FIG. 10. Some of the earlier data of Cd content in rice collected in 1961 from various parts of the affected areas.

than with the nature of the soil or variety of rice; (2) the general average content of cadmium in all the polished Japanese rice included in Fig. 9 was 0.066 ppm wet weight with 0.005 ppm of standard deviation, or in another figure, about 0.07 ppm; (3) the Japanese rice had more cadmium than that from the United States (Arkansas, Texas, and Louisiana, with the exception of Mississippi), Taiwan, Thailand, Burma, Egypt, and Spain; and (4) the rice bran and the embryo bud contained more cadmium than the polished rice.

In Fig. 9, however, it is illustrated that among 200 samples collected from all over Japan only eight samples contained more than 0.2 ppm of cadmium. Later, some of them were found to have been affected by factories. Therefore, it is quite reasonable to insist that the safety standard for cadmium in rice should be lower than that set by the Ministry of Health and Welfare (1970) at 1.0 ppm for unpolished and 0.9 ppm for polished rice. An experiment on mice within the limits of the present standard showed that the more cadmium there was in the rice for their diet, the more cadmium concentrated in the kidneys and livers (Kobayashi et al. [46]).

CLOSER OBSERVATIONS AT THE ENDEMIC AREA

In 1967, Fukushima et al. [47] analyzed the geographical distribution of cadmium concentration in the surface soil (to the depth of 13-18 cm) collected from 34 paddy fields irrigated by the Jinzu River and 16 adjacent fields irrigated from other streams (Fig. 11). They found a striking correlation [47,30] between the distribution pattern of cadmium in the soil and the prevalence rate of patients and patient suspects of the itai-itai disease (see Table 5 and Fig. 12) found in the endemic area. The higher concentration of cadmium found in the surface soil and near the irrigation inlet of each field than in the deeper soil or at the end farthest from the inlet suggested that the cadmium in the fields had been deposited after being transported in the form of suspended particles rather than in solution through the irrigation water from the mining area upstream of the Jinzu River.

In 1968, Fukushima et al. [49] analyzed rice samples collected from 77 farmhouses in different parts of the endemic and control areas, and the distribution of cadmium concentration in rice as shown in Fig. 13 was found to be similar to that shown in Figs. 11 and 12. Thus, a correlation [50] between the geographical distribution of the prevalence of the disease and the cadmium contents in the soil and rice in the endemic area was established

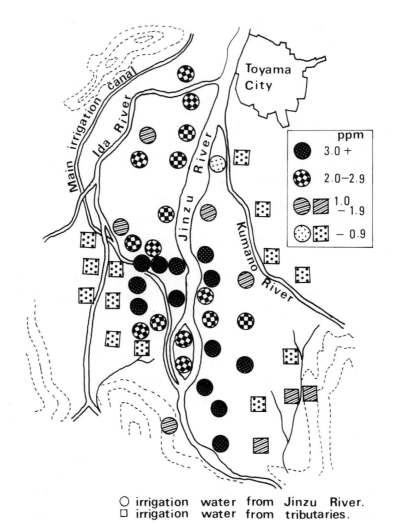

○ irrigation water from Jinzu River.
□ irrigation water from tributaries.

FIG. 11. Distribution of Cd in rice field soil (surface layer) (from Fukushima et al. [47] or Yamagata and Shigematsu [30]).

by their studies in 1967-1968, even though by that time the pollution by
poisonous metals had been considerably decreased.

As described in the third section, the urinary findings in the patients
of the itai-itai disease had shown proteinuria, glucosuria, and aminoaciduria. These are considered signs of renal dysfunction. An increase in
cadmium excretion was always found. Urinary calcium excretion was about
normal while urinary phosphate was reduced. So the urinary Ca/P ratio in
the patients was higher than normal. After an epidemiological investigation

TABLE 5. Results of a 1967 Epidemiological Study[a,b]

		Total	I[c]	i[d]	(i)[e]	Oob[f]
	Total	234	50	17	31	136
District	Fuchu-machi	132	28	11	14	79
	Toyama City	61	16	3	8	34
	Osawano-machi	31	6	3	6	16
	Yatsuo-machi	10	0	0	3	7
Sex	Male	42	1	1	3	37
	Female	192	49	16	28	99
Age	30-44	2	0	0	0	2
	45-59	36	6	0	2	28
	60-74	158	40	10	22	86
	75-	38	4	7	7	20
	Mean age		66.6	72.2	68.3	66.1

[a]From Ishizaki and Fukushima [48].

[b]In later years, the numbers have been modified.

[c]I = a clear case of the itai-itai disease (typical bone signs on x-ray).

[d]i = deeply suspected (bone signs but not typical).

[e](i) = suspected (slight bone signs).

[f]Oob = need for further observation (urinary and/or blood signs).

carried out in 1967, Ishizaki [51] reported that the excretion of protein-uria and glucosuria and the simultaneity of these two excretions among people of older age groups were more common in the endemic than in the control area. This trend of simultaneous excretion of proteins and glucose serves as an indicator for renal tubular dysfunction. They also noticed that the urinary Ca/P ratio in the inhabitants over 30 years of age was higher in the polluted than the unpolluted area.

By separating the urinary proteins excreted from the patients and suspected patients by electrophoresis, gel filtration, and other methods, Fukushima and Sugita [52,50], Kubota et al. [53], and others obtained a typical tubular protein pattern like that which Piscator [54] demonstrated in workers exposed to cadmium, as cited by Friberg et al. [55]. The main part of the urinary proteins had molecular weights less than that of albumin.

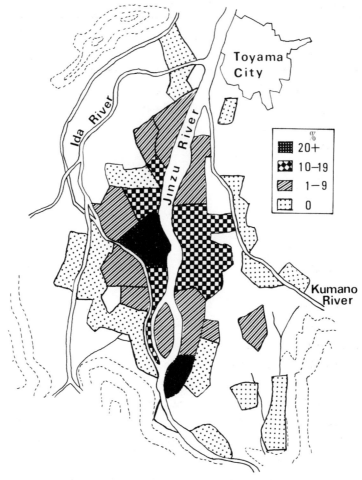

FIG. 12. Distribution of itai-itai patients (percentage of women over 50 years of age) (from Ishizaki and Fukushima [48] or Yamagata and Shigematsu [30]).

        The increased urinary excretion of different kinds of amino acid was also tested in the itai-itai disease patients. Toyoshima et al. [56] reported that the increases of proline and hydroxyproline are the most noticeable among the urinary amino acids found in the patients. Fukushima [50] observed a considerably increased excretion of proline in some inhabitants of the endemic area even with no bone changes observed on x-ray.

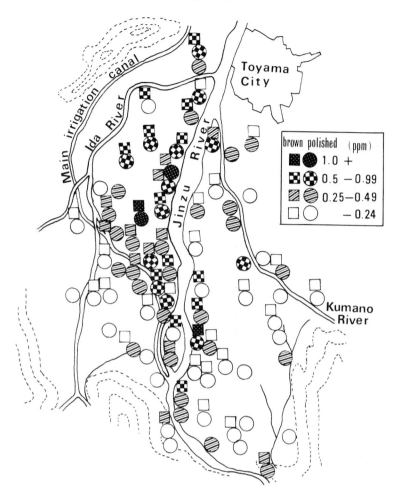

FIG. 13. Distribution of Cd in ordinary (nonglutinous) rice (from Fukushima et al. [49] or Fukushima [50]).

DAILY INTAKE OF CADMIUM IN AFFECTED
AND NONAFFECTED AREAS

How much cadmium is ingested daily in a typical Japanese diet, and how does this figure compare with the amounts of cadmium that were ingested by inhabitants in the polluted areas?

A study group conducted by Shigematsu [30] established the daily cadmium intake in a typical Japanese diet at about 60 μg, based on the results of the National Nutrition Survey in 1966 and on available data of cadmium

TABLE 6. Normal Concentration of Cadmium in Japanese Foodstuffs[a]

Category	Item	No. of Samples	Locality	Cd (ppm/fresh)	Cd (ppm/dry)
Cereals	Wheat flour	2	Unknown[b]	0.020	0.023
	Bread, white	6	Unknown[b]	0.046	0.072
Vegetables	Potato	3	Hokkaido and Kanazawa	0.038	0.18
	Sweet potato	3	Kanazawa and others	0.015	0.044
	Onion	2	Hokkaido	0.007	0.11
	Carrot	3	Aichi and Ishikawa	0.041	0.42
	Cabbage	3	Kanazawa and others	0.009	0.12
	Chinese cabbage	3	Kanazawa and others	0.022	0.51
	Cucumber	3	Kanazawa and others	0.007	0.19
	Tomato	4	Kanazawa and others	0.032	0.63
	Beans	7	Hokkaido	0.026	0.029
Fruit	Apple	5	Aomori	0.004	0.030
	Mandarin orange	2	Fukuoka and Wakayama	0.003	0.026
Meat, eggs, and milk	Beef	2	Unknown	0.054	0.16
	Beef liver	2	Tochigi	0.097	0.33
	Pork	2	Unknown	0.016	0.05
	Pork liver	2	Unknown	0.055	0.17
	Chicken	3	Unknown	0.027	0.093
	Chicken liver	3	Unknown	0.20	0.55
	Chicken viscera	3	Unknown	0.15	0.53
	Egg yolk	3	Unknown	0.029	0.055

Category	Item		Origin		
	Egg white	2	Unknown	0.002	0.024
	Milk	3	Kanazawa	0.003	--
Meat of fish and shellfish	Sea fish	19	Japan Sea and Pacific	0.010	0.040
	Freshwater fish	8	Kanazawa	0.003	0.067
	Cuttlefish, dried	3	Japan Sea	0.075	0.31
	Cuttlefish with viscera (Shiokara)	4	Unknown	0.38	0.99
	Octopus	2	Japan Sea	0.005	0.032
	Clam	2	Ibaragi and Chiba	0.19	0.84
	Short-necked clam with viscera (Tsukudani)	2	Osaka and Kagawa	0.83	1.55
	Oyster with viscera	3	Ishikawa	0.62	3.05
	Shrimp	3	Japan Sea	0.032	0.18
	Crab	3	Japan Sea	0.054	0.23
Seaweed	Laver, dried	2	Unknown	0.11	0.13
	Tangle, dried	2	Hokkaido	0.11	0.16
Miso		3	Kanazawa and others	0.068	0.14
Drinks	Sake	6	Kanazawa	0.002	--
	Beer	2	Unknown	0.0005	--
	Orange juice	2	Unknown	0.001	--
	Cola	2	Unknown	0.0005	--
	Tomato juice	4	Unknown	0.013	--

[a]From Ishizaki et al. [57].
[b]From imported wheat (?).

content in various kinds of foodstuff obtained by Ishizaki et al. [57] as
tabulated in Table 6. A breakdown of the 60 μg of cadmium comes from a
mean intake of 335 g of polished rice. The normal concentration of cadmium
in polished rice from all areas in Japan has shown an average value of 0.07
ppm (see the fifth section).

By assuming the cadmium content of rice harvested in the affected area
of the Jinzu River basin to be 1.0 ppm and other food items which were produced and consumed locally to be contaminated to as high as 10 times the
normal values, the study group suggested the daily intake of cadmium during
recent years in the endemic area must have been 600 μg or 10 times the normal intake. Yamagata and Shigematsu [30] concluded that the daily intake
of this amount of cadmium would never cause the itai-itai disease, as they
supported the safety standard for cadmium set by the National Government
at 1.0 ppm for unpolished and 0.9 for polished rice.

However, Friberg et al. [55] suggested that 600 μg cadmium per day
would cause renal dysfunction after about 10 years with an assumed body
retention rate of 5% of the ingested cadmium, and so the cadmium exposure
in the endemic area, even now, is sufficient to cause renal damage. This
estimate was based on their conclusion that a characteristic symptom of
renal damage is urinary excretion of tubular-type proteins with low molecular weights, and that when a cadmium level of about 200 ppm wet weight is
reached in the renal cortex, the first sign of tubular dysfunction (tubular
proteinuria) may appear in sensitive persons.

Furthermore, it must be noticed that the dietary intake of calcium
which has a protective effect against cadmium has been very low in Japan.
Since polished rice, which forms more than half of the Japanese diet on a
dry basis, contains only 6 mg [58] of calcium in 100 g, the daily intake
of calcium from 300 g of polished rice is only 18 mg, while that of cadmium
from the same amount of normal rice is 21 μg. The ratios of Ca to Cd and
Ca to Zn in normal rice are about 900 and 3.5, respectively, and in polluted
rice with 1.0 ppm of cadmium, the former ratio comes down to 60. Thus,
calcium content is very low in rice.

To investigate the influence of calcium deficiency on the accumulation
of cadmium in organs, Kobayashi et al. [46] gave normal rice (0.1 ppm of
cadmium) and polluted rice (0.6 ppm) to mice on low and normal calcium intakes. After 75 weeks of experiment, it was found that animals on a low-
calcium diet had retained from about 1.5 times (in the polluted rice groups)
to 3-4 times (in the normal rice groups) as much cadmium in liver and kidney
concentrations as those on a normal calcium diet.

TABLE 7. Geographical Variation in Cd and Zn Concentrations in Human Tissues (median values, ppm ash weight, from males 20-59 years of age)[a]

Tissue	Element	United States	Africa	Near East	Far East	Zn/Cd (Far East)
Aorta	Zn	1800	1500	2000	2500	50
	Cd	< 50	< 50	< 50	50	
Brain	Zn	740	680	780	920	--
	Cd	< 50	< 50	< 50	< 50	
Heart	Zn	2500	2000	2200	2800	--
	Cd	< 50	< 50	< 50	< 50	
Kidney	Zn	4500	3100	3300	6000	1.4
	Cd	3200	810	1600	4200	
Liver	Zn	3400	4000	3800	7200	15
	Cd	180	50	140	480	
Lung	Zn	1300	1200	1600	1600	30
	Cd	50	< 50	< 50	50	
Pancreas	Zn	2200	2300	1900	2800	15
	Cd	82	46	76	190	
Spleen	Zn	1300	1300	1500	1700	30
	Cd	< 50	< 50	< 50	50	
Testis	Zn	1300	1900	1700	1600	30
	Cd	< 50	< 50	< 50	50	

[a]From Tipton et al. [59].

Tipton et al. [59] reported that the Far East group of tissue samples collected from Japan and Hong Kong analyzed by emission spectrograph had median values of cadmium significantly higher than groups from the United States, Africa, and the Near East in liver, kidney, pancreas, and other tissues (Table 7). Friberg et al. [55] concluded that the body burden of cadmium at the age of 50 in noncontaminated areas in Europe is probably about 15-20 mg, in the United States 30, and in Japan 40-80 mg. Thus,

even in so-called nonpolluted areas in Japan the amount of cadmium retained
in the body is considerably higher than elsewhere.

These results may be attributed to the higher content of cadmium in
Japanese rice (see Fig. 9) and other home products and to the deficient
intake of calcium. From the results of a pot-culture experiment on rice
and wheat (see the ninth section) and from analytical data on the products
from the polluted area in Annaka near a smelter (see following section),
one might expect the "normal" levels of cadmium in wheat and barley to be
higher than in rice. In Table 8 some of the analytical data on foodstuffs
and a comparison between home-produced and imported wheat are given. In
former days, the people depended much more on rice together with home-produced barley (as mugi-meshi) and wheat (as udon noodles) in addition to a
lower daily intake of calcium (probably less than 300 mg). Such dietary
conditions must have contributed to the higher body burden of cadmium.
However, in recent years, when dietary habits have been much influenced by
Western standards and most of the wheat and barley are imported, a lower
intake of cadmium together with an increased intake of calcium, proteins,
and vitamins can be expected. Actually, it is reported that the mean daily
intake of calcium in Japan [60], which was 389 mg in 1960, has risen to
536 mg in 1970. This amount is about twice that of earlier days up to
roughly 1950.

Fukushima in 1972 [61] estimated the average daily intake of cadmium
in nonpolluted areas of Japan to be 47 µg instead of the 60 µg which had
been reported by Yamagata and Shigematsu [30]. Analyses of cadmium in
foodstuffs (see Table 6) have shown that cadmium concentrations on a wet-
weight basis are generally less than that of rice (0.066 ppm). Fish meat,
most of which comes freshly from oceans far from Japan, and flours from
imported wheat would have the same cadmium concentrations as those of other
countries. Shellfish, such as oysters and clams, contain appreciable amounts
of cadmium, but their approximate yearly total consumption is only 140,000
tons. Compared to rice (10 million tons), flour from imported wheat (4
million tons), and fish meat (5 million tons), the contribution of shellfish
to the daily average cadmium intake in Japan would be negligible, except in
certain localities. So at present the cadmium intake in nonpolluted areas
may have decreased so as to be almost the same as that of Americans or
Europeans if we take into consideration the fact that the quantity of diet
should be proportional to body weight.

TABLE 8. Normal Concentrations of Cd and Zn in Foodstuffs (Imported Wheat Included)

Category	Item	Locality and Description	Cd	Zn	Refs.
Cereals (ppm wet weight)	Wheat, imported	Canada, ICW	0.045	22.0	F. Morii (unpublished)
		U.S.A., American soft	0.028	15.3	
		U.S.A., H.W.	0.053	19.0	
		Argentine, ARG	0.013	19.6	
		Australia, FAQ	0.015	13.4	
		Mean	0.031	17.9	
	Wheat, home produced	Kitami-shi, Hokkaido	0.029	14.3	
		Nishinasuno-cho, Tochigi Prefecture	0.070	19.3	
		Higashiibaragi-gun, Ibaragi Prefecture	0.037	15.9	
		Konosu-shi, Saitama Prefecture	0.094	26.5	
		Konosu-shi, Saitama Prefecture	0.139	27.7	
		Sanyo-cho, Okayama Prefecture	0.125	31.5	
		Kayo-cho, Okayama Prefecture	0.118	44.3	
		Kurashiki-shi, Okayama Prefecture	0.172	49.2	
		Fukuyama-shi, Hiroshima Prefecture	0.054	18.0	
		Fukuyama-shi, Hiroshima Prefecture	0.049	20.0	
		Kishima-gun, Saga Prefecture	0.083	22.8	
		Mean	0.088	26.3	
	Flour from imported wheat	Canada, ICW, for bread	0.027	4.6	
		U.S.A., American soft, for cake	0.017	3.4	

TABLE 8 (Continued)

Category	Item	Locality and Description	Cd	Zn	Refs.
		U.S.A., H.W., for bread	0.022	4.5	
		Australia, FAQ, for <u>udon</u> noodles	0.012	2.8	
		Mean	0.020	3.8	
	Flour from home-produced wheat	Kōnosu-shi, Saitama Prefecture	0.075	7.7	
		Kōnosu-shi, Saitama Prefecture	0.075	7.3	
		Fukuyama-shi, Hiroshima Prefecture	0.036	5.6	
		Fukuyama-shi, Hiroshima Prefecture	0.056	9.4	
		Mean	0.061	7.5	
Green vegetables (ppm dry weight)	Cabbage, inner leaves	Okayama Prefecture	0.11	54[a]	Kobayashi et al. [71]
	Same cabbage, outer leaves	Okayama Prefecture	0.13	54[a]	
	Chinese cabbage, inner leaves	Okayama Prefecture	0.55	78[a]	
	Same cabbage, outer leaves	Okayama Prefecture	0.62	55[a]	
	Welsh onion	Okayama Prefecture	0.11	67	
	Onion	Okayama Prefecture	0.17	41	
Root crops (ppm dry weight)	Potato	Okayama Prefecture	0.10	17	
	Carrot	Okayama Prefecture	0.28	48	
	Taro potato	Okayama Prefecture	0.38	18	
	Scallion	Okayama Prefecture	0.27	35	

Fruit vegetables (ppm dry weight)				
Tomato	Okayama Prefecture	0.30	37	Kobayashi et al. (unpublished)
Pumpkin	Okayama Prefecture	0.20	22	
Fruit of eggplant	Okayama Prefecture	0.93	36	
Persimmon	Okayama Prefecture	0.06	29	
Shellfish (ppm dry weight)				
Oyster with viscera	Hiroshima (Tanna) Prefecture	1.83	1920	
Oyster with viscera	Hiroshima (others) Prefecture	4.84	1640	
Oyster with viscera	Iwate Prefecture	3.73	840	
Seaweed (ppm dry weight)				
Laver, dried	Fukuoka and Saga Prefectures	1.46	52	Kobayashi et al. (unpublished)

a From the same sample.

Another approach to the cadmium intake is possible by determining the amount of cadmium excreted in feces and urine, with which 90-95% of ingested cadmium would be excreted. Since Japan is lagging behind the West in flush toilets and sewerage systems, except for the central parts of cities, the "night soil" which is a mixture of feces and urine is stocked in each household for several weeks to be collected by tank trucks with vacuum pumps for treatment by activated sludge plants. Kobayashi et al. [62] collected the mixed night soil once a month, 12 times throughout the year 1968, from the treating plants of eight towns located in the northeast part of Japan. Among the various kinds of mineral components determined, the data on cadmium and zinc are shown in Table 9. The average content of cadmium was found to be 36 µg/l, while that of zinc was 10 mg/l; thus the ratio of Zn/Cd was 290. The next experiment [63] was carried out in 1974. We chartered vacuum tank trucks to collect the night soil from six areas of Okayama Prefecture, including a city and farm villages, and obtained the same mean value on cadmium, 36 µg/l (see Table 10). As the night soil is stocked at the rate of about 1.0-1.2 liters per person per day, and as several percent of ingested cadmium might be retained in the body, the mean

TABLE 9. Cd and Zn Concentration in the Night Soil Collected 12 Times in 1968 from the Northeast of Japan

Locality of Collection	Cd (µg/l)	Zn (mg/l)	Zn/Cd
Miyagi Prefecture			
Kurokawa	24	7.5	310
Furukawa	33	11	330
Kesennuma	13	4.6	350
Shizukawa-Utazu	52	18	350
Iwate Prefecture			
Ichinoseki	31	10	320
Miyako	40	8.7	220
Akita Prefecture			
Yuzawa	53	11	210
Yokote	39	8.9	230
Average	36	10	290

TABLE 10. Cd and Zn Concentration in the Night Soil Collected in 1974 from Okayama Prefecture[a]

Locality	Date of Collection (1974)	No. of Households	No. of People	Cd (μg/l)	Zn (mg/l)	Zn/Cd
Kurashiki City, residence sections	Feb. 6	20	69	32	6.1	190
Kurashiki City, farming area with side jobs	Feb. 20	12	50	38	6.6	170
Aba-son, farming village	Apr. 15	11	54	37	12.4	340
Yatsuka-son, farming village	Apr. 16	11	56	36	8.6	240
Osa-cho, farming village	Apr. 17	13	70	38	11.6	310
Average				36	9.1	250

[a]From Kobayashi and Morii, unpublished.

daily intake of cadmium from the first and second studies would come to approximately 42 μg, and thus about one half of this must have come from normal rice. From the results of Fukushima and Kobayashi et al., it was found that the recent daily intake has markedly decreased due to the fact that most of the food materials, except for rice and vegetables, are imported and, in addition, due to the reduced intake of rice, especially among younger people influenced by Western diet.

What different result might be obtained if such a study of night soil were extended to the polluted areas? Table 11 gives some of the results of a study carried out by Kobayashi [64,69] in the older endemic area of the Jinzu River basin as well as in the new polluted Annaka area by the largest zinc smelter of Japan. The average concentration of cadmium in samples collected in the Jinzu River basin from 10 farmhouses (51 families in total) which had itai-itai disease patients was 280 μg/l, so the daily cadmium intake was assumed to be about 320 μg. This figure, which is seven times as much as that of the nonpolluted areas, was found to be about one half of the 600 μg estimated by the study group conducted by Shigematsu [30]. This dissimilarity may be explained by the fact that cadmium-polluted products in the endemic area were limited to rice and soya beans since paddy

TABLE 11. Cd and Zn Concentration in the Night Soil Collected in 1969 and 1970 from Inhabitants Exposed to Cd Pollution

Area	Collection Date	No. of Farmhouses	No. of People	Cd (μg/l)	Zn (mg/l)	Zn/Cd
Jinzu River basin, Toyama Prefecture	June 1969	10	51	280	15.0	54
Regions south of the smelter, Annaka City, Gunma Prefecture	June and Oct. 1969	12	62	690	52.2	76
Regions south of the smelter, Annaka City, Gunma Prefecture	June 1970	9	57	270	24.4	90
Regions east of the smelter, Annaka City, Gunma Prefecture	June and Oct. 1969	9	40	910	58.1	64
Regions east of the smelter, Annaka City, Gunma Prefecture	June 1970	8	33	620	44.5	72

fields were the only fields that had been polluted by irrigation from the Jinzu River. No crops or vegetables were grown in the paddy fields there except rice and soya beans. Soya beans were planted at the edge of the fields. Therefore, the estimation by Shigematsu, in which various kinds of local products were assumed to be contaminated to as high as 10 times the normal values, was of course an overestimate. During World War II, however, when Kobayashi [17] observed the great damage to rice crops and the marked chlorosis on rice leaves, the daily cadmium intake from rice and from drinking the poisonous river water was by far more serious.

In contrast, the Ministry of Health and Welfare [65] had underestimated the daily cadmium intake in the recently but very rapidly contaminated farm area in Annaka City of Gunma Prefecture, where the cadmium and zinc concentrations in soil and products (see the following section) are much higher than elsewhere among the polluted areas of Japan. This pollution has come from the biggest zinc smelter in Japan, which has rapidly increased its production in recent years. By neglecting the airborne pollution which is the primary cause of contamination in the vicinity of the smelter, the study group (conducted by Shigematsu) [66] selected a "control" area and affected areas in paddy fields up- and downstream from the smelter; however, since this control area was actually affected by polluted air, little difference was found in cadmium content of rice between control and affected areas. The Ministry [65] without any studies on the serious contamination of crops and vegetables occurring in the nonpaddy fields which extend over the hilly regions adjacent to the smelter, calculated the daily cadmium intake to be 400 µg by a method similar to that reported by Yamagata and Shigematsu [30], and it concluded that this level of intake would not induce the itai-itai disease in the immediate future. However, this report was discussed in the newspapers [67] and in the House of Councilors of the Japanese Diet [68] for underestimating the daily intake by neglecting airborne cadmium.

The night soil analysis performed in 1969 by Kobayashi and Muramoto [64,69] gave higher results, as shown in Table 11. The average content of cadmium in samples collected in 1969 from nine farmhouses within 1 km of the smelter to the east was 910 µg/l, and that of 12 farmhouses to the south was 690 µg/l. Thus, it was demonstrated that farmers in the polluted area in Annaka City were to a serious degree ingesting more cadmium from oral and pulmonary routes than that estimated by the Ministry, and that the pollution was higher to the east of the smelter, where it was blown by the west wind, than in the south.

Again in June 1970, a similar investigation [64,70] was repeated in Annaka. The average cadmium content for eight farmhouses to the east of the smelter was 620 µg/l, and 270 µg/l for nine farmhouses to the south (a detail is given in Table 12). Compared with the previous year, it was observed that the cadmium ingestion had decreased. This must be due to the effect of warning the inhabitants not to eat the heavily polluted agricultural products and also to improvements and reconstruction made inside the factory to prevent air pollution. At present, a further reduction in the intake of cadmium would be expected, as the farmers are more cautious about cadmium and the dust collection in the smelter has been more improved. However, farmers in the vicinity of the smelter are still ingesting much more cadmium than elsewhere in Japan since the cadmium, zinc, and lead concentrations in wheat, barley, cabbages, etc., on which they are depending even more than on rice, are seriously high as described in the next section. In any case, though there have been no symptoms of itai-itai reported in Annaka, more complete epidemiological studies of the people are obviously needed.

TABLE 12. Cd and Zn Concentration in the Night Soil Collected in June 1970 from Inhabitants Exposed to Cd Pollution in Annaka City, Gunma Prefecture

Area	No. of Farmhouses	Distance (m) and Direction from the Chimney	No. of People	Cd (μg/l)	Zn (mg/l)
Regions south of the smelter	1	600 S	6	440	36.5
	2	800 WS	8	150	14.0
	3	800 SES	7	280	19.1
	4	400 S	6	110	6.2
	5	400 SW	7	580	63.2
	6	450 SSE	2	570	51.6
	7	500 SES	6	260	22.2
	8	500 SE	9	40	3.5
	9	450 S	6	20	3.1
Average				270	24.4
Regions east of the smelter	1	800 E	5	650	57.5
	2	850 E	4	50	8.4
	3	800 EN	2	660	44.2
	4	800 E	5	440	24.6
	5	800 E	6	180	23.2
	6	800 E	2	1020	69.8
	7	850 E	2	730	65.6
	8	900 E	7	1250	62.5
Average				620	44.5

ENVIRONMENTAL POLLUTION BY CADMIUM ATTRIBUTED
TO THE LARGEST ZINC SMELTER OF JAPAN

The discovery of the cause of the disease in the Jinzu River basin led to the examination of other areas that could be contaminated by cadmium. One such area was adjacent to a zinc, lead, and cadmium mine on Tsushima Island (see the second section). Another area chosen for study by Kobayashi et al. [64,71,72] was Annaka City, where Japan's largest zinc smelter is located. Here about 500 samples of soil and 80 samples of agricultural products were analyzed for cadmium, zinc, and lead. A high correlation was found to exist between (1) the content of cadmium and other metals in soil,

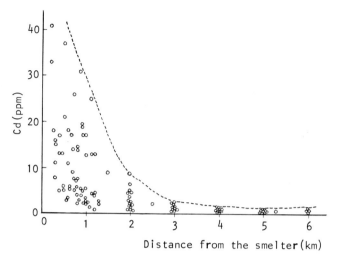

FIG. 14. Relation between the content of cadmium in surface soil (0-10 cm, dried) and the distance in all directions from the smelter, Annaka City. Collected in 1970.

mulberry leaves (the feed of silkworms, which are extensively cultivated by farmers in this district), and wheat flour and (2) the distance from the smelter stack (see Figs. 14-17). In other words, the closer the soil and the vegetation to the stack, the greater the contamination.

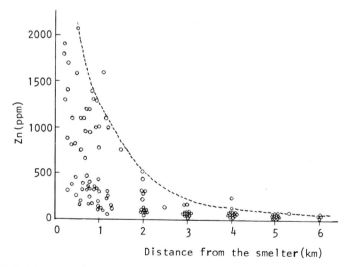

FIG. 15. Relation between the content of zinc in surface soil (0-10 cm, dried) and the distance in all directions from the smelter, Annaka City. Collected in 1970.

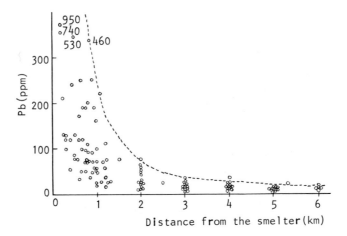

FIG. 16. Relation between the content of lead in surface soil (0-10 cm, dried) and the distance in all directions from the smelter, Annaka City. Collected in 1970.

By comparing the analytical results of the agricultural products in the polluted area (see Table 13) with those of an unpolluted area (see Table 8), the serious contamination by cadmium, zinc, and lead can be clearly seen. The concentrations of these metals in leaf vegetables, such as cabbage, were found to be substantially higher than those in root vegetables, fruits, and cereals. In the same cabbage, the outer leaves had more cadmium than the inner or younger leaves. A variety of moss which grows thick in the gardens of farmhouses was noticed to accumulate large quantities of these poisonous metals. Moreover, flour from wheat and barley contained much more cadmium, zinc, and lead than rice in the same polluted area. Thus, the content of these metals in products varies greatly with the kind and portion of plants.

It was established that the hilly regions covered by nonpaddy fields are affected only by contaminated air, whereas the low-lying rice fields, lower than the factory, are affected by both contaminated air and water. Analytical data on the silkworms and their excrement are given in Table 14. Silkworms, fed on contaminated mulberry leaves and noticeably retarded in growth, excreted feces with a much higher concentration of cadmium and lead than that found in their bodies on a dry-matter basis.

In a hilly region 900 m east from the smelter stack, soil profiles were examined (see Table 15). The upper soil (0-10 cm) was most polluted, and soil deeper than 40 cm was least polluted, showing approximately the same values as unpolluted soil.

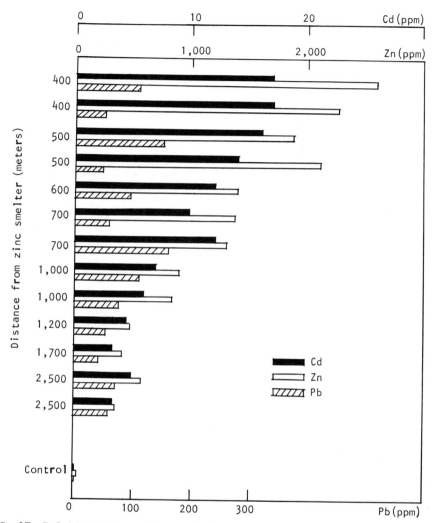

FIG. 17. Relation between the content of Cd, Zn, and Pb (ppm dry weight) in mulberry leaves and distance to the east from the zinc smelter, Annaka City. Collected in October 1968.

At any rate, complete standards on cadmium concentrations for all products including wheat and vegetables need to be set for the health of the people and to place more responsibility on the industries for any pollution they cause. Even the standard 1.0 ppm for brown rice set by the National Government seems to be too lax (see Fig. 9) and to give too free a rein to enterprises which pollute the environment.

Recently, in some of the polluted areas of Japan, decontamination of agricultural land producing food with excessive cadmium content has been

TABLE 13. Content of Cd, Zn, and Pb in Products from the Polluted Area, Annaka City (Collected in October 1968 and June 1969)

Category	Item	Distance (m) and Direction from the Chimney	$\frac{\text{Dry wt}}{\text{Wet wt}} \times 100$ (%)	ppm in Dry Matter Cd	ppm in Dry Matter Zn	ppm in Dry Matter Pb	Paddy or Nonpaddy
Cereals	Rice, polished	500 NE	--	0.61	24	<0.2	Paddy (C)[a]
	Rice, polished	1000 NE	--	0.95	32	<0.2	Paddy (C)[a]
	Rice, polished	1500 NEN	--	0.72	24	<0.2	Paddy (C)[a]
	Rice, polished	1500 NE	--	0.47	23	<0.2	Paddy (C)[a]
	Rice, unpolished	700 E	--	0.6	59	0.65	Paddy (C)[a]
	Wheat flour	220 ESE	--	5.2	280	15	Nonpaddy
	Wheat flour	250 ESE	--	7.1	360	11	Nonpaddy
	Wheat flour	300 SW	--	6.5	320	15	Nonpaddy
	Wheat flour	450 SSW	--	1.5	110	7.9	Nonpaddy
	Wheat flour	500 SE	--	8.6	370	7.2	Nonpaddy
	Wheat flour	800 E	--	2.5	170	4.2	Nonpaddy
	Wheat flour	800 E	--	2.5	220	4.0	Nonpaddy
	Wheat flour	800 SE	--	1.1	88	1.5	Nonpaddy
	Wheat flour	800 S	--	0.50	59	1.3	Nonpaddy
	Wheat flour	850 SSE	--	0.50	57	1.3	Nonpaddy
	Wheat flour	1250 SE	--	0.55	63	0.8	Nonpaddy
	Wheat flour	1250 S	--	0.23	34	1.2	Nonpaddy
	Wheat flour	1500 ENE	--	6.8	310	3.6	Paddy (I)[b]
	Wheat flour	1500 ENE	--	1.7	110	2.3	Paddy (E)[c] }[d]
	Wheat flour	1500 ENE	--	2.5	190	1.5	Paddy (I)[b]

Wheat flour	1750 ENE	--	1.3	110	1.1	Paddy (I)[b]
Wheat flour	1750 ENE	--	0.9	80	0.7	Paddy (E)[c] }[d]
Wheat flour	1800 ENE	--	1.7	120	1.2	Paddy (E)[c]
Barley flour	1500 ENE	--	3.5	140	14	Paddy (I)[b]
Barley flour	1500 ENE	--	1.7	220	10	Paddy (C)[a] }[d]
Corn	850 E	--	0.03	74	3	Nonpaddy
Green vegetables	Cabbage, inner leaves	450 SE	7.9	2.6	270	0.7	Nonpaddy
	Cabbage, inner leaves	1200 E	5.8	3.2	380	4.3	Nonpaddy
	Cabbage, outer leaves	1200 E	7.9	6.9	680	7.0	Nonpaddy
	Chinese cabbage	450 SE	3.7	41	1300	30	Nonpaddy
	Greens for pickling	800 E	--	32	1800	140	Nonpaddy
	Greens, severely damaged	800 E	--	56	3380	260	Nonpaddy
	Rape	1200 E	2.5	9.6	670	19	Nonpaddy
	Welsh onion	500 SWS	7.0	14	1100	4.2	Nonpaddy
	Welsh onion	850 E	10.1	9.4	1000	4.4	Nonpaddy
	Welsh onion	1000 E	--	7.6	1000	9	Nonpaddy
	Leaves of Japanese radish	500 SWS	7.1	2.6	250	13	Nonpaddy
	Leaves of Japanese radish	1200 E	2.5	14	950	23	Nonpaddy
	Asparagus	500 SE	5.5	8.6	220	0.9	Nonpaddy
Root crops	Potato	460 SWS	--	0.91	70	<0.2	Nonpaddy
	Potato	500 SWS	--	1.1	62	<0.2	Nonpaddy
	Potato	850 E	5.8	1.4	100	<0.2	Nonpaddy

TABLE 13 (Continued)

Category	Item	Distance (m) and Direction from the Chimney	Dry wt / Wet wt × 100 (%)	ppm in Dry Matter			Paddy or Nonpaddy
				Cd	Zn	Pb	
	Carrot	850 E	--	9.3	530	21	Nonpaddy
	Radish	500 SWS	--	1.5	140	1.5	Nonpaddy
	Radish	1200 E	4.0	3.2	250	1.3	Nonpaddy
	Taro potato	850 E	--	17	180	63	Nonpaddy
	Scallion	750 E	51.8	15	880	11	Nonpaddy
	Scallion	500 SWS	--	2.2	50	<0.2	Nonpaddy
	Garlic	500 SWS	62.9	1.2	67	<0.2	Nonpaddy
Fruit Vegetables	Tomato	800 E	--	1.7	150	11	Nonpaddy
	Pumpkin	800 E	--	0.4	76	4	Nonpaddy
	Fruit of eggplant	800 E	--	8.6	140	11	Nonpaddy
	Adzuki bean	800 E	--	0.6	59	0.4	Nonpaddy
	Field pea	500 SWS	16.5	2.6	160	1.4	Nonpaddy
	Field pea	850 E	13.0	2.2	200	1.6	Nonpaddy
	Persimmon	850 E	--	0.3	29	6	Nonpaddy
Grass for cows		350 SW	--	11	770	11	Nonpaddy
Moss		1000 E	--	61	7000	370	Nonpaddy

[a](C) collected at the center of the paddy field.
[b](I) collected near the inlet for irrigation water.
[c](E) collected at the end farthest from irrigation inlet.
[d]From the same paddy field.

TABLE 14. Cd, Zn, and Pb Concentration in Bodies and Excrement of Silkworms Fed on Contaminated Mulberry Leaves in Annaka and of Those in a Normal Area (Collected in June 1969)

| | | ppm Dry Weight | | |
		Cd	Zn	Pb
Polluted area, Annaka City	Silkworms fed contaminated mulberry leaves	4.2	1200	8.5
	Excrement	23	1500	66
Unpolluted area, Okayama Prefecture	Normal silkworms	0.14	100	2.0
	Excrement	0.22	54	5.9

attempted by changing the surface soil which is the most polluted in soil profiles. However, since the total area of contaminated fields, including the Jinzu River basin and many other areas, will cover about 2000 hectares, it will cost a colossal sum if these fields are decontaminated by changing the soil. In order to spare this expense, Kobayashi et al. [73], who have been carrying out an experiment in which polluted paddy and nonpaddy field soil is treated with a chelating agent, EDTA (see Fig. 18), found that this is a promising method for decreasing the cadmium content in soil and in the rice produced from it.

TABLE 15. Vertical Distribution of Trace Metals in Soil of Hilly Region, Annaka City (Collected in October 1968)

| Depth (cm) | ppm Dry Weight | | | Zn/Cd |
	Cd	Zn	Pb	
0-2	31	1680	510	54
5	44	1590	280	36
10	32	1310	250	41
20	6.9	540	59	78
30	1.4	140	26	100
40	0.4	80	15	200
60	0.3	62	10	207

FIG. 18. Application test of EDTA sprinkled over farm soil.

UPTAKE OF CADMIUM IN RICE AND WHEAT FROM SOIL

Some of the results of pot-culture experiments on rice plants and wheat performed by Kobayashi and Muramoto [13,74,75] in which different amounts of cadmium or zinc were added to the soil are given in Tables 16-18. In Table 16, it is indicated that the more cadmium oxide added to the soil, the greater the uptake of cadmium, especially in wheat, and that the yield of wheat was reduced at levels of 0.003% cadmium in the soil, while in rice a significant decrease in yield was not seen until 0.1%. Furthermore, an addition of zinc (ZnO) of up to 5% to the soil did not significantly decrease the yield of rice, whereas 0.1% zinc resulted in a marked decrease in yield of wheat. Thus, the rice plant has a greater tolerance than wheat for some poisonous elements and a lesser tendency to assimilate them. The increased uptake of zinc in rice due to the addition of zinc is not so marked as that of cadmium when cadmium is added to the soil.

The results of a pot-culture experiment on wheat carried out on limed and unlimed acid soil are given in Table 17. On limed soil an antagonistic effect was seen between the increased addition of cadmium (CdS was used) to the soil and the decreased uptake of zinc by the wheat. However, on acid soil, such as that found in many parts of Japan, such antagonism in wheat was not observed. In this experiment it was also noticed that liming the

TABLE 16. Uptake of Cadmium (ppm dry weight) by Rice and Wheat (1967-1968)

Addition to Soil of Cd or Zn in Oxides (%)		Rice				Wheat	
		Yield (%)	Polished (90%)		Bran (10%)	Yield (%)	Whole Grain
			Cd	Zn	Cd		Cd
Cd	0	100	0.16	19	0.59	100	0.44
	0.001	100	0.28	15	0.79	106	8.27
	0.003	92	0.40	16	0.84	72	15.5
	0.01	92	0.78	11	1.60	16	29.9
	0.03	93	1.37	--	2.68	13	41.4
	0.1	69	1.62	12	2.94	3	60.7
	0.3	32	1.94	9	3.19	3	48.6
	0.6	19	1.73	11	3.94	2	90.8
	1.0	1	4.98[a]	--	--	1	139.0
Zn	0	100	0.16	19	--	100	0.44
	0.1	109	0.10	26	--	38	0.71
	0.3	99	0.04	28	--	19	1.26
	1.0	97	0.02	34	--	43	0.26
	3.0	95	0.05	34	--	0	--
	5.0	95	0.11	32	--	1	0.89
Cd 0.01% + Zn 0.05%		101	0.62	16	--	1	3.70
Cd 0.1% + Zn 0.5%		59	3.07	29	--	2	32.3

[a]Unpolished.

soil is effective both in reducing the damage to the growth and yield of wheat and in restraining the increased uptake of cadmium by the wheat. Another result of a pot-culture experiment in which two kinds of cadmium compounds, sulfide and chloride, were added to the soil is given in Table 18. Chloride of cadmium added to the soil led to a higher uptake of cadmium by the rice than that of sulfide. At levels of 0.05% cadmium added to the soil as chloride, the uptake of cadmium by the rice was 6.2 ppm dry weight, whereas it was 0.64 ppm at the same levels of cadmium added as sulfide. A lower yield of rice was obtained by chloride of cadmium than by sulfide. Thus, chloride of cadmium soluble in water is more poisonous than

TABLE 17. Uptake of Cadmium (ppm dry weight) by Wheat (1968-1969)

Addition of Cd to Soil (%) in CdS	Unlimed Soil (pH 5.9)			Limed Soil (pH 7.4)		
	Yield (%)	Whole Grain		Yield (%)	Whole Grain	
		Cd	Zn		Cd	Zn
0	100	0.51	87	100	0.49	91
0.0005	103	1.33	63	92	0.89	90
0.0015	96	8.5	120	107	1.96	52
0.005	83	18.1	129	91	3.33	52
0.015	66	23.1	90	99	5.20	51
0.05	23	64.0	100	66	14.6	20
0.15	1	--	--	50	57.5	18

TABLE 18. Uptake of Cadmium (ppm dry weight) by Rice (1968)

Addition of Cd to Soil (%)	Added in CdS		Added in CdCl$_2$	
	Yield (%)	Cd in Unpolished Rice	Yield (%)	Cd in Unpolished Rice
0	100	0.16	100	0.16
0.0005	94	0.47	95	0.50
0.0015	97	0.39	89	0.54
0.005	93	0.44	82	1.01
0.015	81	0.41	78	2.05
0.05	82	0.64	65	6.20

sulfide, which is insoluble in water. The cadmium content of 6.2 ppm in rice by the addition of chloride in this experiment is the highest value ever reported for contaminated rice.

EFFECTS OF CADMIUM ON CALCIUM METABOLISM IN RATS

After the analytical findings of cadmium, zinc, lead, and copper accumulated in itai-itai patients (see the second and fourth sections), Kobayashi undertook calcium metabolism experiments [76,77] in order to ascertain whether bone decalcification is caused by cadmium and/or other trace metals. In two experiments, female rats were kept each in a separate metabolism-testing cage; feces and urine were collected separately and the balance of calcium was determined quantitatively every 3 weeks over a period of about

1 year. If decalcification should occur, the calcium excreted should exceed the amount of calcium gained from the food.

Calcium-deficient food was given to rats divided into the following groups:

Experiment	Group and Food	No. of Rats
First (test of young rats)	I. Control	10
	II. 100 ppm Cd was mixed in the food	10
Second (test of 11-mo.-old rats which had given birth to four litters)	A. Control	9
	B. 50 ppm Cd and 500 ppm Zn mixed in the food	9
	C. 30 ppm Cd, 300 ppm Zn, 150 ppm Pb, and 150 ppm Cu mixed in the food	9

Chlorides of cadmium, zinc, lead, and copper were used for these examinations.

Results of the two experiments, i.e., amount of calcium received from the diet and excreted in feces and urine and the balance of calcium calculated every 3 weeks up to 51 weeks, are given in Tables 19 and 20. In the first experiment, groups I and II exhibited a great difference in the total balance of calcium: +223.5 mg and -154.7 mg, respectively. Thus in contrast to the positive balance in control group I, group II, which was fed cadmium, indicated a negative balance, making a total of 378.2 mg difference between the two groups over the period examined. The balance of calcium in the individual rats is shown in Fig. 19. In the second experiment, the balance of calcium was also quite different among the three groups, A, B, and C, resulting in +84.2 mg, -462.4 mg, and -433.1 mg, respectively, in the total balance calculated up to 51 weeks. The results for each individual in this second experiment are shown in Fig. 20.

Because no increased urinary excretion of calcium was found in the animals fed cadmium in either the first or second experiment, the negative balance of calcium metabolism in rats fed on cadmium was due to fecal excretion of calcium in excess of the amount of calcium gained from the diet. Such an imbalance of calcium was observed from the beginning to the end of

TABLE 19. Balance of Calcium Metabolism in Rats in the First Experiment

| Weeks | Group I (Control, Rats Nos. 1-10) | | | | Group II (100 ppm Cd, Rats Nos. 11-20) | | | |
	Ca Intake from Diet (mg)	Ca Excreted Feces (mg)	Urine (mg)	Balance of Ca (mg)	Ca Intake from Diet (mg)	Ca Excreted Feces (mg)	Urine (mg)	Balance of Ca (mg)
16-18	61.9	46.7	3.6	11.6	52.5	55.8	1.4	-4.7
19-21	54.3	29.9	2.3	22.1	47.4	51.8	1.0	-5.4
22-24	55.7	26.9	1.3	27.5	53.0	45.0	1.0	7.0
25-27	51.8	22.7	2.1	27.0	53.8	50.2	0.9	2.7
28-30	56.8	33.1	1.5	22.2	46.4	58.2	0.7	-12.5
31-33	52.8	30.5	1.8	20.5	45.0	83.6	1.0	-39.6
34-36	57.9	35.6	2.9	19.4	47.1	67.4	1.0	-21.3
37-39	54.8	31.7	1.6	21.5	44.9	70.9	0.9	-26.9
40-42	45.9	40.8	2.0	3.1	37.2	74.2	1.1	-38.1
43-45	48.0	34.9	1.9	11.2	39.9	61.5	1.1	-22.7
46-48	55.0	39.0	2.7	13.3	48.9	51.2	0.9	-3.2
49-51	50.6	24.9	1.6	24.1	44.5	33.4	1.1	10.0
Total	645.5	396.7	25.3	223.5	560.6	703.2	12.1	-154.7
(%)	(100)	(61.5)	(3.9)	(34.6)	(100)	(125)	(2.2)	(-27.6)

the experiments, both before and during proteinuria. In Figs 21 and 22 the results of calcium metabolism in the second experiment are illustrated. In drawing these two figures, only the difference in the amount of calcium gained from the diet and excreted in feces was calculated, omitting the urinary excretion of calcium which was so small as to be negligible and was found to decrease through the influence of cadmium.

After the animals had died, the kidney, liver, femur, and humerus were analyzed for wet weight, dry weight, ash, cadmium, lead, zinc, and copper content. The rest of the body was heated in hot water, pH adjusted to weak alkaline, and digested with pancreatin. During the digestion, toluene was added and the containers were kept sealed. All the bones and the nails remaining undissolved were washed, dried, and weighed (see Fig. 23). In the results, the weight of the total dried bones obtained by the pancreatin treatment of rats among different groups differed as follows: I > II and A > B or C, and 99% of the total amount of calcium in the rat body was found in the bones. Thus, both from the experiments of calcium

TABLE 20. Balance of Calcium Metabolism in Rats in the Second Experiment

Weeks	Group A (Control) Rat Nos. 31 - 39				Group B (Cd 50 ppm, Zn 500 ppm) Rat Nos. 40 - 48				Group C (Cd 30 ppm, Pb 150 ppm, Zn 300 ppm, Cu 150 ppm) Rat Nos. 49 - 57			
	Ca Intake from Diet (mg)	Ca Excreted Feces (mg)	Ca Excreted Urine (mg)	Balance of Ca (mg)	Ca Intake from Diet (mg)	Ca Excreted Feces (mg)	Ca Excreted Urine (mg)	Balance of Ca (mg)	Ca Intake from Diet (mg)	Ca Excreted Feces (mg)	Ca Excreted Urine (mg)	Balance of Ca (mg)
1 - 5	85.8	104.2	1.5	-19.9	68.1	139.5	1.1	- 72.5	66.3	166.7	1.4	-101.8
4 - 6	76.2	85.4	1.5	-10.7	72.8	113.3	1.1	- 41.6	72.0	154.2	1.4	- 83.6
7 - 9	66.3	63.9	1.5	0.9	61.8	101.2	1.1	- 40.5	71.0	110.5	1.4	- 40.9
10 - 12	71.5	55.9	1.8	15.8	67.7	79.3	1.3	- 12.9	71.9	85.2	1.3	- 14.6
13 - 15	70.2	44.5	1.8	23.9	66.5	64.6	1.3	0.6	60.9	55.5	1.3	4.1
16 - 18	70.0	44.0	1.8	24.2	65.0	64.5	1.3	0.8	72.0	53.9	1.4	16.7
19 - 21	72.3	51.9	1.7	18.7	72.4	78.0	1.0	- 0.8	64.3	62.0	1.5	0.8
22 - 24	72.9	72.8	1.7	- 1.6	71.4	115.8	1.1	- 45.5	71.1	90.4	1.5	- 20.8
25 - 27	74.9	70.2	1.7	3.0	65.2	98.6	1.0	- 34.4	68.2	97.8	1.5	- 31.1
28 - 30	72.9	76.7	2.3	- 6.1	72.1	102.1	1.1	- 31.1	73.4	107.0	1.1	- 34.7
31 - 33	67.5	65.0	2.3	0.2	68.5	89.6	1.0	- 22.1	84.2	98.4	1.3	- 15.5
34 - 36	77.7	74.5	2.6	0.6	73.9	93.6	0.8	- 20.5	70.1	87.4	1.3	- 18.6
37 - 39	65.2	72.6	1.2	- 8.6	58.4	94.2	0.6	- 36.4	40.9	88.5	1.3	- 48.9
40 - 42	75.5	80.4	1.4	- 6.3	76.0	104.5	0.6	- 29.1	47.3	85.6	1.4	- 39.7
43 - 45	84.7	80.7	1.2	2.8	72.5	101.7	0.6	- 29.8	67.0	78.5	0.8	- 12.3
46 - 48	87.3	80.1	2.9	4.3	77.8	110.5	0.9	- 33.6	67.7	59.4	1.5	6.8
49 - 51	114.9	69.0	2.9	43.0	75.0	79.7	0.9	- 5.6	75.0	72.5	1.5	1.0
Total	1305.8	1189.8	31.8	84.2	1185.1	1630.7	16.8	-462.4	1143.3	1553.5	22.9	-433.1
(%)	(100)	(91.1)	(2.4)	(6.4)	(100)	(138)	(1.4)	(-39.0)	(100)	(136)	(2.0)	(-57.9)

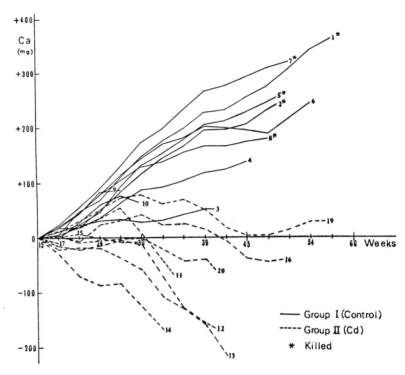

FIG. 19. Balance of calcium metabolism of rats in the first experiment (each line = one individual rat).

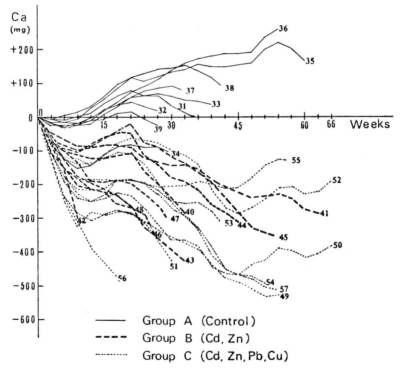

FIG. 20. Balance of calcium metabolism of rats in the second experiment (each line = one individual rat).

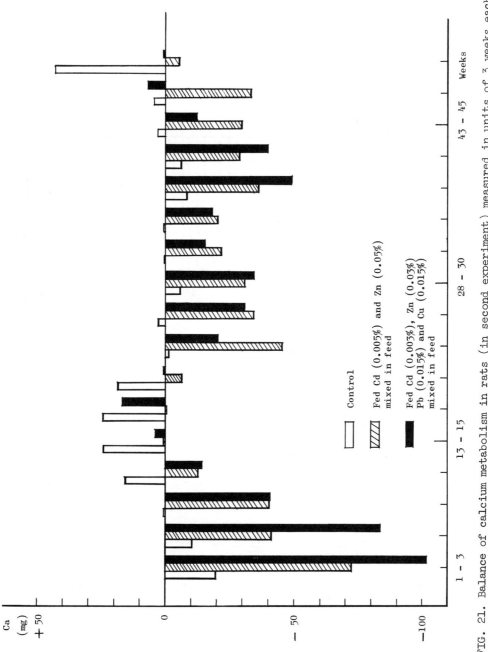

FIG. 21. Balance of calcium metabolism in rats (in second experiment) measured in units of 3 weeks each.

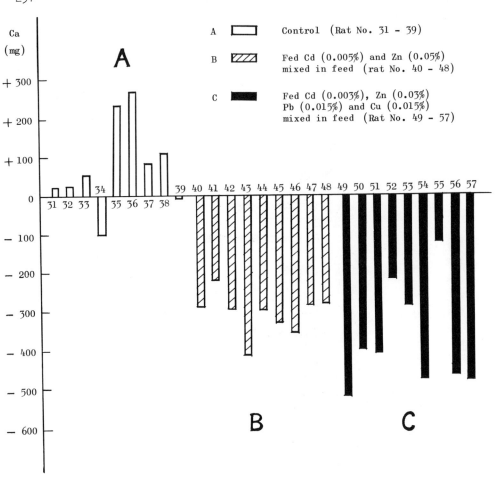

FIG. 22. Total balance of calcium metabolism in each rat in the second rat experiment.

metabolism and analyses of bones, it was clarified that the oral administration of cadmium resulted in decalcifying the bones.

    After the experiments, Kobayashi examined the effect of 1000 ppm zinc
on another group of rats, but no noticeable imbalance of calcium was obtained. Furthermore, in another experiment a marked difference due to demineralization was found in the backbones of three groups of 10 young carp
(see Fig. 24) [78]. In comparison with a control group in normal water,
each of these three groups was kept for 3 months in soft water with various
concentrations of cadmium, 0.01, 0.05, and 0.1 ppm. In each of these three
groups some fish, by the end of 3 months, had severely deformed backbones.
The deformed fish shown in Fig. 24 had been kept in water containing only

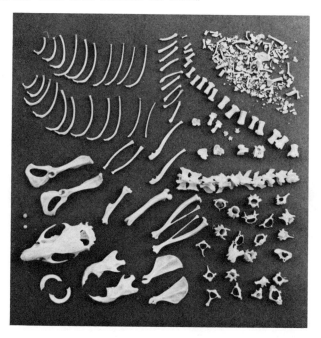

FIG. 23. Complete bones and other hard parts of a rat obtained by pancreatin treatment (minus one femur and one humerus which had been used for determination of ash weight).

0.01 ppm cadmium, which is the standard for our drinking water. Thus, among the metals which had heavily accumulated in the human victims, cadmium was found to be most effective in decalcifying the bones of animals.

Furthermore, the experiments in calcium metabolism indicated that the calcium which was lost from rat bones was excreted in feces instead of in the urine and therefore gave no support to the hypothesis which had been prevalent among medical circles [39,48,79] that the decalcification (osteomalacia) was secondary to a renal dysfunction as in the Fanconi syndrome. Instead they favored the view that the bone changes were induced, like those [80] in the vitamin D deficiency disease, by the low absorption and high excretion of calcium from the gastrointestinal system. So, is the causal mechanism of bone changes in the itai-itai disease due to enhanced deficiency of activated vitamin D? Is it due to some impairment of the intestines as reported by Murata [81,82]? No factual data have been established to answer these questions, which should be very important in the investigation of the itai-itai disease. In any case, it is true that large doses of vitamin D, even though it is not effective in improving the renal dysfunction, is the most effective remedy in reducing the pains and bone changes in the patients.

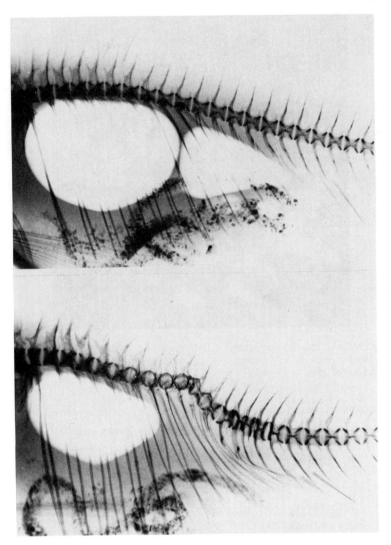

FIG. 24. Soft x-ray photos showing deformation of the backbone of a carp:
(upper) a young carp kept in normal water; (lower) a young carp kept for
3 months in water containing 0.01 ppm cadmium.

REFERENCES

1. S. Matsunaga, M. Kohno, K. Nagai, T. Nakayama, and H. Ohide, Arch. Kohno
    Clin. Med. Res. Inst., $\underline{1}$ (1), (1962) (in Japanese).

2. M. Kohno, H. Sugihara, T. Nakayama, and N. Hagino, Clin. Nutr., $\underline{9}$ (2),
    I (1956) (in Japanese).

3. M. Kohno, H. Sugihara, T. Nakayama, and Y. Kuramoto, Clin. Nutr., $\underline{9}$ (4),
    I (1956) (in Japanese).

4. M. Kohno, H. Sugihara, T. Nakayama, Y. Kuramoto, and N. Hagino, Clin. Nutr., 9 (7), I (1956) (in Japanese).

5. I. Murata, I. Taga, and S. Nakagawa, J. Jap. Soc. Int. Med., 45, 199 (1956) (in Japanese).

6. I. Murata, I. Taga, and S. Nakagawa, J. Jap. Orthop. Ass., 30, 381 (1956) (in Japanese).

7. I. Murata, I. Taga, and S. Nakagawa, J. Jap. Orthop. Ass., 31, 387 (1957) (in Japanese).

8. I. Murata and S. Nakagawa, Clin. Radiol., 2, 637 (1957) (in Japanese).

9. I. Murata and S. Nakagawa, J. Jap. Orthop. Ass., 34, 1365 (1961) (in Japanese).

10. S. Nakagawa, J. Rad. and Phys. Ther., University of Kanazawa, 56, 1 (1960) (in Japanese).

11. T. Nakayama, J. Jap. Orthop. Ass., 34, 1367 (1961) (in Japanese).

12. K. Kajikawa, S. Okuno, K. Ikawa, and R. Hirono, Trans. Soc. Pathol. Jap., 46, 655 (1957) (in Japanese).

13. J. Kobayashi, Mizu No Kenko Shindan (Health Examination of Water) (in Japanese), Iwanami, Tokyo, 1971.

14. J. Kobayashi, On Geographical Relationship Between the Chemical Nature of River Water and Death-Rate from Apoplexy, Ber. Ohara Inst. landw. Biol. Okayama Univ., 2, 12 (1957) (in English).

15. J. Kobayashi, A Chemical Study of the Average Quality and Characteristics of River Waters of Japan, Ber. Ohara Inst. landw. Biol. Okayama Univ., 2, 313 (1960) (in English).

16. J. Kobayashi, Chemical Investigation of River Waters of Southeastern Asiatic Countries (Report I), The Quality of Waters of Thailand, Ber. Ohara Inst. landw. Biol. Okayama Univ., 2, 167 (1959) (in English).

17. J. Kobayashi and K. Ishimaru, The Agricultural Damage in the Jinzu River Basin Caused by the Kamioka Mine (in Japanese), Ministry of Agriculture and Forestry, 1943.

18. Toyama Newspaper, Itai-itai Disease Was Caused by Poisonous Metals from the Mine, reported by Kobayashi (in Japanese), May 14, 1961.

19. N. Hagino and K. Yoshioka, J. Jap. Orthop. Ass., 35, 812 (1961) (in Japanese).

20. J. Kobayashi and N. Hagino, Strange Osteomalacia Itai-itai Disease in the Jinzu River Basin Polluted with Zinc, Lead, and Cadmium (II), private communication, 1961, pp. 1-8.

21. P. Nicaud, A. Lafitte, and A. Gros, Arch. Mal. Prof. Med. Trav. Secur. Soc., 4, 192 (1942).

22. Asahi, Mainichi, and other newspapers, Cadmium Might Be the Cause of the Itai-itai Disease (in Japanese), Oct. 1, 1966.

23. J. Kobayashi, Relation Between the "Itai-itai" Disease and the Pollution of River Water by Cadmium from a Mine, Advances in Water Pollution Research, Vol. I, Proceedings of the 5th International Conference (S. H. Jenkins, ed.), Pergamon Press, San Francisco and Hawaii, 1970, pp. 1-7.

24. N. Hagino and J. Kobayashi, Jap. J. Hyg., 23, 196 (1968) (in Japanese).

25. J. Kobayashi, Jap. J. Hyg., 20, 156 (1965) (in Japanese).

26. M. Hirota, Jap. J. Hyg., 19, 12 (1964) (in Japanese).

27. Japanese Public Health Association, 1966 and 1967 (in Japanese).

28. Parliamentary Records of the Upper House of the Japanese Diet, Special Committee on Industrial Public Nuisances Countermeasures and Traffic Safety Measures, Nos. 7 and 8, 1967 (in Japanese).

29. Japanese Public Health Association, 1968 (in Japanese).

30. N. Yamagata and I. Shigematsu, Cadmium Pollution in Perspective, Bull. Inst. Pub. Health, 19, 1 (1970).

31. All of the main Japanese newspapers, May 9, 1968.

32. All of the main Japanese newspapers, July 1, 1971.

33. J. Takeuchi, Jap. J. Clin. Med., 31, 2048 (1973) (in Japanese).

34. All of the main Japanese newspapers, August 10, 1972.

35. The Record of the Itai-itai Disease Trials, Vols. I-VI (edited by K. Shoriki, Leader of the Group of Lawyers for the Prosecution), Sogo-Tosho, Tokyo, 1971-1974 (in Japanese).

36. N. Hagino, J. Jap. Accident Med. Ass., 17, 300 (1969) (in Japanese).

37. N. Hagino, Modern Science and Pollution, Kubiki, Tokyo, 1972, p. 48, (in Japanese).

38. Ministry of Health and Welfare, Method of Health Examination: A Part of Provisional Countermeasures Against Environmental Pollution of Cadmium, May 19, 1971 (in Japanese).

39. L. Friberg, M. Piscator, G. Nordberg, and T. Kjellström, Cadmium in the Environment, II, The Karolinska Institute, Department of Environmental Hygiene, Stockholm, 1973, pp. 79-92, 97.

40. K. Nogawa, A. Ishizaki, I. Shibata, and N. Hagino, Medicine and Biology, 86, 277 (1973) (in Japanese).

41. I. H. Tipton and M. J. Cook, Health Physics, 9, 103 (1963).

42. A. Ishizaki, M. Fukushima, and M. Sakamoto, Jap. J. Hyg., 24, 515 (1969) (in Japanese).

43. A. Ishizaki, M. Fukushima, and M. Sakamoto, Jap. J. Hyg., 26, 268 (1971) (in Japanese).

44. M. Moritsugu and J. Kobayashi, Study on Trace Metals in Bio Materials II, Cadmium Content in Polished Rice, Ber. Ohara Inst. landw. Biol. Okayama Univ., 12, 145 (1964) (in English).

45. T. Saino and J. Kobayashi, Simultaneous Quantitative Determination of Copper, Lead, Zinc and Cadmium in the Polished Rice by Square-wave Polarography, Nogaku Kenkyu, 49, 189 (1963) (in Japanese).

46. J. Kobayashi, H. Nakahara, and T. Hasegawa, Accumulation of Cadmium in Organs of Mice Fed on Cadmium-Polluted Rice, Jap. J. Hyg., 26, 401 (1971) (in Japanese).

47. M. Fukushima, A. Ishizaki, M. Sakamoto, and E. Hayashi, Jap. J. Hyg., 24, 526 (1970) (in Japanese).

48. A. Ishizaki and M. Fukushima, Jap. J. Hyg., 23, 271 (1968) (in Japanese).

49. M. Fukushima, A. Ishizaki, M. Sakamoto, and E. Kobayashi, Jap. J. Hyg., 28, 406 (1973) (in Japanese).

50. M. Fukushima, New Methods in Environmental Chemistry and Toxicology, Collection of papers presented at the International Symposium on Ecological Chemistry (F. Coulston, F. Korte, and M. Goto, eds.), Susono, November 1973.

51. A. Ishizaki, J. Jap. Med. Soc., 62, 242 (1969) (in Japanese).

52. M. Fukushima and Y. Sugita, Jap. J. Public Health, 17, 759 (1970) (in Japanese).

53. K. Kubota, Y. Fukuyama, and K. Shiroishi, Jap. J. Hyg., 26, 169 (1971) (in Japanese).

54. M. Piscator, Arch. Environ. Health, 12, 335, 345 (1966).

55. L. Friberg, M. Piscator, and G. Nordberg, Cadmium in the Environment, CRC Press, Cleveland, 1971, pp. 61, 84, 90, 113, 134.

56. S. Toyoshima, T. Hoshino, and K. Tsuchiya, Studies of Standardization of Analytical Methods for Chronic Cadmium Poisoning, Japanese Public Health Association, 1972, p. 34 (in Japanese).

57. A. Ishizaki, M. Fukushima, and M. Sakamoto, Jap. J. Hyg., 25, 207 (1970) (in Japanese).

58. The Official List of Food Standards Tables (edited by Resources Survey Committee of the Science and Technology Bureau of the Japanese Government), 1973, p. 18 (in Japanese).

59. I. H. Tipton, H. A. Schroeder, H. M. Perry, and M. J. Cook, Health Physics, 11, 403 (1965).

60. Statistical Tables of Japan's Consumption and Supplies of Foodstuffs for the Year 1972, Society for Statistics on Agricultural and Forestry, 1974, p. 145 (in Japanese).

61. M. Fukushima, Cadmium Concentration in Various Foodstuffs, in Report 11 from the Japanese Association of Public Health, April 1972, p. 72 (in Japanese).

62. J. Kobayashi, F. Morii, S. Muramoto, and S. Nakashima, Chemical Investigation on Inorganic Constituents of Night Soil (Excreta). II, On the Amounts of Ca, Mg, Na, K, P, Cl, S, Mn, Fe, Cd, Cu, Zn, Ni and Pb in the Night Soil of Inhabitants of the Tohoku District, Nogaku Kenkyu, 55, 161 (1976) (in Japanese).

63. J. Kobayashi and F. Morii, unpublished work, 1974.

64. J. Kobayashi, Air and Water Pollution by Cadmium, Lead, and Zinc Attributed to the Largest Zinc Refinery in Japan, Trace Substances in Environmental Health. V (D. D. Hemphill, ed.), Missouri University, Columbia, 1972, pp. 117-127.

65. Ministry of Health and Welfare, Environmental Hygiene Bureau, The Opinion of the Ministry of Health and Welfare as Regards Environmental Pollution by Cadmium and Countermeasures in the Future, March 27, 1969 (in Japanese).

66. Japanese Public Health Association, March 20, 1969 (in Japanese).

67. Asahi and other Japanese newspapers, March 28, 1969.

68. Parliamentary Records of the Upper House of the Japanese Diet. Special Committee on Industrial Public Nuisances Countermeasures and Traffic Safety Measures, No. 7, 1969 (in Japanese).

69. J. Kobayashi and S. Muramoto, Jap. J. Hyg., 25, 80 (1970) (in Japanese).

70. J. Kobayashi, S. Muramoto, S. Nakashima, and K. Hara, Jap. J. Hyg., 26, 164 (1971) (in Japanese).

71. J. Kobayashi, F. Morii, S. Muramoto, and S. Nakashima, Effects of Air and Water Pollution on Agricultural Products by Cd, Pb, Zn Attributed to Mine Refinery in Annaka City, Gunma Prefecture, Jap. J. Hyg., 25, 364 (1970) (in Japanese).

72. J. Kobayashi, F. Morii, S. Muramoto, S. Nakashima, Y. Urakami, and H. Nishizaki, Distribution of Cd, Pb and Zn Contained in Soils Polluted by a Zinc Refinery in Annaka City, Gunma Prefecture, J. Sci. Soil and Manure (Japan), 44, 471 (1973) (in Japanese).

73. J. Kobayashi, F. Morii, and S. Muramoto, Removal of Cadmium from Polluted Soil with the Chelating Agent, EDTA, Trace Substances in Environmental Health. VIII (D. D. Hemphill, ed.), Missouri University, Columbia, 1975, pp. 179-191.

74. J. Kobayashi and S. Muramoto, Jap. J. Hyg., 24, 66 (1969); 26, 170 (1971) (in Japanese).

75. J. Kobayashi, Kagaku (Science), 39, 369 (1969) (in Japanese).

76. J. Kobayashi, Jap. J. Hyg., 22, 155 (1967) (in Japanese).

77. J. Kobayashi, Effects of Cadmium on Calcium Metabolism of Rats, Trace Substances in Environmental Health. VIII (D. D. Hemphill, ed.), Missouri University, Columbia, 1974, pp. 295-304.

78. J. Kobayashi, F. Morii, S. Muramoto, S. Nakashima, K. Hara, and M. Togura, Jap. J. Hyg., 27, 223 (1972) (in Japanese).

79. J. Takeuchi, A. Shinoda, K. Kobayashi, Y. Nakamoto, I. Takazawa, and M. Kurosaki, Int. Med. (Naika), 21, 876 (1968) (in Japanese).

80. K. Ashida, Outline of Nutritional Chemistry (in Japanese), Yokendo, Tokyo, 1974, pp. 233, 268.

81. I. Murata, J. Jap. Med. Ass., 65, 15 (1971) (in Japanese).

82. I. Murata, T. Hirono, Y. Saeki, and S. Nakagawa, Bull. Soc. Int. Chirurgie, 1, 1 (1970).

13

METHYL MERCURY POISONING DUE TO ENVIRONMENTAL
CONTAMINATION ("MINAMATA DISEASE")

Masazumi Harada
Kumamoto University
Kumamoto, Japan

Before the discovery of Minamata disease (Md) in 1956, about 40 cases of organic mercury poisoning had been reported, dating back to the report by G. N. Edwards [5,6]. Before 1956, all the reports were about poisoning caused by direct exposure to organic mercury, as in the cases of a chemist [35], a chemical-plant worker [23,24], and persons in charge of organic mercury-treated seeds and timber (farmers, lumbermen, etc.) [1,22,35] or persons who had mistakenly eaten the disinfected seeds [3,8,50]. Until then there had been no recorded cases of what was later called Minamata disease, in which methyl mercury (MeHg) contaminates the environment, is transmitted through the food chain, and causes mass human poisoning of regional scope. This absence of information made it very difficult to discover the cause of the disease, and many problems remain unsolved even today.

What we now call "Minamata disease" was first observed in communities near Minamata Bay in southwestern Japan. It was officially "discovered" in 1956, and by 1959 it had been demonstrated that the disease was caused by ingestion of fish contaminated by mercury discharged from a chemical plant of the Chisso Corporation, a major chemical manufacturer [33,38]. At first there were efforts to dissuade people from eating fish caught in the bay. But by 1962, it was generally believed that the emergency had been overcome, and inhabitants of the region resumed their ordinary diet, in which fish figured prominently. Meanwhile, however, the Chisso Corporation continued to discharge mercury-polluted effluent. This practice continued until 1968, and even today the bay is contaminated with MeHg [10,17,18,27].

When we began our studies, most of the clinical data on Md were those which had been reported before 1962. However, after that time, we saw patients whose symptoms were worsening and others in whom symptoms appeared for the first time. Also, the area of incidence appeared to expand. The fishermen, fearing that the sale of their catch would be threatened, either deceived themselves about persistent contamination or concealed it; local officials also ignored the danger, and as a result the contamination of the bay did not attract public attention again until 1970 [14,18].

Until 1962, the diagnosis of Md was made only in acute, subacute, and severe cases, all showing symptoms of MeHg poisoning as reported by Hunter and co-workers in 1940 [24,25]. These were indeed typical of Md, and there is no doubt that the main symptoms of the patients resulted from MeHg poisoning. However, excluded from such diagnoses were mild cases, those in which the syndrome was incomplete, and chronic cases in which the full syndrome of MeHg poisoning appeared less perceptibly. Since Md is actually an endemic MeHg poisoning affecting the inhabitants of a large area, it could be argued that the typical, acute cases selected probably represent only the tip of the iceberg and that a much larger number of patients remains undiscovered. This report is a general summary of Md together with some clinical surveys reexamined in the light of subsequent findings.

ENVIRONMENTAL CONTAMINATION IN THE MINAMATA AREA

The Shiranui Sea, located off Japan's southern island of Kyushu, is bounded by Uto Peninsula to the north, the Amakusa Islands to the west, and Nagashima Island to the south. Minamata Bay is an inlet on the eastern shore of the Shiranui Sea (Fig. 1). It is bounded by Myojin Promontory to the northwest, Koiji Island to the west, and Modo Promontory to the southwest. At its south end is Fukuro (pouch) Bay. At the northeast tip of Minamata Bay is the drainage channel of the Chisso plant (Fig. 2).

In 1959, after MeHg was suspected as the cause of the disease, an environmental survey of mercury pollution was conducted. Findings showed that an extraordinarily high level of mercury contamination existed in Minamata Bay. In the sludge near the drainage channel the count reached 2010 ppm (total mercury, wet weight) and declined gradually in proportion to the distance from the channel (see Fig. 2) [26,33].

At that time Chisso Corporation was using a large amount of inorganic mercury in an acetaldehyde and vinyl chloride process. According to a

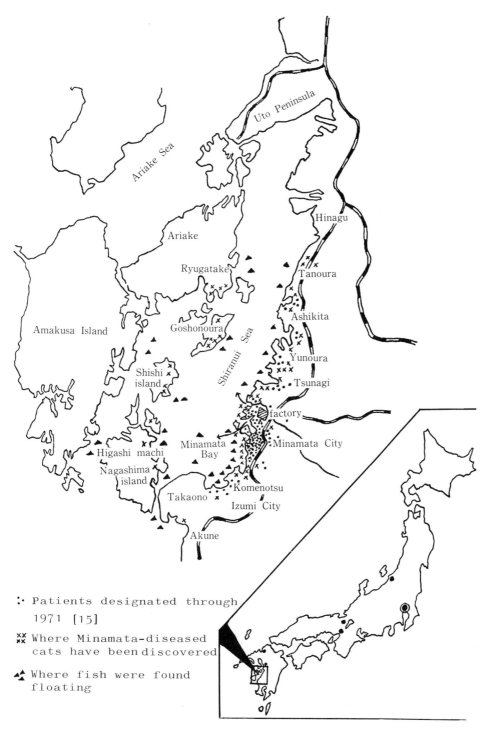

:· Patients designated through
   1971 [15]

×× Where Minamata-diseased
×× cats have been discovered

▲▲ Where fish were found
   floating

FIG. 1. Map of Shiranui Sea.

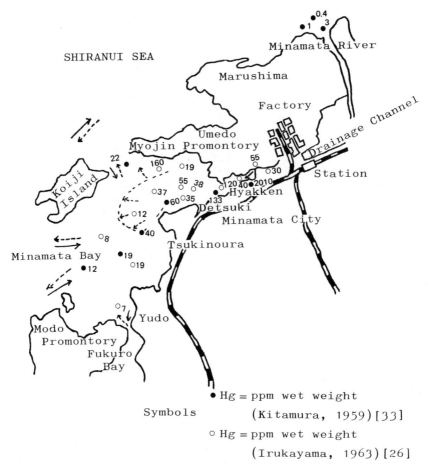

FIG. 2. Minamata Bay and mercury content of sludge at the bottom of the bay.

later estimate the total mercury dumped by Chisso Corporation could be as much as 200-600 tons [17].

In September 1960, Prof. M. Uchida extracted MeHg compound (CH_3HgSCH) in crystalline form from shellfish which had caused Md [64]. However the process by which the **inorganic** mercury changes into organic mercury was still unknown. Again, in the fall of 1962, Prof. T. Irukayama detected MeHg (CH_3HgCl) directly from the sludge of the acetaldehyde process, thereby proving that the factory process itself had methylated the inorganic mercury [34].

In 1973, it was found that Minamata Bay was still contaminated with mercury: 25-400 ppm were detected in sludge from the bottom (Fig. 3).

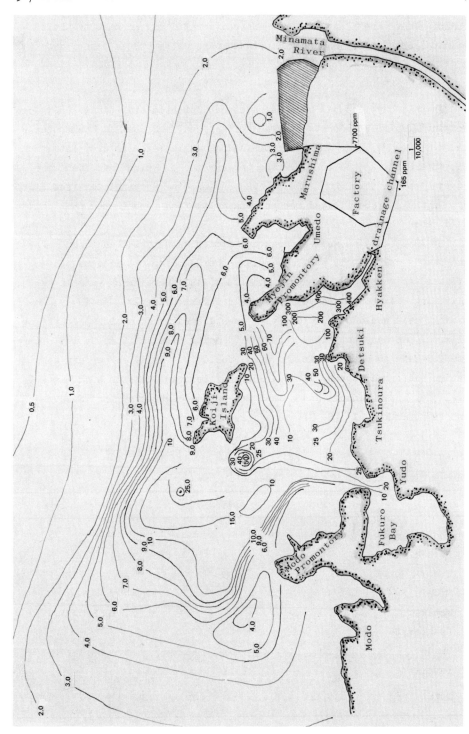

FIG. 3. Recent mercury content of sludge at the bottom of Minamata Bay. Figures = ppm Hg (dry weight) (Environmental Agency 1973).

Although there are plans for dredging and for filling in part of the contaminated area of the bay, no effective measures have been taken so far.

EFFECTS ON LIVING

In this bay, unusual changes have been detected ever since 1950; dead
or dying fish floated on the surface of the sea, and shellfish also per-
ished. In 1952, birds such as crows and seagulls began to drop into the
sea, apparently unable to fly. The sea surface on which dead fish were
floating spread throughout the bay and on into the Shiranui Sea (see Fig.
1). On the surface floated octopus and cuttlefish so weakened that children
could catch them with their bare hands [29,30].

By 1953, cats as well as dogs and some pigs went mad and died. The
inhabitants called this striking malady the "dancing-cat disease" [30].
Afflicted animals would stagger about as if they were drunk (ataxic gait),
salivating. At times, convulsions would strike suddenly. As if by com-
pulsion, cats would whirl violently, then dash away on a zigzag course or
collapse. So many cats died from the "dancing disease" that by 1954 there
were few cats to be found in the area (Table 1). By 1954, mercury contamination had spread throughout the Shiranui Sea, and cats in such distant
localities as Goshonoura and Shishi Island had contracted the "dancing
disease" (see Fig. 1) (S. Kitamura) [29,51].

TABLE 1. Incidence of Minamata Disease in Cats[a]

	1953	1954	1955	1956
Tsukinoura		3	6	7
Detsuki	1	1	7	6
Yudo		6	7	12
Myojin		2	2	2
Mategata		4	1	--
Hyakken		--	1	1
Umedo		1	1	2
Tatara		1	--	--
Total	1	18	25	30

[a]Those discovered through 1956 by S. Kitamura [29].

In 1956, cats brought to Minamata from other areas fell ill within 32 to 65 days on a diet of fish from the bay (K. Sera) [46]. Likewise, 40 very small dried fish from Minamata Bay, weighing about 10 g in all, produced Md symptoms in 51 days when fed to cats three times a day [46].

In 1959, fish and shellfish caught in Minamata Bay showed high mercury values. Hormomya mutabilis contained 11.4-39.0 ppm (wet weight), oyster 5.61 ppm, crab 55.7 ppm (S. Kitamura, 1960) [9,33]. Cats in the area that were spontaneously affected with Md (or dancing disease), and others in which the condition had been induced experimentally by a diet of shellfish from Minamata Bay, showed the following localized mercury levels: liver, 37.0-145.5 ppm (as against 0.9-3.66 ppm in controls); brain, 21.5-70.0 ppm (0.09-0.82 ppm); and hair, 21.5-70.0 ppm (0.51-2.12 ppm) (S. Kitamura, 1960) [33].

Even in 1973-1974, fish and shellfish with high mercury levels were found. For example, Hormomya mutabilis showed 0.28-3.5 ppm (wet weight) S. Tatesu, 1972 [60], and the internal organs of octopus read as high as 13.1 ppm (M. Harada, 1974). At present, however, most of the fish and shellfish taken in this area show values of less than 1.0 ppm.

CONTAMINATION OF HUMAN BEINGS

The mercury content in patients' hair was especially high (96.8-705 ppm) [33] within 1 year after the onset of the disease. A high concentration of mercury was found in the internal organs of those who died of Md. The highest amount was detected in the liver and kidneys [47,54]. Although the amount in the brain was less than that in the liver and kidneys, it is characteristic of MeHg poisoning that the amount deposited in the brain is larger than that seen in the case of inorganic mercury poisoning (Table 2) (T. Takeuchi) [54]. It is notable that 100-191 ppm of mercury were detected in the hair of persons who were thought to be healthy [31,33]. Whether or not symptoms had already appeared by 1960, this fact shows that the inhabitants in the area were all exposed to heavy mercury contamination. Measurement of mercury in the hair was limited to a small sampling, but gave us sufficient data to indicate the extent of mercury pollution among the inhabitants at that time.

High levels in the hair were found not only in people of the Minamata area but also in inhabitants of Tsunagi, Ryugatake, Goshonoura, and Komenotsu as well (Table 3). It is remarkable that 920 ppm of mercury were detected in the hair of those living in the Goshonoura area [17,18,62].

TABLE 2. Mercury Content in Human Organs (ppm wet weight)[a]

Approximate Days from Onset to Death	Autopsy Cases of Minamata Disease			Autopsy Cases of Other Disease		
	Liver	Kidney	Brain	Liver	Kidney	Brain
20	70.5	144.0	9.60	0.18	--	--
25	38.2	47.5	15.4	--	--	0.11
50	34.6	99.0	7.80	0.84	--	--
50	39.5	40.5	8.95	0.45	--	--
60	42.1	106.0	21.3	0.2	--	--
60	38.8	68.2	24.8	0.38	--	--
60	34.7	64.2	7.8	1.06	--	--
90	--	--	9.45	1.02	3.02	--
90	36.2	21.2	4.85	--	0.37	0.11
95	30.0	22.6	4.63	--	0.25	0.08
100	22.0	42.0	2.6	--	1.08	0.12
550	26.0	37.4	5.32	0.07	10.7[b]	0.05
860	6.35	12.8	1.30	--	0.53	0.09
1000	2.05	3.11	0.09	0.60	2.04	0.47
1470	5.44	5.9	2.22	0.97	1.01	1.54

[a]Calculated by S. Kitamura and T. Takeuchi [33,54].
[b]This patient had been treated with a mercury-containing drug.

    After the consumption of fish and shellfish was discontinued, the levels of mercury contamination in patients' hair gradually decreased. However, in the hair of one patient, who had been suffering from the disease more than 3 years, 165 ppm of mercury were found (S. Kitamura, 1960) [32].

    Unfortunately, no data had been gathered before 1959 on the extent of mercury contamination in the area. However, it is a custom among the Japanese to keep the dried umbilical cord of infants, and these were useful as material for studying mercury contamination before 1959. It was found that at least as far back as 1947 human beings had been contaminated with mercury. The mercury levels in the umbilical cords of congenital Md patients were high, but high levels were detected also in the umbilical cords of those who showed no symptoms (Fig. 4) [19].

TABLE 3. Mercury Content in Hair of Healthy Fishermen[a]

ppm	<1	1-10	10-50	50-100	100-150	150-200	200-300	300-	Total
Kuamoto City (control)	4	18	9	0	0	0	0	0	31
Minamata	7	31	100	49	11	1	--	--	199
Tsunagi	--	12	61	23	4	2	--	--	102
Yunoura	--	--	14	9	1	--	--	--	24
Ashikita	1	19	19	1	--	--	--	--	40
Tanoura	--	6	15	11	--	--	--	--	33
Ryugatake	2	22	57	5	--	1	--	--	87
Goshonoura	6	53	334	75	11	1	--	2	482

ppm	<20	20-50	50-100	100-200	200-300	300-	Total
Komenotsu	185	117	105	37	5	1	445
Akune	26	4	1	3	--	1	33
Takaono	2	2	5	--	--	--	10
Higashi Machi	18	32	23	2(142)	--	--	75

[a]This survey was conducted on a voluntary basis by Kumamoto Prefecture in 1960-1961 (M. Harada, 1974) [18,62].

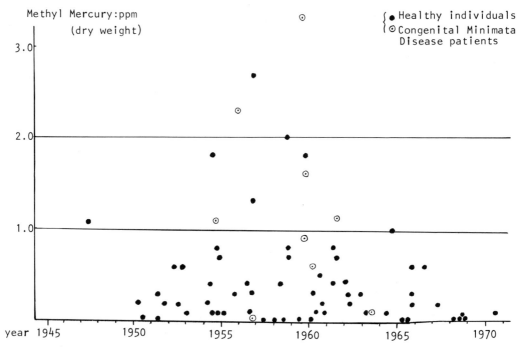

FIG. 4. Methyl mercury found in umbilical cords in the Minamata area.
Calculated in 1972, M. Fujiki and M. Harada [17,19].

In 1973, the mercury content in the hair of inhabitants in the Minamata
area averaged 2.10-14.82 ppm (M. Fujiki) [11]. This is slightly higher than
the level found in other areas of Japan.

METHYL MERCURY POISONING (MINAMATA DISEASE)

Typical Minamata Disease (Postnatal)

Epidemiological and Clinical Findings. Until 1971 it was believed
that the number of adult patients was 98, that they had taken ill between
1953 and 1960, and that the outbreak had occurred in an 80-mile area along
the coast north and south of Minamata (see Figs. 5 and 1, respectively)
[15,63]. Patients were limited to fishermen and others who ate large quantities of fish from Minamata Bay. Therefore, the disease usually affected
families. Also, the number of patients increased in direct proportion to
the increased production of acetaldehyde.

In most cases, the patients began to show symptoms without any apparent signs. The onset began with numbness of the limbs and perioral area,
sensory disturbance, and difficulty with hand movements (such as grasping
things, fastening buttons, holding chopsticks, writing, etc.); also there

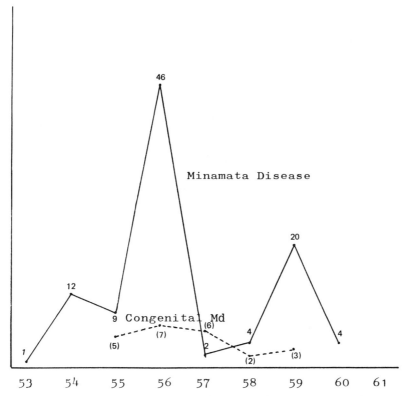

FIG. 5. Annual incidence of Minamata disease cases (designated) through 1971. Figures = number of postnatal patients [15,17]. Figures in paren- theses = number of congenital Md patients.

FIG. 6. Freehand drawing (showing tremor).

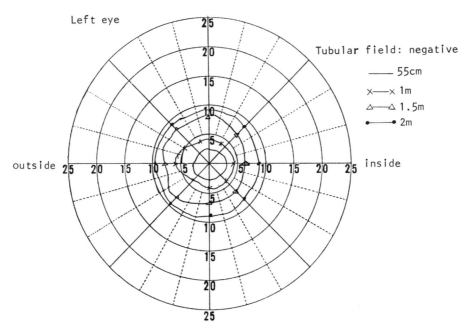

FIG. 7. Constriction of the visual field (21-year-old male).

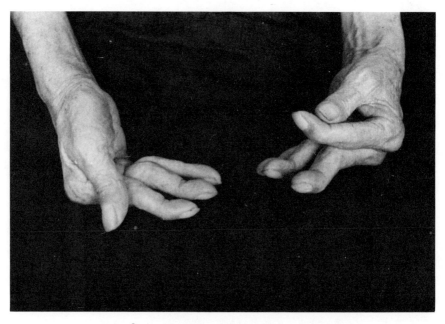

FIG. 8. Deformity of hand in Md patient.

FIG. 9. Infantile Md patient. Miss Kumiko M., born on November 8, 1950,
in the fishing village of Yudo, was taken ill on June 8, 1956. Onset began
with hypersalivation and difficulty in hand movements, followed by tremor
and ataxic gait (June 18) and dysarthria (June 20). These symptoms gradually worsened, and general paralysis ensued within 3 months. No expression
of consciousness was reflected in her face, and there was no response to
stimuli in other parts of her body. She lay in a state of akinesis, with
no voluntary movement, being fed forcibly by mouth. The patient died on
August 26, 1974.

was lack of coordination, weakness and tremor (Fig. 6), dysarthria (speech
slow and slurred), and ataxic gait, followed by disturbances of vision
(Fig. 7), and hearing. These symptoms became aggravated and led to general
paralysis, deformity of the limbs (Fig. 8), difficulty in swallowing, convulsions, and death (Fig. 9).

Among the patients discovered, 21 died within 1 year of onset of the
disease. Of these, 16 died within 3 months and 4 died within 6 months.

The most typical cases presented the so-called Hunter-Russell syndrome
[24]. Therefore, very specific clinical pictures were seen. These cases
had all the main symptoms (75.8-100% of cases) and followed a course which
was considered to be that of the acute or subacute form. It is reported
that the major symptoms developed within 1-3 weeks after the appearance of
the first symptoms: numbness of the limbs and perioral area, clumsiness
of hand movement, etc. (see Table 4, H. Tokuomi, 1960) [61,62].

TABLE 4. Clinical Symptoms of Minamata Disease[a]

Symptoms	%
Disturbance of sensation	
Superficial	100
Deep	100
Constriction of the visual field	100
Dysarthria	88.2
Ataxia	
Adiadochokinesis	93.5
In finger-finger, finger-nose tests	80.6
Romberg's sign	42.9
Ataxic gait	82.4
Impairment of hearing	85.3
Tremor	75.8
Tendon reflex	
Exaggerated	38.2
Weak	8.8
Pathological reflexes	11.8
Salivation	23.5
Mental disturbances	70.6

[a]Thirty-four adult cases of acute and subacute Md (H. Tokuomi, 1959) [61,62].

<u>Pathology of Minamata Disease (T. Takeuchi)</u>. In microscopic examination, the most conspicuous changes were noted in the cerebellum and the calcarine cortex (Table 5) [52,54].

1. Changes in the cerebellum: Granular-cell-type cerebellar atrophy was seen. The disintegration following diffuse loss of granular cells was most severe, while the Purkinje's cells were spared.

2. Focal changes in the cerebral cortex: In all cases the cortex of the calcarine region (Area Striata) was most grossly damaged in both the hemispheres. Also, changes essentially similar to those in the calcarine regions of the precentral cortex, postcentral cortex, temporal transverse cortex, temporal cortex, and insular cortex were often found.

3. Slight changes in the cerebral nuclei: In the cerebral nuclei, brain stem and cord, relatively less severe changes were observed (in acute or subacute cases).

TABLE 5. Pathological Changes of the Nervous System in Acute and Subacute Cases of Human Minamata Disease (H. Takeuchi) [54]

Changes Localization	Meningeal Edema	Perivascular Edema	Perivascular Infiltration	Nerve Cells Degeneration	Nerve Cells Severe Changes	Nerve Cells Loss	Nerve Cells Shrinkage	Glial Proliferation	Glial Degeneration	Myelin Degeneration	Myelin Demyelination
Cerebral hemisphere											
Frontal lobe	+	+	-	++	+	+	+	∓	+	+	∓
Gyrus precentralis	+	++	(∓)	+++	+++	+++	+++	++	++	++	++
Lateral lobe	+	++	-	++	++	++	++	++	+	+	+
Area striata	+	++	(∓)	+++	+++	+++	+++	++	++	++	++
Parahypocampi	+	+	-	∓	∓	-	∓	(∓)	+	+	-
Cerebral nuclei	∓	++	+	++	+	∓	∓	(∓)	+	+	-
Diencephalon	∓	++	+	++	∓	∓	∓	(∓)	+	+	-
Mesencephalon	+	+	-	+	-	-	∓	(∓)	+	+	-
Pons	-	+	-	+	-	-	∓	-	+	+	-
Medulla oblongata	+	+	-	+	+++	-	∓	-	+	+	-
Cerebellum											
Cortex	+	+	-	+++	+++	+++	+++	++	++	++	-
Dentate nucleus	∓	+	-	+	-	-	∓	-	+	+	-
Spinal cord	-	-	-	+	-	-	∓	-	+	+	-

4. Changes in the peripheral nerve: The selective destruction of sensory
   peripheral nerve fibers with demyelination was observed.

5. Changes in other organs: In acute and subacute cases, hypoplasia and
   aplasia of the bone marrow, hypoplasia of lymph nodes, and fatty degeneration of parenchymatous cells in the liver and kidney were found.

Congenital (Fetal, Prenatal) Minamata Disease

Epidemiological and Clinical Studies. From 1955 to 1959, during the
same period as the outbreak of Md, 17 cases of congenital idiocy accompanied
by various neurological symptoms were reported on the coast of Minamata
(Figs. 5, 10, and 11) [13,19,57]. Afterwards, it was concluded that MeHg
passed from the mother to the fetus through the placenta, thereby causing
MeHg poisoning. This disease was diagnosed as congenital Md in 1962 [13].

From the beginning it was suspected that these were MeHg poisoning.
However, none of them had eaten fish or shellfish from Minamata Bay [28,
31,32,41,43,44]. In none of these cases were any abnormal factors observed
during the mother's pregnancy, at delivery, nor as newborns that might have
caused cerebral palsy. Many of the children were breast fed and some had
received mixed feeding. The one common and notable factor was that the
mothers of all these patients had eaten a great deal of fish and shellfish
from Minamata Bay during pregnancy [13,19].

Moreover, the time and locations of incidence of the congenital patients coincided with those of the adult cases. In the Tsukinoura, Yudo,
and Modo areas (the Minamata area most affected by Md), there were 220
births between 1953 and 1955, and 13 of the infants had symptoms of Md, at
the point of 1962.

The figures for that area represent an incidence of 6.9% (M. Harada)
[13], conspicuously higher than the 0.2-0.3% rate of cerebral palsy cases
in other areas of Japan. In 1962, 64% of the congenital patients belonged
to families of which at least one member had typical, acute Md. All the
mothers of those patients were thought to be healthy at that time. However, careful observation revealed certain neurological symptoms in 73% of
the mothers (M. Harada) [13]. After 10 years had passed, slight manifestations of Md were found in almost all the mothers, and the symptoms worsened
(M. Harada) (Table 6) [19].

However, the symptoms of the mothers of the congenital Md patients
were generally mild. Also, those who had had miscarriages and stillbirths
were found to have relatively lighter symptoms than other members of their

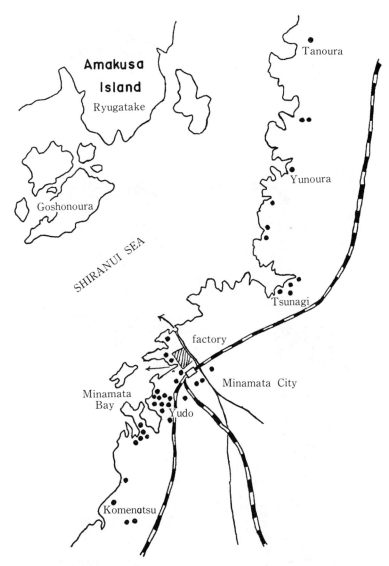

FIG. 10. Distribution of congenital Minamata disease (40 patients discovered up to 1974) [19].

families. It was suggested that MeHg was discharged from their bodies through the placenta into the fetus.

The mercury content in the hair of congenital Md patients born between 1955 and 1958 was still high in 1961. And the mercury content in the hair of their mothers was high also (Y. Harada) (Table 7) [21]. Abnormalities were found at about 6 months of age, when most patients showed instability

FIG. 11. Congenital Minamate disease (M. Harada, 1962).

TABLE 6. Symptoms in Mothers of Congenital Minamata Disease Patients
(M. Harada, 1974)

Symptoms	Cases	%
Sensory disturbance	28	100
Constriction of the visual field	16	57
Dysarthria	12	43
Incoordination (total)	22	79
Ataxic gait	12	43
Finger-finger test	16	57
Adiadochokinesis	22	79
Auditory disturbance	21	75
Tremor	11	39
Muscular rigidity and spasticity	10	35
Hyper-reflexes	11	39
Pathologic reflexes	3	11
Deformities of limbs	4	14
Pain	18	64
Intelligence disturbance	20	71

TABLE 7. Mercury Content in Hair of Congenital Minamata Disease Patients
and Their Mothers (Y. Harada)[a] [21]

	Congenital Minamata Disease		Mother
Sex	Age (year:months)	Mercury Content (ppm)	Mercury Content (ppm)
M	3:5	59.0	72.9
F	2:10	34.3	65.4
F	2:7	61.9	101.0
F	2:7	92.0	--
M	2:7	58.1	--
F	3:7	57.8	191.0
F	2:1	37.7	191.0
F	3:5	26.2	--
F	1:3	35.0	--
M	1:1	5.25	6.56
F	4:4	11.3	12.7
F	3:8	8.14	1.82

TABLE 7 (Continued)

Sex	Congenital Minamata Disease Age (year:months)	Mercury Content (ppm)	Mother Mercury Content (ppm)
F	3:6	10.6	--
M	4:1	100.0	90.0
M	4:3	73.5	16.0

[a]Calculated in 1961 by S. Kitamura [21,33].

of the neck; some showed convulsion or were unable to follow a light with their eyes [13,19,57].

The clinical characteristics of these patients were serious mental retardation, primitive reflexes such as oral tendency and grasping reflex, cerebellar symptoms such as instability of the neck, asynergy, incoordination, ataxia, adiadochokinesis, dysmetry, intention tremor, dysarthria, and nystagmus. They also manifested disturbance of growth, akinesia, hypokinesia, hyperkinesia (chorea, athetasis, etc.), hypersalivation, character disorder, paroxysmal symptoms (such as generalized tonic convulsion, loss of consciousness, myoclonic jerk, etc.), and also strabismus, deformity of limbs, and pyramidal symptoms (M. Harada, 1962) [13,19] (Table 8).

TABLE 8. Comparison of Clinical Symptoms in Congenital Minamata Disease, Exogenous Idiocy, and Infantile Cerebral Palsy (%) (M. Harada, 1962) [13,19]

	Cong. Md	Exog. Idiocy	Inf. C. P.
Intelligence disturbance	100	100	75
(severe)	(100)	(100)	(25)
Cerebellar symptoms	100	33	18
Primitive reflexes	100	61	31
Hyperkinesia	95	32	68
Salivation	95	32	68
Paroxysmal symptoms	82	16	0
Strabismus	77	11	37
Pyramidal symptoms	75	100	43
Dysarthria	100	--	--
Deformity of limbs	100	100	100

TABLE 8 (Continued)

	Cong. Md	Exog. Idiocy	Inf. C.P.
Inhibited growth	100	72	31
Asymmetry of the symptom	0	72	18
Causative factors present	0	56	62
Total no. of cases	17	18	16

Sensory disturbance and constriction of the visual field could not be confirmed at that time because impaired intelligence precluded examination. But later in 1972, constriction of the visual field was confirmed in four cases (H. Moriyama et al.) [42].

The clinical picture of these patients was similar to that of exogenous idiocy, seriously feeble-minded and physically handicapped infants, or cerebral palsy cases. However, when a comparison was made between the symptoms of congenital Md and similar symptoms in other areas of Japan, it was found that in the group of exogenous idiocy cases by other factors cerebellar symptoms and strabismus were rarely observed, but pyramidal symptoms and paralysis were conspicuous [13].

Among cerebral palsy patients in general, as compared to those of Minamata cases, mental retardation was slight and cerebellar symptoms were rarely observed, but hyperkinesia (particularly chlorea and athetosis) was common (M. Harada, 1962) [13].

When each congenital Md patient was observed individually, diagnosis was difficult. However, the characteristic clinical symptoms of congenital Md as compared to cerebral palsy and exogenous idiocy became clear when the patients were observed as groups. Therefore, an epidemiological survey was vital in diagnosing these patients. At present (1974), 40 such congenital cases have been found and still more are expected to appear (Table 9, Fig. 10).

Pathological Changes in Congenital Minamata Disease. In 1961 and 1962, two typical cases died, and the autopsies provided significant findings. Professor Takeuchi reported as follows (1962) [37,53,54,56]:

1. Bilateral cerebral atrophy and hypoplasia.
   a. Decrease and disappearance of the cortical nerve cells with glia proliferation.

TABLE 9. Frequency of Symptoms in Congenital Minamata Disease

	1962		1971		1974	
	Cases	%	Cases	%	Cases	%
Intelligence disturbance	17	100	25	100	37	100
Dysarthria	17	100	24	96	34	91
Extrapyramidal hyperkinesia	16	95	23	92	34	91
Cerebellar symptom	17	100	19	76	29	78
Deformities of limbs	17	100	21	84	23	62
Primitive reflexes	17	100	18	72	25	67
Strabismus	13	77	18	72	25	67
Hypersalivation	17	100	18	72	27	72
Paroxysmal symptom	14	82	9	36	13	35
Pathological reflex	12	75	12	48	17	45
Inhibited bodily growth	17	100	17	68	22	59
Number of cases	17	--	25	--	37[a]	--

[a]Based on a total of 40 cases of congenital Md, of which three died (M. Harada, 1974).

    b. Hypoplasia and malformation of the nerve cells.

2. Cerebellar atrophy and hypoplasia.

    a. Atrophy of the central and granular cell type.

    b. Hypoplasia of granular cell layers and other layers.

3. Abnormality of the cytoarchitecture.

    a. Columnar grouping of the cortical nerve cells.

    b. Remaining matrix cells.

    c. Intramedullary preservation of nerve cells.

4. Hypoplasia of corpus callosum.

5. Dysmyelination of the pyramidal tract.

6. Hydrocephalus externus.

7. Underdevelopment and emaciation of the body and organs, hypoplasia of the bone marrow, etc.

    Furthermore, it was proven by autoradiographical experiments of pregnant animals and histological experiments on their embryos that methyl

mercury passed through the placenta and damaged the central nervous system
of the embryos (N. Morikawa et al., 1961 [40], S. Tatetsu et al., 1969 [59],
E. Fujita, 1969 [12], T. Miyakawa, 1971 [39], H. Shiraki, 1972 [49]). It
has been proven that MeHg concentration is high in the milk of mothers who
have been contaminated with this poison (Y. Harada, 1966) [21] and that
poisoning by MeHg through mother's milk can be caused experimentally (M.
Deshimaru, 1969) [4]. Since the ingestion of milk containing MeHg causes
poisoning, it cannot be denied that these congenital cases may have been
aggravated by mother's milk after birth.

RECENT CLINICAL STUDIES OF MINAMATA DISEASE

Changes of Symptoms in Typical Acute or Subacute Minamata Disease

Even though 10 years have passed since first observation, the main
symptoms of acute and subacute Md found originally can still be seen frequently, indicating that these main symptoms of Md are difficult to improve
(Table 10). However, detailed investigation of those symptoms (not only
their frequency of appearance) indicates that there are considerable differences from symptom to symptom and from case to case. Some symptoms become so mild that they could not be diagnosed unless previous observations

TABLE 10. Changes in Frequency of Symptoms in Minamata Disease (%)[a] (M. Harada, 1972)

	1961	1970
Disturbance of coordination	100	92
Constriction of the visual field	91	91
Auditory disturbance	86	81
Dysarthria	83	83
Sensory disturbance	78	73
Salivation	49	39
Tremor	44	59
Pathological reflexes	7	15
Muscular rigidity and spasticity	12	28
Muscular atrophy	28	31
Number of cases examined	43	3

[a]Only officially designated patients were examined.

had been made, and unless other symptoms had become more prominent. Generally, among adults there has been improvement in cerebellar symptoms such as ataxia, dysarthria, tremor, and hypersalivation, a worsening of pyramidal symptoms (spastic gait, pathological reflex, etc.), muscular atrophy, deformity of fingers, and mental disturbance [16,58].

Among infantile and congenital cases generally, there has been improvement or even a disappearance of cerebral symptoms, hypersalivation, primitive reflex, inhibited bodily growth, and paroxysmal symptoms; but mental retardation, strabismus, and dysarthria remain, and disturbance of the intelligence becomes predominant (Table 11; M. Harada, 1971) [19].

Therefore, with the passing of years these formerly typical cases tend to become incomplete or atypical cases of Md. In some infantile and congenital cases, it became difficult to distinguish between them and ordinary cases of mental subnormality. These facts indicate the existence of slight, atypical, or incomplete cases from the beginning [15,18,19].

TABLE 11. Changes of Symptoms in Congenital Minamata Disease from 1962 to 1972

	1962 (17 Cases)	1971 (15 Cases)		
		Total	Disappeared	Improved
Intelligence disturbance	17(15)[a]	15	0	11
Primitive reflex	17(15)	10	5	3
Cerebellar symptom	15(15)	13	3	5
Extrapyramidal hyperkinesia and hypokinesia	16(14)	14	0	9
Dysarthria	17(15)	15	0	2
Paroxysmal symptom	14(12)	4	8	2
Strabismus	14(12)	12	0	5
Pathological reflex	12(10)	7	3	3
Hypersalivation	16(14)	12	2	6
Deformity of limbs	17(15)	15	0	9
Inhibited growth	17(15)	11	4	7

[a]In 1962, 17 patients were examined and were kept under observation up to the present. Two cases died and the remaining 15 cases were reexamined [19].

Health Survey of the Most Severely
Contaminated Minamata Area

It is a fact that in the Minamata area, whether symptoms had occurred
or not, and whether or not any members of the family had **contracted** typical
Md, virtually all the inhabitants were contaminated with MeHg at a very
high level. This occurred because they had been eating large quantities
of fish daily, as is the custom in the region. Also, a high mercury count
was detected in the hair of inhabitants of this area.

In such a situation, where the entire population of an area is subjected to MeHg contamination, one cannot determine its effect by picking
only those who clearly have all the typical symptoms of MeHg poisoning.

In 1972, surveys of health were conducted [17,60] in the following
three areas of the region (see Fig. 12): (1) Yudo, Tsukinoura, and Detsuki
in the Minamata area, where the contamination was most severe; (2) the Goshonoura area (across the Shiranui Sea from Minamata area), which was undoubtedly contaminated by mercury; although no patients had been discovered
until 1972, the dancing disease in cats had already been recognized; and
(3) the Ariake area, which was used as the control. Since its fishing
grounds are different from those of the other two, the Ariake area shows
little or no contamination.

The investigation showed that the neurological symptoms among the inhabitants of the Minamata area are of distinctly higher frequency than those
among people of the other areas. The rate would be even higher if the 37
patients of Md already officially designated as Md cases were included.
It was significant that the main symptoms of typical Md, including sensory
disturbance, incoordination, dysarthria, auditory disturbance, and constriction of the visual field, were seen to be increasing in proportion to
the degree of contamination. In other words, the greater the level of
contamination, the higher the frequency of symptoms.

The survey concluded that these neurological symptoms must be due to
the effects of MeHg (Table 12) [17,60]. When considering sensory disturbance, which is significantly more frequent, peripheral and perioral sensory
disturbances show an even more remarkable frequency. Therefore, this type
of sensory disturbance can be considered to be the effect of MeHg poisoning
(Table 13) [17]. However, some neurological symptoms such as sensory disturbance of one side or the lower half of the body, muscular atrophy, paroxysmal symptoms, pain in the limbs, pyramidal sign, etc. which were not

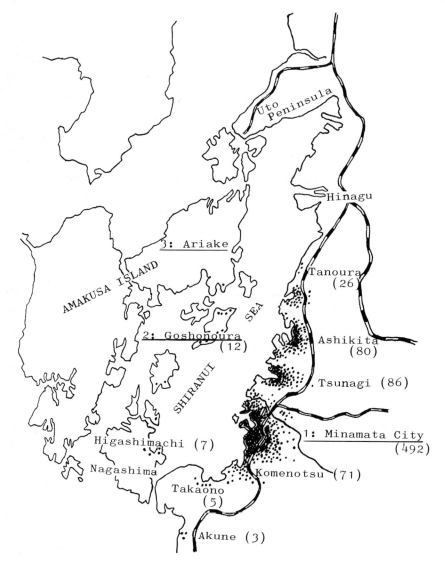

FIG. 12. Distribution of officially designated Minamata disease patients.
Figures in parentheses = number of patients designated through December
1974. See Tables 12 and 13 for tabulated data on areas 1, 2, and 3.

previously attributed to Md, were observed at a high rate among those sur-
veyed in the Minamata area (Table 13).

Even in the Goshonoura area, where no patients had been discovered
until 1972, effects of MeHg contamination on humans were found. Since
1973, 12 Md patients have been officially identified in this area (Fig. 12).
Therefore, even in such limited surveys as have been described above, MeHg

TABLE 12. Frequency (No. and %) of Neuropsychiatric Symptoms in Inhabitants of Minamata, Goshonoura, and Ariake Areas [17,60]

	Minamata	Goshonoura	Control
Sensory disturbance	260(28.0)	132(7.6)	94(10.3)
Incoordination	228(24.7)	193(11.2)	122(13.4)
Ataxic gait	84(9.0)	50(2.9)	20(2.2)
Adiadochokinesia	171(18.4)	101(5.8)	50(5.5)
Finger-nose test	106(11.4)	28(1.6)	11(1.2)
Dysarthria	114(12.2)	63(3.6)	18(1.9)
Auditory disturbance	272(29.2)	156(9.0)	135(14.9)
Constriction of the visual field	127(13.7)	47(2.7)	9(0.9)
Tremor	94(10.1)	87(5.0)	27(2.9)
Pathologic reflexes	56(6.0)	34(1.9)	21(2.3)
Hemiplegia	17(1.8)	9(0.5)	10(1.1)
Pain (limbs)	128(13.7)	92(5.3)	74(8.1)
Epileptic seizure	26(2.8)	19(1.1)	8(0.8)
Muscular atrophy	55(5.9)	13(0.7)	7(0.8)
Parkinsonismus	23(2.5)	13(0.7)	7(0.8)
Deformities of limbs	81(8.7)	104(6.0)	73(7.8)
Intelligence disturbance	211(22.7)	178(10.3)	98(10.8)
Hypertension	218(23.4)	237(27.5)	180(19.1)
Total	928(100.0)	1723(100.0)	904(100.0)
Officially designated Minamata disease	37	0	0

contamination was found, and previously concealed cases of Md were identified at this time (December 1974).

In the whole region there have been 784 officially designated patients of whom 103 have died. About 3000 persons are suspected Md cases and are applying for designation as Md sufferers (Fig. 12). There are approximately 100,000 inhabitants living on or near the shores of the Shiranui Sea. At this time it is impossible to estimate how many have been affected by MeHg contamination.

Moreover, the survey showed that the appearance of the symptoms was not limited to the 1953-1960 period, as had been thought before. After

TABLE 13. Types and Frequency of Sensory Disturbances in Inhabitants on Minamata, Goshonoura, and Ariake Areas [18]

	Total Number Showing Sensory Disturbances	(1)	(2)	(3)	(4)	(5)	(6)	(7)
Minamata	260 cases (28.0%)	70 (7.5)	145 (15.6)	7 (0.7)	51 (5.4)	21 (2.2)	14 (1.5)	16 (1.6)
Goshonoura	132 cases (7.6%)	9 (0.5)	64 (3.7)	2 (0.1)	17 (0.9)	13 (0.8)	2 (0.1)	35 (1.9)
Control (Ariake)	76 cases (8.4%)	1 (0.1)	29 (3.2)	0 (0)	8 (0.8)	5 (0.5)	0 (0)	29 (3.2)

In some cases, type (4) or (5) is counted twice.

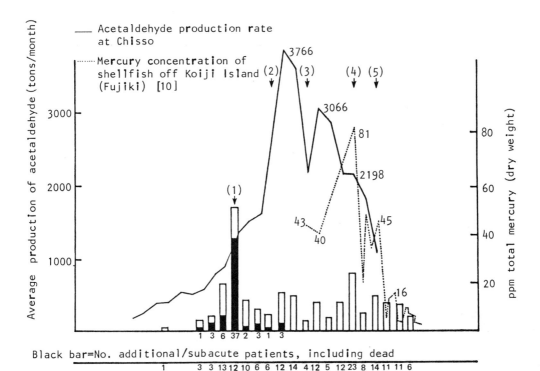

FIG. 13. Comparative chart of acetaldehyde production rate, incidence of
Md cases, and mercury content of shellfish from Minamata Bay. The patients
shown in this chart are limited to the Tsukinoura, Detsuki, and Yudo areas.
(1) Official discovery of patients; (2) cause proven to be MeHg; (3) year-
long labor dispute; (4) circulatory waste-water system installed; (5) operation suspended.

1960, there were people whose symptoms worsened, and others whose symptoms
appeared for the first time. The onset of symptoms was correlated with the
acetaldehyde production of the Chisso Corporation and the mercury content
of shellfish from Minamata Bay (Fig. 13).

     Survey of Healthy Pupils in the Minamata Area

     From our clinical observation of congenital Md patients, we presume
that some slight or incomplete cases of the congenital Md may coincide with
the clinical picture of ordinary mental retardation.

     In 1962 we conducted our first survey, examining children living in
Yudo, Tsukinoura, Detsuki, and Modo. Among these children, born from 1953

TABLE 14. Frequency (No. and %) of Occurrence of Symptoms in Pupils of Lower Secondary School (13-16 Years Old)[a]

	Minamata Area	Control
Sensory disturbance[b]	47(21)	5(3)
Auditory disturbance	19(9)	19(10)
Strabismus	11(5)	10(5)
Nystagmus	11(5)	11(5)
Adiadochokinesia[b]	20(9)	8(4)
Dysarthria[b]	26(12)	4(2)
Pathological reflex	8(4)	6(3)
Constriction of the visual field	2(1)	0(0)
Intelligence disturbance[b]	39(18)	14(7)
Total	223(100)	196(100)

[a]Examined by M. Harada and T. Fujino in 1971 [19].

[b]$P \leq 0.01$.

to 1960, three cases of incomplete or slight congenital Md were found. Surprisingly, we also found mental retardation at the high rate of 29.1%. At first it was thought that this mental subnormality was unrelated to the congenital Md (M. Harada, 1962) [13].

In 1971, we conducted our second survey, covering all pupils in the lower secondary schools in these areas. This revealed that despite their school attendance, some 18% showed mental retardation; and in comparison with the control group, there was a higher rate of slight neurological symptoms such as sensory and auditory disturbance (Table 14). These mentally retarded children aroused interest as part of the "normal" matrix surrounding those diagnosed as having congenital Md [20].

Clinical Characteristic of Chronic Minamata Disease

In the chronic cases of Md that have been discovered recently, the basic symptoms are constriction of the visual field and sensory disturbance. Dysarthria and ataxia (cerebellar symptoms) are relatively slight or non-existent. However, pyramidal symptoms, muscular atrophy and mental disturbance are prominent. Therefore, some such cases have been diagnosed and treated as other diseases. This occurred because the symptoms of Md were concealed by hemiplegia, muscular atrophy, pyramidal symptoms, Parkinsonism, or dementia.

Most characteristic of such chronic cases is their course. Frequently, it takes from 3 to 10 years for the symptoms to worsen (the chronic, slowly progressive course), and in most cases, the first symptoms appeared after 1960 (delayed onset). In some cases, symptoms of MeHg poisoning such as ataxia, sensory disturbance, and dysarthria were precipitated or aggravated by apoplectic-like attack.

In the chronic cases, not all the clinical pictures could be understood solely in terms of the typical Hunter-Russell syndrome [24,25]. In order to clarify the clinical pictures, the chronic cases were classified into seven types [16], depending on the main symptoms (Table 15):

1. Hunter-Russell type: Cerebellar symptoms are conspicuous in cases of this type. The symptoms appeared after 1960 and have a tendency to worsen gradually.

2. Polyneuropathy type: The main symptom is peripheral sensory disturbance, but other symptoms are relatively slight or even nonexistent. Generally, cases of this type are the mildest of the chronic cases, and are found in families of severely affected Md patients.

3. Myopathy type (with muscular atrophy): The main symptom of this type is muscular atrophy accompanied by pyramidal symptoms and sensory disturbance. When constriction of visual fields and other symptoms of Md

TABLE 15. Clinical Types of Minamata Disease[a]

	No. of Cases
Hunter-Russell type	11
Polyneuropathy type	41
Myopathy type	6
Apoplectic type	12
Dementia type	12
Myelopathy type	2
Parkinsonism type	3
Others	3

[a]Patients are those discovered in families of acute or subacute typical cases (M. Harada, 1972) [16].

are not observed, it is difficult to distinguish this type from MLS and
other muscular diseases.

4. Apoplectic type: Here, hemiplegia or sensory disturbance of one side is
   observed. In most cases, the symptoms occur after an attack similar to
   apoplexy. However, on the side opposite the hemiplegia, peripheral sensory disturbance and ataxia are observed; also, constriction of visual
   fields and perioral sensory disturbance are seen.

5. Dementia type: Disturbance of the intelligence is conspicuous and is
   accompanied by dysarthria, ataxia, and tremor. In most cases, these
   symptoms of Md, such as sensory disturbance, ataxia, and tremor, had
   been observed even before the onset of dementia. In most cases of this
   type, dementia prevents verification of constriction of visual fields
   and sensory disturbance.

6. Myelopathy type: Main symptoms are disturbances of the lower half of
   the body accompanied by incontinence of urine. However, dysarthria,
   ataxia of the upper extremities, and constriction of visual fields are
   observed. Such cases are similar to spastic spinal paralysis and other
   diseases of the spinal cord.

7. Parkinsonism type: When constriction of the visual field and peripheral
   sensory disturbance are not observed, diagnosis of this type may be difficult.

The above types are regarded as Md because in addition to the predominant symptoms, other common symptoms of typical MeHg poisoning are observed. Many typical Md patients are found among members of these patients'
families. Moreover, the variety of symptoms in chronic cases coincides with
recent pathological changes in such cases observed by Prof. T. Takeuchi
(1972). In addition to the original pathological changes in Md patients,
some other characteristic changes are seen. Brain damage is less localized
and more diffuse. Especially noticeable are demyelization of the bone marrow, sclerotic atrophy of nerve cells, and lesion of the pyramidal tract
of the spinal cord [26,54,55]. However, mercury value in the hair or blood,
which is considered a decisive factor, is low, according to present findings. Other important criteria are life history, family history, and comparison with other patients.

Survey of Chronic Minamata Disease

In Minamata and surrounding areas, symptoms of chronic Md are becoming
more evident in many individuals and appear for the first time in some.

However, most of the fish and shellfish in this area have an MeHg content
of less than 1.0 ppm, and persons showing these symptoms have low levels
of mercury content in their hair (14.82-2.10 ppm) (M. Fujiki, 1973) [11].
This fact cannot be explained by new poisoning due to intake of mercury at
a high level. There are three plausible explanations for the fact that
symptoms are coming to the surface now:

1. The high level of mercury intake during the period before and after
   1956 is causing the symptoms now [48].

2. Brain damage already caused by MeHg during the period of greatest contamination is now surfacing as a result of aging or other factors or
   complications.

3. In Minamata the inhabitants continued to eat fish which were contaminated with a relatively small amount of MeHg.

Thus, the new outbreak of Md may be due to long-term, low-level ingestion
of MeHg.

At this point it is impossible to state which of these three factors
is causing the present appearance of symptoms. We conducted three surveys
in order to determine the most important factor among those mentioned.

The Long-Term Effect of Resting Mercury. Forty-nine cases who were
living in the Minamata area around 1956, but departed afterward, were examined in the Nagoya and Osaka areas. For limited periods these patients
had been eating fish that was highly contaminated with MeHg. From the
survey it became clear that neurological symptoms are found in many of
those cases and that in some patients the symptoms have a tendency to
worsen (Table 16) (M. Harada, 1973). The syndromes are of the incomplete
or atypical category but suggest that symptoms of Md can appear many years
after ingestion of mercury has been suspended.

The Relation Between Symptoms and Aging in Chronic Cases. It is known
that in aged patients symptoms frequently worsen. However, our survey
showed that age does not play a definite role in generating new patients
(Table 17) [17,48].

Previous Mercury Content of the Hair and Present Clinical Symptoms.
In the years from 1960 to 1961, the mercury content in the hair of fishermen of the Shiranui region was calculated. In the present survey, 137 of
the same persons were reexamined [15]. Forty-eight were from the Minamata
area, 66 from the Goshonoura area, and 23 from the Ashikita area (see
Table 3).

TABLE 16. Long-Term Effects of Residual Mercury (M. Harada, 1972)

Symptoms	Years[a]		
	< 3	< 5	> 5
Sensory disturbance	14	9	13
(Perioral)	(8)	(4)	(4)
Constriction of the visual field	6	4	11
Auditory disturbance	8	8	16
Disturbance of coordination	7	5	11
Dysarthria	8	6	10
Tremor	5	6	8
Pyramidal symptom	3	2	4
Number of cases	16	12	21
Presence of other diseases	2	2	10

[a]After leaving Minamata area.

TABLE 17. Annual Distribution of Patients with Chronic Cases and the Relationship Between Appearance of Symptoms and Age (M. Harada, 1972)

	Age Under 50 Years	Age Over 50 Years
1951	1	4
1952	0	1
1953	0	1
1954	1	1
1955	2	4
1956	9	10
1957	3	3
1958	2	2
1959	1	5
1960	0	6
1961	4	3
1962	3	3
1963	0	3
1964	0	0
1965	3	4

TABLE 17 (Continued)

	Age Under 50 Years	Age Over 50 Years
1966	1	4
1967	2	1
1968	1	3
1969	3	2
1970	6	3

A high rate of the main symptoms of Md was observed (Table 18). According to the previous mercury content in the hair, the sampling was divided into groups of under 10 ppm, under 50 ppm, under 100 ppm, and over 100 ppm. A high rate of symptoms in the under-50-ppm groups was found. Apparently, the present symptoms of the chronic cases are not related to the mercury content of their hair in those days but to the subsequent

TABLE 18. Previous Mercury Content in the Hair and Present Clinical Symptoms (No. and %)[a]

Symptom	< 10	10–50	10–100	> 100
Peripheral sensory disturbance	11(41)	40(51)	26(84)	6(60)
Central sensory disturbance	2(7)	7(9)	1(3)	1(10)
Ataxic gait	6(22)	28(35)	16(52)	5(50)
Adiadochokinesia	8(30)	36(46)	20(65)	5(50)
Dysmetria	8(30)	23(29)	15(48)	4(40)
Constriction of the visual field	9(33)	23(29)	13(42)	4(40)
Impairment of hearing	10(37)	37(47)	24(77)	6(60)
Dysarthria	10(37)	31(39)	18(58)	4(40)
Tremor	5(19)	17(22)	10(32)	1(10)
Deformity of limbs	5(19)	12(15)	5(16)	2(20)
Pain in limbs, neuralgia	8(30)	24(30)	14(45)	3(30)
Total	27(100)	79(100)	31(100)	10(100)

[a]Cases tested from November 1960 to the beginning of 1961 (see Table 3). Data from clinical survey (1972) of survivors among inhabitants tested in 1970-1971 (M. Harada, 1972) [15].

consumption of contaminated fish and shellfish. Indeed, in Minamata Bay
mercury contamination persisted over a long period of time (see Fig. 3).
From these three surveys I concluded that mercury ingested long ago may
cause the later appearance of symptoms of Md. Such delayed appearance may
have something to do with aging, or may be related to the ingestion of MeHg
over a long period. Any or all of these factors may be involved.

CONSIDERATION OF THE SAFETY LEVEL

The effect of MeHg on the human body in terms of the degree of contamination may be thought of as follows: When MeHg enters the body in
large doses, there are symptoms of acute brain damage such as aberrations
of consciousness, convulsions, and paralysis, followed by death. When the
MeHg intake is lower, its effects appear as a subacute case with typical
symptoms as reported by Hunter and Russell [24,25]. Still lower doses of
MeHg cause mild, atypical, or incomplete cases as seen in chronic Md or
may appear as a nonspecific disease.

Previously, it was thought that the harmful effects of MeHg were confined to the nervous system. However, it has become apparent that effects
on other organs must be considered also [17]. This is especially true in
chronic Md or when there is low-level contamination by MeHg over a long
period. For example, disturbances of liver function, diabetes, and hypertension are abnormally prevalent in chronic cases, recent examinations
show (Table 19) (T. Fujino, 1974). Indeed, pathological findings on Md by
autopsy have shown damage by MeHg to the liver, kidneys, pancreas, and bone
marrow (T. Takeuchi) [54].

Concerning the effects of MeHg on mother and child, it has been observed that when the female's intake of the poison is large and she becomes
ill with Md, pregnancy does not occur. When the dosage is smaller, pregnancy occurs but the fetus is aborted spontaneously or is stillborn. An
even smaller dosage permits conception and live birth, but the baby will
suffer from congenital Md, as mentioned previously, with severe neurological
symptoms.

A dosage too small to cause noticeable neurological symptoms in the
child may still cause congenital mental deficiency. But in any of these
cases, the mother's symptoms are relatively mild. In Minamata, the levels
of poison were so great that the heavier dosages could be readily detected.
However, it is very possible that in other areas with lower levels of

TABLE 19. Frequency of Abnormal Findings in Biochemical Examinations
(Chronic Minamata Disease)

	Patients
	No. Abnormal/No. Examined[a]
Direct bilirubin	7/152
GPT	22/152
GOT	17/152
Thymol turbidity test	1/152
Zinc sulfate test	18/152
Serum protein	15/152
Alkalic phosphotase	16/152
Serum cholesterol	18/152
LDH	7/152
Serum amylase	3/152
Blood urea nitrogen	21/152
Uric acid	12/152
Total lipoprotein	15/152
Lipoprotein	20/152
Blood sugar level (hunger time, 110 mg)	64/152
Glucose tolerance test (positive)	46/152

[a]By K. Fujino in 1974.

contamination, symptoms may go undetected or be attributed to some cause
other than poisoning. In light of this probability, the worldwide implications of the effects of long-term, low-level poisoning become self-
evident. And this issue involves the present concept of what is considered
a "safe" level of MeHg consumption.

In cases of Md, it has been calculated that an accumulation of 30 mg
of MeHg in a 70-kg adult causes sensory disturbance and that a 100-mg accumulation causes all the typical Md symptoms.

In Iraq, a 51-kg adult developed sensory disturbance with an accumulation of 25 mg of MeHg, disturbance of gait with 55 mg, dysarthria with
90 mg, auditory disturbance with 170 mg, and a 200-mg accumulation caused
death (F. Bakir et al., 1973) [2].

When considering incomplete and light cases, one can assume that an
accumulation of over 25 mg is dangerous. On the basis of this calculation,

the oral intake and the rate of output are significant factors in measuring the total accumulation of MeHg at a given point. Experimentally, the biological half-life of MeHg is calculated at 70 days (L. Eckmann et al., 1968) [78]. Since the accumulation in the brain is directly connected to the appearance of symptoms, the half-life of MeHg in the brain is most significant. It has been found that MeHg can easily enter the brain and that it remains there much longer than in other organs (T. Takeuchi, 1972) [55]. The present safety level has been based on the 70 days half-life for calculation, and acute or subacute typical cases have been used also as the basis for calculating the safety standard. Thus, calculations ignore the chronic and atypical case now being found in Minamata. These calculations ignore the higher susceptibility of the fetus to MeHg poisoning, and do not fully consider the stronger effect such poisoning may have on individuals with other illnesses.

It must be noted also that there have been no experiments as yet to test the presently calculated safety level where amounts considered marginally safe are consumed continuously over a period of many years.

The old concept of the safety level is based upon the theory that symptoms do not appear until a certain level of MeHg accumulation is attained. However, it is misleading to think that MeHg does not cause any cell damage until it reaches a certain level in the human body. Although outward symptoms may not appear until such a level is reached, it can be assumed that damage is proportionate, in some degree, to the amount of MeHg consumed. The larger the MeHg intake, the greater the cell damage. The smaller the intake, the fewer the cells that are damaged. Therefore, on the cell level, there is no "threshold." In fact, in the Minamata area, symptoms are becoming more evident in many individuals and are appearing for the first time in some individuals, as mentioned above. At present, most of the fish and shellfish caught in this area have relatively low mercury contents, and people showing Md symptoms have low levels of mercury in their hair. This fact cannot be explained by the mercury "accumulation theory" based upon the concept of the biological half-life.

And finally, even in this day and age of multiple pollutants in the environment, comprehensive safety standards have not been considered. There is a possibility of multiple contamination with mercury. When these various factors are considered, it becomes clear that the present approach in calculating the safety level must be seriously reexamined.

REMAINING PROBLEMS

The full range of neurological symptoms caused by the ingestion of
MeHg has not been fully clarified, and follow-up studies are needed to explain what happens within the body after MeHg is ingested. Further research
is needed also into the effects on other organs, treatment of Md, pathology
(especially that of chronic stages), safety levels, effect on chromosomes,
etc.

Most urgent is an investigation into the scope of contamination to
determine how many persons have been exposed to dangerous contamination and
the degree to which each exposed person has been affected by it. At present,
there is not even a reliable estimate of the number of Md sufferers. Research on the conditions and degree of contamination of fish and shellfish
must be conducted continuously, and the mercury-polluted sludge still deposited in Minamata Bay must be removed promptly.

Also urgent is more adequate medical and economic relief for the victims. It must be emphasized that the outbreak of Md has not been concluded
by any means. What we must do is to utilize properly the knowledge gained
so far, not only to clear away the remaining problems of Minamata and other
affected regions in Japan, but also to spare the rest of mankind from a
similar ordeal. There is no way to reach these goals but to make the results of this massive, though unintentional, human experiment entirely
clear. Minamata has become a symbol of environmental pollution throughout
the world. Properly viewed, our experience can be a means of preventing
its tragic consequences elsewhere.

REFERENCES

1. A. Ahlmark, Brit. J. Ind. Med., 5, 117 (1948).

2. F. Bakir, S. F. Damluji, and L. Amin Zaki, Science, 181, 230 (1973).

3. I. A. Brown, Arch. Neurol. Psychiatr., 72, 674 (1954).

4. M. Deshimaru, Psychiatr. Neurol. Jap., 71, 506 (1969) (in Japanese).

5. G. N. Edwards, St. Bartholomew's Hosp. Rep. (London), 1, 141 (1865).

6. G. N. Edwards, St. Bartholomew's Hosp. Rep. (London), 2, 211 (1866).

7. L. Ekmann, Nord. Med., 79, 450 (1968), in G. Löfroth, Kagaku, 39, 658
    (1969) (in Japanese).

8. G. Engleson and T. Herner, Acta Pediatr., 41, 289 (1952).

9. M. Fujiki, J. Kumamoto Med. Soc., 37, 494 (1963) (in Japanese).

10. M. Fujiki, Presented at the Sixth Int. Water Pollution Research,
     Sec. 6, 12 (1972).

11. M. Fujiki, S. Tazima, and S. Oomovi, A Report of the Second Medical Study Group, Kumamoto University, Japan, 1972 (in Japanese).

12. E. Fujita, J. Kumamoto Med. Soc., 43, 47 (1969) (in Japanese).

13. M. Harada, Psychiatr. Neurol. Jap., 66, 429 (1964) (in Japanese).

14. M. Harada, Science, 41, 250 (1972) (in Japanese).

15. M. Harada, Advan. Neurol. Sci., 16, 870 (1972) (in Japanese).

16. M. Harada, Psychiatr. Neurol. Jap., 74, 668 (1972) (in Japanese).

17. M. Harada, Minamata Disease: A Medical Report in Minamata (W. Eugene Smith, words and photographs), Holt, Rinehart, and Winston, New York, 1975, pp. 180-192.

18. M. Harada, Japan Quarterly, 25, 20 (1978).

19. M. Harada, Bull. Inst. Constitutional Med., Kumamoto Univ. Suppl., 25, 1 (1976).

20. M. Harada, Brain and Develop., 6, 378 (1974) (in Japanese).

21. Y. Harada, Minamata Disease, Study Group of Minamata Disease, Kumamoto University, Japan, 1968, pp. 93-117.

22. W. H. Hill, Can. Public Health J., 34, 158 (1943).

23. O. Höök, K. D. Lundgren, and A. Swensson, Acta Med. Scand., 150, 131 (1954).

24. D. Hunter, R. R. Bombord, and D. S. Russell, Quart. J. Med., 9, 193 (1940).

25. D. Hunter and D. S. Russell, J. Neurol. Neurosurg. Psychiatr., 17, 235 (1954).

26. K. Irukayama, F. Kai, M. Fujiki, T. Kondo, S. Tazima, T. Den, H. Satsa, M. Hashiguchi, and S. Ushigusa, Jap. J. Public Health, 11, 645 (1964) (in Japanese).

27. K. Irukayama and M. Fujiki, Jap. J. Public Health, 19, 25 (1972) (in Japanese).

28. T. Kakita, J. Kumamoto Med. Soc., 35, 287 (1961) (in Japanese).

29. S. Kitamura, N. Miyata, M. Tomita, M. Date, K. Ueda, H. Misumi, T. Kojima, H. Minamoto, S. Kurimoto, Y. Noguchi, and R. Nakama, J. Kumamoto Med. Soc., 31 (Suppl. 1), 1 (1957) (in Japanese).

30. S. Kitamura, N. Miyata, M. Tomita, M. Date, K. Ueda, H. Misumi, T. Kojima, E. Honda, and K. Fukagawa, J. Kumamoto Med. Soc., 31 (Suppl. 2), 238 (1957) (in Japanese).

31. S. Kitamura, Y. Hirano, Y. Noguchi, T. Kojima, T. Kakita, and H. Kuwamoto, J. Kumamoto Med. Soc., 33 (Suppl. 3), 559 (1959) (in Japanese).

32. S. Kitamura, T. Kakita, Z. Kojoo, and T. Kojima, J. Kumamoto Med. Soc., 34 (Suppl. 3), 477 (1960) (in Japanese).

33. S. Kitamura, Minamata Disease, Study Group of Minamata Disease, Kumamoto University, Japan, 1968, pp. 257-266.

34. T. Kondo, J. Kumamoto Med. Soc., 38, 353 (1964) (in Japanese).

35. K. D. Lundgren and A. Swensson, Nor. Hyg. Toxicol., 29, 1 (1948).

36. H. Matsumoto, K. Kameda, and T. Takeuchi, Advan. Neurol. Sci., 10, 729 (1966) (in Japanese).

37. H. Matsumoto, G. Koya, and T. Takeuchi, J. Neuropath. Exp. Neurol., 24, 563 (1965).

38. D. McAlpine and S. Araki, Lancet, 1958, 629 (1958).

39. T. Miyakawa, Advan. Neurol. Sci., 16, 861 (1972) (in Japanese).

40. N. Morikawa, Kumamoto Med. J., 14, 87 (1961).

41. H. Moriyama, J. Kumamoto Med. Soc., 41, 506 (1962) (in Japanese).

42. H. Moriyama, Y Harada, and Y. Sakata, Brain and Develop., 5, 420 (1973) (in Japanese).

43. S. Nagano, Y. Harada, S. Nishiura, T. Ueno, K. Hashiguchi, T. Naga-yoshi, Z. Nagata, Z. Oohara, S. Nishida, H. Hosokawa, K. Noda, and T. Kakita, J. Kumamoto Med. Soc., 31 (Suppl. 1), 1 (1957) (in Japanese).

44. S. Nagano, T. Kida, Y. Harada, K. Hashiguchi, T. Ueno, H. Kumabe, T. Nagayoshi, S. Setomoto, K. Inokuchi, T. Ueda, A. Ikezawa, S. Nishida, K. Kawahara, T. Ninomiya, K. Miyatake, T. Sakaguchi, T. Kakita, and M. Fuka, J. Kumamoto Med. Soc., 34 (Suppl. 3), 511 (1960).

45. A. Pentschew, Handbuch d. Speziellen Pathologischen Anatomie u. His-tologie, Bd. 13, Teil 2, Druck d. Univ. Stürz AG, Würzburg, 1958, p. 2007.

46. K. Sera, A. Matsunaga, T. Sasaki, Z. Oonishi, M. Matsukura, H. Mura-kami, K. Sato, H. Yamashita, H. Yoshimovi, T. Miyazaki, K. Shinagawa, and S. Ookubo, J. Kumamoto Med. Soc., 31 (Suppl. 2) 307 (1957) (in Japanese).

47. Y. Shiraishi, J. Kumamoto Med. Soc., 37, 361 (1963).

48. H. Shiraki, Advan. Neurol. Sci., 13, 113 (1969).

49. H. Shiraki, J. Hyg. Chem., 17, 93 (1972).

50. A. Swensson, Acta Med. Scand., 143, 365 (1952).

51. T. Takeuchi, H. Matsumoto, A. Fujii, and H. Ito, J. Kumamoto Med. Soc., 31 (Suppl. 2), 267 (1957) (in Japanese).

52. T. Takeuchi and N. Morikawa, Psychiatr. Neurol. Jap., 62, 1850 (1960) (in Japanese).

53. T. Takeuchi, H. Matsumoto, and G. Koya, Advan. Neurol. Sci., 8, 867 (1964) (in Japanese).

54. T. Takeuchi, Minamata Disease, Study Group of Minamata Disease, Kuma-moto University, Japan, 1968, pp. 141-228.

55. T. Takeuchi, M. Eto, H. Kojima, Y. Ootsuka, T. Miyayama, S. Suko, T. Sakai, N. Sakurama, T. Iwamasa, and H. Matsumoto, J. Kumamoto Med. Soc., 46, 666 (1972) (in Japanese).

56. T. Takeuchi, in Intern. Congr. Environ. Mercury Contamination (1970), Ann Arbor Sci., 1972, pp. 247-301.

57. S. Tatetsu and M. Harada, Advan. Neurol. Sci., 12, 181 (1968) (in Japanese).

58. S. Tatetsu, E. Murayama, and M. Harada, Advan. Neurol. Sci., 13, 76 (1969) (in Japanese).

59. S. Tatetsu, M. Takaki, and T. Miyakawa, Advan. Neurol. Sci., 13, 130 (1969) (in Japanese).

60. M. Tatetsu, M. Harada, E. Hattori, T. Inoue, K. Kabashima, and R. Minami, Minamata Disease, Methylmercury Poisoning in Minamata and

Niigata, Japan (T. Tsubaki and K. Irukayama, eds.), Kodansha Ltd., Tokyo, 1977, pp. 240-267.

61. H. Tokuomi, Psychiatr. Neurol. Jap., 62, 1816 (1960) (in Japanese).

62. H. Tokuomi, T. Okazima, M. Yamashita, and S. Matsui, Advan. Neurol. Sci., 7, 276 (1963) (in Japanese).

63. H. Tokuomi and T. Okajima, Advan. Neurol. Sci., 13, 69 (1969) (in Japanese).

64. M. Uchida, Biochemistry, 35, 430 (1963) (in Japanese).

TOXICITY OF INORGANIC AND ORGANIC MERCURY COMPOUNDS IN ANIMALS

Delmar Ronald Cassidy

National Veterinary Services Laboratories
Ames, Iowa

Allan Furr

Iowa State University
Ames, Iowa

INORGANIC MERCURY

Mercury was probably first used by man in the form of the sulfide ore, cinnabar, as a source of a red pigment for his early artistic efforts. Archeologists have found drawings in the ruins of ancient Egypt and Pakistan that were created using this pigment [1].

The extended use of mercury compounds in medicaments and amalgams is recorded by the Romans prior to and following the time of Christ [1].

Putman [1] writes that Aristotle in the fourth century B.C. was the first to leave a written account of mercury, which was called "liquid silver." Five centuries later Dioscorides recorded that cinnabar was "good for eye medicines" in that it healed burnings and breaking out of pustules. He also noted that it was dangerous if swallowed [1].

The Romans used cinnabar for writing their books and as a pigment to decorate tombs, statues, and walls. They also used elemental mercury as an amalgam to separate gold from other material and as an amalgam to coat gold onto copper [1].

The medieval alchemists must have been fascinated when by heating the red ore cinnabar in an oven a heavy, silvery-white liquid was obtained [1]. The early Hindu wise men thought mercury had aphrodisiacal properties. The sages of ancient China thought it to have immortal attributes.

The Arabian and European alchemists considered it to be one of the two "contraries," the other "contrary" being sulfur, from which all other elements were believed to emerge. The alchemists named the liquid metal after the fleet-footed Greek god, Mercury [1].

As time and science progressed, the various compounds of mercury, both inorganic and organic, were developed. Their usage in human and veterinary medicine gradually increased. The initial application was primarily to aid in control of bacteria, fungi, and other pests considered detrimental to the welfare of man.

As early as 1705 mercuric chloride was used as a wood preservative and by the 1800s was used to control wheat smut in grain [2,3]. It was used as a weak tincture as early as 1882 to control bedbugs. In 1860, it is recorded as being used to control earthworms. In England as early as 1864, it was used to control cabbage maggots. In 1889, a mercuric solution was discovered that would control mold on germinating seed [2,3]. The 1883 Pharmacopoeia of the United States of America [4] lists 12 mercury compounds formulated in that era as medicaments to be prescribed for human ailments.

The inorganic mercury compounds have been used extensively as antiseptics, disinfectants, purgatives, and counterirritants in human and veterinary medicine. Many of these applications are gradually being replaced by other compounds that are more specifically effective yet offer a wider range of safety.

Until the early 1900s, the major toxic problems involving mercury compounds in veterinary medicine had been the occasional poisoning that resulted from the injudicious use of mercurial antiseptics, ointments, and diuretics. In 1915, Germany introduced the use of organic mercury compounds as seed treatment to control fungi [5].

Chemical and Physical Characteristics of Mercury

Mercury compounds are chemically classified as inorganic if either in the form of the elemental metal or as salts in the mercurous or mercuric state. The organic compounds are those associated with carbon in their molecular structure [6].

The elemental metallic form of mercury (Hg), often referred to as quicksilver, has an atomic weight of 200.59 and the atomic number of 80 on the periodic chart. Mercury has a valence of either 1 (mercurous) or 2 (mercuric). The chemical symbol (Hg) is derived from its Latin name, hydraargyrum [6].

The metallic form has the characteristic of being the only metal that is a liquid at ordinary temperatures. It is a silvery-white, mobile, very heavy liquid which at $-39°$ F $(-38.87°$ C) solidifies to a tin-white, malleable mass which can be cut with a knife [6].

The high density (13.59) of this metal makes it uniquely suited to uses that require the mass of a very heavy liquid substance [6].

Mercury is slightly volatile at room temperatures. The volatility increases appreciably at increased temperatures, and it is emitted as an invisible mercury vapor. The volatility of mercury can be subtly hazardous to personnel working in laboratories and industry if careful handling procedures and proper ventilation are not employed.

The metal does not tarnish at room temperatures, but at temperatures near its boiling point of $356.58°$ C it slowly oxidizes to red oxide of mercury [6].

Mercury readily forms amalgams with most metals with the exception of iron. It very readily combines with sulfur; hence the term mercapto, i.e. sulfur-capturing, is used in conjunction with naming many of the sulfur compounds [6].

Commercial and Environmental Sources

Mercury may occur in nature as the metallic form; however, the major commercial source is as cinnabar, mercuric sulfide (HgS), a scarlet-red ore, commonly called vermilion, which has a specific gravity of 8 to 8.2 [6].

Cinnabar occurs abundantly in underground veins in various parts of the world. A major portion of the world's production of mercury is obtained from mines in Spain, Austria, and Italy [1,6]. The supply of mercury for the United States is obtained partially from mines such as those in California, with imported mercury augmenting the domestic supply [1,6-8].

The mine at Alamden, Spain, the earth's richest known mercury deposit, has been mined since the days of the Roman empire. It still produces about 15% of the world's supply of mercury [1].

The crushed cinnabar is heated to about $1500°$ F in furnaces to drive off the mercury as a vapor which is condensed into almost pure elemental mercury [1,6].

The commercial unit of measurement for mercury is the "flask" which contains 76 lb (34.5 kg) by weight or approximately 2.54 liters by volume [1,6-8].

The total production of mercury in the United States in 1972 was
48,072 flasks (about 3,600,000 lb), with 6296 flasks (about 478,000 lb) from
domestic sources, 29,178 flasks (about 2,200,000 lb) being imported, while
400 flasks (about 30,000 lb) were exported [7].

The world production figures for the years 1971 and 1972 are not completed; however, in 1970 the world production was 279,000 flasks (about
21,000,000 lb), while in the same year the total production in the United
States was 45,650 flasks (about 3,400,000 lb) [7].

The New York price for mercury for the years 1965-1969 ranged from a
low yearly average of $441.72 to a high of $570.75 per flask ($5.81-$7.50
per lb) [8].

The major uses of mercury in the United States for the years 1971 and
1972 are listed in Table 1 (adapted from Metal Statistics, 1973; data compiled and reported by the U.S. Bureau of Mines).

The production of electrical apparatus (switches, batteries, lamps,
etc.) accounts for nearly one-third of the annual use of mercury. During
the same period, agriculture used about one-tenth of the amount used by
the electrical industry.

TABLE 1. Major Uses of Mercury in the United States (1971-1972)

Use	1971 Flasks[a]	1971 Pounds	1972 Flasks	1972 Pounds
1. Electrical apparatus	16,646	1,265,096	14,630	1,111,880
2. Chlor-ackali apparatus	12,252	931,152	11,491	873,316
3. Paint, mildew-proofing	8,192	622,592	8,188	622,288
4. Industrial and control instruments	3,906	296,856	5,874	446,424
5. Dental preparations	1,871	142,196	2,550	193,800
6. Agriculture	1,478	112,328	1,835	139,460
7. All other: laboratories, pharmaceuticals, amalgams, catalysts, and redistilled	8,380	636,880	8,632	656,032
Total	52,725	4,007,100	53,200	4,043,200

[a]Flask = 76 pounds.

In consideration of the environmental impact, the amount used does not necessarily correlate with the amount of contamination nearly as well as does the manner of use. The major use by agriculture of mercury compounds is for the control of pests, which necessitates distributing the compounds in low levels over a vast amount of the earth's surface. Thus, even though agriculture has used only about 3% of the annual production of mercury in the United States, it has probably contributed a disproportionate amount to the contamination of the environment. The ban in 1970 on the use of mercurials as seed fungicides should curtail this source of contamination.

The ban has caused a drastic drop in the number of mercury poisonings in livestock, so an effect is already apparent.

The high-grade cinnabar ore is estimated to constitute only about 0.02% of the mercury in the earth's crust [3]. The balance is found distributed in low levels in soils and rocks at an average of 50 ppb. Fossil fuels may contain 20-300 ppm, and as high as 300 ppm is found in tars [2,3]. It is readily apparent that the high consumption of fossil fuels without adequate pollution safeguards would provide a constant source of mercury contamination to the environment.

Mercury compounds in the crust of the earth are believed to degrade to the metallic form, which in turn is volatilized to the atmosphere. The atmosphere turnover is estimated to be about 2 years, which further contributes to the worldwide environmental distribution [2,3]. The ground-air concentration of mercury near mercury deposits at midday may be as high as 16 ppb by weight [2,3].

Rainfall acts as a "scrubber" of the atmosphphere and may deposit up to 500 mg annually per acre on the earth's surface [2,3] (Table 2).

TABLE 2. Environmental Levels of Mercury [2,3,9-13]

A. Air	
1. World average	20 mg/m^3
2. Mineral areas	20,000 mg/m^3
3. Nonmineral areas	3-9 mg/m^3
4. In mines	20,000 mg/m^3
5. Industries using Hg	100 mg/m^3
B. Soils	
1. Swedish soils	50-510 ppb

TABLE 2. (Continued)

2. England and Northern Ireland	5000-15,000 ppb
3. French and Sudan soils	10-60 ppb
4. Northern U.S.A.	40 ppb
C. Water	
1. Rainwater	0.2 ppb
2. Inland groundwater	0.1 ppb
3. Spring water	Up to 80 ppb
4. Northeast U.S.A. inland water	0.6 ppb
5. Ocean water	0.03-2 ppb

The levels of mercury found normally in the air, soils, and water at various locations attest to the wide distribution of mercury in the environment. Plant and animal tissues also have a substantial amount of mercury with a wide range of levels (Tables 3 and 4).

Physiopathology

The inorganic mercury compounds are readily but not efficiently absorbed from the alimentary tract. Studies with rats indicated that about 50% of the dietary intake of mercuric acetate was absorbed [21]. Mercury

TABLE 3. Biological Levels of Mercury [11, 14-16]

1. Apples	New Zealand	11-135 ppb
2. Apples	United Kingdom	20-120 ppb
3. Rice	Japan	227-1000 ppb
4. Rice	United Kingdom (imports)	5-15 ppb
5. Carrots	U.S.A.	20 ppb
6. White bread	U.S.A.	4-8 ppb
7. Whole milk	U.S.A.	3-10 ppb
8. Beer	U.S.A.	4 ppb
9. Haddock	U.S.A.	17-23 ppb
10. Walleye	Canada, St. Clair	0.89-2.43 ppm
11. Walleye	Canada, Lake St. Clair	1.29-5.01 ppm
12. Walleye	Canada, Lake Erie	0.58-0.90 ppm

TABLE 4. Toxicity of Some Mercurial Compounds [2,6,17-20]

Inorganic compounds

Mercuric chloride (corrosive sublimate)	Rat-oral-LD$_{50}$, 37 mg/kg
	Cattle-oral-toxic, 4-8 g
	Horse-oral-toxic, 5-10 g
	Sheep-oral-toxic, 4 g
	Dog-oral-toxic, 0.1-0.3 g
	Dog-oral-lethal, 60 mg/kg
Mercuric cyanide	Rat-oral-MLD, 25 mg/kg
Mercuric iodide (red iodide of mercury)	Rat-oral-LD$_{50}$, 40 mg/kg
Mercuric nitrate	Mouse-IP-LD$_{50}$, 5 mg/kg
Mercurous chloride	Rat-oral-LD$_{50}$, 210 mg/kg
	Cattle-oral-toxic, 8-10 g
	Horse-oral-toxic, 12-16 g
	Sheep-oral-toxic, 1-5 g
	Dog, oral-toxic, 1-2 g
	Dog-oral-lethal, 160 mg/kg
Mercurous iodide	Mouse-oral-LD$_{50}$, 110 mg/kg
Mercurous nitrate	Mouse-IP-LD$_{50}$, 4 mg/kg
Mercury (metal) (inhalation vapors)	Human-inhalation, LTC, 1 mg/m^3
	Rabbit-LTC, 29 mg/m^3

Organic compounds

Ceresan (chloroethyl mercury)	Rat-oral-LD$_{50}$, 30 mg/kg
Ceresan (ethylmercuri-p-toluene sulfonanilide)	Rat-oral-LD$_{50}$, 100 mg/kg
Ethylmercury chloride	Rat-oral-LD$_{50}$, 30 mg/kg
Methylmercury chloride	Guinea pig-oral-LD$_{50}$, 7 mg/kg
Methylmercury dicyanamide	Swine-oral-lethal, 27 mg/kg

vapors are also readily absorbed from the lungs [2,3]. Flatla [17] reported poisoning in cattle from absorption via the uterine and vaginal mucosa following the use of a mercuric chloride solution to treat an infection. He also reported poisoning from the use of mercuric chloride to disinfect deep wounds in animals.

Mercury compounds tend to have an irritating effect on mucous membranes and epidermal tissue. This irritating effect is manifested in the clinical signs and postmortem lesions that are associated with inorganic mercury poisoning.

The clinical signs of acute inorganic mercury poisoning resemble inorganic arsenic poisoning in domestic animals. The characteristic clinical signs observed by the clinician are violent abdominal pains with blood-stained diarrhea. The inorganic mercurials have a coagulative necrotizing effect on the alimentary mucosa and the blood vascular system that causes an extensive hemorrhagic gastroenteritis and blood loss which can lead to shock, collapse, and ultimately death.

The prolonged ingestion of a low level of inorganic mercury can result in a subacute or chronic syndrome of mercury poisoning. The animal that is thus exposed accumulates mercury due to the slow excretion rate. Sublethal exposure can result in excessive salivation, anorexia, oliguria, and a foul breath. The animal becomes emaciated and depressed, has a stiff-legged walk, and is weak. If the animal survives for a week or more, it develops a generalized alopecia with scabby lesions around the mouth, anus, and vulva. The animal develops a pruritis, and the gingival tissue becomes tender and inflamed, which can result in loosening and shedding of the teeth. The animal has a chronic diarrhea that does not respond to treatment.

Postmortem lesions in the acute cases are consistent with a severe hemorrhagic gastroenteritis. Edema, hyperemia, and petechial hemorrhages of the internal organs are found along with swollen liver and kidneys plus congested lungs.

Histologically, the alimentary tract will show extensive areas of coagulative necrosis. The kidneys will contain advanced degenerative lesions of coagulative necrosis.

Diagnosis is based on a good clinical history, the clinical signs, and postmortem lesions. Laboratory analysis of liver and kidney tissue is recommended to confirm the preliminary diagnosis. The highest tissue levels of mercury from inorganic compounds will usually be found in the kidney, more specifically the cortical tissue. The level of mercury in the kidney tissue will generally be above 10 ppm on a wet-weight basis. Actually, any level of mercury above 3 ppm should be viewed with suspicion if the clinical signs and lesions are compatible with mercury poisoning. Mercury levels above 100 mg/l in the urine are indicative of an undue exposure. The analysis of urine provides diagnostic criteria to evaluate a low-level exposure, to aid in diagnosis, and to monitor a treatment regime.

Treatment

Treatment in the acute cases has to be initiated within 15 min of ingestion to be effective. The stomach should be emptied if at all feasible followed by a source of raw protein given orally such as raw egg white. Sodium thiosulfate in an aqueous solution at up to 30 g IV followed by up to 60 g orally in an adult cow at 6-hr intervals for 2-3 days has shown beneficial results in the early stages of poisoning. BAL (dimercaprol) IM at a dosage of 6-7 mg/kg body weight is now recommended. It is usually formulated as a 5% mixture with 10% benzylbenzoate in peanut oil. Repeat this dosage at 8-hr intervals until the animal recovers. Change injection sites as the injections tend to be painful. Abscesses that may be either septic or sterile are a common sequela, so aseptic procedures should be maintained.

Supportive fluid and antibiotic therapy should be maintained until the animal has fully recovered. Vitamin and protein hydrolysates may be indicated depending on the severity of the syndrome [2,18,19,22].

ORGANIC MERCURIALS

Historical

Organic mercury compounds were originally introduced as fungicides for seed disinfection in Germany during 1915 [5]. Hydroxyphenylmercury salts were the first organomercurials used for this purpose. Due to their effectiveness they soon found wide application in both agriculture and industry for fungus control. Because of their more efficient biocidal properties, the alkyloxymethyl and alkylmercury salts soon replaced the hydroxyphenylmercury compounds. The methyl homolog of the alkylmercury salts is now considered by many to be the most effective biocide among the organomercurials [23].

Although not commercially utilized until 1915, the organomercurials had been recognized as lethal agents 50 years earlier in 1869 when Edwards [24] described three cases of chronic human mercurial poisoning. The three individuals affected were laboratory assistants working with a new compound, dimethyl mercury. One patient died and the other two survived as invalids. In a dissertation published in 1870, Prumers [25] described several additional cases of poisoning among chemists due to an ethylmercury compound.

There is little doubt that many domestic and feral animals died of organomercurial intoxication following the introduction of the compounds as seed disinfectants. This was due principally to the practice of feeding excess seed grain to livestock. However, the first report did not appear until 1941 [26]. In this report, Boley et al. described poisoning in a group of steers and two pigs fed seed grain that had been treated with an organic mercury insecticide. A number of chemical analyses were performed on liver and kidney tissues from the calves. Mercury levels were found to be as follows: liver, 3.2 to 7.2 ppm; kidney, 0.5 to 4.0 ppm. The chemical method used was only roughly quantitative. According to analytical chemists, the amount of mercury detected probably represented only one-third to one-half of the mercury present. In a later report, Taylor [27] described the death of a 400-lb sow and nine younger pigs weighing approximately 90 lb each after having been fed corn treated with ethylmercury phosphate for a period of approximately 2 months. Toxicologic analyses demonstrated "appreciable" quantities of mercury in the liver tissue of one animal.

The danger of organomercurials as toxicants in some species has been questioned by a number of workers. In 1922, Hagemann [28] found that grain treated with an organomercurial could be fed to hens with no apparent untoward results. Other workers in Germany [29-31] confirmed his findings and investigated the problem in livestock. Carnaghan and Blaxland [32], working in England with pigeons and pheasants, also confirmed Hagemann's work in 1957. However, Borg [33], in Sweden, published in 1938 results of feeding several species of birds on grain treated with an alkylmercury compound. He reported that death occurred after about 1 month with mercury levels in the visceral organs ranging from 90 to 100 ppm. He also reported that mercury could be detected in the body organs of nonfatal cases for as long as 6-7 months after ingestion of the mercury-treated grain had stopped. Egg production of pheasants fed the treated seed for 9 days was not affected but hatchability of the eggs was reduced. Muscle tissue from these birds contained 8 ppm of mercury.

In Minamata, Japan, from 1953 to 1960, a plastics-manufacturing operation discharged methylmercury chloride into Minamata Bay and river. Levels of methylmercury chloride reached 50 ppm in fish and 85 ppm in shellfish obtained from the contaminated areas. One hundred and twenty-one persons were poisoned, 46 fatally, from eating the contaminated fish. Dogs, cats, pigs, rats, and birds living around the bay developed classical clinical signs and many died [34].

Organomercurial poisoning of a dairy herd fed oats to which a commercial seed disinfectant containing ethylmercuri-p-toluene sulfonanilides had been added was reported by Herberg [35] in the United States in 1954. Seven of 16 dairy cattle were acutely affected and two animals died within 24 hr. The five other animals made an uneventful recovery. In 1955, Japanese veterinarians encountered a mysterious disease in cattle from the Monbetsu District in Hokkaido, Japan [36,37]. The cases of the disease appeared from early February to May. Studies revealed that 171 cattle were fed linseed meal treated with a mercurial fungicide containing 3.6% organomercury compounds (chlorophenylmercuric chloride, phenyldimercuric chloride, and methoxyethyl chloride), 79% talc, 10% caorin, 2% mineral oil, and 6% iron oxide. Twenty-nine of the 171 animals were affected, 10 of which died or were slaughtered. Prominent clinical signs were fever, depression, anorexia, decline in milk production, lymph node swelling, catarrhal bronchitis, petechiation of visible mucous membranes, and cardiac irregularities. Hematologically, there was anemia, neutrophilia with a left shift in the severe cases and lymphocytosis in the mild cases. These workers emphasized that in mercurialism in cattle, onset varied with sensitivity of individual animals to mercury. (Note: The variation of incidence, clinical signs, and pathomorphic changes observed in the affected Japanese cattle are typical of the unpredictable response of many species after ingesting organomercurial fungicides.)

After the appearance of several additional published descriptions of fatal intoxications among swine ingesting oats treated with organomercurials [38,39], Australian workers [40] in 1960 reported that two pigs fed a ration containing wheat treated at a rate of 2 ounces of ethylmercuri-p-sulfonanilide per 100 lb of grain thrived well for a period of 90 days. The pigs showed no clinical ill effects, and no carcass abnormalities were seen when the animals were examined after slaughter. However, analysis of kidneys and livers revealed mercury levels as high as 300 ppm.

Nine years after the Australian work was published, the hideous results of considering tissues of clinically normal pigs, previously fed organomercurial compounds, as fit for human consumption became evident in a farm family in Alamogordo, New Mexico [24,41-44]. On December 4, 1969, an 8-year-old farm girl living near Alamogordo, New Mexico, developed ataxia, visual disturbances, and a reduced state of consciousness which progressed to coma within a period of 21 days. Two weeks after the onset of her illness a 13-year-old male sibling developed similar clinical signs and, like

his sister, became comatose within a 3-week period. By the end of the same
month, a 20-year-old sister developed similar clinical signs and became
semicomatose. At the time the pork containing the high levels of mercury
was being eaten by the 10 members of the farm family, the mother was 3
months pregnant. She did not eat any of the meat after her sixth month of
pregnancy. Clinical examinations during the last 2 months of her pregnancy
disclosed only normal findings; however, her urine contained high levels of
mercury. The pregnancy terminated with delivery of a 6.7-lb male infant.
At birth he manifested intermittent trembling of the extremities which persisted for several days; however, he was otherwise normal. Electroencephalograms, electromyograms, blood electrolytes, calcium, magnesium, glucose,
and bilirubin remained normal during the first 6 weeks. Marked elevation
of urinary mercury was present during the first 6 weeks; however, after
that time urinary mercury was no longer detected. Electroencephalograms
became slightly abnormal at 3 months of age; by 6 months of age they were
markedly abnormal, and generalized myoclonic jerks had developed. By 6
months of age, the infant had nystagmoid eye movements without evidence of
visual fixation, was hypotonic and irritable. All of these clinical signs
and other physical examination findings have been observed in Japanese
children born to parents who consumed various fish caught from Minamata
Bay and surrounding areas. The poisoning of these children presumably resulted from transplacental poisoning with organic mercury [42]. This phenomenon has been demonstrated with animal experimentation [45]. The devas-
tating results of organic mercury intoxication on the human nervous system
has been well documented [46-52].

Six months after the initial appearance of clinical signs among members of the farm family in New Mexico, their condition remained essentially
unchanged. The 8-year-old girl and 13-year-old boy remained comatose; however, the 20-year-old sister continued to improve and was able to speak and
walk with difficulty. The neonate's condition remained unchanged [44].

Subsequent epidemiologic studies revealed that the patients' father
and five farmer neighbors had been given waste seed grain by a local grain
storage facility. The grain had been treated with an organomercurial fungicide, either a formulation containing methylmercuric dicyandiamide or
cyano-methylmercuriguanadine as the active ingredient, or a group of formulations consisting of ethyl-, methyl-, or methoxyethylmercury compounds as
the active ingredients. Both fungicides had been used at various times for

treating the seed grain given the six farmers which was later used in rations fed to hogs on their farms. Late in August or early September, the
father of the affected children began feeding his pigs the treated grain.
Two to 3 weeks later, one hog which had been fed 60% more grain than the
others was slaughtered and the family consumed the meat during the next
3-1/2 months. The animal manifested no clinical signs of poisoning or
other illness at the time of slaughter. Assay of muscle tissue from this
animal by atomic absorption spectrophotometry demonstrated 29.4 parts per
million (ppm) of mercury. Similar analysis of feed grain fed the animals
disclosed 32.8 ppm of mercury. (Note: The high degree of correlation between amounts of organic mercury fed and that present in the muscle attests
to the tendency of organomercurials to accumulate in various mammalian body
tissues without eliciting clinical signs of toxicosis.) Approximately 6
weeks later, 14 other pigs kept on a similar diet but fed smaller quantities
of the treated grain became blind, incoordinated, and developed posterior
paralysis. Twelve of the 14 pigs died during the next 3 weeks. The survivors remained blind and stunted; however, the locomotor impairment improved. When epidemiologic investigations of the incident began, there
were 215 live hogs on the farm of the six individuals who were given the
treated seed grain. Most of the animals were voluntarily sacrificed and
the meat destroyed. The remaining animals were embargoed by the appropriate
state and federal regulatory agencies. Subsequently, all but one carcass
was found free of mercury and released. The contaminated carcass was destroyed.

In 1956 in northern Iraq, over 100 people were poisoned by eating flour
mixed with wheat seed treated with a fungicide containing 7.7% ethylmercuri-
p-toluene sulfonalide. Fourteen, and probably more, of the intoxicated
people died. They had fed the treated seed to chickens for several days
and after observing no ill effects had eaten it themselves. In addition to
central nervous system manifestations, a number of other clinical signs
were observed including frequent polydypsia, polyuria, weight loss, severe
proteinuria, deep musculoskeletal pain refractive to analgesics, and pruritis of the palms, soles, and genitals. The authors attribute the other
clinical signs to the prevalence of ancylostomiasis and dietary deficiencies
of protein and vitamins [53].

Four years later during the winter and spring of 1961, an additional
100 people were poisoned by flour and wheat seed treated with a fungicide

containing 1% mercury as ethylmercury chloride and phenylmercury acetate.
Four of 34 patients hospitalized died; however, the authorities believed
others probably died after refusing hospitalization or signing out against
medical advice. A combination of clinical signs was observed reflecting
insult to the central nervous system by the alkyl (ethylmercury) fraction
and to the renal and gastrointestinal organs apparently due to the aryl-
(phenylmercury) component. All but three of the patients treated with Dimercaprol (BAL) improved significantly; however, when optic nerve degeneration occurred, blindness was frequently permanent [54].

During the wheat-planting seasons of 1963-1965, numerous cases of a
disease suspected of being a viral encephalitis occurred in and around
Panorama, Guatemala. Forty-five cases were observed, over 50% of which
occurred in children, and 20 of these died. Autopsy and subsequent tissue
analyses disclosed high levels of methylmercury dicyandiamide in brain,
liver, and kidney tissue of one victim. Being too poor to buy enough food
to survive, the victim had eaten wheat seed treated with a fungicide containing 1.5% mercury as methylmercury dicyandiamide [55].

In the United States on February 19, 1970, as a result of the Alamogordo, New Mexico, incident, the Pesticide Regulation Division, U.S. Department of Agriculture, suspended the registration of cyano(methylmercuri)-
guanadine or methylmercury dicyandiamide for use as a seed treatment. On
March 9, 1970, the suspension was extended to include all alkylmercury
seed-treatment compounds [44,56,57]. Subsequently, one of the manufacturers
sought and successfully obtained a preliminary injunction against the suspension [58], allowing shipment and sale of the fungicide. The latter
apparently constitutes a shining example of corporate conscience.

Studies by the Swedish Natural Science Research Council [59] from
1964 to 1968 established mercury levels in six different foods: eggs, pork
chops, beef, bacon, pig liver, and ox liver. On February 1, 1966, alkyl-
mercury-containing seed fungicides were banned in Sweden, to be replaced
initially by alkoxyalkylmercury compounds and eventually by fungicides
containing no mercury. After the ban, the mercury levels in the six foods
mentioned above declined to one-third of those detected before the ban.

The impact of organomercurials on the human population as described
above is only one of the many facets of their existence, which probably
predates that of man on this planet. Additional features of their effects
on the world and its inhabitants constitute the basis for discussion in the
remainder of this chapter.

Organomercurials and the Environment

Mercury is a rare element in the earth's crust comprising less than 30 of each billion of its parts. There are only 15 elements present in smaller amounts in the earth than mercury. Only a small number of sites where it occurs in more than trace amounts are known. In elemental form in liquid state it is nontoxic, and a human being might ingest up to a pound without significant adverse effects [60].

What is the normal distribution of mercury in nature? In rocks, soils, water, and tissues of plants and animals, it is present only in fractions of a part per million. Generally speaking, the atmosphere and both fresh and sea waters contain mercury only in levels of one or more parts per billion. However, both plants and animals tend to store, or concentrate, mercury. Examples of this phenomenon are some sea algae which contain 100 times more mercury than is present in the ocean water around them, and certain marine fish which contain over 120 parts per billion of mercury while their aqueous environment contains only one or less [60].

In nature, mercury is distributed by a complex cycle involving atmosphere, hydrosphere, and lithosphere. From the lithosphere (rocks and soils) mercury is released into the atmosphere from the hydrosphere by evaporation of waters. Mercury is returned to the lithosphere by sedimentation from water and precipitation from the atmosphere. Man has been and currently is throwing this cycle out of balance by concentrating the amount of mercury released into certain waters and the atmosphere from certain industrial processes which pollute the environment with several forms of mercury [60]. Major contamination sources include (a) pulp and paper factories using phenylmercuric acetate (PMA); (b) chlorine factories using mercury electrodes; (c) electrical industries using mercury; (d) combustion of fossil fuels containing mercury; (e) wastewater treatment plants whose products are used as soil fertilizer; (f) use of mercury-containing catalysts for the manufacture of vinyl chloride; (g) industries extracting the metal by heating cinnabar and other mercury ores; (h) chemical industries manufacturing paints, pesticides, and fungicides containing mercury; (i) in agriculture, the chlorides of mercury, both mercurous and mercuric, and some organic compounds of mercury used to protect and disinfect seeds, to control diseases of tubers, bulbs, corms, and to prevent fungus diseases of seeds, fruits, and vegetables, including growing plants [61].

In general, it is safe to state that inorganic mercurials are not significant problems in contamination of the general environment. Of primary

concern with regard to the environment are the organic mercurials. There
are two major classes of organic mercurials, the aryl compounds which contain a substitute aromatic ring and the straight carbon chain and the alkyl
compounds. It is the latter groups which pose the greatest threat to animal
life by virtue of their ability to inflict a wide range of damage from producing congenital mental retardation to chromosome abnormalities. Alkyl
mercury appears to be especially dangerous because the extremely high degree
of stability of the bond between the carbon and mercury atoms results in the
molecule not being degraded. This permits the molecule to maintain its
destructive action for from weeks to years. Aryl mercurials are much less
stable; consequently, the injuries which they and inorganic mercury cause
are almost invariably reversible.

    Where do the aryl- and alkylmercury compounds in the environment orig-
inate? In the case of Minamata Bay, apparently the polluting effluent itself
contained both methyl mercury and elemental mercury. The elemental mercury
was, subsequently, methylated by microorganisms in the mud on the bottom of
the bay [60]. In 1967, Swedish workers reported that microorganisms were
capable of methylating inorganic mercury [62]. The same authors later
showed that dimethyl mercury was also formed, depending on the pH, either
as an intermediate compound during the reaction in which monomethyl mercury
was the final product, or as the final product in a reaction in which monomethyl mercury was an intermediate compound [63]. Other workers demonstrated
that biological methylation of inorganic mercury is a nonenzymatic process
involving vitamin B12. Extracts from a methane-forming bacterium isolated
from a culture of Methanobacterium emelianskii and in vitro solutions of
methylcobalamin were able to methylate mercury [64]. Similar results were
obtained by Yamada and Tonomura using Clostridium cochlearis [65] and
Landner [66] using Neurospora crassa. The intermediary dimethyl mercury is
volatile, lipophilic, and decomposes to methyl mercury at acidic pH levels.
The latter complexes with anionic groups, one of which, with sulfhydryl
groups, is extremely stable. At pH 4.6, a substantial amount of decomposition from dimethyl mercury to monomethyl mercury occurs. Consequently,
methylation of mercury is a normal biologic process which may occur in
anaerobic ecosystems which are inevitably associated in an industrialized
society with its polluted waters, e.g., sewage water and factory effluents.
Whenever a mildly acidic environment accompanies such a situation, decomposition of the resultant dimethyl mercury occurs and the highly stable and

toxic methyl mercury is formed. Since the pH of sea water is about 8.1 [61] and since an alkaline pH favors production of the more volatile dimethyl mercury, this would favor a larger discharge of methylated mercury by evaporation into the atmosphere. A more acid pH would then, as previously pointed out, favor a higher proportion of the less volatile monomethyl mercury, as well as decomposition of the dimethyl mercury to the monomethyl form with less loss to the atmosphere and greater retention by acid waters. Jacobs and Keeney [67] have confirmed the pH relationship of the waters of two U.S. rivers to the formation of methyl mercury in a simple but elegant study. These mechanisms have been cited as possible factors leading to the contamination of remote acid lakes with high levels of mercury in the fish in the total absence of known sources of mercury contamination [68].

That inorganic mercury can be formed from other mercury compounds, such as metallic mercury, phenyl mercury, and alkoxyalkyl mercury, has been well documented and established [61]. Bouveng [69] has demonstrated that mercury levels in fish from water polluted by pulpwood and paper mill industries discharging PMA are much higher than levels in similar fish from waters contaminated by chlorine factories using mercury electrodes. He postulated that the combination of PMA and the organic material promotes the accelerated formation of methyl mercury. It is pertinent to point out that up to 100% of the mercury found in all types of fish investigated in both fresh water and that of the Baltic and Atlantic is methyl mercury. Methyl mercury also often comprises up to 99% of the mercury in other animal tissues. Westoo [70,71] found that hen chickens fed seeds treated with inorganic mercury, methoxyethyl mercury, or phenyl mercury all laid eggs with methyl mercury in the egg white. He also demonstrated that liver hemogenates can methylate mercury. Jernelöv [72] has suggested that all forms of mercury appear to be directly or indirectly capable of conversion to methyl mercury. He has also indicated that PMA-containing slimicides are more efficient sources of mercury for methylation than is inorganic mercury. He also pointed out that although methylation of mercury occurred anaerobically, it appears to occur more efficiently in aerobic systems [68].

It should be pointed out that microbiologic systems can also degrade organic mercury compounds. Tonomura et al. [73] and Tonomura and Kanzaki [74] isolated a Pseudomonas species which was capable of decomposing phenylmercuric acetate, ethylmercuric phosphate, and methylmercuric chloride to metallic mercury, benzene, ethane, and methane [75].

Aside from effluents of factories manufacturing vinyl chloride, the largest single source of alkyl organomercurials in the environment was the use of seed-treatment materials. Smart [76] reported that 200 metric tons of mercury were used in various pesticide applications in Japan and the United States in 1968.

Aryl organic mercurials have also been added to the environment, chiefly rivers and lakes, in the form of phenylmercuric acetate from the pulp and paper industry in sizable amounts. In Sweden, above 24,000 kg of PMA were used in 1960 in these industries. There are 90 sites where fishing is banned in Sweden because of a high alkyl organomercurial content of the fish, of which 10 are pulp or paper mill locations.

The amount of mercury added to the atmosphere from combustion of coal alone has been estimated at 3000 metric tons per year [77]. This mercury is in the gaseous phase.

Weiss et al. [78] reported that the flux of mercury from continents to oceans by way of rivers is much less than that from the continents to the atmosphere. They attribute the doubling of mercury in recent years in Greenland's ice sheet to man's greater exposure of the earth's crust through alteration of terrestrial surfaces, allowing more mercury vapor to enter the atmosphere [75].

In summary, by far the most dangerous of the organic mercury compounds are the alkyl group, the aryl and inorganic forms being much less persistent and toxic. The differences are due in large part to the stability of the mercury-to-carbon bond in the alkylmercurials. Pollution by numerous industries utilizing mercury in various forms for a variety of purposes currently is the greatest threat to life on this planet from mercury. We would be remiss to ignore the contribution to the load of mercury in the environment released through the natural processes of leaching and volatilization. This uncontrollable source of mercury must be considered when evaluating concentrations in air and water used by man. It may account entirely for high concentrations of mercury found in isolated bodies of water which have no known man-made sources of contamination.

Toxicogenic Mechanisms of Organomercurials

The organic radical of the organomercurial compounds largely determines their physical, chemical, and biological properties, e.g., solubility in water, vapor pressure, and pathways by which they are metabolized in the animal or plant. The ability to produce c-mitosis and other chromosomal

alterations of structure and function as well as their selective destruction
of specific brain cells in animals have brought the organomercurials the
greatest amount of notoriety.

It has been known since the latter part of the 1930s that organomercurial compounds have a c-mitotic effect; that is, they produce inactivation
of the mitotic spindle mechanism in much the same way as colchicine does.
The c-mitotic effect in its ultimate form is a division of chromosomes without a concurrent or subsequent division of the somatic components of the
cell. A predictable result is a cell with twice the normal number of chromosomes. Studies using root tips of an onion plant, <u>Allium</u> <u>cepa</u>, were
conducted using methylmercury hydroxide, methylmercury dicyandiamide, phenylmercury hydroxide, and methoxyethylmercury chloride. The onion roots
were treated by being placed in water containing the mercury compound to be
tested. One of the main purposes was to establish the lowest c-mitosis-
producing dose of the organomercurial. Methyl- and phenylmercury compounds
were active at dosages of 0.1 ppm or less, which indicated that methyl- and
phenylmercury compounds have lower threshold values for producing c-mitosis
than any other known c-mitotic substances including colchicine. Methyl and
phenyl mercury are, in fact, about 1000 times more effective than colchicine
in producing c-mitosis [61]. The alkoxyalkyl compound was only one-fourth
as active a c-mitosis producer as the phenyl- and methylmercury compounds.
Sublethal amounts of the mercurials also produced other chromosomal changes
associated with c-mitosis including strong contraction of chromosomes at
metaphase, delayed division of the centromere, and no anaphase movement.
As reduced dosages were used, spindle function was resumed in a much more
gradual manner than observed with equal concentrations of colchicine. At
the intermediate dosages, a large number of multipolar spindles formed and
separation of chromosomes during mitosis became highly irregular producing
various numbers of chromosomes in each cell. In addition, the orientation
of the spindle was affected, and the root tips developed c-mitotic tumors
and other irregular deformities. Another cytologic effect observed exclu-
sively with the aryl mercurial was chromosome breakage [79]. In spite of
the latter, the predominant effect of the mercurials is traceable back to
their action on the mitotic spindle. A wide array of chemicals are capable
of producing the c-mitotic effect. Those which have a low threshold value
for the condition almost inevitably have it in combination with high solubility. Only colchicine and the mercurials are exceptions to this relation-

ship which, according to Ramel, points to the more specific action of these agents.

What is the precise mechanism of the c-mitosis-producing agents? Colchicine has been studied intensively in this regard, and recently an interesting mechanism of action has been suggested by Borisy and Taylor as cited by Ramel [79]. They found that colchicine binds specifically to a 6S protein which constitutes the unit forming the microtubules and related structures. The integration of these protein groups into the spindle fibers apparently involves hydrogen bonding which is prevented by colchicine.

Many features of the formation of the spindle indicate that linkages other than hydrogen bonding are involved, namely oxidation of sulfhydryl groups to disulfide bridges. It is likely that this process is required in the formation of smaller proteins into the 6S protein, whose integration into the unit forming the microtubules is prevented by colchicine through its effect on hydrogen bonding. In view of the affinity of mercurials for sulfhydryl groups, it is obvious how they could interfere with spindle formation at the early stage involving formation of the 6S protein. Experiments to test this concept using 2,3-dimercaprolethanol (BAL) and phenylmercury hydroxide were conducted with the aforementioned onion root tips. If mercurials exert their effects on the spindle by interference in the sulfhydryl-disulfide reaction synthesizing protein to be utilized in forming the mitotic spindle, then BAL should act as an antagonist to the mercurials. The experimental results clearly demonstrated a very pronounced effect of BAL in this respect. The conclusions of the experiments were that organomercurials apparently inhibit polymerization of spindle fiber proteins which require the sulfhydryl-disulfide oxidation reaction.

Ramel, utilizing the fruit fly Drosophila melanogaster, found that phenyl and methyl mercury increased significantly the incidence of nonseparation of sex chromosomes. Females manifested this effect to a much greater extent than males [79]. Because methylmercury hydroxide had been previously shown to bind to DNA causing an irreversible denaturation, it was suspected that the alkyl mercurial might induce mutations. This possibility was investigated by means of monitoring recessive lethal males produced in a mercury-treated population of Drosophila melanogaster. A significant increase at the 5% level of recessive lethals in the mercury-treated flies led to the conclusion that methyl mercury was mutagenic but that this ability to induce mutations must be considered slight. Similar treatments with aryl mercurials gave results pointing in the same direction; however,

no significance in the data was observed. Methoxyethylmercury chloride and
inorganic mercury chloride had no significant effect on chromosome segrega-
tion in the Drosophila melanogaster.

What effects do the organomercurials have on the genetic material of
mammals? Frölén and Pamel, cited by Löfroth [61], investigated the effects
of methylmercury dicyandiamide (MMD) on mice. When inbred CBA mice received
intraperitoneal injections of 0.1 mg of MMD and paired immediately with untreated animals at a ratio of 1 male to 4 females, there was no increase in
dead fetuses; however, an increased incidence of sterility was noted in
both sexes. Pregnant mice given the same treatment on the 10th day of
pregnancy produced a high frequency of resorbed litters and an increase in
percentage of dead fetuses. Although the dose of MMD was only one-third to
one-fourth of an LD_{50} dose for an adult mouse, the fetuses were severely
affected. This finding is consistent with intoxications of pregnant women
observed in the Minamata Bay and Alamogordo, New Mexico, incidents involving
methyl mercury in various forms.

How would the chromosomal damage observed in the fruit fly be manifested
in humans and other mammals? Probably in man the results would be the sex
chromosomes anomaly known as Klinefelter's syndrome or Down's syndrome (mon-
golism). In domestic animals, the tortoiseshell male cat is one prominent
manifestation of the lack of disjunction of the sex chromosomes; however,
no increase in this abnormality has been observed in the offspring of methyl
mercury-treated cats. One of the few instances of genetic material damage
in mammals as a result of intoxication with organomercurial compounds that
has come to our attention is that of Oharazawa [80]. Single high-level
doses of methylmercury phosphate given to female rats on day 10 of pregnancy
resulted in a 31.6% incidence of cleft palate and reduced body weight of the
offspring. A report concerning methyl mercury-induced chromosome damage in
man was published in 1974 [81]; however, the results are difficult to interpret. The work involved culturing of lymphocytes from 23 persons with a
history of prior exposure to methyl mercury through ingestion of fish from
contaminated waters and 16 nonexposed subjects. The cells were studied
cytogenetically. Statistically significant relationships were observed
between blood cell mercury levels and the frequencies of cells with chromatid-type aberrations, "unstable" chromosome-type aberrations, and aneuploidy.
A statistically significant relationship between variations in blood cell
mercury and "stable" chromosome-type aberrations could not be detected.
While the blood cell/plasma mercury ratio was high, which it character-

istically is in short-chain alkylmercury compound-intoxicated subjects as
opposed to the low ratio observed in intoxication with inorganic mercurials,
none of the "exposed" subjects showed clinical evidence of methylmercury
poisoning [81]. In general, the stable chromosome-type aberrations in
lymphocytes are produced in vivo, while the unstable, chromatid-type anoma-
lies arise primarily during development in culture. Because of the many
variables involved in this type of study, additional studies seem necessary
before the value of the blood cell mercury and lymphocyte chromosome abnormality relationship can be evaluated as a meaningful criterion of organomercurial toxicosis in man. In addition, the frequency of aberrations observed in lymphocytes from methyl mercury-exposed human subjects is much
less than that recorded in lymphocytes cultured from patients exposed to
therapeutic doses of x-ray radiation [82], virus infection [83], or lead
intoxication [84]. As we have discussed above, methyl mercury causes chromosome breakage in plant root cells [85]; however, no similar changes were
produced in human white blood cells exposed in vitro to a monomethylmercury
compound [86] or dimethyl mercury [87]. Little is known concerning the
medical significance of somatic cell chromosomal aberrations. Obviously,
most such affected cells will die unless the aberrations are so minor as to
be consistent with life and reproduction. Carcinogenesis arising from such
cells has, without any supporting evidence, often been discussed. The effects of transplacentally passed methyl mercury on the fetus have already
been mentioned; however, careful cytogenetic studies have not been made.
A pertinent question would be, "Does germ cell damage occur in mammals?"
It has already been shown that methyl mercury is distributed to the mammalian gonads [88]. An equally ominous report is the positive dominant
lethal test described in rats in 1973 [89]. The results were attributed
to a reduction in viable embryos due to incompatibility of sperm-to-im-
plantation events with mercury treatment of the parent male rat. The
acrosome of the maturing spermatid with the disulfide bond in the perinu-
clear material beneath the acrosome was considered as a possible site of
interference by the mercury compound. In summary, the organomercurials
apparently do affect the germ cells of mammals. Because of the extreme
difference in susceptibility between species and the few pertinent studies
which have been made in this area, it is not presently possible to evaluate
this hazard in man and the domestic animals.

The affinity of the alkylmercurials for the central nervous system,
which became evident in the Minamata incident [34], focused the attention

of the medical community on the entire family of organomercurials. How do
they enter the body; and after gaining entrance, what unique properties
enable them to selectively damage neurons in certain areas of the CNS?
Most cases of organomercurial intoxication in man, cattle, and swine occur
by ingestion of contaminated food; however, occupational exposures by inhalation and through the skin occur occasionally. Aryl mercurials penetrate
the skin and mucous membranes easily, while both alkylorganic and -inorganic
mercury compounds penetrate the skin slowly, if at all [90].

When mercurials enter the blood stream, the organomercurials are mainly
bound to the red blood cells, while inorganic mercurials are chiefly bound
to serum proteins. Intestinal absorption of the mercurials differs sharply
between compounds. Absorption by the intestine ranges from 90% for methyl
mercury in man and rat to 50% for mercuric acetate in the rat, to 50-80%
for phenylmercuric acetate also in the rat [75]. The variation in metabolism, detoxification, and excretion of the different types of mercurials
is no less remarkable. Phenylmercuric acetate is absorbed unchanged into
the circulation of rats from which it is removed predominantly by metabolism
in liver and kidneys, from whence it is excreted via feces and urine. It
is detectable in unchanged form for only approximately 96 hr, and less than
10% of the original dose is excreted unchanged in urine. Ethyl mercury, in
sharp contrast, when administered to rats is present intact in the liver
and kidneys for 21 days. No appreciable amount of the unchanged compound
is excreted in the urine. Urinary excretion in general is slower than with
phenyl mercury. In addition, the fecal excretion by rats of ethyl mercury,
for 7 days following administration, is only 1/20 of that observed with the
phenylmercuric acetate. There is general agreement among workers in this
area that the inorganic and arylorgano mercurials have similar excretion
patterns due to the rapid metabolism of the latter compounds to inorganic
mercury.

As noted earlier, mercury may occur in the enviroment as either monomethyl or dimethyl mercury. Mice exposed to dimethyl mercury, either by
injection or inhalation, exhale the majority of the exposure dose as intact dimethyl mercury with smaller amounts being metabolized to methyl mercury with corresponding distribution and excretion.

Inorganic mercury poisoning results in large amounts of mercury in the
kidneys with very little accumulating in the brain. In poisoning by alkyl
mercurials, large amounts of mercury are present in both organs, as well as
the liver. However, animal experimentation demonstrated that the alkylmer-

curic compounds do not accumulate to the same extent in the kidneys as do
the inorganic salts and arylorganomercurials. This is illustrated by a
study in which groups of 4 rats received subcutaneous injections of 100 μg
of one of four mercury compounds every other day for 2 weeks. Mercury
levels were determined on blood, liver, kidney, and brain. Animals receiving methylmercury hydroxide (MMH) had blood mercury levels 100 times
higher than those receiving mercuric nitrate (MN) and methoxyethylmercury
hydroxide (MEH) and ten times higher than recipients of phenylmercuric
hydroxide (PMH). Liver mercury levels in animals receiving MMH and PMH
were approximately equal but almost twice those of livers from animals receiving MEH and MN. Kidney mercury levels from PMH- and MEH-dosed animals
were about ten times those of MMH-dosed rats and about 25% higher than
kidney mercury levels in MN-dosed rats. Brain mercury levels in MMH-dosed
rats were 6.5, 19.4, and 17.2 times those observed in MN-, PMH-, and MEH-
dosed rats. These figures should be considered while keeping in mind the
radical differences in metabolizing mercurials which exist between species
[75]. Similar studies with the same groups of compounds administered subcutaneously to rats disclosed several significant trends. Alkylmercury
compounds are excreted slowly and remain in the body longer and at higher
concentration than aryl or inorganic mercurials. Marked accumulation of
alkylmercury compounds in the blood erythrocytes was characteristic of them.
The distribution and excretion of alkylmercurials is profoundly influenced
by their carbon chain length. Alkylmercurials have ratios of mercury concentration in the brain to that in plasma which are much larger than those
of aryl or inorganic mercury compounds. Among alkylmercurials, the ratios
for ethylmercury compounds are larger than for butylmercurials. When
ethylmercury compounds were administered, the concentration of mercury in
the brain was two to three times higher than that in plasma, while mercury
levels in brain and plasma were approximately equal when the N-butylmercury
compound was administered. In summary, the highest neurotoxicity appears
to be a special property of short-carbon-chain alkylmercury compounds.

REFERENCES

1. J. J. Putman, National Geographic, 142 (4), 507-527 (1972).

2. W. B. Buck, G. D. Osweiler, and G. A. Van Gelder, Clinical and Diag-
    nostic Veterinary Toxicology, 1st Ed., Kendall/Hunt, Dubuque, Iowa,
    1973, pp. 203-212.

3. J. F. Dewey and R. F. Pendleton, Agrichemical Age, 14 (4), 8-11 (1971).

4. Pharmacopeia of the United States of America, by Authority of U.S. Pharmacopeial Convention, Inc., Bethesda, Md., 1883, pp. 175-181.

5. U. Ulfvarson, Svensk. Kem., 73, 533 (1971).

6. P. G. Stecher (ed.), The Merck Index, 8th Ed., Merck, Rahway, N. J., 1968.

7. Metal Statistics, 66th Annual Ed., Metallgeseilschaft Aktienge Sell-schaft, Reuterweg, Germany, 1973, p. 173.

8. Chemical Statistics Hand Book, 7th Ed., Chemical Industries Association, London, 1971, p. 307.

9. A. Stock and F. Cucuel, Naturwissenschaften, 22, 390 (1934).

10. E. Eriksson, Oikos Supplementum, 9, 13 (1967).

11. I. J. Selikoff (ed.), Environ. Res., 4, 1-69 (1971).

12. R. A. Wallace, W. Fulkerson, W. D. Shults, and W. S. Lyon, Oak Ridge National Laboratories, DOC-ORNL-NSF-EP-1, 1971.

13. A. L. Hammond, Science, 171, 788-789 (1971).

14. N. A. Smart, Agr. Residue Rev., 23, 1-36 (1968).

15. L. J. Goldwater, J. Roy. Inst. Public Health and Hyg., 27, 270-274 (1964).

16. N. Fimreite, W. N. Holsworth, J. A. Keith, P. A. Pierce, and I. M. Gruchy, Can. Field Naturalist, 85, 211 (1971).

17. J. C. Flatla, Int. Encyl. Vet. Med., 3, 1874-1877 (1966).

18. E. G. C. Clarke and M. Clarke, Garner's Veterinary Toxicology, 3rd ed., Bailliere-Tindall & Cassell, London, 1967, pp. 101-104.

19. H. P. Hoskins, J. U. Lacroix, and Kane Mayer, eds., Canine Medicine, 2nd Ed., Amer. Vet. Publ., Santa Barbara, Calif., 1962, pp. 730-731.

20. U.S. Department of Health, Education, and Welfare, Toxic Substances List, 1973.

21. O. G. Fitzhugh, E. P. Lang, and E. P. Nelson, A. M. A. Arch. Ind. Hyg. Occup. Med., 2, 433-442 (1950).

22. D. C. Blood and J. A. Henderson, Veterinary Medicine, 3rd Ed., Williams & Wilkins, Baltimore, Md., 1968, pp. 771-772.

23. U. Ulfvarson, Int. Arch. für Gewerbepathologie und Gewerberhygiene, 19, 412-422 (1962).

24. G. N. Edwards, St. Bartholomew's Hosp. Rep., 1, 141 (1865).

25. V. Prumers, Über das Quecksilberathyl chlorid Diss Berlin, 1870. (This is a dissertation used in its entirety.)

26. L. E. Boley, C. C. Morrill, and R. Graham, N. Amer. Vet., 22, 161 (1941).

27. E. L. Taylor, J. Amer. Vet. Med. Ass., 111, 46-47 (1947).

28. O. Hagemann, Dtsch. landiv. Pr., 49, 378 (1922).

29. J. T. de Haa, Versl. Pl. Ziekt, Dienst Wageninger, 61, 47 (1937).

30. M. Sy, Z. Pfl. Krankh., 48, 11 (1938).

31. G. Rothes and G. Havermann, Nachr. Schadl. Bekampf. Leverkusan, 14, 47 (1939).

32. R. B. A. Carnaghan and J. D. Blaxland, Vet. Res., 69, 324 (1957).

33. K. Borg, Proc. 8th Nordic Vet. Congr. Helsinki, 1938, p. 394.

34. T. B. Eyl, N. Engl. J. Med., 284, 706-709 (1971).

35. W. W. Herberg, Vet. Med., 49, 401-402 (1954).

36. M. Sonodas, R. Nakamura, T. Kinehiko, A. Matsuhashi, H. Ishimoto, R. Sasaki, K. Ishida, and M. Takahashi, Jap. J. Vet. Res., 4, 5-16 (1956).

37. Y. Fujimoto, K. Oshima, H. Satoh, and Y. Ohta, Jap. J. Vet. Res., 4, 17-35 (1956).

38. E. F. Ferrin, H. C. H. Kernkamp, M. H. Roepke, and M. B. Moore, Minnesota Farm and Home Science, 6, 7 (1949).

39. K. McEntee, Cornell Veterinarian, 40, 143-147 (1950).

40. M. R. Gardiner, J. Dept. Agr. West. Aust., 1, 237-238 (1960).

41. A. Curley, V. A. Sedlak, E. F. Girling, R. E. Hawk, W. F. Barthel, P. E. Pierce, and W. H. Likosky, Science, 172 (3978), 65-67 (1971).

42. R. D. Snyder, N. Engl. J. Med., 284, 1014-1016 (1971).

43. Morbidity and Mortality, U.S. Dept. H. E. W., P. H. S., 19, 25-26 (1970).

44. Morbidity and Mortality, U.S. Dept. H. E. W., P. H. S., 169-170 (1970).

45. T. Suzuki, N. Matsumoto, and T. Miyama, Indiana Health, 5, 149-155 (1967).

46. T. Takeuchi, Kumamoto Univ. Study Group of Minamata Disease, 1968, pp. 229-252.

47. R. A. P. Kark, J. D. Bullock, and D. C. Poskanzer, Neurology (Minnesota), 20, 401 (1970).

48. L. T. Kurland, S. N. Faro, and H. Siedler, World Neurology, 1, 370-395 (1960).

49. D. McAlpine and S. Araki, Lancet, 2, 629-631 (1958).

50. H. Tukuomi, Kumamoto Univ. Study Group of Minamata Disease, 1968, pp. 37-72.

51. Y. Harada, Kumamoto Univ. Study Group of Minamata Disease, 1968, pp. 73-91.

52. D. Hunter and D. S. Russell, J. Neurol. Neurosurg. Psychiatr., 17, 235-241 (1954).

53. M. A. Jalili and A. H. Abbasi, Brit. J. Ind. Med., 18, 303-308 (1963).

54. I. U. Haq, Brit. Med. J., 1, 1579-1582 (1963).

55. J. Ordonex, J. Carrillo, and M. Miranda, Bull. Sanit. Panam., 60, 510-519 (1966).

56. U.S. Department of Agriculture, 576-570, release dated February 19, 1970.

57. U.S. Department of Agriculture, P.R. Notice 70-7, dated March 9, 1970.

58. U.S. District Court, 70 C 838, Chicago, April 21, 1970.

59. G. Lofroth, Methylmercury. A Review of Health Hazards and Side Effects Associated with the Emission of Mercury Compounds into Natural Systems, Ecological Research Committee Bull. No. 4, 1st Ed., Swedish Natural Science Research Council, 1969.

60. L. J. Goldwater, Sci. Amer., 224, 15-21 (1971).

61. G. Lofroth, Methylmercury — A Review of Health Hazards and Side Effects Associated with the Emission of Mercury Compounds into Natural Systems, Ecological Research Committee Bull. No. 4, 2nd Ed., Swedish Natural Science Research Council, 1970.

62. S. Jensen and A. Jernelöv, Nordforsk Biocidalinformation, 10, 4 (1967).

63. S. Jensen and A. Jernelöv, Nordforsk Biocidalinformation, 14, 5 (1968).

64. J. M. Wood, F. S. Kennedy, and C. G. Rosen, Nature, 220, 173 (1968).

65. M. Yamada and K. Tonomura, J. Ferment. Technol., 50, 159-166 (1972).

66. L. Landner, Nature, 230, 452-454 (1971).

67. L. W. Jacobs and D. R. Keeney, J. Environ. Quality, 3, 121-126 (1974).

68. N. Nelson, Environ. Res., 4, 1-69 (1971).

69. H. Bouveng, Mod. Kemi, 3, 45 (1968).

70. G. Westöö, Var. Föda, 19, 121 (1967).

71. G. Westöö, Acta Chem. Scand., 22, 2277 (1968).

72. A. Jernelöv, Chemical Fallout (M. W. Miller and G. G. Berg, eds.), C. C. Thomas, Springfield, Ill., 1969, p. 72.

73. K. Tonomura, K. Maeda, and F. Futal, Nature (London), 217, 644-646 (1968).

74. K. Tonomura and F. Kanzaki, Biochem. Biophys. Acta, 184, 227-229 (1969).

75. I. J. Selikoff, Environ. Res., 4, 1-69 (1971).

76. N. A. Smart, Residue Rev., 23, 1-36 (1968).

77. O. I. Joensun, Science, 172, 1027 (1971).

78. H. V. Weiss, K. Minorw, and E. D. Goldberg, Science, 171, 692-694 (1971).

79. Ramel, Jap. Med. Ass. J., 61, 1072-1076 (1969).

80. H. Oharazawa, J. Jap. Obstet. Gynecol. Soc., 29, 14-79 (1968).

81. S. Skerfving, K. Hansson, C. Mangs, J. Lindsten, and N. Ryman, Environ. Res., 7, 83-98 (1974).

82. K. E. Buckton, P. A. Jacobs, W. M. Court Brown, and R. Doll, Lancet, 2, 676-682 (1962).

83. W. W. Nichols, Hereditas, 50, 53-80 (1963).

84. G. Schwanitz, G. Lehnert, and E. Gebhart, Deut. Med. Wochenschr., 95, 1936-1941 (1970).

85. C. Ramel, Hereditas, 61, 208-230 (1969).

86. G. Fiskesjo, Hereditas, 64, 142-146 (1970).

87. D. Kasputis, R. L. Neu, and L. I. Gardner, Arch. Environ. Health, 24, 378 (1972).

88. L. Albanus, L. Frankenberg, C. Grant, U. von Haartman, A. Jernelöv, G. Nordberg, M. Rydalv, A. Schutz, and S. Skerfving, Environ. Res., 5, 425-444 (1972).

89. K. S. Khera, Toxicol. Appl. Pharmacol., 24, 167-177 (1973).

90. A. Swensson and U. Ulfvarson, Occup. Health Rev., 15, 5-11 (1963).

TOXICITY AND RESIDUAL ASPECTS OF
ALKYLMERCURY FUNGICIDES IN LIVESTOCK*

Fred C. Wright, Jayme C. Riner, and Maurice Haufler
U.S. Department of Agriculture
Kerrville, Texas

J. S. Palmer and R. L. Younger†
U.S. Department of Agriculture
College Station, Texas

Seed for planting is commonly treated with fungicides to reduce loss from plant diseases caused by organisms associated with the seed or present in the soil. Mercury compounds, first used in Germany in 1914, were found to control fungal diseases more effectively than other chemicals. The organic mercury compounds were widely used in agriculture in contrast to lesser usage of inorganic mercury compounds [1]. One of the mercurial fungicides used in the past was Ceresan M, which contains 7.7% active ingredient of N-(ethylmercuri)-p-toluene sulfonanilide; the mercury equivalent is 3.2% of the total. Another previously used mercurial fungicide was Panogen 15, which contains 2.2% active ingredient of cyano(methylmercuri)guanidine; the mercury equivalent is 1.5%.

On March 9, 1970, the U.S. Department of Agriculture suspended the use of alkylmercury compounds. Panogen 15, at that time, was still registered for use on cotton as a liquid or granular formulation applied in-furrow and covered at planting or for nongrazed grass areas. As of March 22, 1972, all uses of alkylmercury compounds were canceled or suspended by the U.S. Environmental Protection Agency. The chances that animals will be exposed

---

*This chapter reflects the results of research only. Mention of a proprietary product or a pesticide in this chapter does not constitute an endorsement or a recommendation of this product by the USDA.
†Deceased.

to grain treated with either Ceresan M or Panogen 15 since the cancellation
of these compounds by the EPA are small, but the possibility still exists
because of the fact that grain in storage that was previously treated or
exposed to these alkylmercury fungicides may still be available.

Most known cases of poisoning by mercurial fungicide-treated grain
have been with swine; however, naturally occurring mercurialism has been
reported in cattle [2-4]. Experimentally induced poisoning has been reported in swine [5], cattle [3], and sheep and chickens [6]. Adult chickens, however, appear to be able to tolerate large quantities of seed treated
with cyano(methylmercuri)guanidine [7,8]. Residues of mercury have been
determined in tissues of chickens fed seed treated with Panogen 15 [9].
Residues of mercury have also been determined in eggs and certain tissues
of chickens that were accidentally [10] or purposely [7,8,11] fed seed
treated with Panogen 15. Mercury poisoning in livestock can normally be
diagnosed by chemical analysis of the tissues and feed samples, supplemented
by observation of histopathologic changes in the kidney [12]. Residues of
mercury in excess of 15 ppm in the kidney are indicative of mercury poisoning in livestock [12].

In a study to determine toxicity and residues of mercury, cattle,
sheep, turkeys, or chickens were given daily oral doses of either Panogen
15 or Ceresan M for fixed lengths of time. All doses of the mercurial fungicides were comparable with those that the animals might consume if accidentally exposed to grain treated with the fungicides in the previously
recommended manner. During these studies, toxicological observations were
made, and samples of blood, hair, eggs, or tissues were collected for analysis of mercury content as required. All samples were analyzed for mercury residues by atomic absorption spectrophotometry with either the flame
or the flameless technique.

SELECTION OF TEST ANIMALS AND DOSAGES

All of the cattle and sheep used in these studies were obtained at
nearby auctions. They were of mixed breed and sex, 1-2 yr old, and considered to be clinically normal. Turkeys used were of mixed sex and were
17 weeks old at the start of the study. All chickens were White Leghorns
of mixed sex and were either 6 weeks old or 8 months old at start of the
study.

The dosages and equivalents of the fungicides are shown in Table 1.
References to dosages will be to active ingredients and mercury equivalents.

TABLE 1. Dosage of Livestock with Mercurial Fungicides

Dosage Rate (mg/kg/day)			
Total Formulation	Active Ingredient	Total Mercury	Species Treated

Ceresan M[a] [N-(ethylmercuri)-p-toluene sulfonanilide]

28	2.156	0.896	Sheep
21	1.617	0.672	Sheep
20	1.540	0.640	Chickens
15	1.155	0.480	Cattle, sheep
14	1.078	0.448	Sheep
10	0.770	0.320	Chickens
5	0.385	0.160	Chickens, turkeys

Panogen 15[b] [cyano(methylmercuri)guanidine]

33	0.726	0.500	Chickens
15	0.330	0.225	Cattle, sheep
10	0.220	0.150	Chickens
5	0.110	0.075	Chickens

[a]7.7% active ingredient and 3.2% mercury equivalent. Obtained from E. I. du Pont de Nemours & Co., Wilmington, Del.

[b]2.2% active ingredient and 1.5% mercury equivalent. Obtained from Nor-Am Agricultural Products, Inc., Chicago, Ill.

Dosages were given daily by oral capsule until death or until the end of the study.

STUDIES WITH N-(ETHYLMERCURI)-p-TOLUENE SULFONANILIDE

Experimental Details

Treatment of Animals. Study I: Toxicity [6]. (1) Three sheep were dosed daily, one each with the amount of fungicide that would be contained in 0.45, 0.68, and 0.91 kg of seed that had been treated according to the manufacturer's recommendations. These dosages are equivalent to an average of 1.078, 1.617, and 2.156 mg of active ingredient/kg of body weight daily, respectively. Dosage continued until death. Weights were recorded just before the start of the test and weekly throughout the test period to evaluate general health by weight gain or loss. (2) Six chickens were randomly paired and each pair was given the fungicide daily at dosages of 0.385,

0.77, or 1.54 mg/kg of body weight. Two chickens served as a control. Chickens were weighed weekly throughout the study so that the dose could be adjusted to compensate for weight gain or loss. Dosage continued until death.

Study II: Toxicity and residues of mercury [13]. (1) Eight yearling cattle were given the fungicide at a dosage of 1.155 mg/kg (mercury equivalent of 0.48 mg/kg) daily until death or until dosage was stopped (27 days). One yearling served as a control. (2) Eight sheep were dosed at the same rate as the cattle until death or until dosing was stopped (24 days). Two sheep served as controls. (3) Nine turkeys were dosed with the fungicide at a dosage of 0.385 mg/kg (mercury equivalent of 0.16 mg/kg) daily until death or until dosing was stopped (42 days). Cattle, sheep, and turkeys were weighed before the start of the study and then weekly. At the end of the study, liver, kidney, muscle, and brain were analyzed for mercury.

Study III: Retention of mercury in tissues [14]. (1) Six cattle were dosed for 7 days and another 6 cattle were dosed for 12 days at 1.155 mg/kg/day. Two cattle served as controls. (2) Five sheep were dosed for 4 days at 1.155 mg/kg and 6 sheep were dosed at 1.155 mg/kg for 7 days. Two sheep were untreated and served as controls. Samples of blood and hair were obtained from all animals before the start of the study and weekly for 20 weeks posttreatment. One animal from each dosed group of cattle and sheep was killed 1 day after the last dose and then monthly during the study. Tissues analyzed for mercury included kidney, liver, muscle, and brain.

Residue Techniques. The analysis for residues of mercury in the various samples consisted of a wet-acid digestion followed by partitioning into methyl isobutyl ketone and final analysis by atomic absorption spectrophotometry [15]. Recovery of mercury from fortified tissues averaged 85% of the total mercury added; the sensitivity of the method was 0.1 ppm. Residues of mercury reported in this chapter have not been corrected for percentage of recovery.

Results: Toxicity and Residual Aspects

Cattle [13,14] were acutely poisoned after 27 daily doses of N-(ethylmercuri)-p-toluene sulfonanilide at a dosage of 1.155 mg/kg (mercury equivalent of 0.48) [13]. Some of the animals became clinically affected 3-11 days after the final dose (Table 2). No signs of poisoning were observed in cattle given fewer than 27 doses. Initial signs of poisoning in cattle

TABLE 2. Number of Doses, Days Until Toxicological Signs, and Fate of Cattle Given Daily Oral Doses of N-(Ethylmercuri)-p-toluene Sulfonanilide at 1.155 mg/kg[a]

No. of Animals	No. of Doses	No. of Days to Initial Signs[b]	Fate of Cattle
1	7	NS	Killed day 7 of treatment.
6	7	NS	One yearling each killed 1, 28, 56, 84, 112, and 140 days posttreatment.
6	12	NS	One yearling each killed 1, 28, 56, 84, 112, and 140 days posttreatment.
1	24	NS	Killed day 24 of treatment.
1	27	27	Killed 1 day posttreatment.
1	27	27	Killed 5 days posttreatment.
1	27	30	Killed 8 days posttreatment.
1	27	30	Killed 10 days posttreatment.
1	27	34	Died 15 days posttreatment.
1	27	38	Died 24 days posttreatment.

[a]Mercury equivalent of 0.48 mg/kg.

[b]NS means no signs of toxicosis.

were partial anorexia and the development of an incoordinated gait and infrequent diarrhea. Excessive salivation and progressive weakness followed by prostration and complete incoordination were observed. Spasms of the facial muscles were observed before death.

After 27 doses, two yearlings had initial and progressive signs of poisoning and became increasingly weaker and incoordinated. One was slaughtered for tissue samples 1 day after the last dose, and the other was slaughtered 5 days after the last dose when prostration and complete incoordination had developed. Four other cattle, which also received 27 doses, became affected 3-11 days after the last dose. Complete anorexia occurred after progressive incoordination prevented the intake of feed. At necropsy, some loss of hair, enlarged, pale kidneys, congestion of the cerebral vessels, and inflammation of the intestinal mucosa were observed.

The residues of mercury in the blood and hair of cattle given 7 or 12 daily oral doses of the fungicide at 1.155 mg/kg are shown in Table 3. Residues of mercury in the blood initially (1 day posttreatment) were about

TABLE 3. Residues of Mercury (ppm) in Blood and Hair of Cattle Given Daily Oral Doses of $\underline{N}$-(Ethylmercuri)-$\underline{p}$-toluene Sulfonanilide at 1.155 mg/kg[a]

| Time Post-treatment Days | Average Residues (ppm) in Cattle Given Indicated Number of Doses[b-d] | | | |
| | Blood | | Hair | |
	7	12	7	12
1	2.4	2.7	0.1	0.8
7	1.5	2.0	1.2	1.6
14	0.6	2.0	18.3	8.0
21	0.6	1.3	15.6	13.8
28	0.2	0.8	16.8	14.0
35	0.3	0.9	27.5	15.5
42	0.2	0.7	38.5	29.7
49	0.5	0.8	35.5	35.3
56	0.8	0.8	30.0	41.9
63	0.4	0.5	26.0	31.3
70	NA	NA	29.9	51.2
77	0.2	0.7	27.7	55.9
84	0.3	0.2	26.3	43.3
91	0.1	0.2	19.9	50.5
98	0.1	0.5	21.5	43.8
105	0.4	0.9	15.7	38.2
112	0.2	0.2	19.4	41.0
119	ND	ND	8.4	53.6
126	0.5	ND	6.0	39.0
133	0.1	ND	7.2	31.0
140	ND	0.1	6.4	19.4

[a]Mercury equivalent of 0.48 mg/kg.

[b]Values for days 119-140 were from one animal.

[c]NA indicates no sample available.

[d]ND indicates none detected.

2.5 ppm. These residues gradually decreased as the posttreatment period increased. The half-life of mercury in blood of cattle given 6 doses was between 7 and 14 days, whereas the half-life for the cattle given 12 doses was 21 days. Residues were still detectable after 112 days (16 weeks) posttreatment. Residues of < 0.5 ppm appeared in some blood samples even at 126, 133, and 140 days posttreatment. No detectable residues of mercury were in the blood of control cattle.

The residues of mercury found in the hair of dosed cattle (Table 3) may be misleading because new areas on the animals were sheared each week. Therefore, the reported values are the sum of the old residues of mercury already present in the hair plus that deposited during the last week. Residues of mercury in the hair of cattle dosed seven times peaked at 42 days posttreatment and in the hair of cattle dosed 12 times, at 77 days posttreatment. After these peaks were reached, the residues gradually decreased, but were still evident at 140 days posttreatment. Residues of mercury in the hair of cattle indicate prior exposure to a mercurial compound. No detectable residues of mercury were in the hair of control cattle.

The residues of mercury found in the tissues of 20 cattle dosed with $\underline{N}$-(ethylmercuri)-$\underline{p}$-toluene sulfonanilide at 1.155 mg/kg are shown in Table 4. Almost without exception, the residue in the tissues decreased from kidney to liver, to muscle, and to brain. Residues of mercury in the kidneys of the cattle dosed 7 days were greatest at 1 day posttreatment and then gradually decreased to 14.8 ppm at 112 days posttreatment. The half-life for mercury in kidneys of cattle dosed seven times was about 84 days. Cattle dosed 12 days had initial residues (1 day posttreatment) of twice the quantity found in the kidneys of cattle dosed 7 days. All cattle dosed 27 days had > 73 ppm of mercury in the kidneys even 24 days posttreatment; this finding indicated the slow elimination of this body burden.

The residues of mercury (Table 4) in liver were the greatest 1 day posttreatment. In most cases, the residues of mercury in liver decreased from 1 day posttreatment. After 140 days (20 weeks), the residues of mercury in liver had decreased to about 3.0 ppm in cattle dosed either 7 or 12 times. In one animal given 27 daily doses, the residues were 73.5 ppm in the kidney and 74.5 ppm in the liver at 15 days posttreatment. These amounts appear to be an exception to the rule.

Levels of mercury in muscle were greatest in cattle given 27 daily doses of the fungicide. These residues remained near this level (18 ppm) through 24 days posttreatment. In the cattle given either 7 or 12 daily doses, the greatest residues were found at 1 day posttreatment and then decreased gradually over the 140-day study period.

The pattern of elimination of the residues in brain was similar to that in muscle, except that the greatest amounts in brain were approximately one-half those in muscle.

TABLE 4. Residues of Mercury (ppm) in Tissues of Cattle Given Daily Oral Doses of N-(Ethylmercuri)-p-toluene Sulfonanilide at 1.155 mg/kg[a] at Various Periods of Time After Exposure

No. of Doses	Time (Days Post-treatment) of Sample Collection[c]	ppm Hg in Tissues[b]			
		Kidney	Liver	Muscle	Brain
7	0	69.5	6.8	4.0	4.2
7	1	71.5	25.5	4.3	0.9
7	28	50.7	24.5	2.4	1.1
7	56	58.8	16.0	2.0	1.0
7	84	39.9	5.9	0.6	0.7
7	112	14.8	3.2	0.5	0.3
7	140	18.4	2.9	0.1	0.4
12	1	146.2	40.9	7.3	3.0
12	28	71.5	37.0	3.5	2.8
12	56	34.2	27.2	3.3	1.6
12	84	64.4	24.4	2.1	1.2
12	112	47.8	4.7	1.3	1.0
12	140	18.4	3.1	1.0	0.3
24	0	109.0	47.0	19.4	6.5
27	1	113.0	66.5	14.0	10.8
27	5	109.0	48.8	19.2	8.2
27	8	105.5	49.0	18.0	8.6
27	10	88.0	62.5	19.4	11.4
27	15	73.5	74.5	18.2	8.8
27	24	112.0	46.5	22.2	10.0

[a]Mercury equivalent of 0.48 mg/kg.

[b]Each value is average of residues in duplicate samples from one animal.

[c]0 indicates sample taken on last day of treatment.

Sheep [6,13,14] were acutely poisoned after 12-23 doses at 1.155 mg/kg [13]. The fate of animals and time when signs of poisoning first occurred are shown in Table 5. Sheep given fewer than 13 doses of the fungicide did not show signs of poisoning except for one animal that received 12 doses at 2.156 mg/kg. After only 6 days of treatment, this animal showed the general signs of toxicosis: loss of weight associated with anorexia, diarrhea, stilted gait, lameness, and a clear mucous nasal discharge. These signs

TABLE 5. Number of Doses, Days Until Toxicological Signs, and Fate of Sheep Given Daily Oral Doses of $\underline{N}$-(Ethylmercuri)-$\underline{p}$-toluene Sulfonanilide at 1.155 mg/kg[a]

No. of Animals	No. of Doses	No. of Days to Initial Signs[b]	Fate of Animal
5	4	NS	One sheep each killed 1, 28, 56, 84, and 140 days post-treatment.
6	7	NS	One sheep each killed 1, 28, 56, 84, 112, and 140 days posttreatment.
1	9	NS	Killed day 9 of treatment.
1	12[c]	6	Died day 12 of treatment.
1	13	13	Died day 13 of treatment.
1	22[d]	16	Died day 22 of treatment.
1	24	12	Killed 4 days posttreatment.
1	24	22	Died 6 days posttreatment.
1	24	27	Died 3 days posttreatment.
1	24	23	Died 5 days posttreatment.
1	24	23	Died 7 days posttreatment.
1	33[e]	32	Died day 33 of treatment.

[a]All sheep given 1.155 mg/kg/day except where noted.

[b]NS means no signs of toxicosis.

[c]Average dosage for this sheep was 2.156 mg/kg.

[d]Average dosage for this sheep was 1.617 mg/kg.

[e]Average dosage for this sheep was 1.078 mg/kg.

became progressively worse until the animal died on day 12. The sheep given 1.617 mg/kg had initial toxicosis at 16 days but improved for 3 days, and then signs of toxicosis worsened on day 19, and the sheep died on day 22. Loss of weight was evident in both of these sheep dosed at the higher levels. The sheep dosed at 2.156 mg/kg had a 26.8% weight loss in 12 days and the sheep dosed at 1.617 mg/kg had a 12.4% weight loss in 22 days. Gross pathologic changes found at necropsy, typical of metallic poisoning, were catarrhal enteritis, ulceration and sloughing of the rumen, pulmonary edema, and inflammation of the lungs, kidneys, and spleen.

At the lower dosage rate (1.155 mg/kg), the initial signs of toxicosis were anorexia and loss of weight. Continued dosing produced diarrhea after

12 days. Dyspnea, anorexia, and frequent urination were observed with the occasional diarrhea. Forced exercise of these animals hastened prostration and a moribund condition.

After 13 doses, one sheep developed signs of toxicosis and died 8 hr later. Initial signs began in two sheep after 22 doses, in two after 23 doses, and in one after 27 doses. All sheep in this group died within 2-7 days posttreatment. At necropsy, the kidneys, liver, and cranial vessels were congested, and the intestinal mucosa were reddened.

The sheep given 33 doses at 1.078 mg/kg had no evidence of poisoning except for a slight weight loss after 2 weeks, until the day before death. At that time, signs of toxicosis associated with colic were observed. Weight loss during the 33-day study period was only 4.1%.

The residues of mercury found in the blood and wool of sheep given four or seven daily doses of the fungicide at 1.155 mg/kg are shown in Table 6. At 1 day posttreatment, levels of mercury in blood were 1.5-2.0 ppm; residues gradually decreased. No detectable residues of mercury were in the blood of control sheep.

As with the residues in the hair of cattle, the residues of mercury in the wool of sheep are misleading because new areas were sheared each week. Residues of mercury in the wool of sheep dosed four times reached their greatest peak at 91 days posttreatment, and those of sheep dosed seven times reached their greatest peak at 112 days posttreatment. Residues of mercury in wool of sheep did not reach the levels found in the hair of cattle. Again, this difference is probably related to the quantity of old wool already on the sheep when an area was sheared for a representative sample. No detectable residues of mercury were found in wool of control animals.

The residues of mercury in the tissues of sheep given daily oral doses of the fungicide are given in Table 7. The residues were generally greater in the kidney than in the liver. Liver had greater residues than muscle, and brain had the least amount over all. Residues of mercury in excess of 200 ppm were in the kidneys of three animals. Two of these animals had been dosed only seven times with the fungicide, and samples were taken 2 days posttreatment. We have no explanation for the quantity found in the kidneys of the two animals dosed only seven times. Another animal in that same group had a level of 22.2 ppm at 1 day posttreatment.

The residues in the liver of animals dosed four or seven times were greater initially at 1 day posttreatment and gradually decreased to <10 ppm

TABLE 6. Residues of Mercury (ppm) in Blood and Wool of Sheep Given Daily
Oral Doses of N-(Ethylmercuri)-p-toluene Sulfonanilide at 1.155 mg/kg[a]

| Time Post-treatment Days | Average Residues (ppm) in Sheep Given Indicated Number of Doses[b-d] | | | |
| | Blood | | Wool | |
	4	7	4	7
1	1.5	2.0	ND	0.4
7	0.7	1.5	0.1	0.1
14	0.6	1.0	0.3	0.3
21	0.1	0.6	0.3	0.7
28	0.1	0.6	0.2	1.5
35	0.2	0.6	0.1	4.2
42	ND	0.3	1.0	1.8
49	ND	0.7	4.0	5.4
56	ND	0.1	3.8	8.4
63	0.1	0.1	3.5	10.4
70	NA	NA	3.6	5.5
77	0.2	0.3	4.5	4.9
84	0.2	0.1	3.2	4.3
91	0.3	ND	8.1	9.3
98	0.4	ND	3.1	5.7
105	0.3	ND	3.6	8.3
112	0.4	ND	5.0	14.5
119	0.5	0.1	3.7	5.7
126	0.1	ND	5.1	5.3
133	0.1	ND	3.3	8.7
140	NA	0.5	4.7	8.7

[a]Mercury equivalent of 0.48 mg/kg.

[b]Values for days 119-140 were from one animal.

[c]ND indicates none detected.

[d]NA indicates no sample available.

after 84 days posttreatment. However, in sheep given 24 doses, the residues
of mercury in the livers remained rather stable (105-121 ppm) for up to 7
days posttreatment.

Sheep given four doses of the fungicide had residues of mercury in
muscle and brain that did not exceed 2 ppm. Animals given seven doses had

TABLE 7. Residues of Mercury (ppm) in Tissues of Sheep Given Daily Oral Doses of N-(Ethylmercuri)-p-toluene Sulfonanilide at 1.155 mg/kg[a] at Various Periods of Time After Exposure

No. of Doses	Time Posttreatment Days[b]	ppm Hg in Tissue[c-e]			
		Kidney	Liver	Muscle	Brain
4	1	18.2	22.4	1.8	1.0
4	28	NA	15.4	1.6	1.4
4	56	83.3	3.1	ND	0.3
4	84	72.7	7.8	0.3	0.5
4	112	57.4	NA	ND	ND
7	1	22.2	31.7	6.4	1.8
7	28	NA	19.3	0.9	2.1
7	56	228.8	16.3	0.1	1.0
7	84	208.2	9.6	ND	0.4
7	140	NA	NA	0.5	NA
9	0	63.2	43.2	8.4	2.6
13	0	145.5	27.9	9.3	1.7
24	0	181.1	118.0	20.5	9.7
24	2	212.0	121.0	19.0	9.8
24	3	194.5	116.0	19.9	10.7
24	5	117.5	105.0	15.3	9.2
24	6	137.0	106.8	22.8	9.8
24	7	135.0	119.0	19.0	10.1

[a]Mercury equivalent of 0.48 mg/kg.

[b]0 indicates sample taken on last day of treatment.

[c]Each value is average of residues in duplicate samples from one animal.

[d]NA indicates no sample available.

[e]ND indicates none detected.

a high of 6.4 ppm in muscle at 1 day posttreatment and 2.1 ppm in brain at 28 days posttreatment. The levels of mercury in muscles of sheep given 24 doses remained almost stable (15.3-22.8 ppm) for 7 days posttreatment. The picture of results in brain were similar to that in muscle except the level in brain was lower. Levels ranged from 9.2 to 10.7 ppm in samples of brain collected within 7 days posttreatment from animals dosed 24 times. Tissues from control sheep had residues of mercury that were < 0.2 ppm.

TABLE 8. Number of Doses, Days Until Toxicological Signs, and Fate of Turkeys Given Daily Oral Doses of N-(Ethylmercuri)-p-toluene Sulfonanilide at 0.385 mg/kg[a]

No. of Birds	No. of Doses	No. of Days to Initial Signs[b]	Fate of Turkeys
1	9	8	Killed 2 days posttreatment.
1	11	10	Killed 35 days posttreatment.
1	17	NS	Killed 1 day posttreatment.
1	31	13	Killed 1 day posttreatment.
1	42	31	Killed 1 day posttreatment.
1	42	42	Killed 1 day posttreatment.
3	42	NS	One killed at 34 days posttreatment; two killed at 43 days posttreatment.

[a]Mercury equivalent of 0.16 mg/kg.

[b]NS means no sign of toxicosis

Initial signs of toxicosis in turkeys [13] were weakness, an incoordinated gait, and pale wattles. Ataxia was often associated with movement that resulted in prostration. No deaths resulted from continued exposure, but convulsions and paralysis developed in one turkey. Three of the nine turkeys had no apparent ill effects after the full course of exposure (Table 8).

One turkey, after nine doses, was killed after showing initial signs of toxicosis. Another turkey developed signs after 10 doses but after one additional dose was maintained for 35 days posttreatment with a decrease in the severity of toxicosis and a slight increase in weight. Before slaughter, only a slight incoordinated gait was observed.

After 13 doses, one turkey had an incoordinated gait, but dosing was continued for a total of 31 doses before the bird was slaughtered. The severity of the signs of toxicosis did not increase during this time. Another turkey developed an incoordinated gait after 31 doses, but the severity of toxicosis did not increase with continued dosage. A sudden development of paralysis followed by prostration and convulsions after the final (42nd) dose developed in another turkey, but the other three turkeys were not visibly affected after 42 doses, as judged by the absence of clinical signs.

The residues of mercury found in the tissues of turkeys dosed daily for various lengths of time at 0.385 mg/kg (mercury equivalent of 0.16 mg/kg)

TABLE 9. Residues of Mercury (ppm) in Tissues of Turkeys Given Daily Oral Doses of N-(Ethylmercuri)-p-toluene Sulfonanilide at 0.385 mg/kg[a] at Various Periods of Time After Exposure

No. of Doses	Time Posttreatment Days[b]	ppm Hg in Tissues[c,d]			
		Kidney	Liver	Muscle	Brain
9	2	7.6	10.9	0.8	1.6
11	35	9.7	5.8	1.0	ND
17	0	7.8	14.1	1.8	3.6
31	0	15.7	23.6	2.4	NA
42	0	2.1	19.5	8.7	NA
42	0	39.5	59.0	2.5	NA
42	34	NA	21.4	2.6	NA
42	43	3.5	20.1	1.5	NA
42	43	7.2	10.7	2.1	1.7
0	Variable[e]	0.2	0.1	0.1	0

[a]Mercury equivalent of 0.16 mg/kg.

[b]0 indicates sample taken on last day of treatment.

[c]ND indicates none detected.

[d]NA indicates no samples available.

[e]Control turkeys; one each killed days 10, 18, and 32.

are shown in Table 9. The residue pattern of mercury in tissues of turkeys is the reverse of the pattern of cattle and sheep. Greatest residues were found in the liver and then in kidney, muscle, and brain.

Dosages and fate of treated chickens [6] are summarized in Table 10. All the dosed chickens died. Control birds had normal gains in weight and size.

Signs of toxicosis in the dosed chickens were meager except for anemia manifested as bleached combs, depression, and partial or complete paralysis in two chickens before death. A weight loss was recorded for one chicken receiving a 0.77-mg/kg dosage, but the other chickens gained weight, although not as pronounced as gains of the control chickens. The two chickens given 0.385 mg/kg had an average weight gain of 67% compared with an average weight gain for the control chickens of about 94%.

Gross pathologic changes observed at necropsy included catarrhal enteritis, ulceration of the crop, icterus, and inflammation and edema of the

TABLE 10. Number of Doses, Days Until Toxicological Signs, and Fate of Chickens Given Daily Oral Doses of N-(Ethylmercuri)-p-toluene Sulfonanilide

No. of Chickens	Dosage Rate (mg/kg) Active Ingredient	Hg Equivalent	No. of Doses	Fate of Chickens
1	0.385	0.16	30	Died day 30 of treatment.
1	0.385	0.16	26	Died day 26 of treatment.
1	0.77	0.32	10	Died day 10 of treatment.
1	0.77	0.32	12	Died day 12 of treatment.
2	1.54	0.64	12	Died day 12 of treatment.

kidneys, liver, and spleen. The atria had hemorrhage areas, and the blood was brownish.

STUDIES WITH CYANO(METHYLMERCURI)GUANIDINE

Experimental Details

Treatment of Animals. Study I: Accumulation of mercury in tissues [15]. (1) Ten cattle were dosed daily at 0.33 mg/kg (mercury equivalent of 0.225 mg/kg) for 10 weeks. Three cattle served as controls. (2) Twelve sheep were dosed daily at 0.33 mg/kg for 10 weeks. One sheep served as a control. (3) Twelve chickens were dosed daily at 0.11 mg/kg (mercury equivalent of 0.075 mg/kg) for 12 weeks, and 12 chickens were dosed daily at 0.22 mg/kg (mercury equivalent of 0.15 mg/kg) for 12 weeks. Two chickens served as controls. At least one dosed animal or bird was killed weekly during the test period. Kidney, liver, muscle, and brain were collected from both cattle and sheep, and kidney, liver, breast muscle, and thigh muscle were collected from the chickens. Undosed animals that served as controls were killed at the end of the study; then tissue samples were collected for analysis.

Study II: Elimination of mercury in eggs from treated chickens [16]. One group of three roosters and 10 hens received daily oral doses of 0.726 mg/kg (mercury equivalent of 0.5 mg/kg) for 100 days or until death. Another group of three roosters and 10 hens were sham treated and served as controls. Dosages for each chicken were calculated weekly, according to weight changes. Semen was collected from the roosters and was pooled within each group. Hens were artificially inseminated three times a week with

pooled semen from roosters in the same group. Feed consumption, egg production, and weekly weight of each chicken were recorded. Eggs collected
from each group of hens were combined into weekly groups; some were analyzed
for residues of mercury, and the rest were used in hatchability studies.
Some of the chicks hatched from eggs laid by dosed hens were killed for
tissue analysis. Adult chickens were slaughtered at 1, 15, 50, and 100 days
after completion of the 100 doses. Samples of tissues collected included
liver, kidney, spleen, gizzard, heart, breast muscle, leg muscle, brain,
ovary, oviduct, and testicle.

    Residue Techniques. The analysis for residues of mercury in tissues
of animals was mentioned earlier. Samples of eggs were digested with a
combination of concentrated nitric acid, vanadium pentoxide, and hydrogen
peroxide. Final analysis was by flameless atomic absorption spectrophotometry.

    Results: Toxicity and Residual Aspects. Of the cattle [15] dosed at
0.33 mg/kg (mercury equivalent of 0.225 mg/kg), one yearling was mildly
poisoned after 56 doses and another after 65 daily doses of cyano(methylmercuri)guanidine. Signs of poisoning included muscular incoordination,
stiffness in the hindquarters, and an unsteady gait. None of the other
dosed cattle had any visible signs of toxicosis even after 70 doses at this
level.

    During the 70-day treatment period, the residues of mercury increased
in the blood to 4.5 ppm at 35 days and in the hair to 67 ppm at 42 days.
After these peaks were reached, the residues remained below that level
through the study.

    The accumulation of the residues of mercury in tissues of cattle is
shown in Table 11. Cattle were dosed for different lengths of time. As
with the cattle given N-(ethylmercuri)-p-toluene sulfonanilide, the greatest residues of mercury were found in the kidneys. Liver had the next
greatest residue, followed by muscle and then brain. The residues of mercury in the kidneys increased rapidly until the peak was reached after 63
doses. After that time, the residues in the kidneys ranged from 134 to
165 ppm until the end of the study. The peak residue of mercury was found
in liver after 63 doses, in muscle after the entire 70 doses had been given,
and in brain after 63 doses.

    There appears to be a very definite accumulation of mercury in the
tissues of cattle dosed in this manner. The residues of mercury in the
tissues of a control did not exceed 0.1 ppm.

TABLE 11. Residues of Mercury (ppm) in Tissues of Cattle Given Daily Oral Doses of Cyano(methylmercuri)guanidine at 0.33 mg/kg[a]

No. of Doses[b]	Hg in Tissues (ppm)[c,d]			
	Kidney	Liver	Muscle	Brain
0	0.1	ND	ND	0.1
7	18.1	7.8	2.1	0.6
14	45.8	14.3	3.2	1.5
21	68.3	21.5	8.6	2.1
28	76.0	20.0	12.5	2.4
35	102.3	19.8	14.2	4.6
42	126.8	20.3	16.1	8.0
49	155.5	29.5	12.0	9.4
56	136.5	31.5	16.7	10.0
63	165.0	39.5	19.5	11.8
70	133.8	39.3	23.2	11.7

[a]Mercury equivalent of 0.225 mg/kg.

[b]Animals were killed 1 day after last dose.

[c]Each value is average of residues in duplicate samples from one animal.

[d]ND indicates none detected.

Sheep [15]. Three sheep were mildly poisoned during the treatment period. One exhibited signs of toxicosis after 42 doses, another after 56 doses, and the third after 59 doses. Two other sheep given more doses than the affected ones had no apparent adverse effects. Signs of toxicosis were muscular incoordination, stiffness, and an unsteady gait.

The residues in the blood of sheep given 0.33 mg/kg increased to near 10 ppm after 63 days of treatment; the greatest (35.7 ppm) in wool was also at this same time period. It is interesting to note that the residues of mercury in the blood of sheep are approximately twice those found in the blood of cattle at the same dosage. However, the reverse is true in a comparison between wool of sheep and hair of cattle. Residues in the hair reached a peak almost twice that reached in wool.

The residues of mercury in the tissues of sheep are shown in Table 12. Again, the residues of mercury were greatest in kidneys and then in liver, muscle, and brain. The residues of mercury in the kidneys reached a peak after 42 doses. in liver after 63 doses, and in muscle and in brain after

TABLE 12. Residues of Mercury (ppm) in Tissues of Sheep Given Daily Oral
Doses of Cyano(methylmercuri)guanidine at 0.33 mg/kg[a]

No. of Sheep	No. of Doses[b]	Hg in Tissues (ppm)[c,d]			
		Kidney	Liver	Muscle	Brain
1	0	0.1	0.1	ND	0.1
1	7	6.6	9.0	2.4	1.0
1	14	36.0	26.0	3.8	4.0
2	21	59.5	26.9	5.8	4.8
1	28	55.8	28.5	10.0	5.8
1	35	62.5	41.3	10.6	6.0
1	42	109.8	35.0	11.2	10.2
1	49	108.5	40.0	10.8	8.6
2	56	95.4	45.2	14.2	12.4
1	63	100.0	54.0	13.6	10.6
1	70	76.0	47.5	12.2	11.4

[a]Mercury equivalent of 0.225 mg/kg.

[b]Animals killed 1 day after last dose.

[c]Each value is average of residues in duplicate samples from one animal.

[d]ND indicates none detected.

56 doses. The increase in residues with an increase in number of doses was
more erratic in sheep than in cattle, and the levels reached were not as
great in sheep as those in cattle, except in the liver. The residues of
mercury in the tissues of a control did not exceed 0.1 ppm.

Chickens [15,16]. Chickens were dosed at three levels, i.e., 0.11,
0.22, or 0.726 mg/kg, for various lengths of time. No ill effects were
visible in dosed chickens given 0.11 or 0.22 mg/kg of cyano(methylmercuri)-
guanidine for 84 days. In the chickens given dosages of 0.726 mg/kg (mercury equivalent of 0.5 mg/kg), signs of toxicosis were more apparent. The
two hens in the group that died had 42 and 90 doses and became anorexic for
3 days before death. Subsequently, the hens rested continuously on their
sternums during this period. At the time of necropsy, both showed a severe
weight loss. No clinical signs of toxicosis were observed in any of the
other chickens given the fungicide for 100 days.

About 50% of the chicks hatched from eggs laid by the treated hens
(0.726 mg/kg) after the third week of the study manifested clinical signs

of mercurial poisoning at the time of hatch or developed clinical signs 3-5 days after hatch. Affected baby chicks had or developed weakness, incoordination, and tremors of peripheral musculature.

No gross lesions were observed in baby chicks that hatched from eggs of dosed hens (0.726 mg/kg) or in embryos from eggs that failed to hatch. Hatchability of fertile control eggs averaged 80%, but by the third week the hatchability of fertile eggs of dosed hens had decreased to 30%. This low level of hatchability continued throughout the 100-dose period. No attempts were made to correlate the low hatchability with loss of fertility in either the hens or roosters.

Egg production of the dosed hens (0.726 mg/kg) was 73-90% of the pretreatment production during the first 7 weeks. The egg production than gradually declined to 19% of the pretreatment production at 14 weeks. Mean egg production for dosed hens was highly significantly lower ($P < 0.001$) than that for the controls.

Mean feed consumption did not differ significantly between hens given cyano(methylmercuri)guanidine (0.726 mg/kg) and control hens, but mean feed consumption for dosed roosters (0.726 mg/kg) was significantly lower ($P < 0.05$) than that for control roosters. At the first and last week of the study, mean body weight of dosed hens and roosters were not significantly different from those of the respective controls.

The residues of mercury found in the blood of chickens dosed at 0.11 mg/kg (mercury equivalent of 0.075 mg/kg) ranged from 2.2 to 4.4 ppm during the study. The peak was after 56 doses. In the blood of chickens dosed at 0.22 mg/kg (mercury equivalent of 0.15 mg/kg), the residues ranged from 2.0 to 11.8 ppm during the study. The peak was after 70 doses. The residues in the blood of chickens dosed at the higher level increased gradually and those in the blood of chickens dosed at the lower level were erratic.

The residues of mercury in the tissues of chickens dosed at 0.11 or 0.22 mg/kg are shown in Table 13. As with turkeys dosed with N-(ethylmercuri)-p-toluene sulfonanilide and as with cattle and sheep dosed with either cyano(methylmercuri)guanidine or N-(ethylmercuri)-p-toluene sulfonanilide, the greatest residues of mercury were found in the liver rather than in the kidney. The next greatest amount of mercury was deposited in the kidney and then in the breast muscle, and then in thigh muscle. The peak residues in the tissues of chickens dosed with 0.22 mg/kg were in kidney after 63 doses, in liver after 77 doses, in breast muscle after 63 doses, and in

TABLE 13. Residues of Mercury (ppm) in Tissues of Chickens Given Daily Oral Doses of Cyano(methylmercuri)guanidine for Various Periods of Time

No. of Birds	Dosage Rate (mg/kg)[a]	No. of Doses[b]	Hg in Tissues (ppm)[c,d]			
			Kidney	Liver	Breast Muscle	Thigh Muscle
4	0	0	0.1	0.1	0.1	ND
2	0.11	14	1.8	3.0	0.4	0.7
1	0.11	21	3.3	4.6	0.6	1.2
1	0.11	28	NA	6.8	1.3	1.9
1	0.11	35	6.8	7.8	2.2	1.4
1	0.11	49	4.5	7.6	2.1	1.8
1	0.11	56	8.8	7.6	3.4	2.0
1	0.11	63	5.6	6.2	1.8	1.6
1	0.11	70	3.8	4.8	2.2	1.7
1	0.11	77	4.8	5.2	1.6	0.7
1	0.11	84	9.6	12.2	4.6	4.0
3	0.22	14	2.0	4.1	0.7	1.0
1	0.22	42	9.8	10.4	4.2	2.7
1	0.22	49	13.8	15.2	4.1	3.3
1	0.22	56	10.4	11.0	5.0	2.9
1	0.22	63	16.4	17.0	5.8	5.0
1	0.22	70	11.2	12.3	4.1	3.1
1	0.22	77	14.0	17.4	5.2	3.9
1	0.22	84	11.6	16.0	6.3	5.3

[a]0.11 and 0.22 have mercury equivalents of 0.075 and 0.15 mg/kg, respectively.

[b]Chickens were killed 1 day after last dose.

[c]ND indicates none detected.

[d]NA indicates no sample available.

thigh muscle after 84 doses. The average residues of mercury in tissues of control chickens did not exceed 0.1 ppm.

Residues of mercury in tissues of chickens given daily oral doses of cyano(methylmercuri)guanidine at 0.726 mg/kg (mercury equivalent of 0.5 mg/kg) were greatest in kidney and liver (Table 14). In these chickens, the residues of mercury in the kidney were generally higher than those in

TABLE 14. Residues of Mercury (ppm) in Tissues of Chickens Given Daily Oral Doses of Cyano(methylmercuri)guanidine at 0.726 mg/kg[a] for 100 days[b]

Sex	Female	Female	Female	Male	Female	Female	Male	Female	Male
Days after initial dose	42	90	100[c]	100[c]	115	150	150	200	200
Liver	114.0	67.0	23.0	38.0	50.9	0.1	10.7	1.5	0.1
Kidney	192.0	78.0	22.0	52.0	50.9	0.1	15.8	1.6	3.3
Spleen	72.0	28.6	13.8	19.3	18.3	1.1	5.9	0.6	0.1
Gizzard	34.0	25.5	11.5	15.5	11.1	0.1	1.0	0.1	0.1
Heart	23.3	14.5	6.1	6.9	10.4	0.1	1.6	0.1	0.1
Breast muscle	31.2	30.0	13.6	18.2	23.0	1.9	4.4	0.1	1.0
Thigh muscle	25.8	24.2	9.2	10.0	8.7	0.6	2.6	0.1	0.2
Brain	25.3	19.0	9.7	10.6	11.1	1.6	2.3	0.1	0.1
Ovary	31.2	14.7	8.5	--	8.2	0.8	--	2.3	--
Oviduct	--	37.0	37.5	--	18.5	0.1	--	0.1	--
Testicle	--	--	--	2.8	--	--	0.7	--	0.5

[a] Mercury equivalent of 0.5 mg/kg.

[b] Determined from pooled sample from two hens, but days 42 and 90 represent single sample of hens that died of apparent mercury poisoning. Data for males represent single sample.

[c] Treatment stopped after 100 days.

the liver, but in chickens dosed at 0.11 or 0.22 mg/kg, the residues were
generally higher in the liver. The difference may be due to the variance in
dosage or to age and physiological state of the chickens. Chickens given
0.726 mg/kg were 8 months old at the start of the study, and those given
0.11 or 0.22 mg/kg were 6 weeks old. The residues of mercury in muscle
followed the same pattern as with chickens dosed at lower rates; i.e.,
residues were greater in breast muscle than in thigh muscle.

The residues of mercury in the tissues of chickens dosed at 0.726 mg/kg
agree favorably with those of earlier work of Tejning and Vesterberg [7];
kidney had the greatest residues of mercury, followed by liver, breast
muscle, brain, heart, and ovaries. After dosing had stopped, the residues
in the tissues of both hens and roosters decreased to <4.0 ppm at 100 days
posttreatment. Residues in tissues of controls were <0.4 ppm during the
study.

The residues of mercury in the eggs of hens given cyano(methylmercuri)-
guanidine at 0.726 mg/kg (mercury equivalent of 0.5 mg/kg) are shown in
Table 15. Residues in albumin were much greater than in yolk. An average
of 90% of the total mercury in the egg was found in the albumin and the
rest in the yolk. Smart and Lloyd [11] found 91% in the albumin fraction.
In the present study, the highest percentage of mercury in the yolk was
near the end of the dosing period (12% at weeks 14 and 15). The highest
percentage (94%) of mercury in albumin was in eggs collected during the
first week after initiation of treatment. Residues of mercury reached a
plateau at about 0.8 mg/egg (calculated figure, see Table 15) by 7 weeks
after initiation of treatment and remained near this level throughout the
feeding period. Residues in eggs of control hens were <0.3 ppm during the
study.

The residues of mercury were also determined in tissues of chicks
hatched from eggs laid by treated hens. The tissues of 1-day-old chicks
hatched from eggs collected during the sixth week of dosing (0.726 mg/kg/day)
showed greatest residues (about 32.0 ppm) in kidney and liver. All tissues
analyzed (liver, kidney, spleen, gizzard, breast muscle, thigh muscle, and
brain) had residues of mercury >10 ppm except spleen, which had 5.5 ppm.
Residues of mercury in tissues of 14-day-old chicks hatched from eggs collected during the eighth week of dosing had decreased sharply because of
the fast growth of chicks at this age and possible loss of residues through
excretion. Greatest residues (about 3.8 ppm) were again found in the kidney

TABLE 15. Residues of Mercury (ppm) in Albumin and Yolk of Eggs from Hens Given Daily Oral Doses of Cyano(methylmercuri)guanidine at 0.726 mg/kg for 100 Days

Week Collected	ppm Hg Albumin	ppm Hg Yolk	Total Mercury in Eggs (mg)[a]
1	7.7	0.9	0.277
2	12.4	2.3	0.461
3	13.6	3.2	0.517
4	18.2	4.2	0.690
5	18.1	3.6	0.677
6	19.8	4.1	0.743
7	24.0	4.0	0.884
8	23.8	4.4	0.884
9	20.9	3.6	0.772
10	23.0	4.7	0.862
11	26.4	4.6	0.976
12	26.7	4.9	0.991
13	20.5	5.0	0.782
14	20.6	5.6	0.796
15[b]	21.6	6.1	0.838

[a]Figure calculated on assumed average weight of egg of 56.0 g, eggshell of 5.0 g, yolk of 17.0 g, and albumin of 34.0 g.

[b]Dosing stopped at 100 days.

and liver. Other tissues analyzed had residues of < 2.6 ppm. Residues of mercury were not detected in tissues of chicks hatched from eggs of control chickens.

GENERAL SUMMARY

The general signs of toxicosis in all species dosed with mercurial fungicides were those of typical mercury poisoning. These signs included lack of appetite, loss of weight, muscular incoordination, unstable gait, and lameness. The information reported here should not be misinterpreted to indicate that N-(ethylmercuri)-p-toluene sulfonanilide is more toxic than cyano(methylmercuri)guanidine. The dosages were based on total formulations rather than on active ingredients or mercury equivalents; therefore, the animals given cyano(methylmercuri)guanidine received only 46.9% as much actual mercury as the animals given N-(ethylmercuri)-p-toluene sulfonanilide.

Mercury from mercurial fungicides accumulates rapidly in the tissues of all species studied. Levels of mercury in the blood of cattle and sheep were inconsistent. The greatest residues of mercury in cattle and sheep were in the kidney, followed by liver, muscle, and brain. Greater residues of mercury were in the kidney and liver of sheep given N-(ethylmercuri)-p-toluene sulfonanilide than in the kidney and liver of cattle at the same dosage level (0.48 mg/kg/day of mercury). Residues in muscle and brain of both species were similar. Greater residues of mercury were in the kidneys of cattle given cyano(methylmercuri)guanidine than in the kidneys of sheep given the same fungicide at the same dosage level. However, greater residues were found in liver and muscle of the sheep than in liver and muscle of the cattle.

In chickens and turkeys, at low dosages, the greatest residues were in the liver, then in kidney, and then in muscle. At higher dosages, the greatest residues in chickens were in kidney and then in liver, and then in muscle.

Mercury persists in tissues and hair of livestock for a considerable length of time after intake of the element has ceased. Therefore, care must be taken about the disposal of livestock after exposure to mercurial compounds such as mercurial fungicides. Residues of mercury in tissues after several weeks of nonexposure may still be great enough to be hazardous to humans if consumed.

REFERENCES

1. N. A. Smart, Residue Rev., 23, 1 (1968).

2. L. E. Boley, C. C. Morrell, and R. Graham, N. Amer. Vet., 22, 161 (1941).

3. Y. Fujimoto, K. Oshima, H. Satok, and Y. Ohta, Jap. J. Vet. Res., 4, 19 (1956).

4. W. H. Herberg, Vet. Med., 49, 401 (1954).

5. R. C. Piper, V. L. Miller, and E. O. Dickenson, Amer. J. Vet. Res., 32, 263 (1971).

6. J. S. Palmer, Amer. J. Vet. Med. Ass., 142, 1385 (1963).

7. S. Tejning and R. Vesterberg, Poultry Sci., 43, 6 (1964).

8. S. Tejning, Acta Oecological Scand., 8, 7 (1967).

9. E. Hanko, K. Erne, H. Wanntorp, and K. Borg, Acta Vet. Scand., 11, 268 (1970).

10. J. Howell, Can. Vet. J., 10, 212 (1969).

11. N. A. Smart and M. K. Lloyd, J. Sci. Food Agr., 14, 734 (1963).

12. W. B. Buck, J. Amer. Vet. Med. Ass., 155, 1928 (1969).

13. J. S. Palmer, F. C. Wright, and M. Haufler, Clin. Toxicol., 6, 425 (1973).

14. F. C. Wright, J. S. Palmer, and J. C. Riner, J. Agr. Food Chem., 21, 614 (1973).

15. F. C. Wright, J. S. Palmer, and J. C. Riner, J. Agr. Food Chem., 21, 414 (1973).

16. F. C. Wright, R. L. Younger, and J. C. Riner, Bull. Environ. Contam. Toxicol., 12, 366 (1974).

16

TOXICITY OF INORGANIC
AND ALIPHATIC ORGANIC ARSENICALS

William B. Buck*
Iowa State University College of Veterinary Medicine
Ames, Iowa

Arsenic is one of the Group V elements, which also includes nitrogen, phosphorus, antimony, and bismuth. Arsenic has an atomic number of 33, an atomic weight of 74.91, and its melting point is 814° C. The chemical characteristics of these five elements vary as the atomic weight increases. As would be expected, the chemistry of arsenic and phosphorus are quite similar in many respects.

For the most part, toxicoses caused by inorganic and aliphatic organic arsenicals are manifested by an entirely different syndrome from that caused by the phenylarsonic compounds [1]. Therefore, inorganic and aliphatic arsenicals will be discussed separately.

The toxic effects of arsenic in humans has been known for many centuries. During the Middle Ages, arsenicals were apparently the preferred homicidal and suicidal agents. Arsenic appears to be second only to lead in importance as a toxicant in farm and household animals [1]. It is a ubiquitous element, being present in virtually all rocks, soils, and water. Commercial arsenic is not mined as such but is produced principally as a by-product of mining, mineral beneficiating, smelting, and refining operations. The worldwide production of arsenic trioxide appears to exceed 50,000 tons each year [2]. Many and varied formulations of arsenic are synthesized for uses varying from medicinal to forensic.

---

*Present Affiliation: University of Illinois College of Veterinary Medicine, Urbana, Illinois.

SOURCES

Inorganic arsenical compounds are widely found in soils and ores, usually combined with other elements such as iron and sulfur. Much of the natural arsenic occurs in the form of pyrites ($FeS_2 \cdot FeAs_2$) and sulfides ($As_2S_{2,3}$). During the process of refining metal ores using heat, arsenic trioxide (As_2O_3) is produced, most of which is collected for additional processing and use; but some is carried to the surrounding countryside in dust or smoke [3]. Much of the arsenic presently being produced is used in pesticide and medicinal formulations [4]. Arsenic trioxide (white arsenic) and sodium arsenite are frequently used as herbicides, as are other alkali salts of arsenic such as sodium, potassium, calcium, and lead arsenates. These compounds are used not only as weed and brush killers but also as defoliants, especially for cotton and fruit trees. The use of aliphatic organic arsenical herbicides, monosodium methanearsonate (MSMA), disodium methanearsonate (DSMA), and dimethyl arsenic acid (cacodylic acid) has increased rapidly during the last decade. Copper acetoarsenite (Paris green), arsenic trioxide, and lead arsenate all have good insecticidal qualities.

Other sources of inorganic arsenic include pesticides for use against ants, snails, and rodents; insulation material such as vermiculite and fiberboard; and wood building materials treated with arsenical preservatives. Arsenic is also found in certain paint pigments (e.g., emerald green), detergents, and medicaments, such as Fowler's solution. It has recently been reported that certain ruminatorics were contaminated with arsenic, which, although at levels not thought to be toxic, gave misleading results when rumen contents were tested for arsenic in connection with suspected poisoning cases [1].

Some of the more common sources of arsenic poisoning in foraging animals include grass clippings from lawns previously treated with arsenical crabgrass control preparations, and grass, weeds, shrubbery, and other foliage previously sprayed with arsenical herbicides. Other sources include livestock dipping vats that years before had been charged with arsenic trioxide, and soils heavily contaminated with arsenic, either through the burning of arsenical compounds in rubbish piles or application of arsenical pesticides to orchards and truck gardens.

Common sources of arsenic poisoning in household pets and children include ant and snail baits containing 1-2% arsenic. Certain organic arsenical compounds such as sodium thiacetarsamide are used for the treatment of blood parasites, such as <u>Dirofilaria</u> <u>immitis</u>, in dogs and cats.

Since the synthesis of arsphenamine (salvarsan) in 1907 for the treatment of syphilis, many organic arsenical compounds have been developed for
the treatment of various blood parasitic and infectious diseases in humans.
Some of these compounds, for example acetarsol (3-acetylamino-4-hydroxy-
phenylarsonate) and sodium cacodylate, have been used in veterinary medicine
as general stimulants and for their anti-blood parasitic properties.

OCCURRENCE AND TOXICITY

Man and all lower animals are susceptible to inorganic and aliphatic
organic arsenic, but poisoning is most frequently encountered in the bovine
and feline species, resulting from contamination of their food supply. Frequency of incidence of arsenical poisoning in these two species is closely
followed in other forage-eating animals, such as the sheep and horse. Poisoning by inorganic and aliphatic organic arsenicals occasionally occurs in
the dog, but rarely occurs in swine and poultry.

The toxicity of the various formulations of inorganic arsenicals varies
with the species of animal exposed, the formulation of the arsenical (trivalent arsenicals are more toxic than pentavalent), solubility of the formulation, route of exposure, rate of absorption from the gastrointestinal
tract, and rate of metabolism and excretion by the exposed individual [5,6].
In practice, the most dangerous arsenical preparations are dips, herbicides,
and defoliants in which the arsenical is in a highly soluble trivalent form,
usually trioxide or arsenite. Unfortunately, animals will frequently seek
out and eat materials such as insulation, rodent baits, dirt, and foliage
that have been contaminated with an inorganic arsenical.

Because of the many factors influencing the toxicity of arsenic, as
noted above, there is little point in attempting to state its toxicity on an
mg/kg body weight basis. The lethal oral dose for most species of animals,
however, appears to be from 1 to 25 mg/kg body weight as sodium arsenite
with arsenic trioxide being from 3 to 10 times less toxic [1].

The fact that the toxicity of an arsenical is greatly influenced by its
solubility and particle size and, thus, by its absorbability from the intestinal tract was illustrated by an experiment conducted with swine [1].
Sodium arsenite was given in the feed at levels up to 500 ppm continuously
for 2 weeks. The pigs readily ate the contaminated feed but manifested no
signs of acute arsenic poisoning. When the level was increased to 1000 ppm,
the pigs refused to eat the feed. When an equivalent amount of sodium arsenite was added to their drinking water, severe poisoning and death occurred

within a few hours. It was concluded that the lethal dose via drinking
water of sodium arsenite was 100-200 mg/kg body weight.

Experience with field cases of arsenic poisoning indicates that animals
which are weak, debilitated, and dehydrated are much more susceptible to
arsenic poisoning than the normal animal. This probably is because of reduced rate of renal excretion.

Arsenical poisoning in most animals is usually manifested by an acute
or subacute syndrome. Chronic poisoning, although it has been reported, is
seldom seen and has not been clearly documented. In humans, chronic arsenical poisoning is manifested by symmetrical hyperkeratosis of the hands and
feet, pigmentation of the exposed skin, conjunctivitis, tracheitis, acro-
cyanosis, and polyneuritis. The polyneuritis involves both sensory and
motor functions. Other chronic effects include anorexia, cachexia, hepatic
cirrhosis, and dementia [7]. Some animals appear to develop a tolerance to
arsenic after prolonged oral exposure.

Sodium thiacetarsamide is recommended for the treatment of adult heartworm in dogs at the rate of 0.1 ml of a 1% solution twice daily per pound of
body weight for 2 days (1.6 mg/kg body weight total) or, in cases of heavily
infested and poor-risk individuals, 0.1 ml of a 1% solution daily for 15
days (6 mg/kg total). The safety margin in the dog is quite low. Dogs have
died of arsenic poisoning after five daily doses of 1.8 mg/kg body weight
(a total of 5.4-9.0 mg/kg). However, dogs did not suffer ill effects from
30 daily doses of 0.9 mg/kg body weight (totaling 27 mg/kg) [8]. Apparently,
the most serious untoward reaction from the use of thiacetarsamide is the
sudden death of the adult heartworm dislodging it into the pulmonary capillary bed causing embolism and fatal pulmonary embarrassment. Systemic arsenic poisoning is not uncommon, however, following the routine treatment
of apparently healthy dogs.

In recent years, thiacetarsamide has been used in the treatment of
Hemobartonella felis in cats. It is reasonable to assume, therefore, that
poisoning will occur in a certain percentage of cats treated with this
regimen.

Palmer has studied the toxicity of organic arsenical herbicides, MSMA
and DSMA, in cattle, sheep, and chickens [9]. Cattle were poisoned and
died after receiving five daily doses of 10 mg/kg MSMA and six daily doses
of 25 mg/kg DSMA. Ten daily doses of 5 mg/kg MSMA and 10 mg/kg DSMA produced no ill effects. Sheep were poisoned and died after receiving six
daily doses of 50 mg/kg MSMA and six daily doses of 25 mg/kg DSMA. Animals

receiving ten daily doses of 25 mg/kg MSMA and 10 mg/kg DSMA had no ill
effects. Chickens failed to exhibit ill effects after receiving 10 daily
doses up to 250 mg/kg of either compound. He also studied the toxicity of
cacodylic acid in these three species. Cattle and sheep were poisoned
after receiving 8 and 10 daily doses, respectively, at a rate of 25 mg/kg
body weight. Chickens had reduced weight gains after 10 daily doses of
100 mg/kg body weight.

MSMA is recommended at 2 lb per acre, DSMA at 3 lb per acre, and cacodylic acid at 8 lb per acre. It was concluded that if these arsenicals
were applied at the recommended rate, MSMA would be hazardous for cattle
but not sheep and chickens; and DSMA and cacodylic acid would be hazardous
for cattle and sheep but not chickens [9].

Inorganic arsenicals have been used for the treatment of psoriasis
and other eruptions, and certain organic arsenicals were developed in a
directed effort to find an effective treatment for syphilis. These therapeutic uses of arsenic have since been replaced by antibiotics and other
agents that are considerably less toxic. The extensive use of arsenicals
in human medicine did, however, provide much information with regard to the
chronic toxicity of arsenic, i.e., the fact that cutaneous reactions such
as acute erythemas, pruritic and vesicular eruptions, and hair loss are
among the major manifestations of arsenic toxicosis. The clinical discovery
that latent effects of arsenical administration included, among other cutaneous changes, the development of multiple skin cancers was a major reason why dermatologists abandoned their use.

There have been several epidemiologic surveys of incidents of lung
cancer in miners of gold-bearing ores containing large amounts of arsenic
and among vineyard workers chronically exposed to lead arsenate dust. In
one study among workers involved in preparation and packaging of an arsenical powder containing sodium arsenite, there was a twofold higher proportion of respiratory and skin cancer deaths in the exposed versus nonexposed
populations [10]. In another study dealing with the mortality experience
of over 8000 smelter workers exposed to arsenic trioxide during 1938-1963
there was a threefold higher incidence of mortality from respiratory cancer
in smelter workers as opposed to a nonarsenic-exposed population [11]. The
incidence of respiratory cancer was up to eightfold higher among employees
who worked for more than 15 years and who were heavily exposed to arsenic.
The incidence in respiratory cancer was a function of the degree of exposure to both arsenic and sulfur dioxide. It appears that one to two decades

of very high arsenic exposure may triple the incidence of respiratory cancer and the addition of an associated heavy sulfur dioxide exposure may further increase the incidence of respiratory cancer to an eightfold excess.

There is epidemiologic evidence that relatively high levels of arsenic in drinking water (0.8 ppm) has been associated with an overall increased prevalence of skin hyperpigmentation, keratosis, and cancer. Generally speaking, the prevalence of all three conditions increases steadily with age and with arsenic content of the water [12,13].

An incident in Japan occurred in which wastewater from an arsenic tri-sulfide manufacturing plant seeped into well waters being consumed by humans [14]. The arsenic levels were found to far exceed the 0.05 ppm limit, reaching 1-2 ppm in most cases, with a maximum content of 3.0 ppm. A total of 140 cases of arsenic poisoning were studied. Another Japanese incident that occurred in 1955, in which prepared powdered milk commonly fed to infants was accidentally contaminated with arsenic and resulted in over 12,000 cases of toxicosis, of whom 131 died, serves to exemplify the potential hazard associated with careless handling of toxic materials [15,16].

CLINICAL SIGNS

Peracute and acute episodes of poisoning by inorganic arsenic are usually explosive with high morbidity and mortality over a 2- to 3-day period. Symptoms are manifested by intense abdominal pain, staggering gait, extreme weakness, trembling, excess salivation, vomiting (in humans, dogs, cats, pigs, and perhaps even cattle), diarrhea, fast feeble pulse, prostration, gastric atony, normal to subnormal temperature, collapse, and death.

In subacute arsenic poisoning, animals may live for several days exhibiting depression, anorexia, watery diarrhea, increased urination at first followed by anuria, dehydration, thirst, partial paralysis of the rear limbs, trembling, stupor, cold extremities, subnormal temperature, and death. The watery diarrhea may contain shreds of intestinal mucosa and blood. Convulsive seizures have been reported but are not a usual manifestation [6]. Should the human patient recover from the acute gastrointestinal, renal, and central nervous system effects, grayish semilunar strips about 1 mm wide and extending across the entire base of the nails will appear in about 2 months [7].

Poisoning resulting from arsenical dips usually results in some of the signs noted previously in addition to blistering and edema of the skin followed by cracking and bleeding with associated secondary bacterial infection.

Chronic arsenical poisoning is rarely seen in most domestic animals but has been well documented in man. Reports of chronic arsenic poisoning indicate that a general wasting and unthriftiness with accompanying rough hair coat and brick-red coloration of visible mucous membranes are the principal clinical signs seen [1].

In cases of poisoning by the toxic gas arsine, acute hemolysis and anemia occur in addition to other characteristic signs of arsenic poisoning [17,18].

Many of the trivalent organic arsenicals, such as thiacetarsamide and arsphenamine, apparently act similarly to inorganic arsenicals by poisoning the intracellular sulfhydryl systems. Poisoning by these compounds results in a syndrome very similar to that produced by the inorganic arsenicals. The pentavalent arsenical herbicides DSMA and MSMA apparently also have action similar to inorganic arsenicals.

Small animals, especially dogs, suffering from thiacetarsamide poisoning exhibit vomiting and a diarrhea within 24 hr of intravenous administration. The diarrhea may be projectile and vary in color from reddish to black. The animals usually are listless, depressed, anoretic, and severely dehydrated. Renal damage resulting in oliguria is usually followed by death. Occasionally, a dog suffering from thiacetarsamide poisoning will exhibit gastric tympany, nonproductive retching, and terminal shock [1].

In man, the most common clinical signs of acute or subacute arsenic poisoning include fever, diarrhea, emaciation, anorexia, vomiting, increased irritability, exanthema, and loss of hair, in this order [19]. Frequent signs associated with subacute and chronic arsenic toxicosis in humans can be summarized as dermatologic (keratosis, melanosis, desquamation, white spots, and fingernail changes), anemia, leucopenia, and hepatic swelling [7,14,15,20].

Trivalent organic arsenicals affect the circulatory system, gastrointestinal tract, kidneys, skin, and nervous system much like the inorganic arsenicals. A polyneuritis has been described which may be similar to that seen in swine and poultry poisoned by the phenylarsonic acids [1,21].

Clinical signs of poisoning by the organic arsenical herbicides MSMA, DSMA, and cacodylic acid in cattle and sheep were anorexia, hematuria,

diarrhea, and depression, not unlike the signs of inorganic arsenic toxicosis. Chickens exhibited only reduced weight gains [9].

Sodium arsenite and arsenate have been associated with teratogenic
effects (exencephally, micrognathia, agnathia, anophthalmia, hydrocephalus,
and skeletal defects) in mice and hamsters when injected intraperitoneally
[22-24]. Such effects were not produced following oral exposure to mice and
rats [25], nor have there been documented reports of such effects in humans.

PHYSIOPATHOLOGY

Mechanism of Action

Soluble forms of arsenic, such as sodium arsenite, are readily absorbed
from all body surfaces. Arsenic trioxide and other less soluble arsenicals
are poorly absorbed through the skin or digestive tract and are largely excreted unchanged in the feces. Once absorbed, pentavalent arsenic is
readily excreted by the kidneys, whereas trivalent arsenic is more readily
excreted into the intestine via bile [21].

It is generally considered that regardless of whether an inorganic or
aliphatic organic arsenical is introduced into the body as trivalent or
pentavalent arsenic, most of the major actions can be attributed to the
trivalent form. Arsenic is believed ultimately to exert most of its effects
by reacting with sulfhydryl (-SH) groups in cells. As a result, sulfhydryl
enzyme systems essential to cellular metabolism are inhibited. Thus, the
net effect is the blocking of fat and carbohydrate metabolism and cellular
respiration. The affinity of trivalent arsenic for the -SH radical provides
the rationale for the use of dimercaprol (BAL) as a specific antidote. The
thioarsenite formed by the reaction between BAL and arsenic provides for
its rapid removal from tissue and excretion by the kidneys. The arsenate
or pentavalent ion is capable of uncoupling phosphorylation. The importance
of this phenomenon is not known, although it could account for some of the
effects on peripheral nerves and spinal cord reported in humans. Arsine
(AsH_3), a highly toxic industrial gas, combines with hemoglobin and is oxidized to a hemolytic compound that does not appear to act by sulfhydryl
inhibition [17,18].

Arsenic affects those tissues rich in oxidative systems, primarily the
alimentary tract, kidney, liver, lung, and epidermis. It is a potent capillary poison; and although injury involves all beds, the splanchnic area
is most commonly affected. Capillary damage and dilatation result in

transudation of plasma into the intestinal tract and sharply reduced blood volume. Blood pressure usually falls to shock levels, and heart muscle becomes depressed, contributing to circulatory failure. The capillary transudation of plasma results in vesicles and edema of the gastrointestinal mucosa, eventually leading to epithelial sloughing and discharge of the plasma into the gastrointestinal tract.

Toxic arsenic nephrosis is commonly seen in small animals and man. Capillaries in the glomeruli dilate, allowing the escape of plasma which results in swelling and varying degrees of tubular degeneration. The anhydremia resulting from the loss of fluid through other capillary beds and the low blood pressure contribute to oliguria characteristic of arsenic poisoning. The urine usually contains protein, red blood cells, and casts.

Following percutaneous exposure, capillary dilatation and degeneration may result in blistering and edema, after which the skin may become dry and papery. At this latter stage, the skin may crack and bleed, providing a choice spot for secondary invaders [6].

Pathologic Changes

Gross lesions characteristic of inorganic arsenic poisoning include reddening of the gastric mucosa (abomasum in ruminants), which may be localized or general; reddening of the small intestinal mucosa (often limited to the first few feet of the duodenum); fluid gastrointestinal contents which are often foul smelling; soft, yellow liver; and pulmonary congestion. In peracute cases of poisoning, occasionally no gross postmortem changes are evident. The inflammation is usually followed by edema, rupture of the blood vessels, and necrosis of the mucosa and submucosa. Sometimes the necrosis progresses to perforation of either the stomach or intestine. The fluid gastrointestinal contents may contain blood and shreds of mucosa. Hemorrhages on all surfaces of the heart and on the peritoneum may occasionally be observed [1].

Postmortem lesions associated with poisoning by the organic arsenical herbicides MSMA, DSMA, and cacodylic acid include inflammation, hemorrhages, and edema of the gastrointestinal mucosa in cattle and sheep [9]. Organic arsenicals used as therapeutic agents in man, such as arsphenamine, and organic compounds such as thiacetarsamide, used for the treatment of heartworms in dogs, also produce lesions characteristic of inorganic arsenicals [8].

Histopathologic changes include edema of the gastrointestinal mucosa and submucosa, necrosis and sloughing of mucosal epithelium, renal tubular

degeneration, hepatic fatty change and necrosis, and capillary degeneration
in vascular beds of the gastrointestinal tract, skin, and other organs.

Lesions associated with chronic arsenosis in humans are primarily in
the skin, liver, peripheral nerves, and bone marrow. As poisoning persists,
selective edemas, especially the eyelids and ankles, associated with conjunctivitis, inflammation of the inner membranes of the nose and throat,
and dermatitis (including pigmentation and hyperkeratoses) appear. In severe cases, renal and hepatic damage, alopecia, blood dyscrasia (chiefly
aplastic anemia), encephalitis, polyneuritis, and exfoliative dermatitis
may develop [26].

Arsenic Concentrations in Tissue and Blood

While most textbooks report that arsenic is accumulated in the tissues
and slowly excreted, this phenomenon appears to be true only in rats. Most
species of livestock and pet animals apparently rapidly excrete arsenic
[1,27]. This phenomenon is very important when one considers arsenic levels
in tissues as a means of confirming suspected poisoning. Experience with
field cases in the Veterinary Diagnostic Laboratory, Iowa State University,
indicates that if an animal lives several days after consuming a toxic level
of arsenic, liver and kidney tissues may be below the level ordinarily considered diagnostic [1]. Other laboratories have reported similar findings
[28].

In peracute, acute, and subacute poisoning, arsenic tends to be concentrated in the liver and kidneys. Normal animals usually have a background level of arsenic in these tissues of less than 0.5 ppm. Animals
dying of acute or subacute arsenic poisoning may contain from 2 to 100 ppm
arsenic on a wet-weight basis in these two organs with the kidney usually
having a higher concentration than the liver. Levels above 10 ppm on a
wet-weight basis would be considered confirmatory of arsenic poisoning [1].
Hatch and Funnell have reported the amount of arsenic found in livers and
kidneys of 21 cattle which died with signs and lesions of arsenic poisoning
averaged 14.0 and 13.3 ppm, respectively [29]. The liver levels ranged from
2.1 to 38.0 ppm, while the kidney levels ranged from 1.5 to 37.0 ppm.

The normal arsenic content in urine of cattle usually ranges less than
0.05 mg/l [30]. Following dipping in an arsenous oxide solution containing
0.18-0.22% arsenous oxide, urine levels increase by 10-fold (5.0 mg/l) in
apparently healthy animals. In addition, water deprivation following the
dipping of cattle in an official arsenical solution may result in urinary

levels as high as 60 mg/l in animals evidencing no symptoms of arsenical poisoning.

The arsenic content of whole blood from humans with no known arsenic exposure has ranged from 0 to 113 µg/100 g (1.13 ppm) [31]. The statistically calculated mean whole blood concentration of arsenic in 146 nonexposed individuals was 6.6 µg/100 g of blood, and it was projected that the upper limit for 95% of the population would be 22 µg arsenic/100 g of blood.

Urinary arsenic concentrations have been studied by several individuals. In one study the daily normal body elimination of arsenic in man via the urine ranged from 0.008 to 0.85 mg/l (0.8-85 µg/100 ml) [32]. In another report, a median value of 0.10 mg/l in 24 patients and a mean of 0.126 mg/l in 13 patients, all on 24-hr specimens with no known exposure to arsenic, were reported [33]. In another survey of 124 persons actively working in industry with no arsenic exposure, an average urinary value for arsenic was 0.13 mg/l, and 88% of the samples had less than 0.2 mg/l [34]. In a study of 223 Japanese individuals with no known arsenic exposure, the minimum 24-hr urinary excretion of arsenic was 12 µg and the maximum was 927 µg/day (0.8 mg/l) [31].

Only about one-fifth of the absorbed arsenic is promptly excreted by man, mostly in the urine, with some being eliminated in the feces, sweat, nails, hair, and by exhalation of methylated metabolites. The remaining four-fifths is stored widely in the body. A single dose may require 10 days for complete elimination [35]. In cases of arsenic poisoning of varying degrees of severity, the 24-hr urine specimens showed an arsenic level of from 0.1 to 6 mg, with the majority of persons with clinical evidence of arsenical intoxication having values greater than 0.1 mg/l per 24-hr urinary excretion [36]. Others have suggested that urinary values of more than 0.2 mg/l is suggestive of arsenic poisoning [33]. In 835 spot urine samples taken from 348 men exposed to arsenic trioxide, however, the average urinary arsenic value was 0.8 mg/l and the median was 0.58 mg/l. The only symptoms in any of these employees were dermatitis and nasal septum erosions.

Normal values of arsenic in human hair are less than 0.1 mg/100 g of hair (1 ppm). In chronic cases of arsenic poisoning, the value may go up to 0.5 mg/100 g of hair (5 ppm), and in acute cases to as much as 1-3 mg/100 g of hair (10-30 ppm) [36]. As with domestic animals, there is probably no correlation between the amount of arsenic in hair and the clinical signs of poisoning [14]. Arsenic levels in the feces have a wide range even in apparently nonexposed individuals, but usually are less than 1.5 ppm [31].

Diagnosis of Arsenic Poisoning

Whenever an episode of illness occurs in domestic animals, especially ruminants, cats, and dogs, that is characterized by rapid onset of gastro-enteritis with only minor signs of central nervous system involvement resulting in weakness, prostration, and rapid death, arsenic poisoning should be considered. A diagnosis can be further substantiated by finding excessive fluid in the gastrointestinal tract, together with various degrees of inflammation and necrosis of the gastrointestinal mucosa. Liver, kidney, stomach and intestinal contents, and urine should be obtained for arsenic analyses. A modified Gutzeit method has provided satisfactory results [37]. This method involves the digestion of 5 g of wet tissue in nitric-perchloric-sulfuric acid (or ashing in a muffle furnace) and utilizing an arsine generator and silver diethyldithiocarbamate arsenic-sensitive color reagent. Depending upon many factors, such as time since exposure, route of exposure, type of arsenical formulation, and others as mentioned above, renal and hepatic tissue will usually contain greater than 8 ppm arsenic on a wet-weight basis in acute poisoning. If several days have elapsed since exposure, however, the liver tissue may contain only 2-4 ppm, whereas the kidney tissue may contain diagnostically significant levels (8-10 ppm). Levels of arsenic in the gastrointestinal contents and urine will aid in determining the route and degree of arsenic exposure; but because of their variability and dependence upon water intake by affected animals, they cannot be relied upon for confirmation of arsenical toxicosis.

Diseases frequently confused with arsenic poisoning, especially in the ruminant, include hypomagnesemia, urea poisoning, organophosphorus insecticide poisoning, bovine virus diarrhea, and poisoning by plants containing nitrates, cyanide, oxalates, selenium, or alkaloids. Sometimes lead poisoning in the bovine results in sudden death and could be confused with arsenic poisoning. However, in most instances, the central nervous system signs such as blindness, circling, and convulsive seizures are more prominent in lead poisoning.

Conditions which may be easily confused with arsenic poisoning in dogs and cats include other heavy metal intoxications, such as thallium, mercury, and lead, and ethylene glycol poisoning. Arsenic poisoning is considerably more acute, however, than the syndromes associated with other heavy metals. Enteric infections causing vomiting, diarrhea, and collapse can also resemble arsenic poisoning.

Arsenical poisoning in the human is often more subacute or chronic as compared to domestic animals. The following arsenic levels are usually associated with poisoning: urine, 0.5-5 mg/l; blood, 0.1-5 mg/l; hair, 20-100 ppm; and nails, 0-500 ppm [26,38].

As with the diagnosis of any toxicosis, the tissue levels of arsenic should be interpreted in light of the circumstances, clinical signs, physiopathologic changes, and clinical pathologic and hematologic data. This is especially true in cases of chronic human arsenical toxicoses in which changes occur in the skin, fingernails, red blood cells (anemia), and thoraxic and abdominal organs.

THERAPY OF ARSENIC POISONING

The key to successful treatment of acute inorganic or aliphatic organic arsenic poisoning is early diagnosis. Even so, the prognosis should be heavily guarded. In the ruminant and horse, which do not readily vomit, large doses of saline purgative in an attempt to remove the unabsorbed material from the gastrointestinal tract may be indicated. Demulcents may be given to coat the damaged gastrointestinal mucous membrane. Sodium thiosulfate should be given orally and intravenously. Adult horses and cattle should be given 20-30 g orally in approximately 300 ml of water and 8-10 g in the form of a 10-20% solution intravenously. Sheep and goats should receive about one-fourth of this amount. BAL is a sulfhydryl-containing specific antidote for trivalent arsenic. Its value as a therapeutic agent for arsenic poisoning in large animals is questionable. There may be disappointing therapeutic results with this compound in large animals because veterinarians have not repeated the treatment every 4 hr for the first 2 days, four times on the third day, and twice daily for the next 10 days until recovery is complete, as has been recommended. BAL should be given as a 5% mixture in a 10% solution of benzyl benzoate in arachis oil at a rate of 3 mg/kg body weight [5]. It is important to give supportive therapy such as electrolytes to replace body fluids and to provide plenty of drinking water.

In small animals, if there is an opportunity for treatment early in the course of the syndrome, the stomach should be emptied before the arsenic can pass into the intestine and be absorbed. Gastric lavage with warm water or a 1% solution of sodium bicarbonate is preferred, although emetics such as apomorphine may be used early in the treatment regimen. When signs of

arsenic poisoning are already present, gastric lavages or emetics should
not be used. BAL should be given intramuscularly at a dosage of 6-7 mg/kg
body weight three times daily until recovery. Fluids should be administered
parenterally to rehydrate animals which have been vomiting or had diarrhea.
If uremia has developed, lactated Ringer's solution should be used; B-complex
vitamins may be added to the Ringer's solution. Following rehydration, 20
ml/kg body weight of 10% dextrose solution should be administered, and this
should result in diuresis. The urinary bladder should be catheterized to
determine the rate of urine flow. If flow increases considerably following
the administration of 10% dextrose and the urine contains considerable sugar,
the uremia may be controlled by alternately administering lactated Ringer's
solution and 5 to 10% dextrose. If acidosis is present, 50% sodium lactate
may be added to the lactated Ringer's solution at the rate of 2.5-5.0 ml/
1000 ml.

Protein hydrolysates may be added to supply amino acids, but they must
be given slowly to avoid inducing more vomiting. B-complex vitamin should
be injected daily, and whole blood should be transfused when indicated by
the occurrence of anemia or shock.

There should be no effort to administer drugs or food orally during the
period when the animal is vomiting. When emesis has stopped, kaolin-pectin
preparations can be given orally to aid in controlling diarrhea. Antibiotics are indicated to prevent secondary infections, and meperidine should
be given as needed to lessen abdominal pain. As improvement occurs, a high-
protein, low-residue diet should be fed and other supportive therapy discontinued [39].

SUMMARY

Arsenic is ubiquitous, being present in virtually all rocks, soils, and
water. It is not mined as such, but is produced as a by-product of the production of other minerals.

Most of the arsenic produced is used for the manufacture of various
pesticide formulations and desiccants for cotton and fruit trees, although
a considerable amount is used for the production of organic arsenicals for
livestock feed additives and for therapeutic agents in both human and veterinary medicine. A small amount of arsenic is used by the electronics
industry.

Man and all lower animals are susceptible to poisoning by inorganic
and aliphatic organic formulations of arsenic. Ruminants and other foraging

animals are most frequently poisoned because of contamination of their food supply resulting from the use of arsenicals as pesticides. Inorganic arsenic poisoning in animals is usally acute and fulminating, often resulting in death, whereas poisoning in humans is more often a result of chronic insidious occupational exposure.

The toxicity of arsenic varies greatly with the formulation (and hence the valence of arsenic), species involved, solubility, and route of exposure, all of which directly affect the rate of absorption, metabolism, and excretion. The most dangerous arsenical preparations are the highly soluble trivalent arsenites. The use of arsenicals for the treatment of blood parasites has not been without hazards to the patients in both human and veterinary medicine.

The inorganic and aliphatic organic arsenical formulations tend to have similar effects in both animals and man, whereas the phenylarsonic compounds used primarily for feed additives apparently have a different mechanism of action and a different toxic effect. For this reason the phenylarsonic compounds are discussed in another chapter.

It is generally considered that regardless of whether an inorganic or aliphatic organic arsenical is introduced into the body as a trivalent or pentavalent compound, most of the major actions can be attributed to arsenic in its trivalent form. Arsenic is believed ultimately to exert most of its effects by reacting with sulfhydryl groups in cells, thus tying up sulfhydryl enzyme systems essential to cellular metabolism. Arsenic affects those tissues rich in oxidative systems, such as the alimentary tract, kidney, liver, lung, and epidermis. It is a potent capillary poison producing injury throughout the capillary system, especially in the splanchnic area.

The acute effects of arsenic poisoning in both man and animals are predominantly associated with the gastrointestinal and circulatory systems. The primary clinical signs include intense abdominal pain, staggering gait, extreme weakness, trembling, excess salivation, vomition, diarrhea, fast feeble pulse, prostration, dehydration, collapse, and death. Chronic arsenic poisoning is rarely seen in domestic animals, but is most frequently seen in humans as a result of industrial or occupational exposure. Dermatologic lesions such as keratosis, melanosis, desquamation, hyperpigmentation, fingernail changes, and cancer have all been epidemiologically associated with long-term exposure to arsenic.

While it is popularly thought that arsenic is accumulated in the body tissues and slowly excreted, this phenomenon appears to be true only in the rat. In fact, most species of livestock and pet animals rapidly excrete arsenic via urine and feces. Animals dying of acute arsenic poisoning may contain from 2 to 100 ppm arsenic on a wet-weight basis in the liver and kidneys, whereas animals not known to have been exposed to arsenic usually contain less than 0.5 ppm. Levels above 10 ppm in the liver and kidney should be considered confirmatory of arsenic poisoning. Because of the relatively rapid excretion of arsenic from body tissues, it is possible for severe signs of poisoning to exist with body tissues containing only 1-2 ppm arsenic. Similarly, urinary levels of arsenic only indicate the amount of arsenic that has been recently excreted and not necessarily the toxic effect. The amount of arsenic in human and animal hair does not necessarily correlate with clinical signs of arsenic poisoning. The diagnosis of arsenic poisoning should therefore be made using varied sources of evidence, including history, clinical signs, pathologic changes, and chemical and clinical pathologic analyses.

The key to successful treatment of acute inorganic or organic aliphatic arsenic poisoning is early diagnosis. Efforts should be made to immediately remove or neuralize the poison in the gastrointestinal tract or on the skin. Supportive therapy, especially with the objective of maintaining fluid and electrolyte balance, is of primary importance. Dimercaprol (BAL) is a sulfhydryl-containing specific antidote for trivalent arsenic. Its value in the treatment of animals, however, has been disappointing.

REFERENCES

1. W. B. Buck, G. D. Osweiler, and G. A. Van Gelder, Clinical and Diagnostic Veterinary Toxicology, 2nd Ed., Kendall-Hunt, Dubuque, 1976, pp. 281-288.

2. U.S. Bureau of Mines, Minerals Yearbook, U.S. Government Printing Office, Washington, D. C., 1972.

3. E. Browning, Toxicity of Industrial Metals, 2nd Ed., Appleton-Century-Croft, New York, 1969, pp. 39-60.

4. Environmental Protection Agency, Office of Pesticides Program, Arsenical Pesticides, Man and the Environment, Report, 1972.

5. E. G. C. Clarke and M. L. Clarke, Garner's Veterinary Toxicology, 3rd Ed., Williams and Wilkins, Baltimore, 1967, pp. 44-54.

6. R. D. Radeleff, Veterinary Toxicology, 2nd Ed., Lea and Febiger, Philadelphia, 1970, pp. 158-161.

7. J. J. Chisolm, Jr., Pediatr. Clin. N. Amer., 17, 591-615 (1970).

8. G. F. Otto and T. H. Maren, Amer. J. Hyg., 51, 353-395 (1950).

9. J. S. Palmer, Toxicity of 45 Organic Herbicides to Cattle, Sheep, and Chickens, U. S. D. A.-A. R. S. Res. Report No. 137, 1972, pp. 22-23.

10. A. B. Hill, E. L. Faning, K. Perry, R. G. Bowler, H. M. Buckell, H. A. Druett, and R. S. F. Schilling, Brit. J. Ind. Med., 5, 1-15 (1948).

11. A. M. Lee and J. F. Fraumen, Jr., J. Nat. Cancer Inst., 42, 1045-1052 (1969).

12. W. P. Tseng, H. M. Chu, S. W. How, J. M. Fong, C. S. Lin, and Shu Yeh, J. Nat. Cancer Inst., 40, 453-463 (1968).

13. J. M. Borgono and R. Greiber, in Proc. Conf. on Trace Elements in Environmental Health, University of Missouri, 1971.

14. T. Hideo, K. Kazuo, S. Tsuomu, S. Hideaki, S. Heiichiro, F. Katsuro, S. Chūkichi, Y. Yoshiro, H. Shigeru, W. Giichi, H. Kazuo, O. Tatsuo, and S. Chūkichi, Jap. J. Clin. Med., 18, 118-127 (1960).

15. S. Satake, Jap. Public Health, 2, 22-24 (1955).

16. O. Masahiko and A. Hideyasu, Jap. J. Hyg., 27, 500-531 (1973).

17. W. J. Levinsky, R. V. Smalley, P. N. Hillyer, and R. L. Shindler, Arch. Environ. Health, 20, 436-440 (1970).

18. D. T. Teitelbaum and L. C. Kier, Arch. Environ. Health, 19, 133-143 (1969).

19. K. Okamura, T. Ota, K. Horiuchi, H. Hiroshima, T. Takai, Y. Sakurane, and T. Baba, Shinryo (Diagnosis and Therapy), 9, 240-249 (1956).

20. O. Neubauer, Brit. J. Cancer, 1, 192-215 (1947).

21. S. C. Harvey, in The Pharmacologic Basis of Therapeutics (L. S. Goodman and A. Gilman, eds.), 4th Ed., Macmillan, New York, 1970, pp. 958-986.

22. V. H. Ferm, A. Saxon, and B. M. Smith, Arch. Environ. Health, 22, 557-560 (1971).

23. P. D. Hood, Bull. Environ. Contam. Toxicol., 7, 216-222 (1972).

24. P. E. Holmberg and V. H. Ferm, Arch. Environ. Health, 18, 873-877 (1969).

25. H. A. Schroeder, Arch. Environ. Health, 23, 102-106 (1971).

26. R. C. Wands, Chronic Toxicity of Arsenic Ingestion, Report of the Advisory Center on Toxicology, NRC-NAS, 1969.

27. S. A. Peoples, N. Y. Acad. Sci., 111, 644-649 (1964).

28. J. W. Moxham and M. R. Coup, N. Z. Vet. J., 16, 161-165 (1968).

29. R. C. Hatch and H. S. Funnell, Can. Vet. J., 10, 112-119 (1969).

30. E. E. Mass, J. Amer. Vet. Med. Ass., 110, 249-250 (1947).

31. N. Iwataki and K. Horiuchi, Osaka City Med. J., 5, 209-217 (1959).

32. Ohio River Valley Water Sanitation Commun., Interim Report, Arsenic: Physiological Aspects of Water Quality Criteria with Regard to Man, March, 1957.

33. R. A. Kyle and G. L. Pease, N. Engl. J. Med., 273, 18-23 (1965).

34. S. S. Pinto and C. M. McGill, Ind. Med. Surg., 22, 281-287 (1953).

35. Public Health Service, Drinking Water Standards, Revised, U.S. Government Printing Office, Washington, D. C., 1962, pp. 25-27.

36. A. Heyman, J. F. Pfeiffer, R. W. Willett, and H. M. Taylor, N. Engl. J. Med., $\underline{254}$, 401-409 (1956).

37. W. B. Buck, J. Amer. Vet. Med. Ass., $\underline{155}$, 1928-1941 (1969).

38. Y. Kitamura, M. Miyao, K. Kawazoe, Y. Kinoshita, H. Hiroshima, C. Toi, E. Katsura, and H. Matsuda, Shikoku Acta Medica, $\underline{8}$, 205-225 (1956).

39. D. G. Low, in Current Veterinary Therapy V (R. W. Kirk, ed.), Saunders, Philadelphia, 1974, pp. 138-139.

17

TOXICITY OF ORGANIC ARSENICALS IN FEEDSTUFFS

Arlo E. Ledet and William B. Buck[*]
Iowa State University College of Veterinary Medicine
Ames, Iowa

Organic arsenicals have been widely used as feed additives for poultry and swine since the mid-1940s. The arsenic cation incorporated in the various organic formulations may be either in the trivalent or pentavalent form. Trivalent forms are referred to as arsenoso compounds and the pentavalent forms as arsonic acids. The organic arsenical feed additives are primarily pentavalent phenylarsonic acids and their salts. Structural formulas and names of five forms used as feed additives are given in Fig. 1.

RECOMMENDED AND TOXIC LEVELS IN FEED

The most widely used compounds include arsanilic acid and its salt, sodium arsanilate, and 3-nitro-4-hydroxyphenylarsonic acid. Other organic arsenical feed additives include 4-nitrophenylarsonic acid and p-ureido-

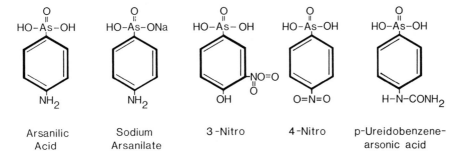

FIG. 1. Structural formulas of organic arsenical compounds.

[*]Present Affiliation: University of Illinois College of Veterinary Medicine, Urbana, Illinois.

benzenearsonic acid. These compounds are all considered to improve weight
gain, feed efficiency, and to aid in the prevention and control of certain
enteric diseases of swine and poultry [1-4].

Arsanilic acid and sodium arsanilate are recommended at 50-100 ppm
(0.005-0.01%) in swine and poultry feeds for increased weight gains and
improvement of feed efficiency. They are recommended at 250-400 ppm (0.025-
0.04%) in swine feed for a duration of 5 to 6 days for the control of swine
dysentery [5]. The margin of safety for arsanilic acid and its salt is
quite wide in normal animals. However, the effective level and the chronically toxic level may impinge upon one another under certain conditions.
The health status of the exposed animals and management practices, especially those involving availability of water, are important contributing
factors for adverse reactions resulting from the addition of organic arsenicals to feed. Animals suffering with a diarrhea are usually dehydrated
and, thus, are excreting very little urine. Since the organic arsenicals
are excreted via the kidneys, their toxicity becomes greatly increased when
given to animals with diarrhea. Usually the morbidity is high and mortality
very low.

Roxarsone or "3-nitro" is recommended at 25-50 ppm (0.0025-0.005%) for
chickens and turkeys and at 25-75 ppm (0.0025-0.0075%) for swine for improving weight gains and feed efficiency. It is also recommended at a level
of 200 ppm (0.02%) for 5 to 6 days for the control of swine dysentery [5].
Animals may exhibit clinical signs after consuming 250 ppm in the feed for
3 to 10 days, and swine have been chronically poisoned on 3-nitro at 100 ppm
for two months [6,7].

The compound 4-nitrophenylarsonic acid ("4-nitro") has been recommended
for chickens and turkeys at 188 ppm for the improvement of weight gains and
feed efficiency. It has not been recommended for ducks or geese and, at the
present time, has only limited use as a feed additive.

ABSORPTION, METABOLISM, AND EXCRETION

In an attempt to explain the toxicity of pentavalent organic arsenicals
the idea of degradation and reduction to trivalent forms has been perpetu-
ated [8-10]. It has also been postulated that trivalent forms were oxidized
to pentavalent states prior to excretion and that the toxicity of these
compounds was related to the rate of oxidation and excretion [11]. The
evidence for the various oxidations and reductions within the animal body

is primarily indirect and certainly contrary to reported findings involving
the metabolism of the arsenic compounds [12-17].

The fate of arsanilic acid and acetylarsanilic acid has been determined
in chickens. Arsanilic acid was excreted unchanged with no evidence that
arsanilic acid was reduced or converted to inorganic arsenic [12]. Metabolic stability of arsanilic acid in chicks was studied through the use of
doubly labeled arsanilic-[1-^{14}C-^{74}As] acid. Arsanilic acid again was found
unchanged in excreta and neither metabolic products nor reduced forms of
arsanilic acid were found in the tissues [16,17].

In a study of the localization and type of arsenic excreted and retained by chickens fed ^{74}As-labeled arsanilic acid, sodium arsenate, and
sodium arsenite, it was found that arsanilic acid was excreted faster and
had less affinity for body tissues than inorganic forms. The biological
half-life of arsanilic acid in blood was about 36 hr compared to more than
60 hr for arsenate [15].

Since the effectiveness of a drug is dependent upon the length of time
during which it is retained in the body, the rate and route of arsenical
elimination is important. In general, from the literature, it would appear
that the organic arsenicals are poorly absorbed from the digestive tract,
but once absorbed they are quickly excreted primarily by the kidneys. When
these same compounds are given parenterally, they are rapidly eliminated in
urine and to a lesser extent in feces [10].

Six hours after intravenous administration of pentavalent organic arsenical to rats 80 to 90% of the compound was excreted in the urine [10,11].
In rabbits similar experiments resulted in a 40% urinary excretion 1 hr
post injection. Following intravenous injection of four forms of organic
arsenicals in rabbits the ratio between total amounts of arsenic excreted
in the urine and feces varied widely, but the urinary excretion predominated.
Arsonic acids were cleared more rapidly than the corresponding arsenoxides
[11].

Intravenous administration of tritium labeled arsanilic acid has been
used in the pig to study rate of clearance from the blood, routes of excretion, and tissue distribution [18]. Samples of blood and urine were collected every hour for 12 hr following the administration of 1.5 mC radioactivity in 100 mg arsanilic acid. Table 1 contains the data obtained from
liquid scintillation counting of radioactivity [18].

Inspection of data in Table 1 would suggest that arsanilic acid is
rapidly cleared from the blood. It would also appear that the kidneys are

TABLE 1. Blood Clearance and Urinary Excretion of Tritiated Arsanilic Acid Administered Intravenously

Post Injection (hr)	Blood (dpm/ml)	Urine		
		Volume (ml)	(dpm/ml)	Initial Dose[a] (%)
1	16,900,000	17	3,060,000	1.56
2	5,320,000	10	115,800,000	34.70
3	4,220,000	37	9,400,000	10.45
4	350,000	46	3,800,000	5.26
5	380,000	20	4,800,000	2.88
6	180,000	14	14,800,000	6.05
7	150,000	50	1,520,000	2.27
8	147,000	26	2,930,000	2.27
9	107,000	64	400,000	0.79
10	100,200	26	321,000	0.25
11	83,200	31	730,000	0.68
12	117,600	20	1,720,000	1.03
Totals		361		68.19

[a]Initial dose contained 1.5 mC radioactivity (3.33×10^9 disintegrations per min).

TABLE 2. Distribution of Radioactivity at Necropsy

Location	Radioactivity
Whole blood	141,000 dpm/ml
Clotted blood[a]	153,000 dpm/ml
Serum	144,000 dpm/ml
Cerebrospinal fluid	42,600 dpm/ml
Spinal cord	38,200 dpm/g
Bile	2,795,000 dpm/ml
Urine[b]	5,180,000 dpm/ml
Liver	362,000 dpm/g
Kidney	348,500 dpm/g
Composite urine[c]	11,070,000 dpm/ml

[a]After removal of serum.

[b]Collected from bladder.

[c]Four hundred and fifty-one milliliters collected over 24 hr.

a primary route of excretion with approximately 36% of the initial dose
cleared by 2 hr post injection.

The tissue distribution of radioactivity was determined 12 hr after a
second injection of 1.5 mC radioactivity in 100 mg arsanilic acid. Table 2
contains the data obtained from liquid scintillation counting of radioactivity. The highest levels of radioactivity were detected in urine and
bile. Urinary excretion of radioactive materials during the 24-hr period
accounted for nearly 75% of the total dose administered. Renal and hepatic
tissue contained similar levels of radioactivity per g weight. At necropsy
the kidneys weighed a total of 12 g and the liver weighed 170 g. Therefore,
on a total weight basis the liver contained approximately 12 times as much
radioactivity as the kidneys [18].

CLINICAL SYNDROME OF TOXICOSIS

Poisoning of swine by high levels of arsanilic acid is very common
and is second only to water deprivation-sodium ion toxicity in frequency
of occurrence. Poisoning of swine with 1000 ppm arsanilic acid in the feed
produces a very typical sequence of developing clinical signs [19]. Pigs
develop a rough hair coat by the third day followed by a mild diarrhea 24
to 36 hr later. After 2 to 3 days of diarrhea, pigs become constipated
passing firm feces coated with mucus. The transient constipation is usually
followed by near normal stools being passed. Cutaneous hyperemia and hyperesthesia are evident by the fifth day. Hyperesthesia is manifested by
squealing and excitement upon handling. Mild incoordination develops in
the hyperesthetic pigs by the seventh day. Incoordinated animals move with
a swaying gait described as a "drunken sailor" syndrome.

Feeding high levels of arsanilic acid does not seem to affect the
animal's appetite. However, with the onset of incoordination many animals
have difficulty with prehension of feed and water. Incoordination becomes
progressively worse and by the 15th day animals may develop posterior paresis. These animals can rise to a standing position only with difficulty
and then only if forced to do so. They prefer to rise on the forelimbs and
remain in a sitting position while eating and drinking. As the paresis
progresses to quadriplegia, the characteristic repose is with forelimbs
extended posteriorly and hindlimbs drawn anteriorly (Fig. 2). Animals so
affected will remain bright and alert and will eat or drink if feed and
water are made available.

The sequential development of clinical signs in relation to duration
of ingestion of toxic levels of arsanilic acid are presented in Table 3.

TABLE 3. Sequence of Onset of Clinical Signs of Toxicosis in Pigs Fed Arsanilic Acid at 10 Times the Recommended Prophylactic Level

	Days Fed Arsanilic Acid at 900 g/Ton Ration																		
	2	3	4	5	6	7	8	9	10	11	12	13	14	15	16	17	18	19	20
Rough hair coat		+	+	+	+	+	+	+	+	+	+	+	+	+	+	+	+	+	+
Diarrhea			+	+	+	+	+												
Firm feces						+		+	+	+	+	+	+	+	+	+	+	+	+
Cutaneous hyperemia				+	+	+	+	+	+	+	+	+	+	+	+	+	+	+	+
Hyperesthesia				+	+	+	+	+	+	+	+	+	+	+	+	+	+	+	+
Incoordination					+	+	+	+	+	+	+	+	+	+	+	+	+	+	+
Posterior paresis														+	+	+	+	+	+
Quadriplegia																	+	+	+

FIG. 2. Pig with quadriplegia following 18 days on ration containing 1000 ppm arsanilic acid.

The clinical course of arsanilic acid toxicosis is reversible if the drug is removed from the diet upon observation of toxic incoordination [20, 23]. However, pigs which have posterior paresis generally develop a progressive paralysis with little or no clinical improvement upon removal of the drug [19]. This suggests that the onset of posterior paresis marks a point in the toxic syndrome at which demyelination becomes progressive and irreversible. Table 4 lists the results of a clinical recovery trial illustrating the progressive nature of the condition.

TABLE 4. Clinical Recovery Trial from Arsanilic Acid Toxicosis

Pig	Days Fed Drug	Signs at Drug Withdrawal	Days Off Drug	Signs Prior to Necropsy
1[a]	18	Posterior paresis	10	Quadriplegia
2	21	Posterior paresis	12	Unchanged
3	15	Posterior paresis	17	Unchanged
4	21	Quadriplegia	20	Unchanged
5	20	Quadriplegia	32	Unchanged
6	21	Posterior paresis	38	Quadriplegia

[a]Suppurative cystitis and bilateral hydronephrosis found at necropsy.

TABLE 5. Arsenic Levels[a] in Tissues of Pigs Fed 1000 ppm Arsanilic Acid in the Ration

Days	0	4	7	10	13	16	19	21	23	27	Means[b]
Kidney	T[c]	5.92	6.27	6.67	15.70	8.38	8.33	6.30	6.13	10.03	8.192
Liver	T	5.50	7.73	7.57	10.93	9.83	9.67	4.57	5.63	5.81	6.360
Muscle	N[d]	0.41	1.17	0.56	0.67	0.61	0.92	0.48	0.44	0.50	0.640
Blood	T	1.65	1.70	1.81	2.98	2.10	1.94	1.03	1.23	1.73	1.797
Rib	N	0.35	0.51	0.37	0.89	0.77	0.46	0.34	0.32	0.53	0.504
Peripheral nerve	T	0.60	1.20	0.97	1.06	1.53	1.57	1.01	0.96	1.15	1.117
Spinal cord	T	0.26	0.36	0.42	0.59	0.99	0.74	0.84	0.79	0.75	0.638
Brain stem	T	0.37	0.38	0.69	0.93	1.06	1.04	1.09	0.89	1.14	0.863
Cerebellum	T	0.24	1.03	0.68	0.74	1.27	1.23	1.45	1.11	1.00	0.972
Cerebrum	N	0.18	0.34	0.49	0.63	0.74	0.82	1.14	1.10	0.93	0.597
Time period means		1.547	2.068	2.023	3.513	2.728	2.672	1.826	1.861	2.359	

[a]Data in each group represent a mean tissue level in parts per million from three pigs.

[b]Overall tissue means averaged over time periods exclusive of control animals.

[c]Trace, less than 0.02 ppm.

[d]Negative to test.

TISSUE RESIDUES

Mean tissue arsenic content builds up to a peak after 13 days of high-level feeding of arsanilic acid. This represents the period during which arsenic intake exceeds excretion. After 13 days the mean tissue level decreases by approximately one-fourth. This decline coincides with the clinical onset of incoordination and loss of prehensile ability which cause a decreased feed intake. The mean arsenic content of 10 tissues is presented in Table 5 [19].

Arsenic accumulates at different rates and to different levels in the various tissues [19-22]. Nervous tissue accumulates arsenic more slowly than the rest of the body tissues, reaching a maximum level after 19 to 21 days. The delayed attainment of maximum mean arsenic level in nervous tissue coincides approximately with the development of lesions of demyelination.

Arsenic retention by various tissues following prolonged high-level feeding of organic arsenicals has public health significance. Arsenic levels in edible tissues decrease rapidly with discontinuance of arsanilic acid in the diet. Table 6 illustrates the rate of release of arsenic from

TABLE 6. Arsenic (ppm)[a] Retained in Tissues After Feeding Arsanilic Acid (900 g/ton) for 19 Days

	Controls[b]	Days After Arsanilic Acid Withdrawal			
		0	3	6	11
Kidney	Trace[c]	8.33	2.90	2.24	1.90
Liver	Trace	9.67	3.10	1.65	1.75
Muscle	Negative	0.92	0.29	0.29	0.31
Blood	Trace	1.94	0.25	0.19	0.45
Rib	Negative	0.46	0.18	0.24	0.08
Peripheral nerve	Trace	1.57	1.17	1.06	0.61
Spinal cord	Trace	0.74	0.76	0.80	0.25
Brain stem	Trace	1.04	0.90	0.91	0.62
Cerebellum	Trace	1.23	1.58	1.10	0.85
Cerebrum	Negative	0.82	1.09	0.84	0.51

[a]Mean parts per million in tissues of three pigs at day 0 and two pigs each at days 3, 6, and 11.
[b]Three pigs.
[c]Trace, less than 0.02 ppm.

the various tissues. Skeletal muscle retained less than one-third of 1 ppm
of arsenic 3 days after drug withdrawal. Kidney and liver, however, retained
approximately 3 ppm 3 days after withdrawal and still retained close to 2 ppm
11 days after withdrawal [19].

PATHOLOGIC CHANGES

Gross lesions attributable to arsanilic acid toxicosis are minimal and
nonspecific. An occasional pig with prolonged paresis will develop atony
of the urinary bladder and have urine retention [19,20]. Atrophy of skeletal
muscle may occur in chronic toxicosis but is not a common finding [7].

Detectable microscopic lesions are confined to myelinated nervous tissue
with the peripheral nerves, optic nerves, and optic tracts most consistently
affected [19,20]. The earliest changes are condensation and contraction of
myelin sheaths. These changes are followed by fragmentation of the myelin
sheath into granules and globules and finally disruption of axis cylinders.
Advanced lesions contain ovoids of myelin and axonal fragments as in Waller-
ian degeneration of nerve fibers. With the appearance and progressive development of demyelination there is an influx of inflammatory cells, mono-
nuclear and polymorphonuclear, first within small blood vessels and later
among the damaged nerve fibers [19].

In peripheral nerves the more heavily myelinated fibers are affected
first and have the most severe lesions. Schwann cells associated with
damaged fibers appear in various stages of degeneration. Some are swollen
and have eosinophilic cytoplasm while others have pyknotic nuclei or were
karyorrhectic. Schwann cells of unmyelinated fibers do not appear to be
affected. These changes would suggest that there is a difference in sensitivity and response, on the part of the neuron soma, axon, and myelin-
supporting cells, to arsanilic acid.

In spite of severe axonal and myelin sheath damage in the nerves and
nerve tracts, no lesions referable to these changes can be demonstrated in
the neurons or retinal cells [19,20]. Peripheral neuropathy in the absence
of recognizable pathological changes in the neuron soma suggests that the
toxic action of arsanilic acid is directed toward the cells or mechanisms
which support the myelin sheath and axons.

Figure 3 illustrates the histological appearance of a peripheral nerve
from a control animal. Notice the neurokeratin network of the myelin sheaths

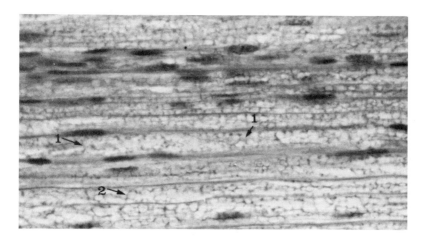

FIG. 3. Sciatic nerve control. Axon appears as faint gray line between
arrows at 1. Darker stained neurokeratin network (arrow 2). Harris
hematoxylin and eosin Y stain (x480).

surrounding the faintly gray axons. Nuclei of Schwann cells are rather large
and oval shaped. Figure 4 illustrates the early lesions in peripheral
nerves associated with arsanilic acid toxicosis, and Fig. 5 the advanced
ones. Notice the presence of many darkly staining nuclei in Figs. 4 and 5
as compared with Fig. 3. A pyknotic Schwann cell nucleus can be seen in
Fig. 5 between the arrows labeled with a number 3.

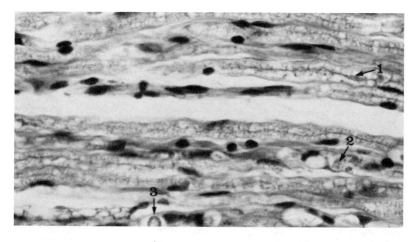

FIG. 4. Sciatic nerve day 16 on 1000 ppm arsanilic acid in ration. Contracted myelin around intact axon (arrow 1). Myelin fragment (arrow 2)
and myelin ovoid (arrow 3). Harris hematoxylin and eosin Y stain (x480).

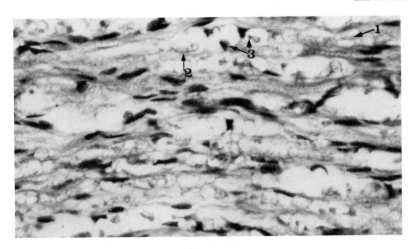

FIG. 5. Sciatic nerve day 27 on 1000 ppm arsanilic acid in ration. Contracted myelin around intact axon (arrow 1), fragment of myelin (arrow 2), and fragments of myelin and axon (arrows at 3). Note pyknotic Schwann cell between arrows at 3. Approximately 60% of the nerve fibers were visibly damaged. Animal was quadriplegic. Harris hematoxylin and eosin Y stain (x480).

Figure 6 illustrates peripheral nerve from a control animal stained by Bodian's method. In Fig. 7, Bodian's method was used to demonstrate axonal disintegration.

Demyelination in the optic nerves and tracts is less obvious, but essentially parallels lesions in the peripheral nerves. Affected optic tracts and nerves appear more cellular than controls. The appearance of increased

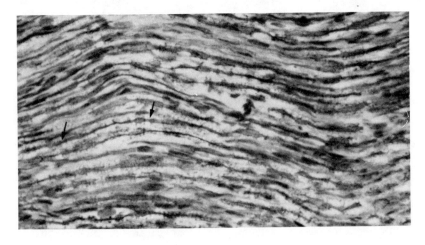

FIG. 6. Sciatic nerve control. Note intact axon between arrows. Bodian's stain (x375).

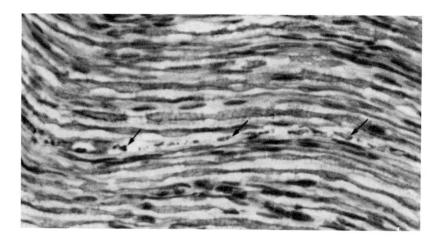

FIG. 7. Sciatic nerve day 16 on 1000 ppm arsanilic acid in ration. Note fragmented axon (arrows) between intact axons. Bodian's stain (x375).

cellularity is in part due to swollen neuroglial cells causing them to be more prominent.

The histological appearance of an optic tract and optic nerve from a control animal can be seen in Figs. 8 and 10; and specimens from an animal which had ingested toxic levels of arsanilic acid for 27 days are seen in Figs. 9 and 11. Histopathologic changes in the animal receiving arsanilic acid are pyknotic nuclei, fragmentation of myelin sheaths, and the presence of many vacuoles in the tissue.

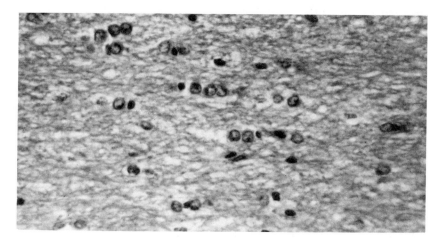

FIG. 8. Optic tract control. Harris hematoxylin and eosin Y stain (x415).

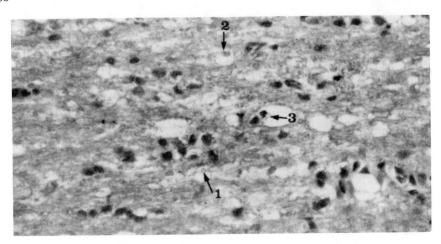

FIG. 9. Optic tract day 27 on 1000 ppm arsanilic acid in ration. Contracted
myelin around intact axon (arrow 1), fragment of myelin (arrow 2), and myelin ovoid containing piece of axon (arrow 3). Harris hematoxylin and eosin
Y stain (x415).

Pathological alterations are evident in the heavily myelinated fibers
of peripheral nerves by days 10-13. These alterations progress to definite
demyelination by day 16. The earliest degenerative change seen in Schwann
cells associated with initial demyelination is nuclear swelling and deeply
eosinophilic cytoplasm. With more advanced demyelination, pyknosis and
eventually karyorrhexis and karyolysis of Schwann cells is noted. The
Schwann cells of nonmyelinated fibers do not appear to be affected in a

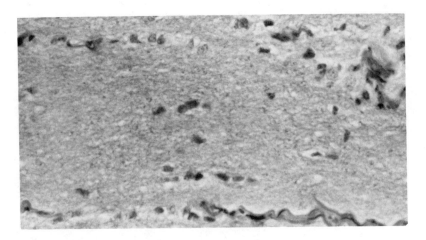

FIG. 10. Optic nerve control. Harris hematoxylin and eosin Y stain (x415).

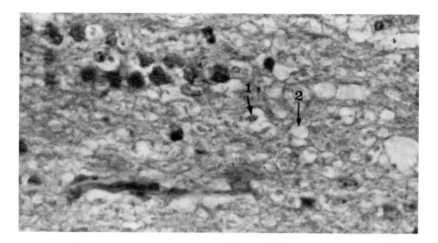

FIG. 11. Optic nerve day 27 on 1000 ppm arsanilic acid in ration. Myelin contracted around axon (arrow 1) and vacuolation (arrow 2). Harris hematoxylin and eosin Y stain (x480).

similar manner. Demyelination is less apparent in the fibers within the dorsal root ganglia.

SUMMARY

The phenylarsonic compounds used as animal feed additives are metabolically and toxicologically distinct from the inorganic and aliphatic organic arsenicals. A tentative diagnosis of organic arsenical poisoning can be made on the basis of clinical signs and the presence of the drug in the feed or water. Arsenic levels in tissue of suspected cases are rarely diagnostic alone. Phenylarsonic forms are rapidly excreted via the urine and bile; thus, if the animal has not been ingesting toxic levels continuously, the tissue levels decrease drastically. Generally, liver and kidney levels of 3-10 ppm arsenic on a wet-weight basis or blood levels of 1-2 ppm would be considered diagnostic if clinical signs of toxicity are present. Microscopic examination of longitudinal sections of peripheral nerves is an important procedure for confirmation of a diagnosis of organic arsenical poisoning. One should keep in mind, however, that lesions of demyelination occur only after 2-3 weeks of exposure to toxic levels of the drug [19,20].

Feed levels of 250 and 100 ppm arsanilic acid or 3-nitro, respectively, should be viewed as potentially toxic if the animals ingesting it are suffering from diarrhea or have limited water intake.

Paralysis of extremities without central nervous system involvement, high morbidity with low mortality, alert animals which continue to eat and drink, and little or no gross changes visible at necropsy strongly suggest organic arsenical poisoning.

REFERENCES

1. D. V. Frost, Arsenicals in Biology: Retrospect and Prospect, Fed. Proc., 26, 194-208 (1967).

2. N. F. Morehouse and O. J. Mayfield, The Effect of Some Aryl Arsonic Acids on Experimental Coccidiosis, Infection of Chickens, J. Parasit., 32, 20-24 (1946).

3. N. F. Morehouse, Accelerated Growth of Chickens and Turkeys Produced by 3-Nitro-4-hydroxyphenylarsonic Acid, Poultry Sci., 28, 375-384 (1949).

4. H. R. Bird, A. C. Groschke, and M. Rubin, Effect of Arsonic Acid Derivatives in Stimulating Growth of Chickens, J. Nutr., 37, 215-226 (1949).

5. Animal Health Institute, Feed Additive Compendium, Miller, Minneapolis, 1974.

6. W. B. Buck, Laboratory Toxicologic Tests and Their Interpretation, J. Amer. Vet. Med. Ass., 155, 1928-1941 (1969).

7. W. B. Buck, Untoward Reactions Encountered with Medicated Feeds. Use of Drugs in Animal Feeds, Pub. 1679, Nat. Acad. Sci., Washington, D. C., 1969.

8. H. Eagle and G. O. Doak, The Biological Activity of Arsenosobenzenes in Relation to Their Structure, Pharmacol. Rev., 3, 107-143 (1951).

9. L. S. Goodman and A. Gilman, The Pharmacologic Basis of Therapeutics, 3rd Ed., Macmillan, New York, 1966, pp. 944-945.

10. C. Voegtlin and J. W. Thompson, Rate of Excretion of Arsenicals, a Factor Governing Toxicity and Parasiticidal Action, J. Pharm. Exp. Ther., 20, 85-105 (1923).

11. A. C. Chance, T. B. B. Crawford, and G. A. Levvy, The Fate of Arsenic in the Body Following Treatment of Rabbits with Certain Organic Arsenicals, J. Exp. Phys., 33, 137-147 (1945).

12. J. P. Moody and R. T. Williams, The Fate of Arsanilic Acid and Acetylarsanilic Acid in Hens, Food and Cosmetics Toxicol., 2, 687-693 (1964).

13. J. P. Moody and R. T. Williams, The Fate of 4-Nitrophenylarsonic Acid in Hens, Food and Cosmetics Toxicol., 2, 695-706 (1964).

14. J. P. Moody and R. T. Williams, The Metabolism of 4-Hydroxy-3-nitrophenylarsonic Acid in Hens, Food and Cosmetics Toxicol., 2, 707-715 (1964).

15. L. R. Overby and R. L. Fredrickson, Metabolism of Arsanilic Acid, II. Localization and Type of Arsenic Excreted and Retained by Chickens, Toxicol. Appl. Pharm., 7, 855-867 (1965).

16. L. R. Overby and L. Straube, Metabolism of Arsanilic Acid, I. Metabolic Stability of Double-Labelled Arsanilic Acid in Chickens, Toxicol. Appl. Pharm., 7, 850-854 (1965).

17. L. R. Overby and R. L. Fredrickson, Metabolic Stability of Radioactive Arsanilic Acid in Chickens, J. Agr. Food Chem., 11, 378-381 (1963).

18. A. E. Ledet, Clinical, Toxicological and Pathological Aspects of Arsanilic Acid Poisoning in Swine, Ph.D. Thesis, Iowa State University, Ames, 1970.

19. A. E. Ledet, J. R. Duncan, W. B. Buck, and F. K. Ramsey, Clinical, Toxicological, and Pathological Aspects of Arsanilic Acid Poisoning in Swine, Clin. Toxicol., 6 (3), 439-457 (1973).

20. J. D. J. Harding, G. Lewis, and J. T. Done, Experimental Arsanilic Acid Poisoning in Pigs, Vet. Rec., 83, 560-564 (1968).

21. L. E. Hanson, L. E. Carpenter, W. J. Aunan, and E. F. Ferrin, The Use of Arsanilic Acid in the Production of Pigs, J. Animal Sci., 14, 513 (1955).

22. S. F. Scheidy, P. W. Wilcox, and A. A. Creamer, Residual Arsenic in Swine Tissues Following Oral Administration, J. Amer. Vet. Med. Ass., 123, 341-342 (1953).

23. W. T. Oliver and C. K. Roe, Arsanilic Acid Poisoning in Swine, J. Amer. Vet. Med. Ass., 130, 177-178 (1957).

18

BIOLOGICAL EFFECTS OF SELENIUM

James R. Harr
Pennwalt Corporation
Rochester, New York

INTRODUCTION

Selenium is an essential nutrient, an integral part of the enzyme glutathione peroxidase, and may have a role in other biologically active compounds [24,71,94]. Possible roles involve selenotrisulfide linkage, thiol and sulfhydryl groups, vitamin E, antioxidation, ubequinones, and the host defense mechanism [24,107]. Demographic and experimental evidence associates selenium with lower than anticipated incidences of cancer and with optimal functioning of the vascular endothelium [99,106].

Changes in the valence state of selenium from 2-, through 0, 2+, 4+, and 6+ are associated with its geologic distribution, redistribution, and use [20,90]. Valence also affects selenium's toxicity and deficiency in the biosphere. Radioactive nucleotides of selenium are used as biological tracers and in radiologic diagnostic procedures for visualization of relative rates of protein synthesis and vascularization [68].

Soluble selenates (6+) occur in alkaline soils, are slowly reduced to selenites, and are readily taken up by plants [20,29,31]. Primary accumulators (Astragolus, Oonopsis, Stanleya, and Zylorhiza) may concentrate selenium to several mg/g [90]. Selenites (4+) are less soluble than the corresponding selenates and are easily reduced to elemental selenium. Ferric selenite and combinations of ferric oxides with selenites are quite insoluble and largely unavailable to the biosphere [29,77].

Selenium dioxide (2+) is formed by combustion of elemental selenium present in fossil fuels or rubbish [118]. Generally sulfur dioxide is formed concurrently and reduces the selenium to elemental selenium. Elemental selenium (valence 0) is associated with copper production. Major industrial uses of selenium are in electronic components, photocopying, steel and chromic alloys, glass clearing, pigments, and heat reflective glass. Elemental selenium is quite insoluble and is largely unavailable to the biosphere [42]. However, Handreck and Godwin have supplied the nutritional requirements of sheep for selenium from metallic pellets placed in the rumen [39].

Hydrogen selenide (2-) is highly toxic (1-4 µg/l air), unstable, acidic, and irritative [22,42]. Selenides of mercury, copper, and cadmium are very insoluble [103]. Their insolubility may be the basis for reported detoxification of methyl mercury by dietary selenite, and for decreased heavy metal toxicity associated with selenite. Metallic selenides thus form a biologically important sequestering of both selenium and heavy metals in a largely unavailable form.

Selenate and selenite are absorbed by plants, reduced, and incorporated in amino acids synthesis [71]. Monogastric animals can reduce selenate to selenite but cannot incorporate selenides into amino acids.

The first step in metabolism of selenium in monogastric animals may be formation of seleno-trisulfide with selenite and the sulfhydral groups of amino acids, peptides, and proteins [28]. Selenium metabolism in ruminants is further complicated by rumen physiology and the concurrent effects of microflora, a strong reducing environment, and rough, highly complex feedstuffs. The major excretory products of selenium metabolism are trimethylselenoneium ion in urine and elemental or metallic selenides in fecal material.

The net effect of soil, plant, and animal metabolism of selenium is to convert selenium to inert and insoluble forms: elemental selenium, metallic selenides, or complexes of selenite with ferric oxides [1]. In areas of acid or neutral soils, the amount of biologically available selenium should steadily decline. The rate of decline may be increased by active agricultural or industrial practices. In dry areas with alkaline soil and oxidizing conditions, elemental selenium and selenides in rock and volcanic soils may oxidize rapidly enough to maintain the availability of biologically active selenium [90]. This balance may be disturbed by intensive agricultural practices or by human activity of redistributing metallic compounds, etc. The role of biological oxidation of selenium is largely unknown [29].

EXPOSURE

The concentration of selenium in animal and plant foods is dependent upon the geographic region where the product was produced, on agricultural practices, and on the selenium requirements and tolerance of the species [1,54,66,90]. In general, forages from the Rocky Mountain and high central plains of the United States may contain 200 to 300 ng/g of selenium, while those from the Pacific Northwest and southeastern areas of the United States may contain 1/10 as much (20-30 ng/g). However, there are considerable variations (to 10-fold) from one specific location to another.

Primary selenium accumulator plants require 1-50 ppm selenium in either soil or water for growth and may contain 100-10,000 μg/g selenium as a glutamyl dipeptide or selenocystanthionine [9,90]. Primary accumulator plants include species of the genera Astragolus (24 species can contain 1000 μg/g Se), Oönoposis (800 μg/g Se), Stanelya (700 μg/g Se), Zylorhiza (120 μg/g Se), and Machaeranthera.

Secondary accumulator plants grow in either seleniferous or nonseleniferous soil and may contain 25-100 μg/g selenium: Astor (72 μg/g Se), Gutierrezia (60 μg/g Se), Atriplex (50 μg/g Se), Grindelia (38 μg/g Se), Castilleja, and Comandra [42,90]. Nonaccumulator plants growing on seleniferous soils contain 1-25 μg/g selenium, and meat and eggs from seleniferous areas may contain 8-9 μg/g selenium. Most foods of animal origin contain 0.01 to 1.00 μg/g of selenium on a wet-weight basis. Tissues from animals maintained on high-selenium feeds generally contain less than 3 to 5 μg/g of selenium. Some tissue from animals killed with selenium intoxication may contain as much as 20 μg/g of selenium.

Fossil fuels may contain 1 to 10 μg/g of selenium [84]. The annual release of selenium from this source in the United States is estimated to be 1500 tons. In addition, industrial losses are estimated as 2700 tons and municipal waste at 360 tons. Of these 4600 tons of selenium released in fuel, industrial, and refuse processing, about 25% is in atmospheric emissions and the balance is in the ash.

Air and surface water generally contain less than 10 ng/g of selenium [21,54]. There are no known significant increases in the ambient concentration of selenium in specific areas apart from industrial environments and ground water leaching from seleniferous soils. In neither case have these instances produced occupational or exposure poisoning in workers or animals.

Biological use of selenium has included the insecticide selenocide to control mites in citrus groves, grapes, and ornamentals [104]. This use has been discontinued. The dandruff shampoo Selsun contains 1% selenium sulfide. Similar preparations are used to control dermatitis and mange in dogs. The largest single biological use of selenium is the transportation of grains from seleniferous areas to deficient areas as animal and human food incidental to grain production in the seleniferous central plain states [51]. The greatest current and direct use of selenium in biology is in the supplementation of animal feeds to provide nutritional requirements for selenium [43,53,71]. Methods used to provide nutritionally required selenium include parenteral injection (1 mg/50 kg), addition to complete feeds (0.1-0.2 μg/g), and application of small amounts to the soil or foliage [16]. The latter requires larger quantities of selenium and is expensive both in direct cost and in maintaining control over the rate of administration to the animal.

METABOLISM

Absorption of oral radioselenite by rats is 95 to 100%, regardless of the concentration of selenium in the feed (0.02 or 4.02 μg/g) [13,24]. Absorption may be less in ruminants because of the reduction of natural forms of selenium in feeds to elemental selenium or by complexing with other components of feed such as ferric oxides.

A single dose of selenium is concentrated in the pancreas, intestine, erythrocytes, liver and kidney, and testes [13,32,44]. Tissue distribution of selenium from chronic exposure is similar [49]. Subcellular distribution of selenium favors nuclear and mitochondrial fractions of testes and kidney [7]. Uptake into erythrocytes from serum is dependent on the basal dietary sufficiency of selenium [120]. The ratio of the concentration of selenium in the liver and kidney of shoats poisoned with selenium was 1.0 to 4.0 in swine that died of the selenium poisoning and 0.3 to 0.6 in normal animals [92,114].

The major excretory pathway for selenium is the urine [14,24]. Urinary excretion is related to the basal exposure to selenium in the feed, and to the amount of additional selenium given to the animals. Rats fed 4 ng/g selenium excreted 6% of an injected submicrogram dose, while those fed 1 μg/g selenium excreted 67% of the injected dose [13]. Burk et al. postulated a threshold dietary concentration of selenium between 54 and 84 ng/g of selenium [14]. Selenium intake greater than this threshold was excreted in proportion to the intake, and below the threshold selenium was conserved by the body. This hemostatic mechanism promotes rapid excretion of excess selenium resulting in a plateauing of selenium in tissue and preventing accumulation of selenium in edible tissues.

Excretion of selenium in feces, bile, saliva, and hair appears to be relatively constant regardless of the amount of exposure [13,37,63]. However, pulmonary excretion is active only when selenium intake is greater than 1 mg/kg. Passage of selenium compounds through the lungs may reduce the evidence and severity of bronchitis [40]. The amount of selenium in the hair is integrative and thus may indicate the degree of chronic exposure [119].

Selenium is an integral part of the enzyme glutathione peroxidase. This selenotrisulfide derivative of glutathione reacts further to form selenopersulfide [60]. Other proteins containing thiol groups apparently can also form selenotrisulfides. Selenium may also be involved in the metabolism and action of ubiquinones [24].

Dietary sulfate, arsenic, linseedmeal, and methionine partially protect against selenium poisoning [38,76]. Sulfate was effective against selenate poisoning in rats and was associated with increased urinary excretion of selenium. Arsenic decreased exhalation of volatile selenium compounds and increased biliary excretion of selenium up to 10-fold. Linseed meal probably changes the chemical form of selenium in tissues and thus reduces signs of toxicosis but increases the amount of selenium held in the tissue. The effect of methionine was dependent upon ample fat-soluble antioxidants and was associated with hepatic lesions and a decreased concentration of selenium in liver and kidney.

Selenomethionine but not selenocystine or selenite was absorbed against a concentration gradient [62]. Methionine inhibited the transport of selenomethionine, but sulfite and cystine did not compete with selenite or selenocystine. The methionine/selenomethionine transport competition may account for the protection of high-protein diets in selenium toxicity [61]. These interactions are variable and in some instances may be additive rather than protective.

Selenium protects animals from toxic doses of cadmium, mercury, and methyl mercury probably by binding with the agent [80]. This may be a normal chemical defense of the individual against environmental heavy metals.

The association of plants with the selenium content of soils is involved
[90]. Grasses growing near selenium accumulator plants contain more selenium than those growing nearby but not in association with accumulators
[66,78]. Addition of various proteins, amino acids, or plant extracts to
a culture solution increases selenium uptake by plants. The pH and colloid
content also influence selenium uptake with time from low-selenium soils
[30]. Sulfur reduces selenium uptake and sulfur additions in selenium-
deficient areas can reduce the selenium content of crops and thus increase
the need for selenium supplementation of animal diets [2].

DEFICIENCY

Experimental selenium deficiency has been produced in rats, sheep,
swine, poultry, and squirrel monkeys by use of low-selenium feeds supplemented with α-tocopherol [64,69,70,73]. Naturally occurring selenium responsive disease has been observed in sheep, cattle, and swine [43,113].
These deficiencies were prevented or reversed by parenteral or dietary
supplementation with sodium selenite or selenate (100 ng/g of feed or
parenterally 1 mg/50 kg). Neither parenteral injection nor additional
dietary supplementation with α-tocopherol prevented or reversed the lesions.
Similar preparations are also widely used in dogs, horses, and breeding
males as therapy for clinical signs of musculoskeletal weakness, lameness,
dermatitis, infertility, and abnormal hair coat [19,24].

The usual route of administration is parenteral. However, in Australia
and New Zealand sodium selenate has been added to mineral supplements, to
fertilizer, or supplied as a supplement for mixed feeds [43]. Supplementation of mixed feeds for young poultry and pigs was approved in Canada on
September 6, 1973, and in the United States on February 7, 1974.

Selenium responsive diseases include nutritional myopathy in sheep,
swine, and cattle (white muscle disease), hepatic necrosis (Hepatosis dietetica), myocardial necrosis and hemorrhage (dietetic microangiopathy or
mulberry heart disease), gasteoesophageal ulceration and hertztod of swine,
and steatitis and exudative diathesis of chicks and poults. Less well defined effects of selenium therapy occur as lameness in dogs, horses, and
breeding bulls and poor growth and reproduction in sheep and rats. The
primary site of injury from selenium deficiency may be the vascular beds.

Myopathy in Sheep and Cattle

Nutritional myopathy (white muscle disease) became prevalent in ruminants in the northeastern and northwestern portions of the United States

and in Australia, New Zealand, and Northern Europe after World War II [68, 69]. It was associated with changes in methods of sheep production from low-yield hay and grass foraging supplemented with grain and protein concentrates to intensified high-yield grass production. The large yield of high-protein grass permitted lambs to be reared and marketed with minimal grain feeding. The major deficiency of plant nutrients was generally sulfur, and the change to high-performance grass management necessitated heavy sulfate fertilization. Hays grown under these conditions often contain less than 20 mg/g of selenium (dry-weight basis).

The incidence of nutritional myopathy in lambs is not entirely related to the concentration of selenium (13-23 ng/g) in the hay. The ability to produce myopathy varies with the location of the field, year, and time of growth. The factors responsible for these variations are unknown but may involve lipid components of the forages and concomitant rumen activity.

Lesions of nutritional myopathy in sheep and cattle are primary calcification and degeneration of skeletal muscle and myocardium [6,67]. The left ventricular wall and interventricular septum are usually affected, while the auricles, right ventricle, and apex are spared. The more active skeletal muscles, especially the abductors and the longissimus dorsi, are also usually affected.

Lesions of myocardial nutritional myopathy have been observed in lambs and calves aborted during the last month of gestation, in stillbirths, and in neonatal deaths. Living neonates may be edematous and have passive congestion, labored breathing, and weak, irregular pulse.

The muscles affected are those with the greatest work requirement. In the neonate these are ventricles of the heart and the abductors of the thigh. Later, the longissimus dorsi and triceps are affected. Mortality may be 65% during the first 10 days of life. Oral or parenteral administration of 1 mg of selenium as selenite may cause remission of clinical signs within a few hours.

Clinical signs of slow growth and weakness of the semitendinosus-semimembranosus muscles and the longissimus dorsi may occur or recur during the initial phase of rapid growth, 3 to 8 weeks after birth. The connective tissue around the distal portion of the muscles may contain a heavily proteinaceous exudate. The concentrations of lactic dehydrogenase, 5-nucleotidase, and glutamic, oxalic, and pyruvic transaminases in serum are increased, and lysosome and lysosomal enzyme changes have been reported [10, 83,122]. Mortality may be 35%. Parenteral administration of 1 mg of selenium as selenite per 50 kg may reverse clinical signs and lesions.

Sheep and cattle that are unthrifty, have muscular weakness or reduced fertility may clinically improve after parenteral administration of 1 mg of selenium per 50 kg of body weight. Annual repetition of selenium adminis- tration, a series of two or three injections during gestation or metallic ruminal pellets of elemental selenium generally prevent selenium deficiency.

The occurrence and initial lesions of nutritional myopathy and the therapeutic response to selenium may differ with climatic, forage, and cropping conditions. Generally in the United States, Australia, and New Zealand, selenium salts are more effective than tocopherols in preventing nutritional myopathy [26,69]. Forages are heavily fertilized, and hay is easily cured or overcured. The initial histologic lesions of nutritional myopathy in these areas are microscopic deposits of calcium midway between the Z bands of the sarcomere [6,67]. The deposits are less than a micro- meter in diameter and may coalesce to form grayish-white, elongated plaques of calcium.

In northern Europe, α-tocopherol has been considered more effective than selenium in preventing nutritional myopathy [74,105]. In this area, nutritional myopathy is associated with poor curing of lightly fertilized native grasses. The primary morphologic lesion of the sarcomere may be de- generation, and calcium deposition is secondary, delayed, or absent. The difference between these syndromes may be associated with changes in the unsaturated fatty acids of the feedstuffs.

Hepatic, Arteriolar, Skeletal, and Myocardial Degeneration in Swine

Selenium and vitamin E are used to prevent a deficiency syndrome of swine characterized by sudden death of feeder pigs, hepatic necrosis, ic- terus, mesenteric edema, fibrinoid degeneration and microangiopathy of the media of the arterioles, and skeletal and cardiac myopathy [92,112,113,116]. The concentration of selenium in feeds associated with this syndrome is 20-60 ng/g [114]. Supplementation with α-tocopherol does not prevent the hepatic, muscular, or vascular lesions, but prophylactic administration of selenium-tocopherol mixtures (100 ng Se/g) is effective. This prophylaxis reduced the incidence of stillborn pigs from 9% to 3%, neonatal deaths from 6.2% to 3.4%, and mortality to 6 months of age (about 225 pounds) from 17% to 11%. The regimen also increased the concentration of selenium in the liver and psoas muscle from 33 and 25 ng/g (wet weight), respectively, to 400 and 65 ng/g.

The selenium concentrations in commercially reared pigs without the disease or therapeutic regimen was 265 (liver) and 83 (muscle) ng/g. The ratio of the concentration of selenium in liver and kidney of the selenium-deficient pigs was about 0.12, and of the treated pigs was 0.35. Experimentally produced lesions of selenium and vitamin E deficiency included distress, ataxia, sudden death, hepatic swelling, necrosis and hemorrhage, irregular Glisson's capsule, hemorrhagic ileitis, and ecchymotic hemorrhages; increased concentrations of plasma transaminases, creatine phosphorinase, α-hydroxybutyric acid dehydrogenase, isocitric dehydrogenase, lactic dehydrogenase; selective destruction of type I skeletal muscle fibers; and a decrease of phosphorylase activity in type II fibers. Muscle lesions appeared to result from primary injury of the myofibrils as reported in sheep [92].

Exudative Diathesis and Myopathy in Poultry

Selenium salts prevent exudative diathesis in chicks and Japanese quail fed synthetic (amino acid) diets that contained little selenium but high levels of α-tocopherol [85]. Lesions include exudative diathesis, poor growth and feathering, and fibrotic degeneration of the pancreas [33]. Death was associated with decreased production of bile enzymes, failure of the lipid-bile salt micelle formation, and adsorption of α-tocopherol [110]. Therapy to improve the absorption of tocopherol prevented exudative diathesis but did not prevent degeneration of the pancreas. The addition of selenium (20-40 ng/g) as selenite to the feed prevented pancreatic degeneration [92].

Selenium deficiency in turkey poults produces a mild form of exudative diathesis, characterized by degeneration of the gizzard muscle [82,97]. Myopathy of the pectoral muscles (25%) and myocardial failure can be prevented by addition of 80-280 ng of selenium as selenite per gram of feed.

Lesions of exudative diathesis are similar to the exudation observed in connective tissue surrounding the digital portion of the semitendenosus and semimembranous muscle of lambs with nutritional myopathy. The myopathy is a degeneration of the muscle fibers with perivascular infiltration and proliferation of histocytes and granulocytic leucocytes, similar to that in sheep and swine.

Rats

Attempts to produce uncomplicated direct selenium deficiency in laboratory rats were unsuccessful until 60 ng/g of α-tocopherol was added to low-selenium (18 ng/g) torula yeast feeds [64]. Rats maintained on these feeds for 2-3 generations or those from the second or third litters of females grown on the Se feeds became selenium depleted.

Selenium-deficient rats grew slowly, had poor hair coats, and were sterile. Vascularization of the subcutis and dermis was incomplete and there were lens cataracts [106]. The tactile hairs (nourished from cavernous blood sinuses) were not affected. Germinal epithelial cells of the skin and endothelial cells of the capillaries and small arteries contained fewer stainable RNA or sulfhydryl groups than in rats on the same feed supplemented with 100 ng/g of selenium as selenite. Addition of 10-100 ng/g of selenium as selenite or selenate to the feed of 60-day-old selenium-deficient rats completely reversed signs of selenium deficiency in 30-90 days.

Selenium-deficient tocopherol-supplemented brown rats fed torula yeast feeds ate 50-60% more feed than did littermates fed the same feed with the addition of 2 µg of selenium as selenite per gram of feed [40]. Feed efficiency (feed eaten/weight gain) was 75% greater in the selenium rats compared to those on the basal diet.

The concentrations of selenium in liver, muscle, and kidney of selenium-deficient rats (18 ng/g Se) were, respectively, 0.4, 1.0, and 2.0 µg/g (dry weight) [41]. Addition of 100 ng/g of selenium as selenite to the feed increased selenium concentration in liver to 2.0 µg/g but did not change the concentration of selenium in the muscle or kidney. Addition of 2500 ng/g of selenium as selenite increased the concentration of selenium in liver to 7.4 µg (dry weight) but again did not increase the concentration of selenium in muscle or kidney. The plateauing of selenium concentration with increased supplementation of the feed with selenium may be the effect of hemostasis or decreased food consumption [14].

Maintenance of the concentration of selenium in rat muscle at the expense of the concentration in the liver was a contrast to observations in sheep, wherein the concentration of selenium in the liver (1.0 µg/g) was maintained at the expense of the concentration in muscle (0.5 µg/g wet weight) [11,13]. The sparing of selenium in the liver or muscle may affect the type of selenium deficiency lesions in these species: myopathy, hepatic degeneration, etc.

Monkeys

Seven adult squirrel monkeys were fed a low-selenium semipurified feed with Candida utilar as the protein source and adequate vitamin E [70].

After 9 months the monkeys developed alopecia, loss of body weight, and listlessness. After one monkey died, three other monkeys were given parenterally 40 µg/g of selenium as selenite at 2-week intervals. The three monkeys given selenium recovered and the other three died or became moribund. Lesions included hepatic necrosis, myopathy (skeletal and myocardial), and nephrosis. The tactile hairs (as in the rat) are supplied by cavernous blood sinuses rather than by capillaries and were not affected.

Human

Selenium is present in human blood, urine, and tissues and in a low-density serum lipoprotein [3,23,25]. Concentrations in the tissues of California residents determined by x-ray emission spectrophotography were generally similar to those in lambs and swine produced on commercial feeds. Selenium levels in tissues of newborn infants in Russia are of the same order as those found in pigs and lambs on low-selenium diets in the United States.

Based on the shape of distribution curves of selenium in human tissues, Liebscher and Smith concluded that selenium is an essential nutrient, although there is no conclusive evidence of human disease which can be attributed specifically to selenium deficiency [12,45,58]. Frost and others estimated that the average human diet in the United States contains 1.8 mg of selenium per month (about 25 µg/kg/month) [24]. This compares to 1 mg per month in a ewe whose lambs develop nutritional myopathy (about 20 µg/kg per month); and to 20 µg per month required for selenium-adequate rat diets [64].

Selenium deficiency in people, if present, may be associated with Kwashiorkor [12,45], peridontal disease [24,43], sudden infant death syndrome [5,65], and increased incidence of cancer [99,101]. The evidence at this time is based largely on extrapolation, lower-than-normal concentration of selenium in the blood of patients, and on demographic investigation.

Other Species

Clinical observations of the efficacy of the selenium-vitamin E mixture in animals with adequate dietary selenium and vitamin E indicates that an additional 1-3 mg/50 kg per month improves functioning of reproductive, skeletal, muscular, and vascular systems [24]. Efficacy is claimed in clinical cases of stiffness, lameness, arthropathy, myositis, hepatic degeneration, infertility, loss of condition of the hair coat, dermatitis, and poor growth [27,43]. A severe strain is put on the musculoskeletal systems of race horses and dogs and in breeding rams and bulls [19]. Veterinarians in selenium-deficient areas believe that selenium therapy produces a clinical improvement in these and in unthrifty or poorly fertile animals [43].

Reproduction

Selenium deficiency interferes with the general health of the individual and thus effects on reproduction may be secondary. In areas where white muscle disease or unthriftiness are prevalent in sheep and cattle, embryonic mortality is high [43,69]. Field use of commercial selenium-tocopheral mixtures at the time of implantation improved reproduction.

Selenium-deficient feeds imposed over successive generations of rats produced animals that grew and reproduced normally [64]. However, their offspring were almost hairless, grew slowly, and were sterile. The females did not reproduce when mated with normal males. Active spermatogenesis was observed in some of the seminiferous tubules of selenium-deficient rats born to females on a selenium-deficient diet; however, the motility of spermatozoa from the cauda epididymis of these males was poor and the majority of sperm had broken fibrils in the axial filaments [123].

An hour after a single subcutaneous injection of a tracer dose of ^{75}Se in mice, the testis was one of the lowest ranking tissues in ^{75}Se uptake [35]. However, 7 days later the testes had the third highest concentration (^{75}Se) after liver and kidney [13]. One or two weeks later the concentration of ^{75}Se increased in epididymis and then was rapidly depleted from the body. This transport pattern is similar to that of ^{65}Zn except that zinc appears in the epididymis one week earlier than selenium and is incorporated into a later phase of spermatogenesis [34].

Approximately 98% of the ^{75}Se distributed to the seminal fluid in rat epididymis is firmly bound in the midpiece of the sperm and is not removable by challenge with large doses of nonradioactive selenium [34]. As mitochondria are present only in this portion of the sperm, selenium may be associated with their function.

Cadmium-induced testicular lesions can be prevented by selenium salts, although selenium caused a marked elevation in uptake of testicular cadmium [34,36]. As cadmium and selenium mutually increase each other's concentration in the testicle and at the same time prevent cadmium toxicity, there may be a binding between them in a form that is inactive against the testicle. Selenium thus transports inactivated cadmium away from its target site to some other and less vulnerable locus within the testis where it is innocuous [17]. The target of cadmium is probably a protein with a molecular weight of 30,000. Cadmium diverted from this target by selenium became attached to a higher molecular weight protein.

Selenium deficiency in Japanese quail causes reduced viability of the newly hatched chicks, but the rate of egg production and fertility are not changed [47]. As selenium crosses the placental barrier and is excreted in the dam's milk, the injection of selenium into pregnant or nursing females prevents neonatal selenium deficiency. Mercuric salts induce a greater retention of selenium in pregnant females and a reduced transfer of selenium to the fetus by both placental transfer and nursing [81]. Thus, increased concentrations of mercury in the environment will alter the amount of selenium reaching the young and will contribute to selenium deficiency.

Vascular

Consideration of vascular endothelium as a dynamic structure altered by selenium imbalance is a relatively new concept [4]. Biological effects of selenium deficiency, toxicity, and interactions support such a concept. The young of about 40 mammalian and bird species cannot tolerate vitamin E and selenium deficiency. The pathology of the various syndromes varies considerably, but all are characterized by effusion, hemorrhage, and necrosis.

The primary effect of selenium deficiency on exudative diathesis in chicks is alteration of capillary permeability [82]. In addition, the vascular endothelium may be the site of the initial lesions in selenium-responsive necrotic liver degeneration, cardiac and skeletal muscle degeneration, kidney necrosis and pancreatic dystrophy, myopathy, hertztod, and microangiopathy.

Initial ultrastructural changes in myocardium of piglets from vitamin E- and selenium-deprived sows appear in connective tissue and capillaries before structural changes in muscle fiber [108]. Similarly, lesions present in selenium-deficient rats were vascular hypoplasia, endothelial thickening, and degeneration in tissues with a marginal vascular supply and considerable oxygen dependence (skeletal and cardiac muscle, testis, and retina) [106]. Lesions were not observed in highly vascular tissues such as nasal tactile area. Selenium- and vitamin E-deficient poults had discoloration and atrophy of flight feathers secondary to hemorrhage in the pulp of the immature feather. The antiinflammatory properties of selenium compounds used in

the treatment of chronic lameness in dogs may also be related to vasculature
insufficiency [19,88].

SELENOSIS

Rosenfeld and Beath reported a series of ancient cases of selenosis
beginning with Marco Polo and plants that probably accumulated selenium
[90]. Horses grazing these plants lost their hair and developed abnormal
hoofs. Other historical incidents of selenosis involved loss of hair,
nails, and teeth, epidermal malformations, chronic dermatitis, lassitude,
and progressive paralysis.

Criteria of toxicity of various exposures to selenium include death,
clinical signs, impaired function, gross or microscopic lesions, and histologic and biochemical changes. Because of these differing criteria and the
duration (hours to years), and method of exposure, it is difficult to describe the amount of selenium that constitutes a toxic, a safe, or a nutritional dose. The characterization of selenosis is further complicated by
the syndromes of blind staggers and alkali disease that occur in cattle and
horses grazing seleniferous plants that are often described as separate
entities rather than as field poisonings by selenium.

Concomitant Conditions

Many conditions alter the toxic effects of selenium. These interactions have affected the direction of research on selenosis and the data
obtained from the experiments and have further complicated establishment
of toxic and no-effect conditions of exposure.

Significant concomitant conditions that affect selenium poisoning
include the method of administration, the chemical form of the selenium,
other feeds and chemicals present, and the age and needs of the animal [8,
90]. In general, selenium in natural feedstuffs or added to natural feeds
is considerably less toxic (about fourfold) than similar exposures in water
or purified feeds [40]. Concurrent ingestion of minerals, rough or high-
protein feeds reduces toxicity. Dietary exposure is less toxic than parenteral or inhalation exposure.

It is also difficult to differentiate significant variations from
normal background variation after chronic or lifetime exposures to effectively nontoxic concentrations of selenium. Selenium generally affects
active tissues and appears to be irritative. Thus, the stromal components

of the tissue may be increased compared to the controls, but is this beneficial or harmful? In some aspects it apparently prolongs life, increases
general defense mechanisms, and improves muscular/skeletal function.

Guidelines of toxic exposures to selenium include respiratory exposure
to hydrogen selenide, 1-4 µg/l of air; parenteral lethal dose, 2-5 mg/kg;
minimal dietary lethal dose on semipurified feed (100 days) of 6-8 µg/g;
toxic dietary exposures in natural feedstuffs, 5-40 µg/g [22,33,42,90].
The minimum rate of lifetime exposure under optimal conditions to produce
lesions in rats is 0.8 to 1.0 µg/g of purified feed, or about 10 times the
threshold requirement for optimal nutrition under similar conditions.

Acute

The acute toxicosis of selenium compounds was recognized during the
1930s. The first fatal case of selenosis occurred in a 3-year-old boy
(1966) who ingested selenious acid in a gun-bluing compound [15]. Lesions
suggested vasculature toxicity. General unawareness of the possibility
and mode of presentation of acute selenium poisoning renders its diagnosis
unlikely. Selenium disulfide, a commercial preparation for the treatment
of dandruff, has been suspected of causing nonfatal toxicity in women. Red
lipstick containing selenium has been suggested as the cause of a nervousness, metallic taste, vomiting, depression, and pharyngitis.

Accidental toxicosis from overtreatment of sheep and cattle with commercial mixtures of selenium salts and vitamin E have been reported in
Australia and New Zealand [43]. The deficiency of selenium or vitamin E in
the treated animals increased the susceptibility to selenosis [13,41].
Some fatal accidental poisonings were produced by subcutaneous doses of 10
mg of selenium as selenite per 50 kg of body weight.

Lesions of acute selenosis include central lobular fatty degeneration
of the liver, nephrosis, capillary permeability and hemorrhage, hemorrhagic
necrosis of the pancreas (especially from selenomethionine) and adrenal cortex, necrosis of lymphoid follicles, neuroangiopathy, depletion of bone
marrow, degeneration of body fat and skeletal muscle, gastroenteritis, and
dermatitis (moist and seborrheic) [42]. The inhibitory responses of the
brain are stimulated, and bradycardia is associated with a shift of the RS
axis to the left and prolonged polarization. The concentration of selenium
in the liver is generally less than 25 µg/g but may be as much as 64 µg/g.

The minimum peracute oral lethal dose of selenium as sodium selenite
in pigs (killed 1/7) was 12-18 mg/kg [92]. Signs were inappetence, emesis,

depression and hemorrhage of heart and renal cortex, and fatty degeneration
of the liver. Two pigs given 3 mg of selenium/kg intravenously as selenite
or selenomethionine with a semipurified torula yeast or dried milk diet died
2.5 and 14 hr after dosing. Lesions were pulmonary edema, myocardial and
renal ecchymosis, and pancreatic necrosis. Two pigs given 1.2 and 2.0 mg/kg
of selenium subcutaneously as selenite died in 4 hr and 5 days. Lesions were
degeneration of skeletal muscle and liver.

Cattle and sheep grazing certain primary selenium indicator plants develop a clinically acute syndrome termed blind staggers, characterized by
anorexia, emaciation, and collapse [42]. Affected cattle wander from place
to place stumbling over obstacles; forelegs are weak, and the tongue and
throat become paralyzed. Lesions include hepatic necrosis, nephritis, and
hyperemia and ulceration of the upper intestinal tract. Signs and lesions
in sheep are similar but less distinct.

Subacute

Subacute selenosis may be produced by selenium salts, selenoaminoacids
(LD_{50} = 10-15 mg/kg in the rat) or seleniferous feed (5 μg/g) [40,42]. As
biological processes tend to reduce selenium, biological oxidation of selenium is uncommon. Ruminants may reduce oxidized selenium salts to less
toxic selenide or elemental selenium. Selenides may also form insoluble
complexes with other ions or feed components. Lesions of subacute selenosis
include increased pancreatic weight and metabolic rate, and hemolysis (hypo-
chronic anemia and increased serum bilirubin). Rabbits become unable to
conjugate bilirubin. Selenium concentration in the pancreas was 7 to 9
times greater than in the liver or kidney (about 27 μg/g). South African
workers suggested a relationship between subacute selenosis and the ovine
hemolytic syndrome.

Shoats poisoned with selenite or selenomethionine for 30-90 days had
edema of the subcutis, pancreas, spinal colon, mesenteries, and meninges;
poliomalacia; degeneration and necrosis of the liver; hyaline degeneration;
and fibrinoid necrosis of arterioles and myofibrils [114]. The concentration of selenium in the liver and kidney of four pigs fed 20 μg/g was 8 to
36 μg/g (wet weight).

Rats fed semipurified feeds containing 4-6 μg/g of selenium as selenite
and with or without added methionine died after 60-100 days [40]. The animals were small and had poor hair coats. Lesions included toxic hepatitis
and increased incidence of visceral degeneration and inflammation. The

same amount of selenium added to natural feeds did not produce lesions or death.

Alkali disease is a subacute form of organic selenosis of cattle, sheep, and horses caused by feed that contains 25-50 ppm selenium [42]. It does not usually result from the consumption of highly seleniferous "primary" or "secondary" selenium accumulators but is caused by consumption of grasses that contain between about 5 and 50 µg/g selenium. Pastures where alkali disease is common generally contain very few or none of the selenium absorbers.

Affected individuals eat well but lose weight and vitality and die (sheep). Lesions include alopecia; elongated, weak, and cracked hooves (horse), and erosion of the articular cartilages (cattle). Animals become lame; the heart, liver, and kidneys are degenerated and fibrotic; conception is reduced; fetuses are reabsorbed; and the tissues contain 5-50 µg/g selenium.

Grains which cause alkali disease in cattle also produced similar signs in rats, dogs, poultry, and swine [75,90]. Similar symptoms may be produced in dogs by feeding sodium selenite. Sodium selenite also produced loss of hair, cracked hooves, emaciation and cirrhosis in growing swine. Chicken eggs from farms where the alkali disease syndrome was observed in cattle, horses, or swine have poor hatchability and produce deformed embryos. Similar lesions are produced by injection of selenate into the air sac of eggs prior to incubation.

Chronic

Chronic selenosis occurs from dietary exposure to selenium as selenite, selenate, or seleniferous feedstuffs at a rate of 1 (rat) to 44 (horse) µg/g of feed; or from water containing 0.5 to 2.0 µg/g selenium [42]. The daily intake of selenium in these conditions may be 0.5 (ruminants) to 1.0 µg/kg. Lesions include follicular skin rash, inflammation of the perivascular lymph channels; loss of hair and nails; hemolytic anemia and serum bilirubin; leukopenia; increased weight of the spleen, pancreas, and liver; fatigue, lassitude, and dizziness; and irregular menses.

Cattle (250 kg) fed 0.5 mg Se/kg three times per week became inappetent and depressed [59]. The concentration of selenium in blood was 3 µg/g: lesions included gastroenteritis and polioencephalomalacia (2/8). Sheep (45 kg) fed up to 75 mg of selenite per day developed myocardial degeneration and fibrosis, and pulmonary congestion and edema.

Lesions in rats fed semipurified feeds that contained 0.5 to 2.0 µg/g of added selenium as selenite or selenate included mild chronic hepatitis, pancreatitis and congestion (basal feed contained 0.1 µg/g selenium) [40]. The estimated minimum toxic concentration of selenium in lifetime exposure under these conditions was 0.35 µg/g based on the incidence of liver changes and 0.75 µg/g estimated from longevity and the incidence of histologic changes in the heart, kidney, and spleen. These concentrations of selenium are about 10 times the nutritional threshold for selenium and 25% of the minimum lifetime exposure to selenium in natural feedstuffs to produce similar effects under the same experimental conditions [14,40].

Reproduction

Reproductive effects may be specific or secondary to emaciation. They are the most significant economic effect of chronic selenosis [90]. Crossbreeding of rats with selenosis with normal animals indicated that the males function normally, but the female failed to conceive [123]. Young born to females with selenosis were emaciated and unable to nurse. Mice given selenium in drinking water reproduced normally until the third generation [94]. Their litters were few and small; the pups were runts, had a high mortality rate before weaning, and were infertile.

The chick embryo is extremely sensitive to selenium [89]. The hatchability of eggs is reduced by concentrations of selenium in feeds that are too low to produce symptoms of poisoning in other species. Poor hatchability of eggs, grossly deformed embryos, and anemia have been aids in locating potentially seleniferous areas [52]. Deformed embryos are characterized by missing eyes and beaks and distorted wings and feet. Inherited abnormalities, such as the Creeper mutation in hens, exaggerated the developmental derangements caused by selenium [52,90].

Selenosis also causes congenital malformation in rats, swine, and cattle. Selenosis in sheep affects the same primordium as in chicks (eyes and joints) resulting in deformed legs and impaired locomotion [90].

Vasculature

The endothelium reacts selectively to injury and may be the primary site of damage [4]. Certain nephrotoxic snake venoms, mercuric and uranium salts, and chromates have been shown to exert their renal damage by means of selective damage to the glomerular vasculature [45]. Vascular lesions of selenosis include edema, congestion, petechial to diffuse hemorrhage;

degeneration, inflammation, and fibrosis of the liver, intestine, lung, heart, kidney, spleen, and cuticle.

Arsenic interference with capillary permeability can be prevented by sulfhydryl compounds [24]. Arsenic is also an antagonist to many of the toxic effects of selenium. The overwhelming necrosis caused by subcutaneously administered cadmium salts is due to selective damage to the internal spermatic artery pampiniform plexus complex and its branches [34]. Although large amounts of zinc, sulfhydryl compounds, and estrogens protect against this vascular damage, the most potent known protector is selenium [35]. Cadmium also evokes overwhelming vascular reactions in the female including acute hemorrhagic necrosis of the placenta (rats and mice), acute, and transient hemorrhage of the prepuberal ovary (rat), which can be blocked by zinc and selenium [94,96].

Hepatic

Addition of 4-16 μg/g of selenium as selenite or selenate to semipurified feeds prevented growth and normal hair coat and produced small livers (under 2 g) and death in 10-180 days [40]. Hepatic lesions were acute toxic hepatitis. After 8-12 weeks of exposure, lesions were hyperplasia of hypatocytes and microscopic foci of toxic hyperplasia. Addition of 2-4 μg/g selenium to these feeds for 6 to 9 months produced hyperplasia of the bile and pancreatic ductal cells, new cholangeoles, and infiltration of the perivascular tissues with reticulo-histocytic cells. These lesions persisted for up to nine months after selenium was no longer added to the feed. Lesions did not become progressive or form cirrhotic, metaplastic, anaplastic, or neoplastic lesions. The effect of dietary selenium on the nails, hooves, and hair is similar in that the development of the cutaneous adnexa is altered and the effect persists after selenium is removed but is not productive or anaplastic. The physiological effect of concentrations of dietary selenium that produce lesions that persist after the insult has been removed, but do not progress, is unknown.

Addition of 0.5 to 2.0 μg/g selenium produced chronic toxic hepatitis in aged rats including degeneration of the central lobular hepatic cells and proliferation of the peripheral hepatocytes. Fibroblasts around the central vein, between the hepatic lobules, and around the bile ductal cells were proliferative, but there was no cirrhosis or fibrosis even after 30 months of exposure.

In animals that die of selenosis, increases in the concentration of selenium in the cutaneous tissues and liver are minor compared to the increase in exposure to selenium [41]. Both the liver and the germinal layer of the epidermis from animals poisoned with selenium contain more stainable sulfhydryl groups than tissues from selenium normal animals [106].

Therapy

Frost lists the toxic concentration of selenium salts in feed as 2-3 μg/g, and the lethal dose as 5-10 μg/g [24]. Schroeder and Harr found similar effects in rats by addition of selenium salts to water and commercial feed, respectively [40,95]. South Dakota workers report that forages that contain 5-10 μg/g Se cause the naturally occurring selenium poisoning known as alkali disease [90]. Repeated dosage of sheep and cattle with 250-500 μg/lb body weight caused death in 6 to 23 weeks. In these instances there was an increased concentration of selenium in the blood, urine, muscle, hair, liver, kidney, and other tissues of poisoned sheep and cattle.

Therapeutic measures for acute selenosis are symptomatic and similar to the recommendations for arsenic poisoning [42]. The use of bromobenzene is controversial and BAL is counterindicated. For cattle, recommendations include several small doses of strychnine sulfate (20-30 μg/kg) at 2- to 3-hr intervals, with lavage with tepid water two to three times a day and neostigmine (25-40 μg/kg) with intravenous glucose for 2-3 days.

Selenium poisoning from ingestion of seleniferous feed rapidly responds to supportive therapy once the exposure has been removed. Recommendations include a high-protein (25%) ration, addition of arsenicals to the feed (50-100 μg/g), and compounds that are metabolized to mercapturic acid. Sodium arsenite (5 μg/g) in drinking water modifies the reproductive and hemolytic effects of selenium and reduces renal concentration and excretion; however, the liver concentration of selenium is increased.

NEOPLASIA, PHYSIOLOGY, AND PHARMACOLOGY

Identification of chemical aspects of carcinogenesis and anticarcinogenesis and knowledge of the histochemical, morphologic, and biochemical effects of selenium have suggested possible association between optimum selenium nutrition and pharmaceutical amounts of selenium and prophylaxis and health of biochemically stressed cells.

Neoplasia

Nelson et al. reported selenium-induced hepatic cirrhosis and neoplasia
in 53 of 126 principals and 14 of 18 control rats that survived a low-pro-
tein (12%) feed for 18-24 months [72]. The feed of the principals contained
4.3 μg/g of added selenium in the form of seleniferous grain, potassium
ammonium selenide, and potassium sulfur selenide. The mixed selenides were
used as insecticides at that time and are no longer manufactured. Of the
53 principals, 11 had hepatic adenomas and 42 contained areas of adenomatous
hyperplasia. Adenomatous, neoplastic, or cirrhotic lesions were not present
in the 14 control rats. The spontaneous incidences of hepatic tumors in
18- to 24-month-old rats in the colony at that time was 0.5%. The tumors
were not carcinomas and did not metastasize.

An extensive bioassay of selenium carcinogenesis was reported by Harr
et al. [40,111]. A semipurified ration containing 0.5 to 16 μg/g added
selenium as selenate or selenite was fed to 1437 rats for up to 30 months.
Twenty to thirty percent of the rats fed the control feed or that supplemented with 0.5 to 2.0 μg/g selenium lived 24 to 30 months. Selenium did
not produce cirrhosis, fibrosis, amaplasia, or neoplasia in any rat. Experimental variables included oxidation state of selenium; concentration of
protein (22%, 12% with 0.3 DL methionine); concentration of added dietary
selenium, and husbandry regimens (continuous or partial feeding). The
basal ration contained 0.1 μg/g selenium and the control groups included a
placebo, a commercial ration, and 100 to 150 μg/g of N-2-fluorenylacetamide
(FAA), a known hepatic carcinogen.

The 12% protein feed was similar to that used by Nelson, except that
it was composed of semipurified components, rather than natural feedstuffs;
and the added selenium was an oxidized inorganic salt rather than organic
or an insecticide. Cirrhosis was not present in 119 rats fed the 12% protein feed with 4-8 μg/g added selenium. Hepatic lesions in the principals
were acute and chronic toxic hepatitis and toxic or regenerative hyperplasia. There were no metoplastic, anoplastic, or neoplastic lesions.
The incidence of hepatic carcinoma in the FAA-induced (positive control)
rats was 30%.

Volgarev and Tscherkes reported three experiments on the carcinogenic
potential of selenium [115]. They fed 200 rats in three series of experiments 4.3 or 8.6 μg/g selenium as selenate. The feeds contained 12 to 30%

casein with addition of riboflavin, methionine, α-tocopherol, cystine,
nicotinic acid, and choline in appropriate groups. There were no negative
or placebo controls included in the experiment.

In the first series (40 rats), 23 rats lived 18 months or longer; four
developed sarcomas, three hepatic carcinomas (two with metastases), and
three hepatic adenomas. Other lesions included cholangiofibrosis, oval
cell (bile duct cell) proliferations, and biliary cysts characteristic of
chronic toxic hepatitis. In the second series 60 rats were fed selenium
and 12 or 30% casein. One liver carcinoma, one hepatic adenoma, and three
sarcomas were reported. The 100 rats in series three did not have neoplasms
or precancerous lesions.

The 15 neoplasms observed occurred in approximately 120 rats that lived
18 to 24 months, and only in the first two series of experiments. They included seven sarcomas, four hepatic carcinomas, and four hepatic adenomas.
The seven sarcomas (six lymphomas and one ductal) were unrelated to reported
induction of hepatic cancers. Six of the eight liver cancers and four of
the seven sarcomas occurred in the 23 old rats from the first series. Thus,
the incidence of hepatic neoplasia in series one was 6 of 23 (26%) of the
old rats; and in series two and three it was 2 of 40 (5%) and 0 of 57 (0%).

The fourth bioassay of selenium carcinogenesis was reported by Schroeder
and associates [94-96]. Mice and rats were treated with 2 to 3 μg/g selenium as selenate or selenite in the drinking water. The 90% survival time
of rats supplemented with selenate was 1113 days compared to 1118 days for
the nonsupplemented rats. The selenium-treated rats were 3-7% heavier than
the nontreated controls. Despite the heavy supplementation with near-lethal
amounts of selenium salts (about 25 times basal intake), the concentration
of selenium in kidney, liver, heart, lung, spleen, and erythrocytes of
selenate-supplemented rats was only one-half to twice as much as in tissue
from control rats.

Approximately 70% of the rats were autopsied and 65% of those were
sectioned. Hepatic cirrhosis or neoplasia were not present. Fifty of the
rats did have neoplasms, 20 in the control group and 30 in the selenate-
treated group; 21 mammary carcinomas (10 control, 11 treated), 25 sarcomas
(8 control, 17 treated), and 6 leukemia (2 control, 4 treated). The increase
in sarcomas was probably caused by the degenerative effects of selenium or
aftereffects of chronic murine pneumonia rather than by the neoplastic potential of selenium.

The occurrence of hepatic tumors in the Nelson and Volgerev experiments apparently resulted from concomitant or unique factors related to cirrhosis rather than to exposure to selenium. The sarcomas may have been granulomatous growths caused by an epizootic of chronic murine pneumonia 12-18 months before autopsy.

Despite initial reports of selenium as a cirrhotic and carcinogenic agent, chronic experimental exposure of rats and mice fed several diets containing selenium salts did not induce cirrhosis or neoplasia, in spite of considerable effort over more than 15 years. During the same period of time, selenium salts have been used prophylatically and therapeutically in ruminates, omnivores, and carnivores throughout the world [43,68]. In addition, sections of the world, including the north central and Rocky Mountain regions of the United States, are geologically rich in selenium, and forage crops and plants in these areas may contain more than 10 $\mu g/g$ selenium [90]. Ruminants and horses in these areas may die from consuming enough selenium in forages to develop selenium toxicity. Despite these widespread experimental, field, and natural chronic and toxic exposures to selenium in feed, water, shampoo, and industrial plants, there has been no increase in the incidence of neoplasia in any of the treated or exposed species, and in New Zealand the incidence of intestinal cancer in treated sheep has decreased [26].

Anticarcinogenicity and Pharmacology

Demographic and experimental observations of Schamberger and associates [98-102] support the concept of pharmacologic and medical uses of selenium salts. They found an inverse correlation between the incidence of cancer deaths, the concentration of selenium in the patient's serum, and in the geographic area of the city. The selenium contents of diets of 17 paired human males with and without gastric cancer were compared and related to dietary antioxidants and food preservatives. Patients with gastrointestinal cancer or metastases to gastrointestinal organs had significantly lower levels of selenium in the blood than normal patients. No elevations of selenium in the blood of cancer patients were noted. The authors postulated that selenium acted to prevent attachment of the carcinogen to DNA sites.

The unique ability of selenium to reduce methylene blue was reported by Schrauzer and Rhead, who suggested that this ability might provide a

basis for the presence or susceptibility to testing for cancer [93]. In
studies of lipid therapy based on the types of lipid imbalance in cancer
patients, Revici reported that the most satisfactory and reproducible pal-
liative effects were obtained by using synthetic lipids containing bivalent
selenium [86].

The demonstration of the relationship of selenium to human cancer is
limited to demographic studies and comparisons of blood levels of selenium
in patients with and without malignancies. However, Weisberger and Suhrland
discussed the effect of selenium cystine on leukemia, and Chu and Davidson
listed selenium compounds among potential antitumor agents [18,117].
Shamberger also reported on the effect of adding sodium selenide to cancer-
inducing preparations of anthracene compounds or adding sodium selenite to
the feed of rats exposed to anthracene compounds [98]. Rats fed dietary
selenite and those treated with preparations of anthracene compounds with
added selenide developed fewer skin papillomas than rats treated with anthracene compounds without added selenide. Johnston studied the effect of
selenium on the induction of cancer by 2-N-fluorenyl-acetamide (FAA) and
diethyenitrosamine (DEN) in selenium-depleted rats over a restricted exposure period [40]. Because of widely varying rates of feed consumption by
the principal and control groups and the high incidence of neoplasia in all
the exposed groups, results were confusing.

Nutritional requirements for dietary selenium are 50 to 200 µg/g in
the diet or 1 mg/month per 50 kg body weight [14,40]. Additional but nontoxic amounts of selenium will affect the function of cellular membranes,
the availability of metallic ions and electron-deficient radicals, vascularization, and erythrocyte, muscle, and reproductive physiology.

Assay of the concentration of selenium in tissues from selenium-deficient, supplemented, and poisoned individuals indicates that there is a
strong hemostatic mechanism for selenium and that the concentration of selenium in tissue does not increase more than threefold within the capacity
of the individual to resist fatal poisoning [14,41].

Clinical observation of the efficacy of parenterally administered mixtures of selenium and vitamin E to animals with adequate dietary selenium
and vitamin E indicates that pharmacologic amounts of selenium and an additional 1-3 mg/50 kg per month improve functioning of reproductive, muscular,
and vascular systems [19,50,80]. Efficacy is claimed in clinical cases
of stiffness, lameness, myositis, hepatic degeneration, infertility, loss

of condition of the hair coat, and poor growth [43]. In lameness and stiffness in racehorses and dogs and in breeding rams and bulls, a severe strain
is put on the musculoskeletal system and in particular on the animal's
joints. Veterinary practitioners in selenium-deficient areas believe that
a definite improvement is seen in those animals given supplements. Theoret-
ically this could be related to better vascularization of the tissues or to
the role of glutathione peroxidase in maintenance of cellular membranes.

Physiology

The historical relationship between vitamin E and selenium suggested
that selenium has a similar function as an antioxidant [109]. Mechanisms
postulated for the antioxidant activity of selenium compounds include peroxidation inhibition, peroxide decomposition, free radical scavenging,
repair of damaged molecular sites, and catalysis of sulfhydryl compounds.
However, as these functions do not explain some observations, selenium
probably has a more subtle biological function. As examples, dietary selenium does not protect rats exposed to chronic whole-body irradiation, or
to lipid peroxidation induced in mitochondria by ascorbate, oxidized plus
reduced glutathione, or iron; and there are selenium-responsive diseases
(myopathy) in animals adequately supplied with vitamin E. Several other
hypotheses have been suggested to account for the effects of selenium.
These include the following.

The observation that selenium protected erythrocytes against oxidative
hemolysis characteristic of vitamin E deficiency indicated that selenium
was involved in glutathione metabolism but had no effect on the generation
of glutathione. The fault was in the utilization of glutathione in selenium deficiency in that the enzyme glutathione peroxidase contained selenium [91].

The biologically active form of selenium may be a selenide at the
active site of nonheme iron proteins. Significant portions of the selenium
in the mitochondrial and microsimal fractions of rat liver in animals given
adequate vitamin E are at the selenide valence. In contrast, there is
little selenide in the subcellular fractions of rats deficient in vitamin E.
Thus, vitamin E may protect an unstable selenide from oxidation.

The effect of dietary vitamin E or selenium on rat liver mitochondria
swollen by various chemical agents in vitro suggests that selenium plays a
role in the electron transport chain at the level of cytochrome c [56,57].

Dietary vitamin E, but not selenium, protects mitochondria against swelling promoted by lipid peroxidation. Selenium was a highly effective catalyst for the reduction of cytochrome c by glutathione in a chemically defined system. Selenium may function in vivo by facilitating the transfer of electrons from glutathione or other sulfur compounds into the cytochrome system. A possible role for selenium in biologic electron-transfer reactions is also supported by the observation that a selenoprotein was a component of the clostridial glycine reductase system. Moreover, Whanger et al. have found a selenoprotein in lamb muscle that contains a heme group identical to that of cytochrome c [121].

These theories of selenium function are closely related and may be variations of a single action. The role for selenium in glutathione peroxidase could explain the antioxidant effects of selenium and mitochondrial swelling and contraction, nonheme iron proteins, and electron transfer. Future work may contribute a great deal of information about these and other possible functions of selenium in such areas as ubiquinones and aging.

REFERENCES

1. W. H. Allaway, Selenium in the Food Chain, Cornell Vet., 63, 151-170 (1973).

2. W. H. Allaway, Sulphur-Selenium Relationships in Soils and Plants, Sulfur Inst. J., 6 (3), 3-5 (1970).

3. W. H. Allaway, J. Kubota, F. Losee, and M. Roth, Selenium, Molybdenum, and Vanadium in Human Blood, Arch. Environ. Health, 16, 342-348 (1968).

4. R. Altschul, Endothelium. Its Development, Morphology, Function, and Pathology, Macmillan, New York, 1954, p. 171.

5. A. J. Bergman, B. Beckwith, and C. G. Ray (eds.), Sudden Infant Death Syndrome, Proc. of the Second International Conference on Causes of Sudden Death in Infants, University of Washington Press, Seattle, 1970, p. 268.

6. E. Bonucci and R. Sadun, Experimental Calcification of the Myocardium: Ultrastructural and Histochemical Investigations, Amer. J. Pathol., 71, 167-192 (1973).

7. D. G. Brown and R. F. Burk, Selenium Retention in Tissues and Sperm of Rats Fed a Torula Yeast Diet, J. Nutr., 103, 102-108 (1973).

8. D. G. Brown, R. F. Burk, R. J. Seely, and K. W. Kiker, Effect of Dietary Selenium on the Gastrointestinal Absorption of $^{75}SeO_3$ in the Rat, Int. J. Vit. Nutr., 42, 588-591 (1972).

9. T. C. Broyer, C. M. Johnson, and R. P. Huston, Selenium and Nutrition of Astragalus. I. Effects of Selenite or Selenate Supply on Growth and Selenium Content, Plant Soil, 36, 635-649 (1972).

10. J. G. Buchanan-Smith, E. C. Nelson, and A. D. Tillman, Effect of Vita-
     min E and Selenium Deficiencies on Lysosomal and Cytoplasmic Enzymes
     in Sheep Tissues, J. Nutr., 99, 387-394 (1969).

11. J. G. Buchanan-Smith, B. A. Sharp, and A. D. Tillman, Tissue Selenium
     Concentrations in Sheep Fed a Purified Diet, Can. J. Physiol. Pharma-
     col., 49, 619-621 (1971).

12. R. Burk, Jr., W. N. Pearson, R. P. Wood II, and F. Viteri, Blood Se-
     lenium Levels and in Vitro Red Blood Cell Uptake of ^{75}Se in Kwashior-
     kor, Amer. J. Clin. Nutr., 20, 723-733 (1967).

13. R. F. Burk, D. G. Brown, R. J. Seely, and C. C. Scaief, III, Influence
     of Dietary and Injected Selenium on Whole-Body Retention, Route of
     Excretion, and Tissue Retention of $^{75}SeO_3^{2-}$ in the Rat, J. Nutr., 102,
     1049-1055 (1972).

14. R. F. Burk, R. J. Seely, and K. W. Kiker, Selenium: Dietary Threshold
     for Urinary Excretion in the Rat, Proc. Soc. Exp. Biol. Med., 142,
     214-216 (1973).

15. R. F. Carter, Acute Selenium Poisoning, Med. J. Austral., 1, 525-528
     (1966).

16. E. E. Cary and W. H. Allaway, Selenium Content of Field Crops Grown
     on Selenite-Treated Soils, Agron. J., 65, 922-925 (1973).

17. R. W. Chen, P. A. Wagner, W. G. Hoekstra, and H. E. Ganther, Affinity
     Labelling Studies with [109]Cadmium in Cadmium-Induced Testicular Injury
     in Rats, J. Reprod. Fertil., 38, 293-306 (1974).

18. S. H. Chu and D. D. Davidson, Potential Antitumor Agents. α- and β-2'-
     Deoxy-6-selenoguanosine and Related Compounds, J. Med. Chem., 15, 1088-
     1089 (1972).

19. M. W. Colton, Selenium-tocopherol Treatment of Chronic Lameness in
     Dogs, Mod. Vet. Pract., 46 (10), 92 (1965).

20. R. G. Crystal, Elemental Selenium: Structure and Properties, in Or-
     ganic Selenium Compounds: Their Chemistry and Biology (D. L. Klayman
     and W. H. H. Günther, eds.), Wiley-Interscience, New York, 1973, pp.
     13-27.

21. W. E. Davis and Associates, National Inventory of Sources and Emissions.
     Barium, Boron, Copper, Selenium, and Zinc. Selenium. Sec. IV, Envi-
     ronmental Protection Agency, Office of Air Programs, Contract No. 68-
     02-0100, W. E. Davis and Associates, Leawood, Kansas, 1972, p. 50.

22. H. C. Dudley and J. W. Miller, Toxicology of Selenium. IV. Effects of
     Exposure to Hydrogen Selenide, U.S. Public Health Rep., 52, 1217-1231
     (1937).

23. R. P. Eaton and D. M. Kipnis, Incorporation of [75]Se Selenomethionine
     into a Protein Component of Plasma Very Low Density Lipoprotein in
     Man, Diabetes, 21, 744-753 (1972).

24. D. V. Frost, The Two Faces of Selenium. Can Selenophobia Be Cured?
     C. R. C. Crit. Rev. Toxicol., 1, 467-514 (1972).

25. J. M. Fuller, E. D. Beckman, M. Goldman, and L. K. Bustad, Selenium
     Determination in Human and Swine Tissues by X-ray Emission Spectometry,
     in Symposium: Selenium in Biomedicine. First International Symposium,
     Oregon State University, 1966 (O. H. Muth, ed.), AVI, Westport, 1967,
     pp. 119-124.

26. B. J. Gabbedy, Effect of Selenium on Wool Production, Body Weight, and
     Mortality of Young Sheep in Western Australia, Austral. Vet. J., 47,
     318-322 (1971).

27. B. J. Gabbedy and R. B. Richards, White Muscle Disease in a Foal, Aus-
     tral. Vet. J., 46, 111-112 (1970).

28. H. E. Ganther, Selenium: The Biological Effects of a Highly Active
     Trace Substance, in Trace Substances in Environmental Health. Pro-
     ceedings of University of Missouri's Fourth Annual Conference on
     Trace Substances in Environmental Health, June 23-25, 1970 (D. D.
     Hemphill, ed.), University of Missouri, Columbia, 1971, pp. 211-221.

29. H. R. Geering, E. E. Cary, L. H. P. Jones, and W. H. Allaway, Solu-
     bility and Redox Criteria for the Possible Forms of Selenium in Soils,
     Soil Sci. Soc. Amer. Proc., 32, 35-40 (1968).

30. P. L. Gile and H. W. Lakin, Effect of Different Soil Colloids on the
     Toxicity of Sodium Selenite to Millet, J. Agr. Res., 63, 560-581 (1941).

31. G. Gissel-Nielsen and B. Bisbjerg, The Uptake of Applied Selenium by
     Agricultural Plants. 2. The Utilization of Various Selenium Compounds,
     Plant Soil, 32, 382-396 (1970).

32. M. W. Glenn, J. L. Martin, and L. M. Cummins, Sodium Selenate Toxico-
     sis: The Distribution of Selenium Within the Body After Prolonged
     Feeding of Toxic Quantities of Sodium Selenate to Sheep, Amer. J. Vet.
     Res., 25, 1495-1499 (1964).

33. C. L. Gries and M. L. Scott, Pathology of Selenium Deficiency in the
     Chick, J. Nutr., 102, 1287-1296 (1972).

34. S. A. Gunn and T. C. Gould, Cadmium and Other Mineral Elements, in The
     Testis, Vol. III. Influencing Factors (A. D. Johnson, W. R. Gomes,
     and N. D. Vandemark, eds.), Academic Press, New York, 1970, pp. 377-
     481.

35. S. A. Gunn, T. C. Gould, and W. A. D. Anderson, Incorporation of Se-
     lenium into Spermatogenic Pathway in Mice, Proc. Soc. Exp. Biol. Med.,
     124, 1260-1263 (1967).

36. S. A. Gunn, T. C. Gould, and W. A. D. Anderson, Mechanisms of Zinc,
     Cysteine, and Selenium Protection Against Cadmium-Induced Vascular
     Injury to Mouse Testis, J. Reprod. Fertil., 15, 65-70 (1968).

37. D. M. Hadjimarkos and T. R. Shearer, Selenium Concentration in Human
     Saliva, Amer. J. Clin. Nutr., 24, 1210-1211 (1971).

38. A. W. Halverson, P. L. Guss, and O. E. Olson, Effect of Sulfur Salts
     on Selenium Poisoning in the Rat, J. Nutr., 77, 459-464 (1962).

39. K. A. Handreck and K. D. Godwin, Distribution in the Sheep of Selenium
     Derived from 75Se-Labelled Ruminal Pellets, Austral. J. Agr. Res., 21,
     71-84 (1970).

40. J. R. Harr, J. F. Bone, I. J. Tinsley, P. H. Weswig, and R. S. Yamamoto,
     Selenium Toxicity in Rats. II. Histopathology, in Symposium: Selenium
     in Biomedicine. First International Symposium, Oregon State University,
     1966 (O. H. Muth, ed.), AVI, Westport, 1967, pp. 153-178.

41. J. R. Harr, J. H. Exon, P. H. Weswig, and P. D. Whanger, Relationship
     of Dietary Selenium Concentration; Chemical Cancer Induction; and
     Tissue Concentration of Selenium in Rats, Clin. Toxicol., 6, 487-495
     (1973).

42. J. R. Harr and O. H. Muth, Selenium Poisoning in Domestic Animals and
     Its Relationship to Man, Clin. Toxicol., 5, 175-186 (1972).

43. W. J. Hartley, Levels of Selenium in Animal Tissues and Methods of Se-
     lenium Administration, in Symposium: Selenium in Biomedicine. First
     International Symposium, Oregon State University, 1966 (O. H. Muth,
     ed.), AVI, Westport, 1967, pp. 79-96.

44. M. Heinrich, Jr. and F. E. Kelsey, Studies on Selenium Metabolism.
     The Distribution of Selenium in the Tissues of the Mouse, J. Pharmacol.
     Exp. Ther., 114, 28-32 (1955).

45. L. L. Hopkins, Jr. and A. S. Majaj, Selenium in Human Nutrition, in
     Symposium: Selenium in Biomedicine. First International Symposium,
     Oregon State University, 1966 (O. H. Muth, ed.), AVI, Westport, 1967,
     pp. 203-214.

46. K. J. Jenkins, R. C. Dickson, and M. Hidiroglou, Effects of Various
     Selenium Compounds on the Development of Muscular Dystrophy and Other
     Vitamin E Dyscrasies in the Chick, Can. J. Physiol. Pharmacol., 48,
     192-196 (1970).

47. L. S. Jensen, Selenium Deficiency and Impaired Reproduction in Japanese
     Quail, Proc. Soc. Exp. Biol. Med., 128, 970-972 (1968).

48. W. K. Johnston, The Effect of Selenium on Chemical Carcinogenicity in
     the Rat, M.S. Thesis, Oregon State University, Corvallis, 1974.

49. G. B. Jones and K. O. Godwin, Studies on the Nutritional Role of Sele-
     nium. I. The Distribution of Radioactive Selenium in Mice, Austral.
     J. Agr. Res., 14, 716-723 (1963).

50. D. L. Klayman, Selenium Compounds as Potential Chemotherapeutic Agents,
     in Organic Selenium Compounds: Their Chemistry and Biology (D. L.
     Klayman and W. H. H. Günther, eds.), Wiley-Interscience, New York,
     1973, pp. 727-761.

51. J. Kubota, W. H. Allaway, D. L. Carter, E. E. Cary, and V. A. Lazar,
     Selenium in Crops in the United States in Relation to Selenium-Re-
     sponsive Diseases of Animals, J. Agr. Food Chem., 15, 448-453 (1967).

52. G. Kury, L. H. Rev-Kury, and R. J. Crosby, The Effect of Selenous Acid
     on the Hematopoietic System of Chicken Embryos, Toxicol. Appl. Pharma-
     col., 11, 449-458 (1967).

53. K. L. Kuttler and D. W. Marble, Prevention of White Muscle Disease in
     Lambs by Oral and Subcutaneous Administration of Selenium, Amer. J.
     Vet. Res., 21, 437-440 (1960).

54. H. W. Lakin and D. F. Davidson, The Relation of the Geochemistry of
     Selenium to Its Occurrence in Soils, in Symposium: Selenium in Bio-
     medicine. First International Symposium, Oregon State University, 1966
     (O. H. Muth, ed.), AVI, Westport, 1967, pp. 27-56.

55. W. Landauer, Studies on the Creeper Fowl. XIII. The Effect of Sele-
     nium and the Asymmetry of Selenium-Induced Malformations, J. Exp. Zool.,
     83, 431-443 (1940).

56. O. A. Levander, V. C. Morris, and D. J. Higgs, Acceleration of Thiol-
     Induced Swelling of Rat Liver Mitochondria by Selenium, Biochemistry,
     12, 4586-4590 (1973).

57. O. A. Levander, V. C. Morris, and D. J. Higgs, Selenium as a Catalyst
     for the Reduction of Cytochrome c by Glutathione, Biochemistry, 12,
     4591-4595 (1973).

58. K. Liebscher and H. Smith, Essential and Nonessential Trace Elements. A Method of Determining Whether an Element Is Essential or Nonessential in Human Tissue, Arch. Environ. Health, 17, 881-890 (1968).

59. D. D. Maag and M. W. Glenn, Toxicity of Selenium: Farm Animals, in Symposium: Selenium in Biomedicine. First International Symposium, Oregon State University, 1966 (O. H. Muth, ed.), AVI, Westport, 1967, pp. 127-140.

60. K. P. McConnell, Distribution and Excretion Studies in the Rat After a Single Subtoxic Subcutaneous Injection of Sodium Selenate Containing Radioselenium, J. Biol. Chem., 141, 427-437 (1941).

61. K. P. McConnell and G. J. Cho, Active Transport of L-Selenomethionine in the Intestine, Amer. J. Physiol., 213, 150-156 (1967).

62. K. P. McConnell and G. J. Cho, Transmucosal Movement of Selenium, Amer. J. Physiol., 208, 1191-1195 (1965).

63. K. P. McConnell and R. G. Martin, Bilary Excretion of Selenium in the Dog After Administration of Sodium Selenate Containing Radioselenium, J. Biol. Chem., 194, 183-190 (1952).

64. K. E. M. McCoy and P. H. Weswig, Some Selenium Responses in the Rat Not Related to Vitamin E, J. Nutr., 98, 383-389 (1969).

65. D. F. L. Money, Cot Deaths and Deficiency of Vitamin E and Selenium, Brit. Med. J., 4, 559 (1971).

66. A. O. Moxon, O. E. Olson, and W. V. Searight, Selenium in Rocks, Soils, and Plants. South Dakota Agricultural Experiment Station Technical Bulletin No. 2, Agricultural Experiment Station, South Dakota State College of Agriculture and Mechanic Arts, Brookings, 1939, p. 94.

67. O. H. Muth, Discussion of Clinical Aspects of Selenium Deficiency, in Symposium: Selenium in Biomedicine. First International Symposium, Oregon State University, 1966 (O. H. Muth, ed.), AVI, Westport, 1967, p. 225.

68. O. H. Muth, The Biomedical Aspects of Selenium, in Symposium: Selenium in Biomedicine. First International Symposium, Oregon State University, 1966 (O. H. Muth, ed.), AVI, Westport, 1967, pp. 3-6.

69. O. H. Muth, White Muscle Disease (Myopathy) in Lambs and Calves. I. Occurrence and Nature of the Disease Under Oregon Conditions, J. Amer. Vet. Med. Ass., 126, 355-361 (1955).

70. O. H. Muth, P. H. Weswig, P. D. Whanger, and J. E. Oldfield, Effect of Feeding Selenium-Deficient Ration to the Subhuman Primate (Saimiri sciureus), Amer. J. Vet. Res., 32, 1603-1605 (1971).

71. National Research Council, Agricultural Board, Committee on Animal Nutrition, Subcommittee on Selenium (J. E. Oldfield, Chairman), Selenium in Nutrition, National Academy of Sciences, Washington, D.C., 1971, p. 79.

72. A. A. Nelson, O. G. Fitzhugh, and H. O. Calvery, Liver Tumors Following Cirrhosis Caused by Selenium in Rats, Cancer Res., 3, 230 (1943).

73. M. C. Nesheim and M. L. Scott, Nutritional Effects of Selenium Compounds in Chicks and Turkeys, Fed. Proc., 20, 674-678 (1961).

74. H. E. Oksanen, Selenium Deficiency: Clinical Aspects and Physiological Responses in Farm Animals, in Symposium: Selenium in Biomedicine. First International Symposium, Oregon State University, 1966 (O. H. Muth, ed.), AVI, Westport, 1967, pp. 215-299.

75. O. E. Olson, C. A. Dinkel, and L. D. Kamstra, A New Aid in Diagnosing Selenium Poisoning, S. Dakota Farm Home Res., 6, 12-14 (1954).

76. O. E. Olson and A. W. Halverson, Effect of Linseed Oil Meal and Arsenicals on Selenium Poisoning in the Rat, Proc. S. Dakota Acad. Sci., 33, 90-94 (1954).

77. O. E. Olson and C. W. Jensen, The Absorption of Selenate and Selenite Selenium by Colloidal Ferric Hydroxide, Proc. S. Dakota Acad. Sci., 20, 115-121 (1940).

78. O. E. Olson, E. I. Whitehead, and A. L. Moxon, Occurrence of Soluble Selenium in Soils and Its Availability to Plants, Soil Sci., 54, 47-53 (1942).

79. I. S. Palmer, R. P. Gunsalus, A. W. Halverson, and O. E. Olson, Trimethylselenonium Ion as a General Excretory Product from Selenium Metabolism in the Rat, Biochim. Biophys. Acta, 208, 260-266 (1970).

80. J. Pařízek, I. Oštádalová, J. Kalousková, A. Babický, and J. Beneš, The Detoxifying Effects of Selenium Interrelations Between Compounds of Selenium and Certain Metals, in Newer Trace Elements in Nutrition (W. Mertz and W. E. Cornatzer, eds.), Dekker, New York, 1971, pp. 85-122.

81. J. Pařízek, I. Oštádalová, J. Kalousková, A. Babický, L. Pavlik, and B. Bibr, Effect of Mercuric Compounds on the Maternal Transmission of Selenium in the Pregnant and Lactating Rat, J. Reprod. Fertil., 25, 157-170 (1971).

82. E. L. Patterson, R. Milstrey, and E. L. R. Stokstad, Effect of Selenium in Preventing Exudative Diathesis in Chicks, Proc. Soc. Exp. Biol. Med., 95, 617-620 (1957).

83. H. W. Pendell, O. H. Muth, J. E. Oldfield, and P. H. Weswig, Tissue Fatty Acids in Normal and Myopathic Lambs, J. Animal Sci., 29, 94-98 (1969).

84. D. G. Penington, Assessment of Platelet Production with [75]Se Selenomethionine, Brit. Med. J., 4, 782-784 (1969).

85. M. M. Rahman, R. E. Davies, C. W. Deyoe, and J. R. Couch, Selenium and Torula Yeast in Production of Exudative Diathesis in Chicks, Proc. Soc. Exp. Biol. Med., 105, 227-230 (1960).

86. E. Revici, The Control of Cancer with Lipids, Dir. Path. Conf. Beth David Hospital, New York, May, 1955.

87. M. Rhian and A. L. Moxon, Chronic Selenium Poisoning in Dogs and Its Prevention by Arsenic, J. Pharmacol. Exp. Ther., 78, 249-264 (1943).

88. M. E. Roberts, Antiinflammation Studies. II. Antiinflammatory Properties of Selenium, Toxicol. Appl. Pharmacol., 5, 500-506 (1963).

89. D. S. F. Robertson, Selenium, a Possible Teratogen?, Lancet, 1, 518 (1970).

90. I. Rosenfeld and O. A. Beath, Selenium: Geobotany, Biochemistry, Toxicity, and Nutrition, Academic Press, New York, 1964, p. 411.

91. J. T. Rotruck, A. L. Pope, H. E. Ganther, A. B. Swanson, D. G. Hafeman, and W. G. Hoekstra, Selenium: Biochemical Role as a Component of Glutathione, Science, 179, 588-590 (1973).

92. G. R. Ruth and J. F. van Vleet, Experimentally Induced Selenium-Vitamin E Deficiency in Growing Swine: Selective Destruction of Type I Skeletal Muscle Fibers, Amer. J. Vet. Res., 35, 237-244 (1974).

93. G. N. Schrauzer and W. J. Rhead, Interpretation of the Methylene Blue Reduction Test of Human Plasma and the Possible Cancer Protection Effect of Selenium, Experientia, 27, 1069-1071 (1971).

94. H. A. Schroeder and M. Mitchener, Selenium and Tellurium in Mice. Effects on Growth, Survival, and Tumors, Arch. Environ. Health, 24, 66-71 (1972).

95. H. A. Schroeder and M. Mitchener, Selenium and Tellurium in Rats: Effect on Growth, Survival, and Tumors, J. Nutr., 101, 1531-1540 (1971).

96. H. A. Schroeder and M. Mitchener, Toxic Effects of Trace Elements on Reproduction of Mice and Rats, Arch. Environ. Health, 23, 102-106 (1971).

97. M. L. Scott, G. Olson, L. Krook, and W. R. Brown, Selenium-Responsive Myopathies of Myocardium and of Smooth Muscle in the Young Poult, J. Nutr., 91, 573-583 (1967).

98. R. J. Shamberger, Relation of Selenium to Cancer. I. Inhibitory Effect of Selenium on Carcinogenesis, J. Nat. Cancer Inst., 44, 931-936 (1970).

99. R. J. Shamberger and D. V. Frost, Possible Protective Effect of Selenium Against Human Cancer, Can. Med. Ass. J., 100, 682 (1969).

100. R. J. Shamberger and G. Rudolph, Protection Against Cocarcinogenesis by Antioxidants, Experientia, 22, 116 (1966).

101. R. J. Shamberger, E. Rukovena, A. K. Longfield, S. A. Tytko, S. Deodhar, and C. E. Willis, Antioxidants and Cancer. I. Selenium in the Blood of Normals and Cancer Patients, J. Nat. Cancer Inst., 50, 863-870 (1973).

102. R. J. Shamberger and C. E. Willis, Selenium Distribution and Human Cancer Mortality, Crit. Rev. Clin. Lab. Sci., 2, 211-221 (1971).

103. L. G. Sillen and A. E. Martell, Stability Constants of Metal-Ion Complexes. Special Publication No. 17, 2nd Ed., The Chemical Society, London, 1964, p. 754.

104. F. F. Smith, Use and Limitations of Selenium as an Insecticide, in Selenium in Agriculture. U.S. Department of Agriculture Handbook No. 200 (M. S. Anderson, H. W. Lakin, K. C. Beeson, F. F. Smith, and E. Thacker, eds.), U.S. Government Printing Office, Washington, D. C., 1961, pp. 41-45.

105. E. Søndegaard, Selenium and Vitamin E Interrelationships, in Symposium: Selenium in Biomedicine. First International Symposium, Oregon State University, 1966 (O. H. Muth, ed.), AVI, Westport, 1967, pp. 365-381.

106. L. H. Sprinker, J. R. Harr, P. M. Newberne, P. D. Whanger, and P. H. Weswig, Selenium Deficiency Lesions in Rats Fed Vitamin E-Supplemented Rations, Nutr. Rep. Int., 4, 335-340 (1971).

107. T. C. Stadtman, Selenium Biochemistry. Proteins Containing Selenium Are Essential Components of Certain Bacterial and Mammalian Enzyme Systems, Science, 183, 915-922 (1974).

108. P. R. Sweeny and R. G. Brown, Ultrastructural Changes in Muscular Dystrophy. I. Cardiac Tissue of Piglets Deprived of Vitamin E and Selenium, Amer. J. Pathol., 68, 479-492 (1972).

109. A. L. Tappel, Free-Radical Lipid Peroxidation Damage and Its Inhibition by Vitamin E and Selenium, Fed. Proc., 24, 73-78 (1965).

110. J. N. Thompson and M. L. Scott, Impaired Lipid and Vitamin E Absorption Related to Atrophy of the Pancreas in Selenium-Deficient Chicks, J. Nutr., 100:797-809 (1970).

111. I. J. Tinsley, J. R. Harr, J. F. Bone, P. H. Weswig, and R. S. Yamamoto, Selenium Toxicity in Rats. I. Growth and Longevity, in Symposium: Selenium in Biomedicine. First International Symposium, Oregon State University, 1966 (O. H. Muth, ed.), AVI, Westport, 1967, pp. 141-152.

112. A. L. Trapp, K. K. Keahey, D. L. Whitenack, and C. K. Whitehair, Vitamin E-Selenium Deficiency in Swine: Differential Diagnosis and Nature of Field Problem, J. Amer. Vet. Med. Ass., 157, 289-300 (1970).

113. J. F. Van Vleet, W. Carlton, and H. J. Olander, Hepatosis Dietetica and Mulberry Heart Disease Associated with Selenium Deficiency in Indiana Swine, J. Amer. Vet. Med. Ass., 157, 1208-1219 (1970).

114. J. F. Van Vleet, K. B. Meyer, and H. J. Olander, Control of Selenium-Vitamin E Deficiency in Growing Swine by Parenteral Administration of Selenium-Vitamin E Preparations to Baby Pigs or to Pregnant Sows and Their Baby Pigs, J. Amer. Vet. Med. Ass., 163, 452-456 (1973).

115. M. N. Volgarev and L. A. Tscherkes, Further Studies in Tissue Changes Associated with Sodium Selenate, in Symposium: Selenium in Biomedicine. First International Symposium, Oregon State University, 1966 (O. H. Muth, ed.), AVI, Westport, 1967, pp. 179-184.

116. M. E. Wastell, R. C. Ewan, M. W. Vorhies, and V. C. Speer, Vitamin E and Selenium for Growing and Finishing Pigs, J. Animal Sci., 34, 969-973 (1972).

117. A. S. Weisberger and L. G. Suhrland, The Effect of Selenium Cystine on Leukemia, Blood, 11, 19 (1956).

118. H. V. Weiss, M. Koide, and E. D. Goldberg, Selenium and Sulfur in a Greenland Ice Sheet: Relation to Fossil Fuel Consumption, Science, 172, 261-263 (1971).

119. B. B. Westfall and M. I. Smith, Chronic Selenosis. IV. Selenium in the Hair as an Index of the Extent of Its Deposition in the Tissues in Chronic Poisoning, Nat. Inst. Health Bull., 174, 45-49 (1940).

120. P. H. Weswig, S. A. Roffler, M. A. Arnold, O. H. Muth, and J. E. Oldfield, In Vitro Uptake of Selenium-75 by Blood from Ewes and Their Lambs on Different Selenium Regimens, Amer. J. Vet. Res., 27, 128-131 (1966).

121. P. D. Whanger, N. D. Pedersen, and P. H. Weswig, Selenium Proteins in Ovine Tissues. II. Spectral Properties of a 10,000-Molecular-Weight Selenium Protein, Biochem. Biophys. Res. Commun., 53, 1031-1035 (1973).

122. P. D. Whanger, P. H. Weswig, O. H. Muth, and J. E. Oldfield, Tissue Lactic Dehydrogenase, Glutamic-Oxalacetic Transaminase, and Peroxidase Changes of Selenium-Deficient Myopathic Lambs, J. Nutr., 99, 331-337 (1969).

123. S. H. Wu, J. E. Oldfield, O. H. Muth, P. D. Whanger, and P. H. Weswig, Effect of Selenium in Reproduction, Proc. Ann. Meet. W. Sect. Amer. Soc. Animal Sci., 20, 85-89 (1969).

124. T. Fukuyama and E. J. Ordal, Induced Biosynthesis of Formic Hydrogen-lyase in Iron-Deficient Cells of *Escherichia coli*, J. Bacteriol., _90_, 673-680 (1965).

19

THE RELATIONSHIP OF DIETARY SELENIUM
CONCENTRATION, CHEMICAL CANCER INDUCTION,
AND TISSUE CONCENTRATION OF SELENIUM IN RATS*

James R. Harr
Pennwalt Corporation
Rochester, New York

Jerry H. Exon, Paul H. Weswig, and Philip D. Whanger
Oregon State University
Corvallis, Oregon

Selenium has been labeled a toxicant [4,13], a carcinogen [11,14,23,24], an anticarcinogen [4,19,20], an essential trace element [8,21,25], and an effective therapeutic agent in treating several mammalian and avian diseases [6,7,9,10,12,16,17]. The well-documented toxicity and the alleged carcinogenic properties of the various organic and inorganic compounds of selenium in the absence of a recognized nutritional requirement has caused the Food and Drug Administration to restrict the use of selenium to prevent contamination of food. Recognition of the nutritional requirements, the inability to verify carcinogenicity [3,5], and the reported anticarcinogenic potential of selenium have altered the basis for restriction of this element.

Current concern is that animals exposed to selenium may accumulate excessive amounts of the element in various tissues. This concern, as well as the economic importance of the therapeutic and prophylactic uses of selenium in livestock [10] and poultry, and the nutritional requirements for selenium, indicates that the effects of various selenium compounds should be studied in more detail. Optimum regulation of selenium compounds

*Technical Paper No. 3325, Oregon Agricultural Experiment Station. Supported in part by the Selenium-Tellurium Development Association, Inc. and Oregon Agricultural Experiment Station.

will be dependent upon correlation of exposure levels with resulting adverse or beneficial effects, and accumulation of selenium in tissues.

In this study, vitamin E-supplemented, selenium-deficient rats were fed a ration containing a known chemical carcinogen, N-2-fluorenyl-acetamide (FAA), and varying amounts of selenite to determine the effect of selenium on cancer induction. Harr et al. [5] reported the histologic observations from this experiment. The incidence of induced cancer was inversely correlated to the selenite concentration in the feed. This paper reports the concentration of selenium in the liver and skeletal muscle of these rats.

METHODS

Eighty female OSU-brown rats were weaned at 35 days of age and divided into four groups of 20 each. The 40 rats in groups I and II were born from parents reared on the selenium depletion regimen of McCoy and Weswig [8]. The 40 rats in groups III and IV were from the second generation maintained on this depletion regimen. Rats in groups I and II were clinically normal. Those in groups III and IV had clinical signs of selenium deficiency [21].

The basal torula yeast feed contained 18 ppb selenium, as determined by analysis, and was supplemented with 60 ppm vitamin E, as described by McCoy and Weswig [8], and with 150 ppm FAA. Groups I, II, and III received additional selenium supplementation of 2.5, 0.5, and 0.1 ppm sodium selenite, respectively (Table 1). Husbandry and experimental protocol were as described previously by Harr et al. [5].

Selenium in skeletal muscle and liver were determined by fluorometry [1], utilizing oxygen-flask combustion, precipitation with arsenic, oxidation with nitric acid, and fluorescence with 2,3-diaminonaphthalene.

The amount of selenium ingested was estimated from the average age of the rats in each group and the average feed consumption (15 g per rat per day) for that age group. Feed intake was determined from previous observations of rats fed similar rations [22].

Skeletal muscle was estimated, from gross dissection and lipid extraction (6-9%), to be 25% of the body weight. The average amount of selenium in the skeletal muscle of the rats of each experimental group was calculated from the average body weight for each group.

TABLE 1. Occurrence of Neoplasms in Rats Fed FAA and Varying Dietary Concentrations of Added Selenium[a]

Days of Exposure to 150 ppm FAA	Group I (2.50 ppm Se)			Group II (0.50 ppm Se)			Group III (0.10 ppm Se)			Group IV (0.00 ppm Se)		
	MC[b]	LC[c]	Died	MC	LC	Died	MC	LC	Died	MC	LC	Died
1-40	-	-	-	-	-	-	-	-	3	-	-	2
41-80	-	-	-	-	-	-	-	-	-	-	-	1
81-120	-	-	-	-	-	-	1	1	1	-	-	1
121-160	-	-	-	-	-	-	4	-	2	2	3	3
161-200	-	-	-	1	1	2	4	3	4	7	-	3
201-240	4	-	-	6	-	1	3	5	7	1	1	4
241-280	5	2	4	4	7	8	-	-	-	1	2	4
281-320	2	6	8	-	4	6	2	3	3	1	-	1
Total	11	8	12	11	12	17	13	12	20	12	6	19
Killed 321 days	1	4	8	-	3	3	-	-	-	1	1	1

[a]Basal diet contained 0.018 ppm Se.

[b]MC, mammary adenocarcinoma.

[c]LC, hepatoma.

RESULTS

The average concentration of selenium in the liver of rats from each experimental group increased (0.40 to 7.35 ppm) with the addition of selenite to the basal ration (Table 2). The concentration of selenium in the liver of group I rats was 1.3 times greater than that of group II rats. The concentration of selenium in the feed of group I animals was 4.9 times greater than that of group II rations. Thus the ratio of the increase in selenium concentration in the liver to feed between groups I and II was 0.27. The same ratios between groups II and III and groups III and IV (Se deficient) were 0.33 and 1.45, respectively. The concentration of selenium in the skeletal muscle of these animals was unaffected by dietary supplementation with selenite.

The portion of the ingested selenium that was retained by the tissues decreased from 110% to 1% with the addition of selenite to the feed. The amount of selenium retained by the rats was 0.50 mg/kg in group I, 0.42 mg/kg in group II, 0.36 mg/kg in group III, and 0.22 mg/kg in group IV (Se deficient).

The concentration of selenium in the liver was inversely related to the incidence of FAA-induced mammary and hepatic cancer and early mortality (Tables 1 and 2). After 200 days on the experimental rations, 38 of the 40 rats in groups I and II (high selenium rations) were alive and no subcutaneous neoplasms had been observed. In groups III and IV (low and selenium-deficient rations), 20 of the 40 rats died and 24 neoplasms were observed.

Histologic lesions of primary selenium toxicity, i.e., toxic hepatitis [3,4], were present in group I. In these rats the concentration of selenium in the liver was three times the concentration in the feed and seven times that found in muscle. Selenium deficiency lesions [21] persisted for 30 days in group III and for 240 days in group IV. In group IV rats, the concentration of selenium in the liver was 20 times that in the feed.

DISCUSSION

The concentration of selenium in the rations ranged from toxic to deficient levels. Lesions of both selenosis (group I) and selenium deficiency (group II) were observed. Although the concentration of selenite in the rations varied 125-fold, the concentration of selenium in the liver varied only 19-fold. The greatest increase in liver selenium (nine times)

TABLE 2. Selenium Concentration and Retention; Cancer Incidence and Longevity in Rats[a]

	Se in Liver[b] (ppm)	Total Se in Liver and Muscle (mg/rat)	Total Se Ingested (mg/rat)	No. of Rats with Hepatic or Mammary Cancer (200 days)	Average Age at Death[c]	Average Body Weight at Death (g)[b]
Group I 2.500 ppm Se	7.35 s 3.30	0.134	13.68	0	332	267
Group II 0.500 ppm Se	5.58 s 1.58	0.110	2.11	2	309	261
Group III 0.100 ppm Se	3.80 s 1.64	0.088	0.24	12	200	244
Group IV 0.000 ppm Se	0.40 s 0.46	0.055	0.05	12	207	247

[a] Basal ration contained 0.018 ppm Se.

[b] Natural or at termination of experiment.

[c] Concentration of selenium in muscle was 1.3 ppm (s = 0.83).

occurred between groups III and IV. Additional supplementation of the basal
ration with 25 (group I) and 5 (group II) times the concentration of selenite added to the group III ration increased the concentration of selenium
in the liver of group I and II rats 1.5 and 1.9 times, respectively. The
inability of an animal fed a selenium-adequate ration to concentrate selenium in proportion to the selenium content in the feed has been postulated
by other investigators [2,18] and was observed in blood studies [18].

This study supports the hypothesis of self-regulation of selenium accumulation in muscle. The concentration of selenium in the muscle was the
same (1.3 ppm) in all groups of rats, whether fed toxic or deficient levels
of selenium. The selenium concentration in muscle of rats fed the selenium-
deficient ration (group IV) was maintained at the expense of the liver concentration of selenium. Group IV rats had lesions of selenium deficiency,
even though they were estimated to have retained all of their ingested selenium.

In group III rats, clinical signs of selenium deficiency were present
for only 30 days after supplementation with 0.10 ppm Se. The concentration
of selenium in the liver of these rats was 2.9 times the concentration in
their muscle. This increase in the concentration of liver selenium can be
attributed to adequate selenium nutrition. These rats (group III) retained
37% of the calculated total ingested selenium.

The amount of selenium retained by rats in groups I, II, and III was
nearly inversely proportional to the content of selenium in the feed. In
previous experience with similar regimens [3,22], the dietary LD_{50} (100
days) of selenium as selenite was 6 ppm. The rats in group I of this experiment were fed 40% of this LD_{50} level and had lesions of selenosis [5].
These observations suggest a primary self-regulating mechanism for the accumulation of selenium in muscle. The regulation of selenium concentrations
in the liver may be more complex.

There is concern that selenium may accumulate in the food chain and
expose people to toxic quantities of the element. Although selenium-tox-
icity dermatitis has been reported in high seleniferous areas [4], there
are no reported cases of selenosis associated with consumption of meat from
these areas. If livestock react to selenium exposure similarly to rats,
the present data indicate that selenium does not accumulate in muscle, even
when the animal is fed toxic selenium concentrations for long periods.
Animals may therefore not be capable of accumulating selenium in tissue in
large enough quantities to be toxic to the next trophic level. It appears

more likely that human and animal food sources might be nutritionally deficient in selenium.

Observations in this study were confounded with different lengths of
exposure to FAA and selenite supplementation. However, the rats that were
fed the highest concentration of selenite lived the longest and had the
least numbers of cancers. Similar studies with mice also indicate increased
longevity on high selenium rations [15]. Further experimentation has been
initiated on laboratory rats using a fixed period of exposure to cancer-
inducing agents. This will avoid confounding drug exposure with the anticarcinogenic action of selenium.

CONCLUSIONS

The concentration of selenium in the liver of rats fed selenite-supple-
mented rations was inversely correlated with the incidence of hepatic and
mammary cancer. Chronic exposure to various concentrations of selenite in
the feed did not increase the concentration of selenium in skeletal muscle
of rats. The concentration of selenium in the liver of rats fed nutritionally adequate amounts of selenite increased slowly, even though a portion
of the exposure regimen included 40% of the LD_{50} dose. In these groups,
the addition of selenite to the basal selenium-deficient ration increased
longevity and decreased cancer induction.

ADDENDUM

Further work on the relationship between selenium consumption and car-
cinogenesis was completed by W. K. Johnston (The Effect of Selenium on
Chemical Carcinogenicity in the Rat, Masters Thesis, Oregon State University,
July, 1973).

The experimental design of this work included three levels of dietary
selenium (0, 0.2, and 2.0 $\mu g/g$) and three carcinogenic exposures (none, FAA,
and DEN). The concluding paragraph from the thesis summary is as follows:

"Selenium appeared to provide protection against tumors when included
in the DEN groups. Incidence was reduced at least 20% with selenium; the
reduction was most apparent in the females. The protective action of selenium did not present itself in those groups exposed to FAA."

A third experiment on this problem is in progress in Drs. Weswig and
Whanger's laboratory. The result of this work will be available later.
available later.

ACKNOWLEDGMENTS

The authors wish to express their appreciation to Mrs. Louise Hogan
for her capable technical assistance and to Dr. D. V. Frost for assistance
in the inception and design of this experiment.

REFERENCES

1. W. H. Allaway and E. E. Cary, Determination of Submicrogram Amounts
   of Selenium in Biological Materials, Anal. Chem., 36, 1359 (1964).

2. D. V. Frost, Significance of Symposium, in Symposium: Selenium in
   Biomedicine (O. H. Muth, ed.), AVI, Westport, 1967, Chap. 2, pp. 7-26.

3. J. R. Harr, J. F. Bone, I. J. Tinsley, P. H. Weswig, and R. S. Yama-
   moto, Selenium Toxicity in Rats II. Histopathology, in Selenium in
   Biomedicine (O. H. Muth, ed.), AVI, Westport, 1967, pp. 153-178.

4. J. R. Harr and O. H. Muth, Selenium Poisoning in Domestic Animals and
   Its Relationship to Man, Clin. Toxicol., 5 (2), 175-186 (1972).

5. J. R. Harr, J. H. Exon, P. D. Whanger, and P. H. Weswig, Effect of
   Dietary Selenium on N-2-Fluorenyl-Acetamide (FAA)-Induced Cancer in
   Vitamin E-Supplemented Selenium-Deficient Rats, Clin. Toxicol., 5 (2),
   187-194 (1972).

6. W. J. Hartley and A. B. Grant, A Review of Selenium-Responsive Diseases of New Zealand Livestock, Fed. Proc., 20 (2), 679-688 (1969).

7. D. D. Magg and M. W. Glenn, Toxicity of Selenium: Farm Animals, in
   Symposium: Selenium in Biomedicine (O. H. Muth, ed.), AVI, Westport,
   1967, pp. 127-140.

8. K. E. McCoy and P. H. Weswig, Some Selenium Responses in the Rat Not
   Related to Vitamin E, J. Nutr., 98, 383-389 (1969).

9. O. H. Muth, White Muscle Disease (Myopathy) in Lambs and Calves, Occurrence and Nature of the Disease under Oregon Conditions, J. Amer.
   Vet. Med. Ass., 126, 355-360 (1955).

10. O. H. Muth, Theme of the Symposium: The Biomedical Aspects of Sele-
    nium, in Symposium: Selenium in Biomedicine (O. H. Muth, ed.), AVI,
    Westport, 1967, Chap. 1, pp. 3-6.

11. A. A. Nelson, O. G. Fitzhugh, and H. O. Calvery, Liver Tumors Follow-
    ing Cirrhosis Caused by Selenium in Rats, Cancer Res., 3, 230-236
    (1943).

12. M. C. Nesheim and M. L. Scott, Nutritional Effects of Selenium Com-
    pounds in Chicks and Turkeys, Fed. Proc., 20 (2), 674-677 (1961).

13. I. Rosenfeld and O. A. Beath, Selenium: Geobotany, Biochemistry,
    Toxicity and Nutrition, Academic Press, New York, 1964.

14. H. A. Schroeder, Effects of Selenate, Selenite, and Tellurite on the
    Growth and Early Survival of Mice and Rats, J. Nutr., 92 (3), 334-338
    (1967).

15. H. A. Schroeder and M. Mitchener, Selenium and Tellurium in Mice,
    Effects on Growth, Survival, and Tumors, Arch. Environ. Health, 24,
    66-71 (1972).

16. K. Schwarz and C. M. Foltz, Selenium as an Integral Part of Factor 3
    Against Dietary Necrotic Liver Degeneration, J. Amer. Chem. Soc., 79,
    3292-3293 (1957).

17. M. L. Scott, Selenium Deficiency in Chicks and Poults, in Symposium:
    Selenium in Biomedicine (O. H. Muth, ed.), AVI, Westport, 1967, pp.
    231-238.

18. M. L. Scott and J. N. Thompson, Selenium Content of Feedstuffs and
    Effects of Dietary Selenium Levels upon Tissue Selenium in Chicks and
    Poults, Poultry Sci., 50 (6), 1742-1747 (1971).

19. R. J. Shamberger, Relation of Selenium to Cancer. I. Inhibitory Effect
    of Selenium on Carcinogenesis, J. Nat. Cancer Inst., 44 (4), 931-936
    (1970).

20. R. J. Shamberger and C. E. Willis, Selenium Distribution and Human
    Cancer Mortality, C. R. S. Crit. Rev. Clin. Lab. Sci., 211-221 (1971).

21. L. H. Sprinker, J. R. Harr, P. M. Newberne, P. D. Whanger, and P. H.
    Weswig, Selenium Deficiency Lesions in Laboratory Rats Fed Vitamin E-
    Supplemented Rations, Nutr. Rep. Intern., 4 (6), 335-340 (1971).

22. I. J. Tinsley, J. R. Harr, J. F. Bone, P. H. Weswig, and R. S. Yama-
    moto, Selenium Toxicity in Rats I. Growth and Longevity, in Selenium
    in Biomedicine (O. H. Muth, ed.), AVI, Westport, 1967, pp. 141-152.

23. L. A. Tscherkes, S. G. Aptekar, and M. N. Volgarev, Hepatic Tumors
    Induced by Selenium, Byulletin' Eksperimental; noi Biologii i Medit-
    siny, 53, 78-82 (Russian) (1961).

24. M. N. Volgarev and L. A. Tscherkes, Further Studies in Tissue Changes
    Associated with Sodium Selenate, in Selenium in Biomedicine (O. H.
    Muth, ed.), AVI, Westport, 1967, pp. 179-184.

25. P. D. Whanger and P. H. Weswig, Selenium Responses in the Rat Inde-
    pendent of Vitamin E, Fed. Proc., 28 (2), 809 (1969).

EFFECT OF DIETARY SELENIUM ON N-2-FLUORENYL-ACETAMIDE (FAA)-INDUCED CANCER IN VITAMIN E-SUPPLEMENTED, SELENIUM-DEPLETED RATS*

James R. Harr
Pennwalt Corporation
Rochester, New York

Jerry H. Exon, Paul D. Whanger, and Philip H. Weswig
Oregon State University
Corvallis, Oregon

Although earlier reports associated seleniferous compounds with neoplasia in aged rats [3,11], an extensive bioassay of selenium carcinogenesis in rats did not produce cancer [1,10]. The basal rations in these three experiments [1,3,11] contained 0.5-1.0 ppm selenium. The principals received the basal rations with 0.5-8 ppm added selenium. Those treatment rations were associated with toxic hepatitis and hyperplasia and cirrhosis of the liver, in addition to the reported tumors. Current reports indicate that rats given selenium through their drinking water developed similar lesions [4,5].

Several recent reports suggest that selenium is anticarcinogenic. Addition of selenium to feed was associated with a reduction in the number of anthracene-induced papillomas [6]. Shamberger and Willis demographically demonstrated an inverse correlation between dietary and blood concentrations of selenium and human cancer mortality [7].

Cancer induction has not been investigated in vitamin E-supplemented selenium-deficient individuals. In this study, selenium-depleted rats were fed adequate vitamin E, supplementary selenite, and N-2-fluorenyl-acetamide (FAA), a known chemical carcinogen for rats.

*Technical Paper No. 3182, Oregon Agricultural Experiment Station. Supported in part by the Selenium-Tellurium Development Association, Inc. and Oregon Agricultural Experiment Station.

METHODS

Eighty female OSU-Brown rats were weaned at 35 days of age, divided into four groups of 20, and fed the vitamin E-supplemented (60 ppm), low-selenium (18 ppb), torula ration of McCoy and Weswig [2]. The 40 rats in groups 1 and 2 (Table 1) were born from parents reared on a selenium depletion regimen. The 40 rats in groups 3 and 4 (Table 1) were from the second generation maintained on the depletion regimen. Rats in groups 1 and 2 were clinically normal. Those in groups 3 and 4 had clinical signs of selenium deficiency [8].

FAA and selenite were added to the basal ration [2] to produce the following diets:

For group 1, 150 ppm FAA and 2.50 ppm added selenium.

For group 2, 150 ppm FAA and 0.50 ppm added selenium.

For group 3, 150 ppm FAA and 0.10 ppm added selenium.

For group 4, 150 ppm FAA and no added selenium.

Each rat was housed in a suspended cage and had free access to food and distilled water. Observations were made daily for clinical condition, presence of subcutaneous mammary adenocarcinomas, and abdominal masses. Subcutaneous tumors were surgically removed and preserved in 10% formalin when they became 5-10 mm in diameter. Moribund rats were necropsied and lungs, liver, and tumors were preserved in 10% formalin. Tissues were imbedded in paraplast, sectioned at 6 μm, stained with hematoxalin and eosin [9], and examined for neoplasia.

The accumulative percentage incidence of mammary and/or hepatic neoplasia in each experimental group was determined for each 20-day period of exposure to FAA (Figs. 1-3). Rats that died without developing cancer were disregarded in determining the incidence of cancer at subsequent exposure periods.

RESULTS

Clinical signs of selenium deficiency [8] in group 3 (0.1 ppm added selenium) were resolved within 30 days after selenium supplementation. Signs of selenium deficiency in group 4 (no added selenium) persisted for 240 days. Groups 1 and 2 (first depletion generation) did not have signs of selenium deficiency.

Mammary adenocarcinomas and/or hepatic carcinomas occurred in 67 of the rats. From 160 to 260 days of exposure, the curves of the accumulative

TABLE 1. Occurrence of Neoplasms in Rats Fed FAA and Varying Dietary Concentrations of Added Selenium

Days of Exposure to 150 ppm FAA	Group 1 (2.50 ppm Se)			Group 2 (0.50 ppm Se)			Group 3 (0.10 ppm Se)			Group 4 (0.00 ppm Se)[c]		
	MC[a]	LC[b]	Died	MC	LC	Died	MC	LC	Died	MC	LC	Died
1-40	--	--	--	--	--	--	--	--	3	--	--	2
41-80	--	--	--	--	--	--	--	--	--	--	--	1
81-120	--	--	--	--	--	--	--	1	1	--	--	1
121-160	--	--	--	--	--	--	4	--	2	2	3	3
161-200	--	--	--	1	1	2	4	3	4	7	--	3
201-240	4	--	--	6	--	1	3	5	7	1	1	4
241-280	5	2	4	4	7	8	--	--	--	1	2	4
281-320	2	6	8	--	4	6	2	3	3	1	--	1
Total	11	8	12	11	12	17	13	12	20	12	6	19
Killed (321 days)	1	4	8	--	3	3	--	--	--	1	1	1

[a] MC, mammary adenocarcinoma.

[b] LC, hepatoma.

[c] Basal diet contained 0.018 ppm Se.

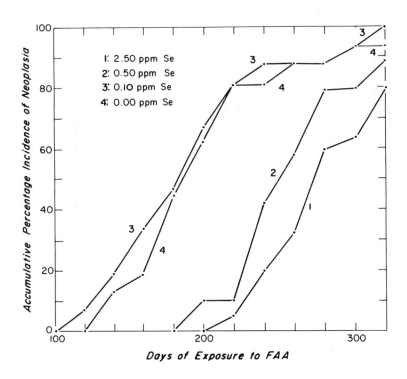

FIG. 1. Accumulative incidence of cancer in rats after dietary exposure to
FAA and varying amounts of added selenite.

incidence of cancer in the four groups were markedly different (see Figs.
1-3). Neoplasms developed more slowly in groups 1 and 2 (0.5-2.5 ppm added
selenium) than in groups 3 and 4 (0.0-0.1 ppm added selenium). In addition,
rats in groups 3 and 4 died at an earlier age than rats in groups 1 and 2
(see Table 1).

Mammary adenocarcinomas in groups 1, 2, and 3 (0.1-2.5 ppm added selenium) occurred predominantly (90%) in the thoracic region. The tumors
were well circumscribed, firm, and easily removed. Those in group 4 (no
added selenium) were predominantly (80%) in the pelvic area and were soft
and fluid, contained little connective tissue, and were invasive.

Toxic hepatitis was present in 18 of the 20 livers from group 1 (2.5
ppm added selenium), but was not present in the other groups. Hepatic
cirrhosis and hyperplasia did not occur in any of the rats.

DISCUSSION

The inverse relationship of dietary selenium concentrations to the
incidence of FAA-induced cancer after 160-260 days of exposure to FAA is

contrary to the suggested carcinogenesis of selenium compounds [3-5,11,12].
However, the group 2, 3, and 4 rats (0.0-0.5 ppm selenium) did not have
cirrhosis or toxic hepatitis as was reported in the previous work [3-5,11,
12]. Previously reported tumors associated with dietary selenium supplementation may have been toxic or hyperplastic nodules rather than neoplasms.
Those that were neoplasms [11] may have been secondary to cirrhosis or
other confounding influences rather than to dietary selenium.

It is difficult to either prevent or evaluate confounding influences
or effects in an experiment involving a relatively few individuals observed
at the end of a lifetime (2-3 yr in the rat). In three of the previous reports on selenium carcinogenesis, negative controls were fortuitous rather
than planned [3,11,12], positive controls were not parallel [3,11,12], and,
in one instance, the ration contained an organic rather than an inorganic
form of added selenium [3].

The observed effect of dietary selenium concentration on FAA-induced
carcinogenesis is consistent with the reported bioassay of selenium carcino-

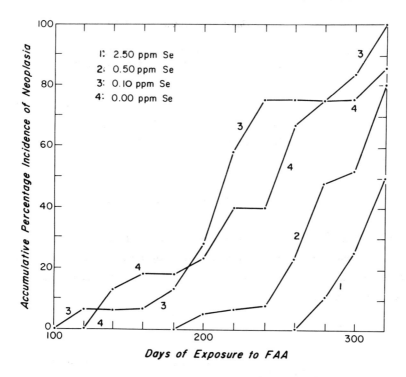

FIG. 2. Accumulative incidence of hepatic neoplasms in rats after dietary
exposure to FAA and varying amounts of added selenite.

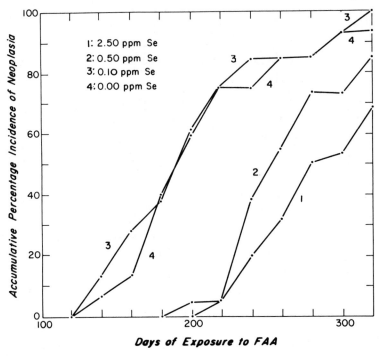

FIG. 3. Accumulative incidence of mammary adenocarcinoma in rats after dietary exposure to FAA and varying amounts of added selenite.

genesis [1], a current demographic study [7], the effects of dietary selenium on the experimental induction of skin cancer [6], and the concept of the metabolic role of selenium in nutrition [8]. The absence of cirrhosis associated with mammary adenocarcinomas in group 4 may result from the inability of connective tissue to proliferate. This failure may be related to hypovascularization associated with selenium deficiency and its effect on membranes [8].

This study did not determine whether the observed effects of added selenite represented a prevention of carcinogenesis or a modification of the rate of induction. This point was confounded by the unequal exposure periods for each rat.

CONCLUSION

Addition of selenite to vitamin E-supplemented, low-selenium, torula rations decreased the effect of cancer induction by FAA in rats.

REFERENCES

1. J. R. Harr, J. F. Bone, I. J. Tinsley, P. H. Weswig, and R. S. Yama-moto, Selenium Toxicity in Rats. II. Histopathology, in Selenium in Biomedicine (O. H. Muth, ed.), AVI, Westport, 1967, pp. 153-178.

2. K. E. McCoy and P. H. Weswig, Some Selenium Responses in the Rat Not Related to Vitamin E, J. Nutr., 98, 383-389 (1969).

3. A. A. Nelson, O. G. Fitzhugh, and H. O. Calvery, Liver Tumors Following Cirrhosis Caused by Selenium in Rats, Cancer Res., 3, 230-236 (1943).

4. H. A. Schroeder, Effects of Selenate, Selenite, and Tellurite on the Growth and Early Survival of Mice and Rats, J. Nutr., 92, 334-338 (1967).

5. Metallic Micronutrients and Intermediary Metabolism, U.S. Clearing-house Fed. Sci. Tech. Inform. No. 708581, 1970, p. 22.

6. R. J. Shamberger, Relation of Selenium to Cancer. I. Inhibitory Effect of Selenium on Carcinogenesis, J. Nat. Cancer Inst., 44, 931-936 (1970).

7. R. J. Shamberger and C. E. Willis, Selenium Distribution and Human Cancer Mortality, C. R. C. Crit. Rev. Clin. Lab. Sci., June, 211-221 (1971).

8. L. H. Sprinker, J. R. Harr, P. M. Newberne, P. D. Whanger, and P. H. Weswig, Selenium Deficiency Lesions in Laboratory Rats Fed Vitamin E Supplemented Rations, Nutr. Rep. Int., 4, 335-340 (1971).

9. S. W. Thompson and R. D. Hunt, Selected Histochemical and Histopathological Methods, Charles C Thomas, Springfield, 1966, 763-772.

10. I. J. Tinsley, J. R. Harr, J. F. Bone, P. H. Weswig, and R. S. Yama-moto, Selenium Toxicity in Rats. I. Growth and Longevity, in Selenium in Biomedicine (O. H. Muth, ed.), AVI, Westport, 1967, pp. 141-152.

11. L. A. Tscherkes, S. G. Aptekar, and M. N. Volgarev, Hepatic Tumors Induced by Selenium, Byulleten' Eksperimental' noi Biologii i Meditsiny, 53, 78-82 (1961).

12. M. N. Volgarev and L. A. Tscherkes, Further Studies in Tissue Changes Associated with Sodium Selenate, in Selenium in Biomedicine (O. H. Muth, ed.), AVI, Westport, 1967, pp. 179-184.

TOXICOLOGY AND ADVERSE EFFECTS OF MINERAL IMBALANCE
WITH EMPHASIS ON SELENIUM AND OTHER MINERALS

Richard C. Ewan
Iowa State University of Science and Technology
Ames, Iowa

Many minerals serve essential functions in mammalian systems, and in the last few years carefully controlled studies have resulted in the addition of tin [1], silicon [2,3], vanadium [4,5] fluorine [6,7], nickel [8-11], and chromium [12,13] to the list of minerals required for normal function of animals. Many of the essential trace elements are among the transition elements in the periodic table. Thus, many of the chemical properties are similar, and there are many interactions that influence the metabolism of the trace elements.

Hill and Matrone [14,15] have pointed out that many of the trace elements have similar electronic configuration and that those with similar electron distribution (i.e., similar orbital arrangement) tend to interact and to have an influence on the metabolism of each other. With this consideration, they suggested that the cuprous ion would have interactions with zinc, cadmium, and mercury while the cupric ion would interact with silver. They were able to demonstrate that zinc enhanced mortality, reduced growth, and decreased hemoglobin levels in chicks fed a copper-deficient diet. Interactions among copper, cadmium, and silver were also demonstrated in chicks. Mercury did not interact with copper, possibly because the electronic orbitals were farther from the nucleus than the other ions. Interactions were demonstrated among Se^{4+}, As^{3+}, Sn^{2+}, and Te^{4+}; between V^{5+} and Cr^{6+}; and among the anions phosphate, vanadate, chromate, arsenate, and selenate. The interactions among the groups of minerals noted have been observed by others, and the theory of Hill and Matrone [14] may provide an explanation of these biological interrelationships.

Since the interactions among minerals are extensive, only the minerals selenium, iron, copper, and zinc will be considered in this review. The effects of inadequate and excess intake will be considered and the metabolism of the elements will be discussed. The interactions among these elements and others in biological systems will be considered.

SELENIUM

Introduction

Selenium was discovered by Berzelius in 1817. The nutritional significance of selenium was first recognized as a toxic effect by Franke [16] in 1934. He discovered that selenium was the cause of a toxicity that occurred in cattle and horses that resulted in loss of hair and sloughing of hoofs. The writings of Marco Polo suggested the possibility of selenium toxicity when he referred to "...a poisonous plant...which if eaten by them [beasts of burden] has the effect of causing the hoofs of the animal to drop off" [17]. Dr. T. C. Madison, a surgeon in the U.S. Army, described the signs of selenium toxicity in horses at Fort Randal in 1857 and used the term alkali disease to refer to the condition [18].

Selenium toxicity occurs because selenium in soils is absorbed and concentrated by plants. The level of selenium in plants and seeds is dependent on the concentration of selenium in the soil and on the type of plant. In general, plants may be grouped based on their ability to absorb and concentrate selenium. Plants that accumulate large amounts of selenium have been designated as indicator plants [19,20]. These plants require selenium for growth [21] and may accumulate several hundred parts per million of selenium on a dry-matter basis. About 24 species and varieties of Astragalus (milk vetch), Xylorhiza (woody aster), Oonopsis (golden weed), and Stanleya (prince's plume) are known to belong to this group of plants [22]. A second classification is plants that accumulate moderate amounts of selenium if grown in soils containing high concentrations of selenium. Species of the genera Aster, Atriplex, Grindelia, Gutierreza, and Manchaeranthera are examples of the plants of this type [23]. The third classification is plants that will accumulate low levels of selenium (0 to 40 ppm) if grown on soils containing available selenium. The normal farm grains and forages are in this class. Selenium content of corn, wheat, and barley grown on seleniferous soils of South Dakota have been reported to contain 25 to 30 ppm [24]. In contrast, samples of corn grown in Michigan, Wisconsin, Indiana, Iowa, and Illinois, where soil levels of selenium are low, have been reported to range from 0.01 to 0.16 ppm [25]. Grains and forages from seleniferous areas can produce selenium toxicity in livestock while those from low selenium areas can result in selenium-vitamin E-related deficiency conditions.

In addition to the soil content of selenium, the management of the crops and the production level have been shown to influence the concentration of selenium in cereal grains. Harmon et al. have reported results of studies of selenium level of corn produced on experimental plots under cultivation since 1876. They found that fertilization and crop rotation increased yield and reduced the selenium in the grain ($r = -0.84$). A negative correlation was also observed between selenium content of the grain and the pH of the soil ($r = -0.64$), indicating that selenium was more available from acid soils [26].

Selenium levels in soils have been identified by studies of the level of selenium in plants. Areas with levels of selenium in soils that produce toxic accumulations in plants have been identified in North and South America, Australia, New Zealand, South Africa, France, and Germany. In the United States, many regions of the western states from North Dakota to Texas produce plants containing toxic amounts of selenium. Areas of the northeastern, midwestern, and far northwestern sections of the United States have soils that are deficient in selenium [27-29].

Toxicity

Selenium is toxic to all animals, but the toxicity occurs naturally primarily in cattle, sheep, and horses when they graze plants containing excessive levels of selenium [24,30,31]. Swine, poultry, and laboratory animals are susceptible to selenium toxicity if they are fed diets that contain excessive levels of selenium [22,32,33]. Selenium toxicity in man has been described in individuals that live in areas with seleniferous soils [34] and industrial workers [35,36]; the toxicity, however, has been observed infrequently in spite of the wide use of selenium compounds in industry [37].

Toxicity from natural sources of selenium (seleniferous grains and forages) can be classified into three types depending on the level of selenium ingested: acute, subacute, and chronic toxicity. The acute form of selenium toxicity results from the ingestion of large amounts of selenium in a single dose. In general, levels of 1 to 5 mg/kg of body weight are

required to produce an acute toxicity [38-40]. Acute toxicity is characterized by changes in movement and posture. The animal may walk a short
distance with an uncertain gait, stop, and stand with the head lowered and
ears drooped. Dark, watery diarrhea may develop. The temperature may be
elevated and the pulse rate may be rapid (90 to 300 per min) and weak.
Respiration may be labored with mucus rales and there may be bloody froth
from air passages. Bloating and abdominal pain are usually pronounced.
The animal usually becomes completely prostrate before death due to respiratory failure. The course of the illness is from hours to a few days
depending on the level of selenium ingested [22,41-47]. The consumption
of seleniferous indicator plants has been reported to be responsible for
acute poisoning of sheep and cattle and occasionally swine and horses.
Many studies have demonstrated the toxicity of selenite and selenate to all
animals [22].

The subacute selenium toxicity is responsible for "blind staggers" in
cattle and is observed when small to moderate amounts of seleniferous plants
are consumed. In the first stages of the disease, cattle have a tendency
to stray from the herd. There is a slight impairment of vision and the animal has difficulty in judging distances of objects in its path. As the
condition progresses, vision is impaired to a greater extent and depraved
appetite may be observed. In the terminal stage, there are varying degrees
of paralysis, evidence of abdominal pain, grating of teeth, salivation, and
grunting. Death usually results from respiratory failure [30,48,49].

Chronic selenium poisoning or alkali disease occurs when animals consume grains and forages from seleniferous areas for long periods of time.
The general symptoms are dullness and a lack of vitality, emaciation, a
rough hair coat, loss of the long hair from the mane and tail of horses and
cattle, and soreness and sloughing of hoofs. Pathologically, atrophy and
cirrhosis of the liver, atrophy of the heart (dishrag heart), and anemia
are observed in chronic toxicity. Animals may die because of the soreness
of the feet and an inability to obtain food and water [22,24].

The biochemical mechanism of selenium toxicity is not known. Selenium
has been shown to inhibit the growth of yeast and the activity of some enzyme systems [50,51]. Selenium from plants is present as organic selenium
such as the amino acids, selenomethionine, selenocysteine, and selenocysta-
thionine. Seleno-amino acids can be incorporated into protein by mammalian
systems [52,53]. Selenite and selenate cannot be substituted for sulfur in

the sulfur amino acids by mammalian systems [54,55], although this can occur in plants and microorganisms [56,57]. The biological effects of substitution of seleno-amino acids for sulfur amino acids in protein has not been determined, so the significance of this route of metabolism cannot be assessed.

Elemental selenium is not available to the animal and apparently is not metabolized in mammalian systems [58]. Selenite and selenate are readily available and are metabolized by rapid binding to protein [59] and reduction to selenide [60,61]. The reduction of selenium occurs in the liver and requires glutathione, reduced triphosphopyridine nucleotide, coenzyme A, adensine-5'-triphosphate, and magnesium for optimal activity [61]. With the ingestion of normal levels of selenium (0.1 to 2 ppm), selenite and selenate are excreted primarily as trimethylselenide in the urine [62,63]. When challenged with a single subacute dose of selenium, the methylation system becomes saturated and the intermediate, dimethylselenide, is formed and lost in the expired air [60,61]. Dimethylselenide has a garliclike odor and is the cause of the garlic odor observed in animals [64] and man exposed to selenium [35,65]. Di- and trimethylselenides are less toxic to animals than inorganic selenium or seleno-amino acids [66,67]; this suggests that the saturation of the detoxification mechanisms results in accumulation of the inorganic forms of selenium and death.

A number of factors are known to modify the toxicity of selenium. As implied by the preceding discussion, the length of exposure, the level of intake, and the form of selenium affect the toxicity. In addition, decreasing the level of protein in the diet increases the toxicity of selenium. Smith [68] observed that 10 ppm of selenium in a 10% protein diet was highly toxic to rats, whereas 10 ppm of selenium in diets containing an additional 20% of protein as casein was tolerated. Linseed meal was shown to be superior to casein in alleviating selenium toxicity. Some fraction other than protein in linseed meal may be involved [69].

Sulfate has been shown to reduce the incorporation of selenate by plants [70,71]. Increasing levels of sulfate in a sulfate-free diet have been shown to relieve the growth inhibition of 10 ppm of selenium in young rats [72]. Sulfate has also been shown to decrease the retention of selenate and to increase the urinary excretion of selenium. Sulfate had much less effect on the retention or excretion of selenite [73]. Halverson et al. have also reported that sulfate has little effect on the toxicity produced by seleniferous wheat [74].

In studies of the effects of inorganic compounds on selenium toxicity, Moxon and DuBois found that 5 ppm of arsenic in the drinking water of rats prevented the toxicity produced by 11 ppm of selenium as seleniferous wheat. Molybdenum, tellurium, vanadium, chromium, fluorine, cadmium, zinc, cobalt, uranium, and nickel were not effective [75,76]. Subsequent investigations indicated that arsenic sulfides (AsS_2 and AsS_3) were ineffective while arsenite and arsenate were equally effective in preventing toxicity of seleniferous wheat, selenite, and selenocystine [77-79]. Organic arsenicals such as arsanilic acid and 3-nitro-4-hydroxyphenylarsonic acid provide partial protection [80,81].

The mechanism by which arsenic alleviates selenium toxicity is not known, but some effects of arsenic on selenium metabolism have been defined. Arsenic has been shown to increase the excretion of selenium in the bile [82] and to reduce the excretion of selenium in respired air [83,84]. Ganther and Hsieh have reported that there are methylation systems in the soluble and microsomal fractions of liver and that the microsomal system is very sensitive to arsenite [85]. These observations suggest that arsenic modifies the detoxification of selenium in the liver and causes an increase in the excretion through the intestine via the bile.

Bromobenzene is detoxified by conjugation with acetylated cysteine to form a mercapturic acid and has been administered to animals receiving seleniferous diets to remove selenocysteine. The administration of bromobenzene to rats, dogs, and steers fed seleniferous diet increased the urinary excretion of selenium [86,87]. Other studies have not observed an increase in urinary loss [88]. The toxicity of bromobenzene, however, limits its use.

Deficiency

The nutritional requirement for selenium was suggested by studies of Schwarz and Foltz [89]. Liver necrosis caused by feeding a vitamin E-deficient diet based on Torula yeast was prevented by supplementation of the diet with selenite. While many studies have been conducted to determine the effect of selenium in vitamin E-selenium-deficient diets, recent studies have clearly demonstrated that the rat, chick, and pig require a dietary source of selenium.

McCoy and Weswig [90] reported the effect of feeding a vitamin E-supplemented Torula yeast diet to female rats. In the second generation, the young failed to grow and did not develop normal hair coats. Supplementation of the diet with selenium restored growth and normal hair coats while additional vitamin E or sulfur amino acids had no effect. The second generation rats were sterile as measured by a lack of breeding of females with fertile males and by lack of viable spermatozoa in the males. Hurt et al. [91] demonstrated that the second and third litters of female rats fed a vitamin E-supplemented Torula yeast diet failed to grow as rapidly as rats pair fed the diet supplemented with selenomethionine. They also observed a decrease in growth rate of commercial weanling rats fed a selenium-deficient amino acid diet for 20 weeks when compared to selenomethionine-supplemented diets.

Thompson and Scott [92] reported that feeding a selenium-deficient amino acid diet to chicks resulted in poor growth, poor feathering, and high mortality. This response was observed when vitamin E was present in the diet and was prevented by supplementation of the diet with 0.1 ppm of selenium. Thompson and Scott [93] reported that selenium deficiency reduced the absorption of vitamin E and lipids and caused atrophy of the pancreas in the chick. Lipid absorption was improved slightly by addition of bile salts to the diet or by use of free fatty acids and monoglycerides as the fat source. The reduction of vitamin E absorption may be due to the decreased pancreatic function and the general decline of lipid absorption.

In evaluating selenium responses, it should be emphasized that the selenium-vitamin E-deficient animal responds in a manner different from the selenium-deficient animal. Exudative diathesis in chicks appears when a diet deficient in vitamin E and selenium is fed [94]. It is characterized by greenish-blue edema of the breast, slow growth, and high mortality. Exudative diathesis can be prevented by selenium or vitamin E [95,96]. Vitamin E deficiency in the presence of selenium results in brain damage and development of encephalomalacia. Encephalomalacia does not respond to selenium supplementation [97,98]. Muscular dystrophy can develop in chicks fed a vitamin E-deficient diet low in sulfur amino acids. The muscle degeneration does not respond to physiological levels of selenium, but the incidence of the condition can be reduced by levels of 1 to 5 ppm of selenium in the diet [99-101].

In lambs, a nutritional muscular dystrophy (NMD) that occurs naturally was demonstrated by Willman et al. [102] to respond to vitamin E administration. The condition is characterized by degeneration of skeletal muscle fibers [103]; a rise in the serum-glutamic oxaloacetic transaminase [104], lactic acid dehydrogenase [105], and 5'-nucleosidase [105,106]; and reduced growth [107,108]. Muth [109] and Hartley and Grant [110] were unable to demonstrate that vitamin E would prevent NMD. Subsequent studies demonstrated that when a vitamin E-selenium-deficient diet was supplemented with selenium, the clinical signs of NMD were prevented and growth was improved, but the increases in serum enzymes were only delayed [111-113].

In swine, Obel [114] described the condition hepatosis diatetica which was demonstrated to respond to vitamin E in the diet. The deficiency resulted in centrilobular liver necrosis, heart and skeletal muscle degeneration, elevated levels of serum enzymes, edema, and sudden death. When unsaturated fish oils are fed, a yellow discoloration of body fat occurs [115]. Either vitamin E or selenium will prevent the liver degeneration, edema, and sudden death, while selenium is less effective than vitamin E in maintaining normal levels of serum enzymes and does not prevent the yellow discoloration of body fat [115,116]. In the presence of adequate amounts of vitamin E, a selenium deficiency in the pig results in reduced growth rate and an increase in the feed required per unit of gain because the digestibilities of dry matter, ether extract, and nitrogen are reduced [117].

Rotruck et al. [118,119] recently reported that dietary selenium in the presence of glucose was effective in preventing the in vitro hemolysis of red blood cells of rats. The reaction steps that enable glucose and selenium to prevent oxidative damage to erythrocytes are (1) phosphorylation of glucose to glucose-6-phosphate, (2) oxidation of glucose-6-phosphate by TPN and glucose-6-phosphate dehydrogenase with generation of TPNH, (3) reduction of oxidized glutathione (GSSG) to 2 moles of GSH by TPNH and glutathione reductase, and (4) utilization of GSH by glutathione peroxidase (glutatuione:H_2O_2 oxidoreductase, E.C.1.11.1.9) to reduce hydrogen peroxide to water. Subsequent studies reported the purification of the enzyme, glutathione peroxidase, from sheep erythrocytes and it was found to contain selenium in stoichiometric quantities. The final preparation was estimated to contain 4 moles of selenium per mole of enzyme protein [120]. Similar stoichiometry has been reported for the bovine enzyme [121] and rat liver enzyme [122]. Since Flohe et al. [121] have demonstrated that bovine glutathione peroxidase has four subunits, it is assumed that there is one selenium atom per subunit.

Studies of glutathione peroxidase activity indicate that the level of activity in the rat, chick, and lamb tissue is directly related to the dietary selenium intake [123-127]. The enzyme provides a convent link between the effects of vitamin E, glutathione, selenium, and sulfur amino acids [Fig. 1].

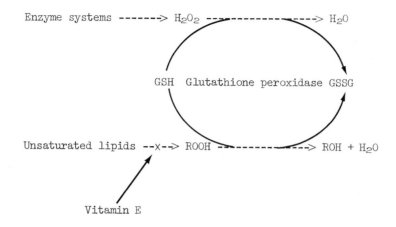

FIG. 1. Relationship between vitamin E and selenium (adapted from Hoekstra [128]).

Both selenium and vitamin E provide mechanisms for the control and harmless destruction of hydrogen peroxide or lipid peroxides; thus if either selenium or vitamin E is supplied in small quantities, the requirement for the other increases, explaining the interdependence of the two nutrients. Sulfur amino acids are required for the formation of glutathione that is essential for enzyme activity [120]. Naguchi et al. have demonstrated that plasma glutathione peroxidase decreases to very low levels in chicks before exudates appear and suggested that the enzyme is necessary to protect the capillary wall from oxidative damage and subsequent leakage of fluid into the subcutaneous tissues [126].

In the rat, chick, and pig, a selenium deficiency has been produced in the presence of vitamin E; but in other species, a clearly defined selenium deficiency has not been demonstrated. Since the signs of selenium deficiency in the rat, chick, and pig differ markedly from those of a vitamin E-selenium deficiency, it is probable that other species will respond differently to a selenium deficiency than to the combined deficiency.

Interactions

Selenium metabolism has been demonstrated to be related to the metabolism of cadmium, mercury, thallium, zinc, silver, and cobalt in addition to the effect of arsenic previously discussed.

Kar et al. [129] and later Mason and co-workers [130,131] reported that small amounts of selenite were effective in preventing testicular necrosis caused by parenteral administration of cadmium salts. Selenium compounds also provide protection for the damage produced by cadmium in the nonovulating ovary [132] or in the placenta [133] or by the selective toxicity of cadmium during the last stages of pregnancy [133]. Selenium compounds also prevent the teratogenic effects of cadmium [134] and decrease the mortality of experimental animals given doses of cadmium lethal to controls [133,135].

The metabolism of cadmium has not been studied in detail so that inferences on the effect of cadmium are difficult. Subcutaneous injection of cadmium prior to selenite into rats results in increased retention of selenium in the liver, carcass, and blood and a decrease in volatile compounds; intestinal and urinary excretion of selenium also decreased [73]. Thus, the presence of cadmium seems to have reduced the ability of the rat to detoxify selenium by methylation. The strong affinity of cadmium for sulfhydryl groups may suggest an involvement with the enzymes required for reduction and methylation of selenium or with the co-factor required for reduction and methylation, glutathione [61]. It is interesting that after subcutaneous injection of cadmium and selenium, cadmium levels in plasma [136], testis, and kidney [137] are elevated. The combined injection has been observed to result in increased levels of the elements at the site of injection suggesting the formation of an insoluble or unavailable complex between selenium and cadmium compounds [84]. Hill [138] has fed the precipitate formed from equimolar amounts of selenium dioxide and cadmium sulfate (assumed to be cadmium selenite) and found it to be less toxic than selenium dioxide.

The toxicity of mercury compounds is well known. Recently, selenium has been reported to decrease mercury toxicity and alter its metabolism. Parizek et al. [139,140] demonstrated in short-term experiments that selenite has a protective effect against renal necrosis and mortality caused by mercuric chloride and that selenite reduced the placental transfer of mercuric chloride to the fetus. Ganther et al. [141] demonstrated that high mercury levels in tuna were accompanied by a high selenium content and that the molar ratio of mercury to selenium was about 1:1. Other observations on liver mercury and selenium levels in seals, dolphins, and porpoises indicate a similar molar ratio [142]. Ganther et al. [141] were also able

to demonstrate that mercury from tuna was less toxic than mercury alone
when fed to rats and that selenite would reduce the toxicity of mercury.
The effect of selenium on mercury toxicity has been confirmed in Japanese
quail [143] and chicks [138]. Selenite is effective in reducing the toxicity
of inorganic mercury [138-140] and the toxicity of methyl mercury [141,143,
144]. The toxicity of selenium to chicks has been found to be reduced by
the presence of mercury in equimolar ratios [138].

The nature of the relationship between mercury and selenium has not
been established, but several suggestions have been proposed. The distribution of mercury administered with selenium is altered with an increase in
liver, red blood cells, and spleen and a decrease in kidney [144,145]. The
shift in distribution is not permanent since after withdrawal of mercury
for 44 days no significant differences in tissue concentrations of mercury
were noted [144].

Since changes in distribution of mercury are observed, the implication
is that there is an interaction between the intermediates of mercury and
selenium metabolism. Sulfate does not protect against mercury toxicity
even though sulfate is chemically similar to selenite and selenate [146].
Selenite and selenate are reduced by mammalian systems while sulfate is not.
It may be suggested that one of the intermediates in the reduction of selenite to selenide may be involved. Dimethylselenide does not protect
against mercury toxicity [147] and is more toxic itself in the presence of
mercury compounds as is trimethylselenide [136]. This may suggest that
mercury and selenium interact in the reduction of selenium between selenite
and dimethylselenide. At this time, however, the specific interaction cannot be defined.

Hollo and Zlatarov [148] reported that selenite injection had a protective effect against thallium toxicity. This was confirmed by Rusieki
and Brzezinski [149], who observed that the lethal effects of thallium
compounds were prevented by oral administration of selenate and that the
thallium content of liver, kidney, and bones increased. Subcutaneous injections of thallium acetate increased the retention of selenium in liver
and kidney, decreased the excretion of volatile selenium and urinary loss
of selenium [82].

Hill [138] has reported that copper (500 ppm) will reduce the toxic
effects of 40 ppm of selenium when fed to chicks. Copper improved growth
and reduced mortality. The precipitate formed when selenium dioxide was

combined with equimolar amounts of cupric sulfate was also less toxic than
selenium dioxide.

McConnell and Carpenter [84] observed that zinc chloride injections
altered the metabolism of subacute doses of selenium. Zinc decreased the
excretion of selenium in urine, feces, and expired air. Marked accumulation
of selenium was found at the injection site when both zinc and selenium were
injected at the same site. They concluded that there was probably a reaction between selenium and zinc that caused selenium to remain at the site
of injection.

Shaver and Mason [150] first described dystrophic lesions, necrotic
degeneration of the liver, and high mortality in rats fed a vitamin E-deficient diet and given silver nitrate or silver lactate in their drinking
water. Dam et al. [151] have also reported that 20 ppm of silver as silver
acetate had a pro-exudative effect in vitamin E-deficient chicks. Diplock
et al. [152] demonstrated that 1500 ppm silver acetate in the drinking water
induced liver necrosis in 2 to 4 weeks in weanling rats fed a vitamin E-
deficient diet. The necrosis could be prevented by vitamin E or partially
prevented by 1 ppm but not by 0.05 ppm selenium. Subsequent studies indicated that 0.05 or 0.1 ppm of selenium would prevent liver necrosis produced by 130 ppm but not by 1000 ppm of silver [153]. Grasso et al. [154]
suggested after histological studies of the development of liver necrosis
that silver may complex with small amounts of selenium present in the livers
of vitamin E-deficient rats, so that selenium is not available for biologically active compounds. Silver has also been reported to reduce the levels
of glutathione peroxidase in rat liver [128] which may support this concept.

Bunyan et al. [155] demonstrated that cobalt was partially effective
in preventing liver necrosis produced by feeding rats a diet containing
bakers' yeast as the protein source. Schwarz et al., however, were unable
to confirm this finding using a diet containing Torula yeast as the protein
source [156]. In New Zealand, which has cobalt- and selenium-deficient
soils, Andrews et al. [157] suggested a relationship between cobalt and
selenium because responses to selenium supplementation occurred in livestock grazed in areas known to have cobalt-deficient soils. They demonstrated that sheep raised on cobalt-deficient diets had more selenium in
the kidney than cobalt-supplemented animals. Wise et al. [158] have reported
a decrease in cobalt level in the kidney of lambs affected with nutritional
muscular dystrophy. They concluded that a metabolic relationship between

cobalt and selenium seemed to exist in lambs with nutritional muscular
dystrophy. No explanation for the effect is apparent.

IRON

Introduction

Iron was recognized in the seventeenth century as a factor in blood
formation by two English physicians who found iron to be of value in treatment of anemia. Iron was identified as a characteristic constituent of
blood, and in 1886, Zinoffsky estimated the iron content of horse hemoglobin
to be 0.335% [159]. The involvement of iron in metabolism was expanded by
the observation of Keilin that iron through its presence in the cytochromes
is essential for the oxidative mechanisms of all living cells. Subsequently,
iron-containing flavoprotein enzymes were discovered [160,161], and iron was
established as a factor in the utilization of oxygen as well as the transfer
of oxygen to the tissue.

Nutritional disorders involving iron in animals are not common except
in the baby pig where iron-deficiency anemia will develop spontaneously if
therapy is not used. A number of significant interactions between iron,
copper, zinc, and manganese have been demonstrated to affect the requirements of animals for iron.

Metabolism

Early investigations revealed that little iron is lost in the urine
even when large doses were injected or ingested. Since such doses were
found to result in accumulation of iron in the epithelial cells of the
small intestine, it was suggested that the intestine was an excretory as
well as an absorptive organ for iron [162]. McCance and Widdowson [163,
164] proposed that the amount of iron in the body must be regulated by
controlled absorption. Intensive studies were centered on the metabolism
and absorption of iron. The "mucosal block" theory was proposed by Hahn
et al. [165] and supported by Granick [166,167] and suggests that iron absorption is controlled by the intestinal mucosal cell. It is well established that factors such as anemia and blood loss that increase the demand
for iron also increase its absorption, but the mechanisms that control the
absorption of iron are still obscure. The absorption of physiological
amounts of iron seem to involve the uptake by the mucosal cell, transport
across the mucosal cell, and transport out of the cell. Within the cell,

iron may be sequestered, not transported across the cell and lost through sloughing of the intestinal mucosal cell [168].

The uptake of iron by the mucosal cell does not seem to be the major control of absorption since in iron excess, significant quantities of iron still enter the mucosal cell. The iron bound or sequestered within the cell seems to be the major variable component within the intestinal cell. The protein apoferritin is known to bind ferric iron in the form of the hydroxide to form ferritin and to be present in the mucosal cell. The significance of ferritin is not clear since Manis and Schachter have estimated that 8% of the trivalent iron in the rat mucosal cell was associated with ferritin. They suggested that ferrous iron was transferred to the serosal surface or to the trivalent pool. The turnover of ferric iron is slower than ferrous iron, and the ferric iron may serve as a storage depot for excess iron [169].

Since ferritin seems to be a storage form of iron and not involved in the transport of iron in the mucosal cell, attempts have been made to find other iron-binding components in the mucosal cell. A large iron-binding protein similar to ferritin has been reported [170], and a small protein (molecular weight of 9000) has been isolated from duodenal segments from chicks [171]. These proteins, however, have not been correlated with the absorption of iron. The presence of an iron-binding protein that is correlated with iron absorption has been reported [168,172-176]. The protein is present in low concentrations in animals receiving an excess of iron and increases in deficiency [168]. The protein has a molecular weight of about 80,000 and has a turnover rate that is much more rapid than the turnover rate of larger iron-binding molecules [174,176]. The results of these studies suggest that the protein may be involved in the transport of iron across the intestinal cell to the blood.

In blood in addition to the iron in hemoglobin, iron is completely bound by the protein transferrin [177,178]. Transferrin is a beta globulin with a molecular weight of 86,000 and has two iron-binding sites, each capable of binding one atom of ferric iron [179]. Transferrin is a carrier for iron and transports iron to the tissues for utilization. More transferrin iron-binding sites are present in the plasma than atoms of iron so that free iron is not present. Free iron in small quantities is toxic, so that in addition to transport, transferrin functions to prevent the occurrence of free iron in the plasma. Similar proteins are present in eggs [180] and milk [181].

Iron is transported to the tissues and is utilized for the iron-containing compounds or stored as ferritin or hemosiderin. The storage compounds are generally present in the highest concentration in liver, spleen, and bone marrow. Crystalline ferritin contains up to 20% iron in a central nucleus surrounded by a shell of protein (apoferritin) and is spherical in shape [182]. Apoferritin is a high-molecular-weight (460,000) globulin [183]. Hemosiderin is a relatively amorphous compound consisting mainly of ferric hydroxide and may contain as much as 35% iron [184]. It can be observed in tissues as a brown, granular, readily stainable pigment. At low level of iron storage, the ratios of ferritin to hemosiderin are relatively constant and ferritin predominates. As the amount of storage iron increases, however, iron is stored primarily as hemosiderin [185].

Toxicity

As implied from the discussion of iron metabolism, the normal individual is able to maintain iron homeostasis, and iron toxicity or overload does not occur. Iron overload in man was first described by Trousseau [186] and was characterized by portal cirrhosis, diabetes mellitus, and skin pigmentation. Subsequent investigations have suggested that the syndrome idiopathic hemochromatosis is a genetic disorder of iron absorption by the intestinal mucosa [187-189]. Iron may accumulate at a rate of 2 to 4 mg/day and is deposited primarily in the liver, heart, stomach, and skin. It usually appears at 50 to 60 years of age and results in progressive, multiple organ dysfunction. Death occurs as a result of cardiac arrhythmia, congestive heart failure, or hepatic failure [190].

Iron toxicity from excessive dietary consumption has been described in the African Bantu population [191,192], Manchurian population [193], alcoholics [194], and individuals consuming excess medicinal iron preparations [195]. Iron overload may also occur in conditions where iron utilization is impaired such as chronic hemolytic anemia [196], pyridoxine deficiency [197], and folic acid deficiency [198].

Acute iron toxicity can occur with ingestion of large quantities of iron supplements. Nausea, vomiting, and diarrhea develop, and in severe cases, hypotension, shock, or coma are observed. Serum iron levels are high and may exceed the total iron-binding capacity of the serum. High levels of serum iron may be reduced by administration of iron chelation agents (deferoxamine intravenously). Oral administration of phosphate

salts inhibits further absorption of iron by forming complexes that are
poorly absorbed [195].

Continuous feeding of high levels of iron in the diet results in a
decrease in the performance of livestock. In calves, as little as 400 ppm
of iron decreased feed intake and rate of gain [199], while levels as high
as 2000 ppm could be tolerated without clinical symptoms of toxicity [200].
In pigs, levels of about 5000 ppm of iron were required to produce a reduction in feed intake and rate of gain [201]. Associated with reduced
performance, high levels of dietary iron increased serum iron, decreased
iron-binding capacity, and decreased serum phosphorus levels [201-203].
Presumably, the high dietary iron levels complex phosphorus, reducing phosphorus availability and the amount of phosphorus available for calcification [201,203].

Injectible iron preparations are routinely used to prevent anemia in
newborn baby pigs. This practice has resulted in acute toxicity when pigs
from sows fed rations high in unsaturated fats and low in vitamin E are
treated [204]. Iron injection results in muscular degeneration similar to
vitamin E deficiency and death from hyperkalaemia in about 2 hr [205].
Patterson et al. have proposed that iron-catalyzed peroxidation is the cause
of the toxicity and that vitamin E-deficient animals are unable to metabolize the peroxide induced by the oxidant, iron [206].

Deficiency

Low iron intake generally results in a reduction of serum iron levels,
an increase in serum iron-binding capacity and anemia with lowered levels
of hemoglobin and myoglobin. Levels of iron-containing enzymes are also
reduced including cytochrome c oxidase, succinic dehydrogenase, and catalase
[207].

Interactions

In 1928, Hart et al. [208] reported the discovery that copper as well
as iron was necessary for normal red blood cell formation in the rat.
Studies of heme synthesis in copper-deficient swine indicated that heme was
synthesized as rapidly by copper-deficient animals as by supplemented animals. The authors concluded that copper was involved in iron transport or
for the synthesis of globulin [209]. Other studies suggested an accumulation of iron in the intestinal mucosa, liver, and spleen, but plasma iron
was lower than in copper-supplemented controls. The data suggest that

copper was necessary for mobilization and transport of iron from the intestine and liver for incorporation into the erythrocyte [210].

The anemia produced by copper deficiency is associated with low plasma copper and with a reduction in the plasma level of ceruloplasmin. Ceruloplasmin is a copper-containing **protein** synthesized by the liver and present in the plasma. It has the enzymatic activity to oxidize p-phenylenediamine and is able to catalyze the oxidation of ferrous ion to ferric ion, which is essential for the formation of transferrin and iron transport [211]. This seems to be the major link between iron and copper metabolism such that copper in ceruloplasmin oxidizes ferrous iron to ferric so that it may combine with transferrin and be transported to the site of utilization. Ceruloplasmin is required for the mobilization of iron from the liver since copper-deficient animals store iron excessively but release it when copper is replenished [212,213]. The releasing agent is ceruloplasmin because this protein, but not copper salts, induces the release of iron from iron-laden livers in perfused liver systems [214-216]. This relationship is also illustrated by the observation that in roosters injected with estrogen there is an increase in ceruloplasmin followed by an increase in serum iron levels but with no change in transferrin level [217].

Iron and copper are both present in the enzyme cytochrome c oxidase. This enzyme functions as the terminal step in the transfer of electrons to oxygen in the hydrogen-electron transport chain. An interaction might be expected since the enzyme contains two copper ions and two heme groups. Copper deficiency reduces the levels of cytochrome c oxidase in rats [218-220], chicks [221], and swine [222]. Iron deficiency also decreased the activity of cytochrome c oxidase in the rat and calf [223-226]. Catalase, another iron-containing **enzyme**, declines in iron deficiency [222-228] and may decline in copper deficiency although experimental results are variable [229,230].

In iron deficiency, the absorption of iron by the intestinal mucosa is increased, and the absorption of manganese, cobalt, **nickel**, chromium, and zinc is also increased [231-236]. These observations suggest that these metals share the same transport system within the intestinal cell. The order of uptake and transfer from the intestine to the iron-deficient animal was cobalt, iron, and manganese [232]. In lambs, high levels of manganese (up to 5000 ppm) have been shown to reduce hemoglobin levels and decrease the iron concentration in the liver, kidney, and spleen [237,238]. In rabbits and pigs fed diets containing 1250 or 2000 ppm manganese, hemo-

globin was depressed and an additional 400 ppm of iron prevented the effect.
With anemic animals lower levels of manganese reduced the rate of hemoglobin
regeneration [239].

In rats, dietary intakes of zinc of 5000 to 10,000 ppm results in hypochromic, microcytic anemia accompanied by high levels of zinc in the liver
and other tissues and reduced levels of copper, iron, cytochrome $\underline{c}$ oxidase,
catalase, and δ-aminolevimulinate dehydrase [240-245]. The anemia and associated biochemical changes can be partially overcome by addition of copper
[241,243] and completely alleviated by supplements of copper and iron [240,
242]. It seems that the anemia of zinc toxicity results in a copper deficiency and, secondarily, an iron deficiency. Wanger and Weswig [246] have
demonstrated that high dietary levels of zinc depress liver copper in rats
and that the major change is in the microsome and soluble fractions. This
emphasizes that the major effect of zinc is on copper metabolism, which in
turn alters the metabolism of iron and its utilization for hemoglobin synthesis.

Chronic cadmium ingestion has been demonstrated to produce anemia in
rats [247-249], pigs [250], and Japanese quail [251]. Iron administration
can prevent the anemia induced by cadmium [248-250]. Wanger [252] has observed that dietary addition of cadmium up to 50 ppm resulted in a linear
depletion of iron from the liver as cadmium increased. The greatest accumulation of cadmium was in the soluble fraction of the liver. Cadmium resulted
in a greater depletion of iron from the microsomal and soluble fractions than
from other subcellular fractions. Cadmium is known to be associated with a
low-molecular-weight, sulfhydryl-containing protein that also binds copper
and zinc [253,254]. At present, the relationship between the various binding
proteins is not clear, but it certainly is involved in the dietary interactions of several trace minerals.

Iron deficiency has been suggested to increase the toxicity of lead.
In rats, iron deficiency increased the retention of lead in liver, kidney,
and bone and decreased the urinary excretion of lead. The urinary excretion
of δ-aminolevulinic acid (a biochemical index of lead toxicity) was greater
for iron-deficient rats than for rats fed normal levels of iron [255].

COPPER

Introduction

While copper was recognized as being present in plant and animal tissues,
the first evidence of the essential role of the element was reported by Hart
et al. [208]. They reported that copper was required in addition to iron for

hemoglobin synthesis. Subsequent investigations have led to the identification of a number of biological compounds that contain copper and that are
significant in metabolism. Many of the compounds that require copper as
part of their molecular structure are directly related biochemically to a
defect observed in copper deficiency. Copper is associated with a number
of enzymes that are essential for normal metabolism, and proteins that
bind copper are also involved in copper transport and metabolism. Copper
metabolism and homeostasis have recently been reviewed [256].

Deficiency

Copper is plentiful in natural foods so that in most areas of the
world, mammals consume diets that contain adequate copper to meet the
minimum requirements [257]. There are areas in the world where livestock
become copper deficient [258]. Copper deficiency in animals results in reduced growth, anemia, defective connective tissue formation and bone abnormalities, loss of pigmentation of hair, loss of crimp in wool, lack of proper
myelination of the spinal cord, and poor reproduction. Many of these defects
can be directly related to a decrease in the activity of one of the copper-
containing enzymes.

The anemia associated with copper deficiency is directly related to
the reduction in plasma ceruloplasmin (ferridoxase I). Ceruloplasmin is a
copper-containing enzyme responsible for mobilization of iron and essential
for erythrocyte formation as discussed.

Defective connective tissue formation causes the death of copper-deficient chicks and pigs because of the rupture of major blood vessels [259-
261]. Hill et al. [262] demonstrated that the fragility of blood vessels
in copper deficiency is caused by a decrease in the activity of monoamine
oxidase. This enzyme contains four atoms of copper per molecule and has
been isolated from aorta tissue of cattle [263] and rabbits [264]. The
enzyme catalyzes the oxidative deamination of peptidyl lysine to form peptidyl α-aminoadipic-δ-semialdehyde. Four of these residues condense to
form desmosine that provides the cross-links in elastin and collagen [265,
266]. Decreases in amine oxidase have been observed in bones from copper-
deficient chicks [267] and suggest that the bone defects and arterial weakness may both be related to the formation of normal connective tissue cross-
links.

Achromatrichia occurs in the copper-deficient rat, rabbit, guinea pig,
cat, dog, goat, sheep, and cattle but has not been observed in the pig
[258]. The enzyme tyrosinase has been demonstrated to require copper [268].
Tyrosinase is essential in the first two steps in the conversion of tyrosine

to melanin pigment [269]. In addition to achromatrichia due to dietary
deficiency of copper, a genetic absence of tyrosinase results in albinism.

Cytochrome c oxidase, the terminal oxidase in the electron transport
chain, requires copper for catalytic activity [270,271]. The oxidase is
essential for cellular metabolism, and a copper deficiency decreases cytochrome c oxidase activity [218-230] and oxidative phosphorylation [220].
Gallagher [272] has suggested that a lack of cytochrome c oxidase caused
the lack of normal myelination in lambs in copper deficiency. He pointed
out that a lack of cytochrome c oxidase activity can lead to demyelination
[273] and that the enzyme is very low in copper-deficient, ataxic lambs.
The decline in enzyme is marked in the central nervous system of copper-
deficient lambs [274-276].

In addition, these enzymes, superoxide dismutase, ascorbic acid oxidase, δ-aminolevulinic acid dehydrase, dopamine-β-hydroxylase, and galactose
oxidase contain copper. The identification of copper in these enzymes has
contributed to the knowledge of the function of copper in biological systems.

Metabolism

Copper absorption was demonstrated by Gittin et al. [277] to involve
more than simple diffusion. An energy-dependent transfer as well as diffusion have been suggested [278,279]. Kirchgessner and Grassmann [280] have
demonstrated that L-amino acids facilitate the intestinal absorption of
copper. Other studies have shown that amino acids facilitate the transfer
of copper across cellular membranes [281,282]. The energy-dependent component of copper absorption may represent the transport of a copper-amino acid
complex [256].

Protein components may be involved in copper absorption. At least two
distinct proteins have been isolated from rat, beef, and chick duodenum.
One protein has the physical properties of superoxide dismutase, and the
second is a sulfhydryl-rich protein with a molecular weight of 10,000. This
protein is found in the cytosol and will bind cadmium and zinc as well as
copper [283-285]. The properties of this protein are very similar to metallothionein isolated from equine kidney [253,254]. It has been suggested
that metallothionein may serve as a supply of binding sites within the intestinal mucosa to insure an adequate supply of the metal for subsequent
absorption. It may also serve as a block to protect against absorption of
toxic amounts of copper and other trace elements [256].

After absorption from the intestine, copper is transported through the portal blood by albumin [283,286]. Ingested copper disappears rapidly from the plasma and is deposited in the liver [287,288]. Albumin binds one copper preferentially before binding other copper ions. This binding site involves an α-amino nitrogen, the imidazole nitrogen from histidine in position three and two peptide bonds [289-292]. Copper is also present in blood as a complex with amino acids [282,291], so that within the plasma ionic copper, copper-amino acid complexes, and albumin-bound copper are in equilibrium.

Absorbed copper is available to all tissues for the synthesis of the copper-containing compounds. The liver, however, seems to be the major organ controlling copper homeostasis. Hazelrig et al. [293] have partitioned copper metabolism in the liver into three distinct processes: (1) preparation of copper for secretion into bile, (2) temporary storage, and (3) synthesis of ceruloplasmin. Bile is the major excretory route for copper [287,294], and it is present in bile associated with amino acids and small peptides and nonspecifically with high-molecular-weight molecules [295,296].

Copper is found in all of the subcellular fractions of liver with the nuclear fraction and the mitochondrial and lysosome fraction containing about 20% each, the microsomal fraction about 10%, and the remainder in the cytosol [297-299]. The copper in the nuclear fraction is probably in association with nucleic acids and basic proteins and may be involved in bridges within polynucleotide strands [300]. Copper in the mitochondrial-lyosomal fraction is sequestered in lyosomes [301,302] and may be the copper that is excreted by the bile. A low-molecular-weight protein high in cystine has been identified in mitochondria [303]. The copper in the microsomal fraction probably represents copper being synthesized into copper proteins, primarily ceruloplasmin [246,304]. Copper in the cytosol is the major portion of the copper in the liver and is in association with a specific metal-binding protein [305,306]. This protein has been shown to have properties similar to metallothionein [283,284].

The copper in the subcellular fractions increases linearly as the intake of copper increases [297,298]. At low to moderate levels of copper intake, the level of copper in the cytosol is the most sensitive to dietary consumption [307]. At high levels of copper loading, the cytosol becomes saturated and marked increases occur in other subcellular fractions [297,

298]. The major increase in the copper-loaded liver is in the lysosomes
and probably is copper that is to be excreted into the bile [302]. In addition to copper status, hepatic copper also varies with age by increasing
during the intrauterine period, reaching a maximum at or before birth and
declining thereafter to adult levels [258].

Toxicity

Copper toxicity has been observed in animals under grazing conditions
but usually occurs through contamination of feeds with copper compounds.
Sheep and cattle are most sensitive, while pigs, rats, and chickens can
tolerate high dietary levels of copper. Chronic copper toxicity is characterized by copper accumulation in the liver, liver degeneration, hemoglobinuria, jaundice, and death [258,308,309]. Copper toxicity is discussed
in detail in Chapter 22 of this volume.

Interactions

A number of elements have been demonstrated to affect the metabolism
of copper including molybdenum and sulfate, zinc, cadmium, silver, and mercury. The relationship between copper-molybdenum and sulfate is discussed
in Chapter 22 of this volume.

High levels of zinc have been observed to produce anemia that can be
corrected by the addition of copper to the diet [14,243,310]. Zinc has
been shown to interact with copper at the site of absorption [311] and to
bind to the metal-binding protein in chick duodenum [285]. It also can
displace copper from metallothionein isolated from bovine duodenum. Similar observations have been made with respect to storage of copper in the
liver. Wanger and Weswig have shown that zinc accumulation in the liver is
associated with decreased concentrations of copper in the microsomes and
cytosol while zinc increases in these fractions [246]. Zinc has been shown
to compete with copper for the sulfhydryl binding sites on metallothionein
from bovine liver [284]. Wanger and Weswig [304] have also presented evidence to indicate that ceruloplasmin synthesis is decreased in animals fed
high-zinc diets which suggests that copper is not available for induction
of the synthesis of the protein or for incorporation into ceruloplasmin.
At high zinc levels, zinc competes for copper binding sites at the level
of absorption, liver storage, and synthesis of ceruloplasmin, resulting in
reduced availability of copper compounds.

Albumin has also been suggested as a site of antagonism between copper
and zinc. Evans and Hahn [312] have demonstrated that copper interferes

with absorption of zinc but does not interfere with the in vivo intestinal uptake of zinc in rats. In vitro observations indicated that copper inhibits zinc binding to albumin, suggesting that the transport of zinc and copper may also be involved in the interaction.

Cadmium and silver have also been shown to compete with copper at the site of absorption and storage for sulfhydryl compounds [311,313]. This would provide an explanation for the antagonistic effect of cadmium on growth and mortality in chicks that can be partially prevented by additional copper. Cadmium has also been demonstrated [304] to reduce the synthesis of ceruloplasmin which may suggest that cadmium interferes with the incorporation of copper into the copper protein at the microsomal level.

Silver has also been shown to depress growth and enhance copper insufficiency [14,304,313]. Silver seems to affect copper metabolism by a slightly different mechanism. Silver had no effect on copper absorption but increased the deposition of radioactive copper in the liver at the expense of blood copper [313]. Ceruloplasmin synthesis has been demonstrated to be depressed when silver is fed [304]. The effect of silver seems to be at the site of synthesis of ceruloplasmin rather than a competition with copper for protein binding sites during absorption.

Mercury has been suggested as a possible element to interact with copper. Attempts to demonstrate an interaction have not been successful [14,304,313].

Rubino et al. [314] reported an elevation in the copper content of erythrocytes in patients suffering from lead poisoning, suggesting an interaction between these elements. Klander et al. [315] reported that rats fed lead were anemic when they were copper deficient but not when a copper-adequate diet was fed. They also observed a reduction in ceruloplasmin and in serum zinc. Erythrocyte lead was inversely related to ceruloplasmin levels. Lead also seems to depress copper metabolism by competing with copper for essential binding sites in protein.

ZINC

Introduction

Raulin [316] in 1869 discovered that zinc was required for the growth of a mold, Aspergillus niger. In 1919, Birckner [317] reported that egg yolk, human milk, and cow's milk contained zinc and suggested that zinc was of nutritive value. Sommer and Lipman [318] demonstrated that zinc was essential for plant growth. Although attempts were made to demonstrate a requirement for zinc by animals, this was not accomplished until 1934 when Todd et al. [319] reported that zinc was essential for growth and normal development of rats. Keilin and Mann [320,321] first demonstrated that zinc was an integral and essential component of carbonic anhydrase from red blood cells. Since then at least 18 metalloenzymes have been shown to contain zinc [322]. The disease of swine, porcine parakeratosis, was demonstrated to be a zinc deficiency by Tucker and Salmon [323] in 1955, which has led to the general practice of supplementing swine feeds with zinc. O'Dell and Savage [324] reported in 1957 that zinc was essential for birds also. At present at least 15 animal species, including man, have been shown to require zinc [325].

In man, Prasad et al. [326,327] described individuals with growth retardation, extreme iron deficiency anemia, and retarded sexual maturation. They suggested that zinc deficiency might be involved and subsequently demonstrated that zinc sulfate significantly increased growth and accelerated sexual development [328-330].

Metabolism

Zinc absorption has not been clearly defined but seems to require uptake by the mucosal cell, transport across the cell, and release into the bloodstream. The small intestine is the active site of absorption with little absorption occurring in the stomach or colon [331]. Uptake of zinc by intestinal segments in vitro has been demonstrated. Metabolic inhibitors gave contradictory results so that a clear demonstration of active transport was not accomplished [332-335]. More recently, Kowarski et al. [336] have demonstrated in vitro that segments of rat jejunum transported zinc from mucosal to serosal sides. The transport was dependent on sodium, oxygen, and the presence of metabolizable hexose. In these studies where initial zinc concentrations were the same on both sides of the intestinal segment, the net transfer was observed to be from the serosal to mucosal side.

Fractionation of the soluble proteins from the jejunum indicated that zinc was bound by a protein with a molecular weight of about 100,000 and that the protein was different from the calcium-binding protein [336]. Starcher [285] demonstrated a protein in intestinal mucosa with a molecular weight of 10,000 that was able to bind zinc. In addition, a low-molecular-weight complex of zinc has been identified in the intestinal mucosa and lumen. This complex was not ionic zinc nor a zinc-amino acid complex [337].

The relationship of these components to zinc absorption is not clear at present.

About two-thirds of the zinc in serum or plasma is loosely bound while the remainder is firmly bound to protein [338]. The tightly bound zinc is primarily associated with an α-2-macroglobulin, and the function of this protein is not known [339]. The loosely bound zinc is associated with albumin and readily exchanges with radioactive zinc [340]. Albumin binding of zinc accounts for about 98% of the loosely bound zinc and amino acid-zinc complexes (primarily histidine-cysteine complexes) compose a small fraction [341].

Absorbed zinc is taken up rapidly by the pancreas, liver, kidney, and spleen and has a rapid turnover in these tissues [342]. Substantial amounts of zinc are also present in bone and muscle. The zinc content of many tissues does not decrease markedly in zinc deficiency [343-345]. In liver, 40 to 50% of the zinc is found in the soluble fraction, 20 to 30% is in the nuclear fraction, and the balance is in the microsomes (15 to 25%) and mitochondria (5 to 10%) [346]. Zinc in the liver cytosol seems to be complexed by metallothionein and may represent a storage form of zinc [246]. The function of zinc in the nucleus is not known, but zinc has been observed to be bound by nucleic acids [347] and to alter nucleic acid metabolism [348-350].

Zinc in the microsome fraction probably represents zinc that is in the process of being incorporated into zinc-containing enzymes. Some of the enzymes that contain zinc are alkaline phosphatase, glutamic dehydrogenase, lactic dehydrogenase, leucine aminopeptidase, and alcohol dehydrogenase [322].

Zinc is excreted primarily by its secretion into the intestine [351]. Studies in which radioactive zinc is injected indicate that zinc appears in the intestine in 10 to 20 min after injection and that it is secreted by the entire length of the intestinal tract. Secretion was maximal after 3 to 6 hr, and at 24 hr 12% of the dose was present in the intestinal tract and 6% had appeared in the feces [331]. Thus, much of the zinc that was secreted was not reabsorbed and seems to be unavailable to the animal. Urinary excretion was very small (0.2%) confirming earlier observations that urine is a minor excretory route [351,352]. In addition to secretion by the intestine, pancreatic secretions contain zinc and contribute to the zinc appearing in the intestine [353,354]. The mechanism that controls

zinc balance is not known, but Evans et al. [355] have suggested that the
amount of zinc in the mucosal cell has a regulating effect on the balance
of absorption and secretion since they observed changes in zinc content of
intestinal mucosa that were inversely related to dietary absorption. Kasar-
kis and Hoekstra [356] were unable to confirm this observation from analysis
of intestinal segments.

Deficiency

Zinc deficiency has been described in a number of species, and the
symptoms can include dermatitis, emaciation, alopecia, retarded growth,
anorexia, ocular lesions, and testicular atrophy [321,323,324,345,357].
The symptoms can be reversed by zinc supplementations with the possible
exception of testicular atrophy in severe deficiency [358]. Malformations
of chick embryos and rat fetuses have been described involving the brain,
mandibles, and limbs [359-361].

Skin lesions are common in zinc deficiency and the healing of wounds
has been observed to be delayed in zinc-deficient rats, cattle, and hamsters [362-365]. Zinc has been reported to improve wound healing in man
[366-369], although all studies have not demonstrated positive responses
[370,371]. Stephan and Hsu [372] found that in zinc-deficient rats, DNA
synthesis was reduced as indicated by reduced incorporation of radioactive
thymidine into the DNA of skin. They suggested that the impaired skin DNA
synthesis with the observation of a decrease in incorporation of radioactive
amino acids into skin proteins [373] may suggest an explanation for the relationship of zinc to healing and dermatitis.

The activities of a number of zinc-containing enzymes are altered in
a zinc deficiency. Alkaline phosphatase in serum of zinc-deficient animals
is decreased [374-376] and is due to the zinc deficiency and not reduction
of feed intake [375,376]. Carbonic anhydrase and lactic dehydrogenase are
normal in zinc-deficient pigs, rats, and chicks, but decreases have been
reported in carbonic anhydrase in blood of zinc-deficient calves and man
[330,345,377]. In tissues, decreases have been observed in lactic dehydrogenase, malic dehydrogenase, alcohol dehydrogenase, alkaline phosphatase,
and carboxypeptidase [343,378-380].

Zinc is necessary for normal reproduction in both sexes. In males,
zinc deficiency affects spermatogenesis and the development of the primary
and secondary sex organs. Atrophy of the seminiferous tubules with a decrease in maturing germ cells has been observed; mitotic activity is retained,

however [358, 381-383]. The transformation of spermatids to spermatozoa seems to be the earliest histological change in the zinc-deficient rat [358].

Severe zinc depletion of the female rat results in disruption of estrus cycles and infertility. Females fed a diet containing 9 ppm of zinc became pregnant when mated and gave birth to malformed or dead young [361]. Zinc deficiency imposed after mating results in difficulty during parturition with excessive bleeding, failure to consume afterbirth, and a lack of maternal behavior [384,385]. Feeding a zinc-deficient diet during lactation reduced milk production and zinc content, and the pups became zinc deficient [387].

Toxicity

Most species are relatively tolerant to excessive intakes of zinc. Rats, pigs, and poultry are able to tolerate levels of 1000 to 2000 ppm without effects [388-390]. Cattle and sheep fed levels of about 1000 ppm grow slowly, consume less feed, and utilize feed less efficiently than controls fed normal levels of zinc [391].

The growth depression that is associated with zinc toxicity is probably a result of reduced intake of unpalatable high-zinc diets [391]. Anemia develops in zinc toxicity and is accompanied by reduced levels of copper, iron, cytochrome oxidase, catalase, and δ-aminolevinate dehydrase [240-245]. The effects can be partially alleviated by supplements of copper [241,243] and can be prevented by copper and iron [240,242]. The major effect of high zinc seems to be a reduction in utilization of copper and affects iron secondarily to copper. These effects are discussed in more detail in the section on iron interactions.

Interactions

Tucker and Salmon [323] first demonstrated that increasing the calcium content of practical swine diets enhanced parakeratosis and reduced growth. Numerous investigations have confirmed a calcium-zinc antagonism when swine are fed diets containing plant protein [392-394]. Experiments with rats and pigs have indicated that when plant proteins are fed, animals have higher zinc requirements than when animal proteins are used [395-397]. The presence of plant proteins reduced the apparent absorption of zinc by rats and chicks [397-400]. Plant proteins contain phytate (the hexaphosphate ester of inositol) that is known to bind many trace elements and calcium in insoluble complexes. Oberleas et al. were able to demonstrate in vitro that ratios of calcium/zinc/phytate of 2:1:1 resulted in precipitation of 97% of the zinc as an insoluble complex [334]. These results suggest that calcium in the presence of phytate precipitates zinc in an unavailable complex in the intestine and reduces absorption of zinc [401]. Studies with casein as the protein source indicate that calcium will also reduce zinc absorption in the presence of high levels of phosphorus [402].

Cadmium and zinc have similar chemical properties and an interaction has been demonstrated between these two elements. Parizek [403] demonstrated that the testicular damage caused by injection of cadmium could be prevented by zinc. This effect has been confirmed [404,405] and extended to alleviation of cadmium toxicity in chicks [406], prevention of cadmium-induced interstitial cell tumors in rats and mice [135,407,408], and reduction in embryo malformation caused by intravenous injection of cadmium salts [409]. The metalloprotein metallothioneine is known to bind both cadmium and zinc. This may suggest that excess cadmium displaces zinc from binding sites while increased zinc allows zinc to compete and prevent cadmium binding.

Zinc and copper have been shown to interact at the sites of absorption and storage. Copper has also been reported to alleviate parakeratosis in swine [410-412], although less effectively than zinc. Low zinc absorption increases the deposition of copper in the liver so that an inverse relationship exists between the two elements in liver [410,411]. Conversely, increased levels of zinc decrease the toxicity of high dietary levels of copper [410-414].

CONCLUSIONS

Many interactions have been noted for the elements selenium, iron, copper, and zinc. The mechanisms that are involved are complex, but some similarities can be noted among the interactions of trace elements. Some of the interactions can be explained by the formation of unavailable inorganic compounds. For example, high dietary iron reduces phosphate availability by forming insoluble iron phosphates; calcium, zinc, and phytate form an unavailable complex stressing zinc status; selenium metabolism is altered by the presence of zinc or mercury that may be the formation of an inorganic complex.

A second type of interaction can be explained by the requirement of one mineral for the formation of a biological compound that is necessary for the

utilization of another element. Copper deficiency results in a reduction
in the synthesis of ceruloplasmin. Since ceruloplasmin is necessary for
the utilization of iron, symptoms of iron deficiency can occur.

The third basic mechanism of trace mineral interactions involves the
competitive binding of metals by macromolecules. A deficiency of a mineral
is the simplest example in which the compounds containing the mineral may
be affected. Prasad has made the following observations on the depletion
of zinc from the enzymes that require it [378]:

> In a growing cell there is a series of apoenzymes, each present
> at some low concentration and, with normal levels of zinc, the
> apoenzymes combine with metal, forming functional zinc enzymes.
> At limiting levels of zinc, the apoenzymes behave as a series of
> ligands, each competing for the available zinc ions according to
> their stability constants...but among apoenzymes or other macro-
> molecular ligands, there will be first one and then others whose
> decreasing activities will limit growth. Even when growth stops
> alogether, we can expect that some of the ligands with the tight-
> est affinities for zinc might still be completely satisfied.

In considering the effects of other trace elements on the compounds
containing a specific element, electronic configuration as proposed by
Matrone and Hill [14,15], the stability of the ligands formed with each
metal and the relative concentration of each metal are factors that will
determine the biological effect. The electronic configuration of the elements determines which elements will interact while the stability of the
complexes formed and the relative concentrations will determine the levels
of intake that are required to observe an effect. This mechanism is probably involved in many of the interactions noted. Some examples are the
copper and zinc antagonism and the interactions among cadmium, zinc, copper,
and selenium. Sulfhydryl-rich proteins are intimately involved in many of
these interactions, although the specific function of these proteins has
not been defined.

As the knowledge of the metabolism of a trace element expands, the interactions between elements can be more clearly defined. Thus, increased
knowledge of the metabolism of tin, nickel, vanadium, fluorine, and chromium
will probably reveal new interactions. The complexity of the interrelations
among elements emphasizes the necessity for defining as accurately and completely as possible the composition of diets consumed by man and animals
during investigations of mineral metabolism.

REFERENCES

1. K. Schwarz, D. B. Milne, and E. Vingard, Biochem. Biophys. Res. Commun., 40, 22 (1970).

2. K. Schwarz and D. B. Milne, Nature, 239, 333 (1972).

3. E. M. Carlisle, in Trace Element Metabolism in Animals, 2 (W. G. Hoekstra, J. W. Suttie, H. E. Ganther, and W. Mertz, eds.), University Park Press, Baltimore, 1974, p. 407.

4. K. Schwarz and D. B. Milne, Science, 174, 426 (1971).

5. L. L. Hopkins, Jr., in Trace Element Metabolism in Animals, 2 (W. G. Hoekstra, J. W. Suttie, H. E. Ganther, and W. Mertz, eds.), University Park Press, Baltimore, 1974, p. 397.

6. K. Schwarz and D. B. Milne, Bioinorg. Chem., 1, 331 (1972).

7. H. H. Messer, W. D. Armstrong, and L. Singer, in Trace Element Metabolism in Animals, 2 (W. G. Hoekstra, J. W. Suttie, H. E. Ganther, and W. Mertz, eds.), University Park Press, Baltimore, 1974, p. 425.

8. F. H. Nielsen and D. A. Ollerich, Fed. Proc., 33, 1767 (1974).

9. F. H. Nielsen and H. E. Sauberlich, Proc. Soc. Exp. Biol. Med., 134, 845 (1970).

10. F. H. Nielsen and D. J. Higgs, in Trace Substances in Environmental Health, Vol. 4 (D. D. Hemphill, ed.), University of Missouri, Columbia, 1971, p. 241.

11. F. H. Nielsen, in Trace Element Metabolism in Animals, 2 (W. G. Hoekstra, J. W. Suttie, H. E. Ganther, and W. Mertz, eds.), University Park Press, Baltimore, 1974, p. 381.

12. K. Schwarz and W. Mertz, Arch. Biochem. Biophys., 85, 292 (1959).

13. W. Mertz, Physiol. Rev., 49, 163 (1969).

14. C. H. Hill and G. Matrone, Fed. Proc., 29, 1474 (1969).

15. G. Matrone, in Trace Element Metabolism in Animals, 2 (W. G. Hoekstra, J. W. Suttie, H. E. Ganther, and W. Mertz, eds.), University Park Press, Baltimore, 1974, p. 91.

16. K. W. Franke, J. Nutr., 8, 597 (1934).

17. M. Polo, The Travels of Marco Polo (Revised from Marsden's translation and edited by M. Komroff), Liveright, New York, 1926, p. 81.

18. T. C. Madison, in Statistical Report on the Sickness and Mortality in the Army of the United States, January, 1855 to January, 1860 (U.S. Congress 36th, First Session), Senate Exch. Doc., 52, 37 (1860).

19. O. A. Beath, J. H. Draize, H. F. Eppson, C. S. Gilbert, and O. C. McCreary, J. Amer. Pharm. Ass., 23, 94 (1934).

20. O. A. Beath, C. S. Gilbert, and H. F. Eppson, Amer. J. Botany, 28, 887 (1941).

21. S. F. Trelease and H. M. Trelease, Science, 87, 70 (1938).

22. I. Rosenfeld and O. A. Beath, Selenium: Geobotany, Biochemistry, Toxicity, and Nutrition, Academic Press, New York, 1964, p. 61.

23. I. Rosenfeld and O. A. Beath, Selenium: Geobotany, Biochemistry, Toxicity, and Nutrition, Academic Press, New York, 1964, p. 81.

24. A. L. Moxon, S. Dakota Agr. Exp. Sta. Bull. No. 311, 1 (1937).

25. G. Patrias and O. E. Olson, Feedstuffs, 41 (42), 32 (1969).

26. B. G. Harmon, S. G. Cornelius, R. W. Bisby, M. G. Oldham, D. E. Ullrey, P. K. Ku, and J. A. Hitchcock, J. Animal Sci., 39, 976 (1974).

27. J. Kubota, W. H. Alloway, D. L. Carter, E. E. Cary, and V. A. Lazar, Agr. Food. Chem., 15, 448 (1967).

28. H. W. Lakin and D. F. Danison, in Symposium: Selenium in Biomedicine (O. H. Muth, ed.), AVI, Westport, 1967, p. 27.

29. W. V. Searight and A. L. Moxon, S. Dakota Agr. Exp. Sta. Bull. No. 5 (1945).

30. J. H. Draize and O. A. Beath, J. Amer. Vet. Med. Ass., 86, 753 (1935).

31. S. G. Knott, C. W. R. McCray, and W. T. K. Hall, Queensland J. Agr. Sci., 15, 43 (1958).

32. A. W. Halverson, T. Ding-Tsair, K. C. Tribwasser, and E. I. Whitehead, Toxicol. Appl. Pharmacol., 17, 151 (1970).

33. A. W. Halverson, I. S. Palmer, and P. L. Guss, Toxicol. Appl. Pharmacol., 9, 477 (1966).

34. R. E. Lemley and M. P. Merryman, Lancet, 61, 435 (1941).

35. A. J. Amor and P. Pringle, Bull. Hyg., 20, 239 (1945).

36. J. R. Glover, Ind. Med. Surg., 39, 26 (1970).

37. W. C. Cooper, in Symposium: Selenium in Biomedicine (O. H. Muth, ed.), AVI, Westport, 1967, p. 185.

38. M. I. Smith and R. D. Lillie, U.S. Public Health Serv., Nat. Inst. Health Bull., 174, 1 (1940).

39. M. I. Smith, E. F. Stohlmann, and R. D. Lillie, J. Pharmacol. Exp. Ther., 60, 449 (1937).

40. K. W. Franke and A. L. Moxon, J. Pharmacol. Exp. Ther., 58, 454 (1936).

41. C. Caravaggi, F. L. Clark, and A. R. B. Jackson, Res. Vet. Sci., 2, 146 (1970).

42. M. W. Glenn, R. Jensen, and L. A. Griener, Amer. J. Vet. Res., 25, 1479 (1964).

43. D. D. Magg, J. S. Osborn, and J. R. Clopton, Amer. J. Vet. Res., 21, 1049 (1960).

44. W. T. Miller and H. W. Schoening, J. Agr. Res., 56, 831 (1938).

45. D. A. Morrow, J. Amer. Vet. Med. Ass., 152, 1625 (1968).

46. H. E. Munsell, G. M. DeVaney, and M. H. Kennedy, U.S.D.A. Tech. Bull., 534 (1936).

47. K. Orstadius, Nature, 188, 1117 (1960).

48. O. A. Beath, J. H. Draize, and C. S. Gilbert, Wyo. Exp. Sta. Bull., 200 (1934).

49. O. A. Beath, H. F. Eppson, and C. S. Gilbert, Wyo. Exp. Sta. Bull, 206 (1935).

50. V. R. Potter and C. A. Elvehjem, Biochem. J., 30, 189 (1936).

51. V. R. Potter and C. A. Elvehjem, J. Biol. Chem., 117, 341 (1937).

52. K. P. McConnell and J. L. Hoffman, F. E. B. S. Lett., 24, 60 (1972).

53. K. P. McConnell, J. M. Hsu, J. L. Herrman, and W. L. Anthony, Proc. Soc. Exp. Biol. Med., 145, 970 (1974).

54. L. M. Cummings and J. L. Martin, Biochem., 6, 3162 (1967).

55. K. J. Jenkins, Can. J. Biochem., 46, 1417 (1968).

56. A. Schrift and T. K. Virupaksha, Biochim. Biophys. Acta, 100, 65 (1965).

57. I. Rosenfeld, Proc. Soc. Exp. Biol. Med., 111, 670 (1962).

58. K. Schwarz, Fed. Proc., 20, 666 (1961).

59. K. Schwarz and E. Sweeny, Fed. Proc., 23, 421 (1964).

60. K. P. McConnell and O. W. Portman, J. Biol. Chem., 195, 277 (1952).

61. H. E. Ganther, Biochem., 5, 1089 (1966).

62. J. L. Byard, Arch. Biochem. Biophys., 130, 156 (1969).

63. I. S. Palmer, R. P. Gunsalus, A. W. Halverson, and O. E. Olson, Biochim. Biophys. Acta, 208, 260 (1970).

64. F. Hofmeister, Arch. Exp. Pathol. Pharmakol., 33, 198 (1894).

65. J. R. Glover, Trans. Ass. Ind. Med. Offic., 4, 94 (1954).

66. K. P. McConnell and O. W. Portman, Proc. Soc. Exp. Biol. Med., 79, 230 (1952).

67. B. D. Obermeyer, I. S. Palmer, O. E. Olson, and A. W. Halverson, Toxic. Appl. Pharmacol., 20, 135 (1971).

68. M. I. Smith, Public Health Rep., 54, 1441 (1939).

69. A. W. Halverson, C. M. Hendrick, and O. E. Olson, J. Nutr., 56, 51 (1955).

70. A. M. Hurd-Karrer, Science, 78, 560 (1933).

71. A. M. Hurd-Karrer, Amer. J. Botany, 25, 666 (1938).

72. A. W. Halverson and K. J. Monty, J. Nutr., 70, 100 (1960).

73. H. E. Ganther and C. A. Baumann, J. Nutr., 77, 408 (1962).

74. A. W. Halverson, P. L. Guss, and O. E. Olson, J. Nutr., 77, 459 (1962).

75. A. L. Moxon, Science, 88, 81 (1938).

76. A. L. Moxon and K. P. DuBois, J. Nutr., 18, 447 (1939).

77. K. P. DuBois, A. L. Moxon, and O. E. Olson, J. Nutr., 19, 477 (1940).

78. M. Rhian and A. L. Moxon, J. Pharmacol. Exp. Ther., 78, 249 (1943).

79. A. L. Moxon, C. R. Paynter, and A. W. Halverson, J. Pharmacol. Exp. Ther., 85, 115 (1945).

80. C. M. Hendrick, H. L. Klug, and O. E. Olson, J. Nutr., 51, 131 (1953).

81. R. C. Wahlstrom, L. D. Kamstra, and O. E. Olson, J. Animal Sci., 14, 105 (1955).

82. O. A. Levander and L. C. Argrett, Toxicol. Appl. Pharmacol., 14, 308 (1969).

83. H. E. Ganther and C. A. Baumann, J. Nutr., 77, 210 (1962).

84. K. P. McConnell and D. M. Carpenter, Proc. Soc. Exp. Biol. Med., 137, 996 (1971).

85. H. E. Ganther and H. S. Hsieh, in Trace Element Metabolism in Animals, 2 (W. G. Hoekstra, J. W. Suttie, H. E. Ganther, and W. Mertz, eds.), University Park Press, Baltimore, 1974, p. 339.

86. A. L. Moxon and O. E. Olson, S. Dakota Agr. Exp. Sta. Rept., 53, 42 (1940).

87. A. L. Moxon, A. E. Schaefer, A. H. Lardy, K. P. DuBois, and O. E. Olson, J. Biol. Chem., 132, 785 (1940).

88. B. B. Westfall and M. I. Smith, J. Pharmacol. Exp. Ther., 72, 245 (1941).

89. K. Schwarz and C. M. Foltz, J. Amer. Chem. Soc., 79, 3293 (1957).

90. K. E. M. McCoy and P. H. Weswig, J. Nutr., 98, 383 (1969).

91. H. D. Hurt, E. E. Cary, and W. J. Visek, J. Nutr., 101, 761 (1971).

92. J. N. Thompson and M. L. Scott, J. Nutr., 97, 335 (1969).

93. J. N. Thompson and M. L. Scott, J. Nutr., 100, 797 (1970).

94. H. Dam, J. Nutr., 27, 193 (1944).

95. E. L. Patterson, R. Milstrey, and E. L. R. Stokstad, Proc. Soc. Exp. Biol. Med., 95, 617 (1957).

96. K. Schwartz, J. G. Bieri, G. M. Biggs, and M. L. Scott, Proc. Soc. Exp. Biol. Med., 95, 621 (1957).

97. H. Dam, G. K. Nielsen, I. Prange, and E. Sondergaard, Experientia, 13, 493 (1957).

98. K. J. Jenkins, L. M. Ewan, and J. D. McConachie, Poultry Sci., 44, 615 (1965).

99. C. C. Calvert, M. C. Neshiem, and M. L. Scott, Proc. Soc. Exp. Biol. Med., 109, 16 (1962).

100. H. Dam and E. Sondergaard, Experientia, 13, 494 (1957).

101. M. C. Nesheim and M. L. Scott, J. Nutr., 65, 601 (1958).

102. J. P. Willman, J. K. Loosli, S. A. Aedell, F. B. Morrison, and P. Olafson, J. Animal Sci., 4, 128 (1945).

103. J. R. Schubert, O. H. Muth, J. E. Oldfield, and L. R. Remmert, Fed. Proc., 20, 689 (1961).

104. C. Blincoe and W. B. Dye, J. Animal Sci., 17, 224 (1958).

105. C. Blincoe and D. W. Marble, Amer. J. Vet. Res., 21, 886 (1960).

106. M. A. Arnold, P. H. Weswig, O. H. Muth, and J. E. Oldfield, Proc. Soc. Exp. Biol. Med., 118, 75 (1965).

107. J. R. Schubert, O. H. Muth, J. E. Oldfield, and L. R. Remmert, Proc. Soc. Exp. Biol. Med., 104, 568 (1960).

108. J. W. McLean, G. G. Thompson, and J. H. Claxton, N. Z. Vet. J., 7, 47 (1959).

109. O. H. Muth, J. Amer. Vet. Med. Ass., 126, 355 (1955).

110. W. J. Hartley and A. B. Grant, Fed. Proc., 20, 679 (1961).

111. E. S. Erwin, W. Sterner, R. S. Gordon, L. J. Machlin, and L. L. Tureen, J. Nutr., 75, 45 (1961).

112. L. L. Hopkins, Jr., A. L. Pope, and C. A. Baumann, J. Animal Sci., 23, 674 (1964).

113. R. C. Ewan, C. A. Baumann, and A. L. Pope, J. Animal Sci., 27, 751 (1968).

114. A. L. Obel, Acta Pathol. Microbiol. Scand. Suppl. XCIV, 1 (1953).

115. R. C. Ewan, M. E. Wastell, E. J. Bicknell, and V. C. Speer, J. Animal Sci., 29, 912 (1969).

116. R. C. Ewan and M. E. Wastell, J. Animal Sci., 31, 343 (1970).

117. L. R. Glienke and R. C. Ewan, J. Animal Sci., 39, 975 (1974).

118. J. T. Rotruck, W. G. Hoekstra, and A. L. Pope, Nature, New Biol., 231, 223 (1971).

119. J. T. Rotruck, A. L. Pope, H. E. Ganther, and W. G. Hoekstra, J. Nutr., 102, 689 (1972).

120. J. T. Rotruck, A. L. Pope, H. E. Ganther, A. B. Swanson, D. G. Hafeman, and W. G. Hoekstra, Science, 179, 588 (1973).

121. L. Flohe, W. A. Gunzler, and H. H. Schock, F. E. B. S. Lett., 32, 132 (1973).

122. W. Nakamura, S. Hosoda, and K. Hayashi, Biochim. Biophys. Acta, 358, 251 (1974).

123. P. J. Smith, A. L. Tappel, and C. K. Chow, Science, 247, 392 (1974).

124. C. K. Chow and A. L. Tappel, J. Nutr., 104, 444 (1974).

125. D. G. Hafeman, R. A. Sunde, and W. G. Hoekstra, J. Nutr., 104, 580 (1974).

126. T. Noguchi, A. H. Cantor, and M. L. Scott, J. Nutr., 103, 1502 (1973).

127. S. T. Omaye and A. L. Tappel, J. Nutr., 104, 747 (1974).

128. W. G. Hoekstra, in Trace Element Metabolism in Animals, 2 (W. G. Hoekstra, J. W. Suttie, H. E. Ganther, and W. Mertz, eds.), University Park Press, Baltimore, 1974, p. 61.

129. A. B. Kar, R. P. Das, and B. Mukerji, Proc. Nat. Inst. Sci. India, 26B (Suppl.), 40 (1960).

130. K. E. Mason, J. A. Brown, O. J. Young, and R. R. Nesbit, Anat. Rec., 149, 135 (1964).

131. K. E. Mason and J. O. Young, in Symposium: Selenium in Biomedicine (O. H. Muth, ed.), AVI, Westport, 1967, p. 383.

132. J. Parizek, I. Ostadalova, I. Benes, and J. Pitha, J. Reprod. Fert., 17, 559 (1968).

133. J. Parizek, I. Ostadalova, I. Benes, and A. Babicky, J. Reprod. Fert., 16, 507 (1968).

134. R. E. Holmberg and V. H. Ferm, Arch. Environ. Health, 18, 873 (1969).

135. S. A. Gunn, T. C. Gould, and W. A. D. Anderson, Proc. Soc. Exp. Biol. Med., 128, 591 (1968).

136. J. Parizek, J. Kalouskovia, A. Babicky, J. Benes, and L. Pavilik, in Trace Element Metabolism in Animals, 2 (W. G. Hoekstra, J. W. Suttie, H. E. Ganther, and W. Mertz, eds.), University Park Press, Baltimore, 1974, p. 119.

137. S. A. Gunn and T. C. Gould, in Symposium: Selenium in Biomedicine (O. H. Muth, ed.), AVI, Westport, 1967, p. 395.

138. C. H. Hill, J. Nutr., 104, 593 (1974).

139. J. Parizek and I. Ostadalova, Experientia, 23, 142 (1967).

140. J. Parizek, I. Benes, V. Prochazkova, A. Babicky, J. Benes, and J. Lener, Physiol. Bohemoslov., 18, 108 (1969).

141. H. E. Ganther, C. Goudie, M. L. Sunde, M. J. Kopecky, P. Wagner, S. Oh, and W. G. Hoekstra, Science, 175, 1122 (1972).

142. J. H. Koeman, W. H. M. Peeters, C. H. M. Koudstaal-Hol, P. S. Tjioe, and J. J. M. de Goeij, Nature, 245, 385 (1973).

143. G. S. Stoewsand, C. A. Bache, and D. J. Lisk, Bull. Environ. Contam. Toxicol., 11, 152 (1974).

144. S. Potter and G. Matrone, J. Nutr., 104, 638 (1974).

145. A. E. Moffitt, Jr. and J. L. Clary, Res. Commun. Chem. Pathol. Pharmacol., 7, 593 (1974).

146. J. Parizek, A. Babicky, I. Ostadalova, J. Kalouskovia, and L. Pavlik, in Radiation Biology of the Fetal and Juvenile Mammal (M. R. Sikov and D. D. Mahlum, eds.), U.S. Atomic Energy Commission, Oak Ridge, 1969, p. 137.

147. J. Parizek, I. Ostadalova, and J. Kalouskovia, in Newer Trace Elements in Nutrition (W. Mertz and W. E. Cornatzer, eds.), Dekker, New York, 1971, p. 85.

148. Z. M. Hollo and Sz. Zlatarov, Naturwissenschaften, 47, 87 (1960).

149. W. Rusiecki and J. Brzezinski, Acta Polon. Pharmacol., 23, 74 (1966).

150. S. L. Shaver and K. E. Mason, Anat. Rec., 109, 382 (1951).

151. H. Dam, G. K. Nielsen, I. Prange, and E. Sondergaard, Nature, 182, 802 (1958).

152. A. T. Diplock, J. Green, J. Bunyan, D. McHale, and I. R. Muthy, Brit. J. Nutr., 21, 115 (1967).

153. J. Bunyan, J. Green, E. Murrell, A. T. Diplock, and M. A. Canthorne, Brit. J. Nutr., 22, 97 (1968).

154. P. Grasso, R. Abraham, R. Hendy, A. T. Diplock, L. Golberg, and J. Green, Exp. Mol. Pathol., 11, 186 (1969).

155. J. Bunyan, E. E. Edwin, and J. Green, Nature, 181, 1801 (1958).

156. K. Schwarz, E. E. Roginski, and C. M. Foltz, Nature, 183, 472 (1959).

157. E. D. Andrews, A. B. Grant, and B. J. Stephenson, N. Z. J. Agr. Res., 7, 17 (1964).

158. W. R. Wise, P. H. Weswig, O. H. Muth, and J. E. Oldfield, J. Animal Sci., 27, 1462 (1968).

159. O. Zinoffsky, Hoppe-Seyler's Z. Physio. Chem., 10, 16 (1886).

160. H. R. Mahler and D. G. Elowe, J. Amer. Chem. Soc., 75, 5769 (1953).

161. D. A. Richert and W. W. Westerfield, J. Biol. Chem., 209, 179 (1954).

162. A. B. Macollum, J. Physiol., 16, 268 (1894).

163. R. A. McCance and E. M. Widdowson, Lancet, II, 680 (1937).

164. R. A. McCance and E. M. Widdowson, J. Physiol., 94, 148 (1938).

165. P. F. Hahn, W. F. Bale, J. F. Ross, W. M. Balfour, and G. H. Whipple, J. Exp. Med., 78, 169 (1943).

166. S. Granick, J. Biol. Chem., 164, 737 (1946).

167. S. Granick, Physiol. Rev., 31, 489 (1951).

168. D. Van Campen, Fed. Proc., 33, 100 (1974).

169. J. Manis and D. Schachter, Amer. J. Physiol., 207, 893 (1964).

170. W. N. Pearson and M. B. Reich, J. Nutr., 99, 137 (1969).

171. C. H. Hill, Proc., 30, 236 (1971).

172. M. Worwood, A. Edwards, and A. Jacobs, Nature, 229, 409 (1971).

173. M. Worwood and A. Jacobs, Life Sci., 10, 1363 (1971).

174. W. Forth, H. Hubers, E. Hubers, and W. Rummel, in Trace Substances in Environmental Health, Vol. 6 (D. D. Hemphill, ed.), University of Missouri, Columbia, 1973, p. 121.

175. H. Hubers, E. Hubers, W. Forth, and W. Rummel, Life Sci., 10, 141 (1971).

176. S. Pollack, T. Compana, and A. Arcario, J. Lab. Clin. Med., 80, 322 (1972).

177. C. G. Holmberg and C. B. Laurell, Acta Chim. Scand., 1, 944 (1947).

178. A. L. Schade, R. W. Reinhart, and H. Levy, Arch. Biochem., 20, 170 (1949).

179. A. Ehrenberg and C. B. Laurell, Acta Chim. Scand., 9, 68 (1955).

180. J. H. Jandl, J. K. Inman, R. L. Simmons, and D. W. Allen, J. Clin. Inv., 38, 161 (1959).

181. P. L. Masson, J. F. Heremans, and Ch. Dive, Clin. Chim. Acta, 14, 735 (1966).

182. A. R. Muir, Quart. J. Exp. Physiol., 45, 192 (1960).

183. A. Rothen, J. Biol. Chem., 152, 679 (1944).

184. A. Shoden and P. Sturgeon, Nature, 189, 846 (1961).

185. A. Shoden and P. Sturgeon, Acta Haematol., 27, 33 (1962).

186. A. Trousseau, Clinique Medicale de l'Hotel Dieu de Paris, 2nd ed., Bailliere, Paris, 1865, p. 812.

187. I. B. Brick, Gastroenterology, 40, 210 (1961).

188. R. Williams, P. J. Scheuer, and S. Sherlock, Quart. J. Med., 31, 249 (1962).

189. L. W. Powell, C. B. Campbell, and E. Wilson, Gut, 11, 727 (1970).

190. F. F. Whitcomb, Jr., F. R. Lummix, Jr., J. A. Mejia, and E. Achkar, Lahey Clinic Foundation Bull., 18, 109 (1969).

191. J. Wainwright, S. African J. Lab. Clin. Med., 3, 1 (1957).

192. T. H. Bothwell and B. A. Bradlow, Arch. Pathol., 70, 279 (1960).

193. K. Huyeda, Jap. J. Med. Sci., 4, 91 (1939).

194. M. E. Conrad, Jr., A. Berman, and W. H. Crosby, Gastroenterology, 43, 385 (1962).

195. D. S. Fischer, R. Parkman, and S. C. Finch, J. Amer. Med. Ass., 218, 1179 (1972).

196. J. T. Ellis, I. Schulman, and C. H. Smith, Amer. J. Pathol., 30, 287 (1954).

197. D. L. Harrigan and J. W. Harris, Advan. Intern. Med., 12, 103 (1964).

198. M. S. Greenberg and N. D. Grace, Arch. Intern. Med., 125, 140 (1970).

199. J. F. Standish, C. B. Ammerman, C. F. Simpson, F. C. Neal, and A. Z. Palmer, J. Animal Sci., 29, 496 (1969).

200. L. J. Koong, M. B. Wise, and E. R. Barrick, J. Animal Sci., 31, 422 (1970).

201. K. Fugugouri, J. Animal Sci., 34, 573 (1972).

202. J. F. Standish and C. B. Ammerman, J. Animal Sci., 33, 481 (1971).

203. P. B. O'Donovan, R. A. Pickett, M. P. Plumlee, and W. M. Beeson, J. Animal Sci., 22, 1075 (1963).

204. N. Lannek, D. Lindberg, and G. Tollerz, Nature, 195, 1006 (1962).

205. D. S. P. Patterson, W. M. Allen, D. C. Thurley, and J. T. Done, Vet. Rec., 80, 333 (1967).

206. D. S. P. Patterson, W. M. Allen, D. C. Thurley, and J. T. Done, Biochem. J., 104, 2P (1967).

207. E. J. Underwood, Trace Elements in Human and Animal Nutrition, 3rd Ed., Academic Press, New York, 1971, p. 36.

208. E. B. Hart, H. Steenbock, J. Waddell, and C. A. Elvehjem, J. Biol. Chem., 77, 797 (1928).

209. G. R. Lee, G. E. Cartwright, and M. M. Wintrobe, Proc. Soc. Exp. Biol. Med., 127, 977 (1968).

210. G. R. Lee, S. Nacht, J. N. Lukens, and G. E. Cartwright, J. Clin. Inves., 47, 2058 (1968).

211. E. Frieden, in Trace Element Metabolism in Animals, 2 (W. G. Hoekstra, J. W. Suttie, H. E. Ganther, and W. Mertz, eds.), University Park Press, Baltimore, 1974, p. 105.

212. H. R. Marston, S. H. Allen, and S. L. Swaby, Brit. J. Nutr., 25, 15 (1971).

213. C. A. Owen, Jr., Amer. J. Physiol., 224, 514 (1973).

214. S. Oksaki and D. A. Johnson, J. Biol. Chem., $\underline{244}$, 5757 (1969).

215. S. Oksaki, D. A. Johnson, and E. Frieden, J. Biol. Chem., $\underline{246}$, 3018 (1971).

216. H. A. Ragen, S. Nacht, G. R. Lee, C. R. Bishop, and G. E. Cartwright, Amer. J. Physiol., $\underline{217}$, 1320 (1969).

217. J. Planas and E. Frieden, Amer. J. Physiol., $\underline{225}$, 423 (1973).

218. M. O. Schlutze, J. Biol. Chem., $\underline{129}$, 729 (1939).

219. M. O. Schlutze, J. Biol. Chem., $\underline{138}$, 219 (1941).

220. G. H. Gallagher, J. D. Judah, and K. R. Rees, Proc. Roy. Soc., London, Ser. B, $\underline{145}$, 134 (1956).

221. G. Matrone, Fed. Proc., $\underline{19}$, 659 (1960).

222. C. J. Gubler, G. E. Cartwright, and M. M. Wintrobe, J. Biol. Chem., $\underline{224}$, 533 (1957).

223. E. Beutler, J. Clin. Invest., $\underline{38}$, 1605 (1959).

224. E. Beutler and R. K. Blaisdell, Blood, $\underline{15}$, 30 (1960).

225. P. R. Dallman and H. C. Schwartz, J. Clin. Invest., $\underline{44}$, 1631 (1965).

226. P. R. Dallman, J. Nutr., $\underline{97}$, 475 (1969).

227. E. Grassmann and M. Kirchgessner, in Trace Element Metabolism in Animals, 2 (W. G. Hoekstra, J. W. Suttie, H. E. Ganther, and W. Mertz, eds.), University Park Press, Baltimore, 1974, p. 523.

228. M. O. Schlutze and K. A. Kuiken, J. Biol. Chem., $\underline{137}$, 727 (1941).

229. D. H. Adams, Biochem. J., $\underline{54}$, 328 (1953).

230. G. E. Cartwright, C. J. Gubler, J. A. Bush, and M. M. Wintrobe, Blood, $\underline{11}$, 143 (1956).

231. M. Diez-Ewald, L. R. Weintraub, and W. H. Crosby, Proc. Soc. Exp. Biol. Med., $\underline{129}$, 448 (1968).

232. A. B. R. Thompson and L. S. Valberg, Amer. J. Physiol., $\underline{223}$, 1327 (1972).

233. S. G. Schade, B. F. Felsher, B. F. Glader, and M. E. Conrad, Proc. Soc. Exp. Biol. Med., $\underline{134}$, 741 (1970).

234. W. Forth, in Trace Element Metabolism in Animals (C. F. Mills, ed.), Livingstone, Edinburgh, 1970, p. 298.

235. W. Forth and W. Rummel, in Intestinal Absorption of Metal Ions, Trace Elements, and Radionucleotides (S. C. Skoryna and D. Waldron-Edwards, eds.), Pergamon Press, Oxford, 1971, p. 173.

236. L. L. Hopkins, Jr. and M. A. Noble, Fed. Proc., $\underline{28}$, 299 (1969).

237. R. H. Hartman, G. Matrone, and G. H. Wise, J. Nutr., $\underline{55}$, 429 (1955).

238. N. W. Robinson, S. L. Hansard, D. M. Johns, and G. L. Robertson, J. Animal Sci., $\underline{19}$, 1290 (1960).

239. G. Matrone, R. H. Hartman, and A. J. Clawson, J. Nutr., $\underline{67}$, 309 (1959).

240. D. H. Cox and D. L. Harris, J. Nutr., $\underline{70}$, 514 (1960).

241. D. R. Grant-Frost and E. J. Underwood, Aust. J. Exp. Biol. Med. Sci., 36, 339 (1958).

242. A. C. Magee and G. Matrone, J. Nutr., 72, 233 (1960).

243. R. Van Reen, Arch. Biochem. Biophys., 46, 337 (1953).

244. R. Van Reen and P. B. Pearson, Fed. Proc., 12, 283 (1953).

245. I. J. Witham, Biochim. Biophys. Acta, 73, 509 (1963).

246. P. D. Whanger and P. H. Weswig, J. Nutr., 101, 1093 (1971).

247. M. Berlin and L. Friberg, Arch. Environ. Health, 1, 478 (1960).

248. R. J. Banis, W. G. Pond, E. F. Walker, Jr., and J. R. O'Connor, Proc. Soc. Exp. Biol. Med., 130, 802 (1969).

249. W. G. Pond and E. F. Walker, Jr., Nutr. Rep. Int., 5, 365 (1972).

250. W. G. Pond, E. F. Walker, Jr., and D. Kirtland, J. Animal Sci., 36, 1122 (1973).

251. M. E. Richardson, M. R. Spivey Fox, and B. E. Fry, Jr., J. Nutr., 104, 323 (1974).

252. P. D. Whanger, Res. Commun. Chem. Pathol. Pharmacol., 5, 733 (1973).

253. J. H. R. Kagi and B. L. Vallee, J. Biol. Chem., 235, 3460 (1960).

254. J. H. R. Kagi and B. L. Vallee, J. Biol. Chem., 236, 2435 (1961).

255. K. M. Six and R. A. Goyer, J. Lab. Clin. Med., 79, 128 (1972).

256. G. W. Evans, Physiol. Rev., 53, 535 (1973).

257. A. H. Schroeder, A. P. Nason, I. H. Tipton, and J. J. Balassa, J. Chronic Diseases, 19, 1007 (1966).

258. E. J. Underwood, Trace Elements in Human and Animal Nutrition, 3rd Ed., Academic Press, 1971, pp. 57-115.

259. B. L. O'Dell, B. C. Hardwick, G. Reynolds, and J. E. Savage, Proc. Soc. Exp. Biol. Med., 108, 402 (1961).

260. G. S. Shields, W. F. Coulson, D. A. Kimball, W. H. Carnes, G. E. Cartwright, and M. M. Wintrobe, Amer. J. Pathol., 41, 603 (1962).

261. W. F. Coulson and W. H. Carnes, Amer. J. Pathol., 43, 945 (1963).

262. C. H. Hill, B. Starcher, and C. Kim, Fed. Proc., 26, 129 (1967).

263. R. B. Rucker, L. F. Roensch, J. E. Savage, and B. L. O'Dell, Biochem. Biophys. Res. Commun., 40, 1391 (1970).

264. R. B. Rucker and W. Goetlich-Riemann, Proc. Soc. Exp. Biol. Med., 139, 286 (1972).

265. W. S. Chou, J. E. Savage, and B. L. O'Dell, Proc. Soc. Exp. Biol. Med., 128, 948 (1968).

266. W. S. Chou, J. E. Savage, and B. L. O'Dell, J. Biol. Chem., 244, 5785 (1969).

267. R. B. Rucker, H. E. Parker, and J. C. Rogler, J. Nutr., 98, 57 (1969).

268. A. B. Lerner, T. B. Fitzpatrick, E. Calkins, and W. H. Summerson, J. Biol. Chem., 187, 793 (1950).

269. A. B. Lerner and T. B. Fitzpatrick, Physiol. Rev., 30, 91 (1950).

270. W. W. Wainio, C. V. Vanderwende, and N. F. Shimp, J. Biol. Chem., 234, 2433 (1959).

271. D. E. Griffiths and D. C. Wharton, J. Biol. Chem., 236, 1850 (1961).

272. C. H. Gallagher, Aust. Vet. J., 33, 311 (1957).

273. E. W. Hurst, Aust. J. Exp. Biol. Med. Sci., 20, 297 (1942).

274. J. McC. Howell and A. N. Davidson, Biochem. J., 72, 365 (1959).

275. C. F. Mills and R. B. Williams, Biochem. J., 85, 629 (1962).

276. B. F. Fell, C. F. Mills, and R. Boyne, Res. Vet. Sci., 6, 170 (1965).

277. D. W. Gitlin, W. J. Hughes, and C. A. Janeway, Nature, 188, 150 (1960).

278. R. F. Crampton, D. M. Matthews, and R. Poisner, J. Physiol., 178, 111 (1965).

279. N. Marceau, N. Aspin, and A. Sass-Kortsak, Amer. J. Physiol., 218, 377 (1970).

280. M. Kirchgessner and E. Grassmann, in Trace Element Metabolism in Animals (C. F. Mills, ed.), Livingstone, Edinburgh, 1970, p. 277.

281. D. I. M. Harris and A. Sass-Kortsak, J. Clin. Invest., 46, 659 (1967).

282. P. Z. Neumann and A. Sass-Kortsak, J. Clin. Invest., 46, 646 (1967).

283. G. W. Evans and W. E. Cornatzer, Fed. Proc., 29, 695 (1970).

284. G. W. Evans, P. F. Majors, and W. E. Cornatzer, Biochem. Biophys. Res. Commun., 40, 1142 (1970).

285. B. C. Starcher, J. Nutr., 97, 321 (1969).

286. A. G. Bearn and H. G. Kunkel, Proc. Soc. Exp. Biol. Med., 85, 44 (1954).

287. J. A. Bush, J. P. Mahoney, H. Marouirtz, C. J. Gubler, G. E. Cartwright, and M. M. Wintrobe, J. Clin. Invest., 34, 1766 (1955).

288. S. B. Osborn, C. N. Roberts, and J. M. Walshe, Clin, Sci., 24, 13 (1963).

289. T. Peters, Jr., Biochim. Biophys. Acta, 39, 546 (1960).

290. T. Peters, Jr. and F. A. Blumenstock, J. Biol. Chem., 242, 1574 (1967).

291. B. Sarker and T. P. A. Kruck, in Biochemistry of Copper (J. Peisach, P. Aisen, and W. E. Blumberg, eds.), Academic Press, New York, 1966, p. 183.

292. T. Peters, Jr. and C. Hawn, J. Biol. Chem., 242, 1566 (1967).

293. J. B. Hazelrig, C. A. Owen, Jr., and E. Ackerman, Amer. J. Physiol., 211, 1075 (1966).

294. C. A. Owen, Jr. and J. B. Hazelrig, Amer. J. Physiol., 210, 1059 (1966).

295. G. W. Evans and W. E. Cornatzer, Proc. Soc. Exp. Biol. Med., 136, 719 (1970).

296. P. A. Farrer and S. P. Mistilis, Birth Defects, 4 (2), 14 (1968).

297. G. W. Evans, D. R. Myron, N. F. Cornatzer, and W. E. Cornatzer, Amer. J. Physiol., 218, 298 (1970).

298. G. Gregoriadis and T. L. Sourkes, Can. J. Biochem., 45, 1841 (1967).

299. H. Porter, W. Wiener, and M. Barker, Biochim. Biophys. Acta, 52, 419 (1961).

300. N. A. Berger and G. L. Eichhorn, Biochem., 10, 1857 (1971).

301. T. Barka, P. Scheuer, F. Schaffner, and H. Popper, Arch. Pathol., 78, 331 (1964).

302. M. A. Verity, J. K. Gambell, A. R. Reith, and W. J. Brown, Lab. Invest., 16, 580 (1967).

303. H. Porter, Biochim. Biophys. Acta, 229, 143 (1971).

304. P. D. Whanger and P. H. Weswig, J. Nutr., 100, 341 (1970).

305. A. G. Morell, J. R. Shapiro, and I. H. Scheinberg, in Wilson's Disease. Some Current Concepts (J. M. Walshe and J. N. Cumin, eds.), Oxford, Blackwell, 1961, p. 36.

306. J. R. Shapiro, A. G. Morell, and I. H. Scheinberg, J. Clin. Invest., 40, 1081 (1961).

307. D. B. Milne and P. H. Weswig, J. Nutr., 95, 429 (1968).

308. S. M. Wolff, Arch. Pathol., 69, 217 (1960).

309. J. R. Todd and R. H. Thompson, Brit. Vet. J., 121, 90 (1965).

310. S. E. Smith and E. J. Larson, J. Biol. Chem., 163, 29 (1946).

311. D. R. Van Campen and P. U. Scaife, J. Nutr., 91, 473 (1967).

312. G. W. Evans and C. Hahn, in Trace Element Metabolism in Animals, 2 (W. G. Hoekstra, J. W. Suttie, H. E. Ganther, and W. Mertz, eds.), University Park Press, Baltimore, 1974, p. 497.

313. D. R. Van Campen, J. Nutr., 88, 125 (1966).

314. G. F. Rubino, E. Pagliardi, V. Paito, and E. Giangrandi, J. Haematol., 4, 103 (1958).

315. D. S. Klauder, L. Murthy, and H. G. Petering, in Trace Substances in Environmental Health, Vol. 6 (D. D. Hemphill, ed.), University of Missouri, Columbia, 1973, p. 131.

316. J. Raulin, Ann. Sci. Natur. Botan. Biol. Vegetale, 11, 93 (1869).

317. V. Birckner, J. Biol. Chem., 38, 191 (1919).

318. A. L. Sommer and C. B. Lipman, Plant Physiol., 1, 231 (1926).

319. W. R. Todd, C. A. Elvehjem, and E. B. Hart, Amer. J. Physiol., 107, 146 (1934).

320. D. Keilin and T. Mann, Nature, 144, 442 (1939).

321. D. Keilin and T. Mann, Biochem. J., 34, 1163 (1940).

322. A. F. Parisi and B. L. Vallee, Amer. J. Clin. Nutr., 22, 1222 (1969).

323. H. F. Tucker and W. D. Salmon, Proc. Soc. Exp. Biol. Med., 88, 613 (1955).

324. B. L. O'Dell and J. E. Savage, Poultry Sci., 36, 459 (1957).

325. J. A. Halsted, J. C. Smith, Jr., and M. I. Irwin, J. Nutr., 104, 345 (1974).

326. A. S. Prasad, J. A. Halsted, and M. Nadimi, Amer. J. Med., 31, 532 (1961).

327. J. A. Halsted and A. S. Prasad, Trans. Amer. Clin. Climat. Ass., 72, 130 (1960).

328. A. S. Prasad, A. Miale, Jr., Z. Farid, H. H. Sandstead, and A. R. Schulert, J. Lab. Clin. Med., 61, 537 (1963).

329. A. S. Prasad, A. Miale, Jr., Z. Farid, H. H. Sandstead, A. R. Schulert, and W. J. Darby, Arch. Intern. Med., 111, 407 (1963).

330. H. H. Sandstead, A. S. Prasad, A. R. Schulert, Z. Farid, A. Miale, Jr., S. Bassilly, and W. J. Darby, Amer. J. Clin. Nutr., 20, 422 (1967).

331. A. H. Methfessel and H. Spencer, J. Appl. Physiol., 34, 58 (1973).

332. W. N. Pearson, T. Schwenk, and M. Reich, in Zinc Metabolism (A. S. Prasad, ed.), Charles C. Thomas, Springfield, 1966, p. 239.

333. D. Oberleas, M. E. Muhrer, and B. L. O'Dell, in Zinc Metabolism (A. S. Prasad, ed.), Charles C. Thomas, Springfield, 1966, p. 225.

334. D. Oberleas, M. E. Muhrer, and B. L. O'Dell, J. Nutr., 90, 56 (1966).

335. B. M. Sahagian, I. Harding-Barlow, and H. M. Perry, Jr., J. Nutr., 90, 259 (1966).

336. S. Kowarski, C. S. Blair-Stanek, and D. Schachter, Amer. J. Physiol., 226, 401 (1974).

337. C. Hahn and G. W. Evans, Proc. Soc. Exp. Biol. Med., 144, 793 (1973).

338. H. P. Wolff, Klin. Wochenschr., 34, 409 (1956).

339. A. F. Parisi and B. L. Vallee, Biochem., 9, 2421 (1970).

340. F. A. Suso and H. M. Edwards, Jr., Proc. Soc. Exp. Biol. Med., 137, 306 (1971).

341. E. L. Giroux and R. I. Henkin, Biochim. Biophys. Acta, 273, 64 (1972).

342. J. P. Feaster, S. L. Hansard, J. T. McCall, F. H. Skipper, and G. K. Davis, J. Animal Sci., 13, 781 (1954).

343. A. S. Prasad, D. Oberleas, P. Wolf, and J. P. Horwitz, J. Clin. Invest., 46, 549 (1967).

344. A. S. Prasad, D. Oberleas, P. Wolf, J. P. Horwitz, E. R. Miller, and R. W. Luecke, Amer. J. Clin. Nutr., 22, 628 (1969).

345. J. K. Miller and W. J. Miller, J. Dairy Sci., 43, 1854 (1960).

346. B. Alfaro and F. W. Heaton, Brit. J. Nutr., 32, 435 (1974).

347. U. Weser, Structure and Bonding, 5, 41 (1968).

348. U. Weser, L. Hubner, and A. Jung, F. E. B. S. Lett., 7, 356 (1970).

349. U. Weser, S. Seeber, and P. Warnecke, Biochim. Biophys. Acta, 179, 422 (1969).

350. U. Weser and H. Moller, Z. Klin. Chem. Klin. Biochem., 8, 137 (1970).

351. R. A. McCance and E. M. Widdowson, Biochem. J., 36, 692 (1942).

352. A. H. Methfessel and H. Spencer, J. Appl. Physiol., 34, 63 (1973).

353. E. B. Miller, A. Sorscher, and H. Spencer, Rad. Res., 22, 216 (1964).

354. M. L. Montgomery, G. E. Sheline, and I. L. Charkoff, J. Exp. Med., 78, 151 (1943).

355. G. W. Evans, C. I. Grace, and C. Hahn, Proc. Soc. Exp. Biol. Med., 143, 723 (1973).

356. E. J. Kasarskis and W. G. Hoekstra, Proc. Soc. Exp. Biol. Med., 145, 508 (1974).

357. E. A. Ott, W. H. Smith, M. Stob, and W. M. Beeson, J. Nutr., 82, 41 (1964).

358. G. H. Barney, M. C. Orgebin-Crist, and M. P. Macapinlac, J. Nutr., 95, 526 (1968).

359. D. L. Blamberg, U. B. Blackwood, W. C. Supplee, and G. F. Combs, Proc. Soc. Exp. Biol. Med., 104, 217 (1960).

360. L. S. Hurley, Amer. J. Clin. Nutr., 22, 1332 (1969).

361. L. S. Hurley and H. Swenerton, Proc. Soc. Exp. Biol. Med., 123, 692 (1966).

362. W. H. Strain, W. J. Pories, and J. R. Hinshaw, Surgeons Forum, 11, 291 (1960).

363. H. H. Sandstead and G. H. Shepard, Proc. Soc. Exp. Biol. Med., 128, 687 (1968).

364. W. J. Miller, J. D. Morton, W. J. Pitts, and C. M. Clifton, Proc. Soc. Exp. Biol. Med., 118, 427 (1965).

365. I. U. Lavy, Brit. J. Surg., 59, 194 (1972).

366. W. J. Pories and W. H. Strain, in Zinc Metabolism (A. S. Prasad, ed.), Charles C. Thomas, Springfield, 1966, p. 378.

367. W. J. Pories, J. H. Henzel, C. G. Rob, and W. H. Strain, Ann. Surg., 165, 432 (1967).

368. W. J. Pories, J. H. Henzel, C. G. Rob, and W. H. Strain, Lancet, I, 121 (1967).

369. T. Hallbook and E. Lanner, Lancet, II, 780 (1972).

370. P. J. Barcia, Ann. Surg., 172, 1048 (1970).

371. M. B. Myers and G. Cherry, Amer. Surg., 37, 167 (1971).

372. J. K. Stephan and J. M. Hsu, J. Nutr., 103, 548 (1973).

373. J. M. Hsu and W. L. Anthony, J. Nutr., 101, 445 (1971).

374. W. J. Miller, W. J. Pitts, C. M. Clifton, and J. D. Morton, J. Dairy Sci., 48, 1329 (1965).

375. R. W. Luecke, M. E. Holman, and B. V. Baltzer, J. Nutr., 94, 344 (1968).

376. E. R. Miller, R. W. Luecke, D. E. Ullrey, B. V. Baltzer, B. L. Bradley, and J. A. Hoefer, J. Nutr., 95, 278 (1968).

377. J. K. Miller and W. J. Miller, J. Nutr., 76, 467 (1962).

378. A. S. Prasad, D. Oberleas, P. Wolf, and J. P. Horwitz, J. Lab. Clin. Med., 73, 486 (1969).

379. C. F. Mills, J. Quarterman, R. B. Williams, A. C. Dalgarno, and B. Panic, Biochem. J., 102, 712 (1967).

380. J. M. Hsu, J. K. Aniline, and D. E. Scanlan, Science, 153, 882 (1966).

381. I. Diamond, H. Swenerton, and L. S. Hurley, J. Nutr., 101, 77 (1971).

382. M. J. Millar, M. I. Fischer, P. V. Elcoate, and C. A. Mawson, Can. J. Biochem. Physiol., 36, 557 (1958).

383. M. J. Millar, P. V. Elcoate, M. I. Fischer, and C. A. Mawson, Can. J. Biochem. Physiol., 38, 1457 (1960).

384. J. Apgar, Amer. J. Physiol., 215, 160 (1968).

385. J. Apgar, Amer. J. Physiol., 215, 1478 (1968).

386. J. Apgar, J. Nutr., 103, 973 (1973).

387. P. B. Mutch and L. S. Hurley, J. Nutr., 104, 828 (1974).

388. V. G. Heller and A. D. Burke, J. Biol. Chem., 74, 85 (1927).

389. W. R. Sutton and V. E. Nelson, Proc. Soc. Exp. Biol. Med., 36, 211 (1937).

390. M. F. Brink, D. E. Becker, S. W. Terrill, and A. H. Jensen, J. Animal Sci., 18, 836 (1959).

391. E. A. Ott, W. H. Smith, R. B. Harrington, and W. M. Beeson, J. Animal Sci., 25, 414 (1966).

392. J. H. Conrad and W. M. Beeson, J. Animal Sci., 16, 589 (1957).

393. P. K. Lewis, Jr., W. G. Hoekstra, and R. H. Grummer, J. Animal Sci., 16, 578 (1957).

394. R. W. Luecke, J. A. Hoefer, W. S. Brammell, and D. A. Schmidt, J. Animal Sci., 16, 3 (1957).

395. W. H. Smith, M. P. Plumlee, and W. M. Beeson, J. Animal Sci., 21, 399 (1962).

396. D. Oberleas, M. E. Muhrer, and B. L. O'Dell, J. Animal Sci., 21, 57 (1962).

397. R. M. Forbes and J. M. Yohe, J. Nutr., 70, 53 (1960).

398. J. E. Savage, J. M. Yohe, E. E. Pickett, and B. L. O'Dell, Poultry Sci., 43, 420 (1964).

399. H. M. Edwards, Jr., Poultry Sci., 45, 421 (1966).

400. D. A. Heath and W. G. Hoekstra, J. Nutr., 85, 367 (1965).

401. B. L. O'Dell, Amer. J. Clin. Nutr., 22, 1315 (1969).

402. D. A. Heath, W. M. Becker, and W. G. Hoekstra, J. Nutr., 88, 331 (1966).

403. S. A. Gunn, T. C. Gould, and W. A. D. Anderson, Arch. Pathol., 71, 274 (1961).

404. S. A. Gunn, T. C. Gould, and W. A. D. Anderson, Amer. J. Pathol., 42, 685 (1963).

405. K. E. Mason and J. O. Young, Anat. Rec., 159, 311 (1967).

406. C. H. Hill, G. Matrone, W. L. Payne, and C. W. Barber, J. Nutr., 80, 227 (1963).

407. S. A. Gunn, T. C. Gould, and W. A. D. Anderson, J. Nat. Cancer Inst., 31, 745 (1963).

408. S. A. Gunn, T. C. Gould, and W. A. D. Anderson, J. Reprod. Fert., 15, 65 (1968).

409. V. H. Fern and S. J. Carpenter, Nature, 216, 1123 (1967).

410. J. A. Hoefer, E. R. Miller, D. E. Ullrey, H. D. Ritchie, and R. W. Luecke, J. Animal Sci., 19, 249 (1960).

411. H. D. Ritchie, R. W. Luecke, B. V. Baltzer, E. R. Miller, D. E. Ullrey, and J. A. Hoefer, J. Nutr., 79, 117 (1963).

412. H. D. Wallace, J. T. McCall, B. Bass, and G. E. Combs, J. Animal Sci., 19, 1153 (1960).

413. N. F. Suttle and C. F. Mills, Brit. J. Nutr., 20, 135 (1966).

414. N. F. Suttle and C. F. Mills, Brit. J. Nutr., 20, 149 (1966).

22

COPPER/MOLYBDENUM TOXICITY IN ANIMALS

William B. Buck*
Iowa State University College of Veterinary Medicine
Ames, Iowa

Copper is essential not only to life in mammals but also to plants and lower forms of organisms. It has varied and numerous biologic effects not only as an essential element but also as a toxicant. These effects are greatly influenced by factors such as form and level of copper exposure, species of organism, diet, disease state, and individuality.

The metabolism of copper is highly complex not only because of its varied functions but also because of its interaction with many other trace elements such as molybdenum, zinc, iron, cobalt, manganese, and the sulfate ion. Several excellent reviews concerning the essentiality, metabolic interrelationships, and toxicity of copper have been published [1-9].

In mammals, a copper deficiency results in anemia, reduced hematopoiesis, and cardiovascular lesions; defects in pigmentation, keratinization, bone formation, reproduction, myelination of the spinal cord, and connective tissue formation; and reduced growth. Chronic excess copper intake results in sudden release of copper from the hepatic storage sites into the bloodstream causing hemolysis, icterus, and anemia with accompanying hepatic and renal necrosis. Much has been written concerning the various dietary and environmental factors that contribute to a net copper imbalance, both deficiencies and excesses.

INTERRELATIONSHIP BETWEEN COPPER
AND OTHER ELEMENTS AND MOLECULES

The metabolism of copper, molybdenum, and the sulfate anion (SO_4^{2-}) is complex and interrelated, especially in ruminant species. Other elements,

*Present Affiliation: University of Illinois College of Veterinary Medicine, Urbana, Illinois.

including zinc and iron, also interact with copper, especially in nonruminant species such as swine, poultry, and laboratory animals.

The first evidence of the relation between copper and molybdenum metabolism was obtained when the drastic scouring disease of cattle known as
teart was shown to be a manifestation of chronic molybdenum poisoning and
could be controlled by treating the cattle with large amounts of copper [1,
10]. Additional evidence of copper-molybdenum interaction was reported by
Dick and Bull [11] when molybdenum was found to be an effective treatment
for copper poisoning in sheep. From these investigations came the realiza-
tion of the profound interrelationship of molybdenum and copper and the
dependence of this interaction upon inorganic sulfate in the diet. Subsequent investigations have quantified these interactions in ruminants [1,6,
12-26].

The combined effects of copper, molybdenum, and sulfate are much less
marked in the nonruminant. Gipp et al. [27] and Hays and Kline [28] were
unable to demonstrate any effect of molybdenum and sulfate on the liver
storage of pigs fed varying levels of copper. Dale [29], who observed similar results, also observed a depression of ceruloplasmin levels when sulfate was added to swine diets containing about 10 ppm of copper. Cromwell
[30] reported that a combination of molybdenum and sulfate were ineffective
in preventing the depressive effects upon weight gains, hematological
changes, and the buildup of liver copper stores in swine associated with
feeding a high (500 ppm) level of dietary copper; however, a combination of
molybdenum and sulfide appeared to be quite effective in preventing excessive copper accumulation in the liver of pigs fed a diet containing a high
level of copper. In rats fed copper-deficient diets, Gray and Daniel [31]
observed greater growth reduction when sulfate was added to high levels of
molybdenum. In diets with adequate copper, no effects were noted.

Zinc and iron have a profound effect upon copper metabolism in nonruminant animals, especially swine. Both zinc and iron have been shown to
protect swine from the adverse effects of high levels (250-750 ppm) of
dietary copper [32-35]. Similarly, zinc and iron deficiency tended to accentuate copper toxicity in swine [35].

In rats, copper was shown to prevent the occurrence of anemia and reduced liver catalase and cytochrome oxidase associated with zinc toxicosis
but did not prevent the growth inhibition produced by zinc [36,37]. Subsequent studies have revealed an influence of zinc on copper metabolism in
rats [38].

There is some indication that the sources or quality of dietary protein may also be a factor in these interrelationships [32]. Suttle and Mills [35] observed severe copper toxicosis in swine receiving whitefish meal but not in those receiving soybean-oil meal with both diets containing up to 425 ppm copper. Other workers did not obtain such decisive results [39,40]. At any rate, there appears to be a regional difference in susceptibility of swine to dietary copper [41]. Significantly, Gregoriadis and Sourkes [42] demonstrated that protein synthesis is required for the removal of copper from the liver of the rat. It is also possible that the effects of dietary protein source upon copper toxicity are related to their concentration of elements such as zinc and iron.

Severe anemia is a prominent manifestation of copper deficiency in swine and other animals [1,43,44]. Copper deficiency is first manifested by a slow depletion of body copper stores, including blood plasma. The type of anemia associated with copper deficiency is identical to that caused by iron deficiency and appears to be a defect in hemoglobin synthesis. Lee et al. [45], however, showed that copper-deficient pigs did not have abnormalities in the heme biosynthetic pathway. On the contrary, as the anemia developed, the activity of heme biosynthetic enzymes increased. Copper, it was concluded, is not a part of heme biosynthesis. For many years it was accepted that iron absorption was unaffected by copper deficiency but that copper was necessary for iron utilization by the blood-forming tissues. Chase et al. [46] showed that copper deficiency in rats reduced their ability to absorb iron, to mobilize iron from the tissues, and to utilize iron in hemoglobin synthesis. Subsequently, Lee et al. [47] confirmed that copper-deficient pigs failed to absorb iron at a normal rate and observed increased amounts of iron in the duodenal mucosa. When radioiron was administered orally, the mucosa of copper-deficient animals extracted iron from the duodenal lumen at a normal rate, but transfer to the plasma was impaired. These and other experiments provided evidence that copper deficiency causes an impaired ability of duodenal mucosa, the reticuloendothelial system, and the hepatic parenchymal cells to release iron into the plasma. This hypothesis is compatible with the suggestion that the transfer of iron from tissues to plasma requires the enzymic oxidation of ferrous iron and that ceruloplasmin is the enzyme (ferroxidase) that catalyzes the reaction. The authors [47] proposed an additional defect in iron metabolism in copper deficiency, residing within the normoblast itself. The excessive levels of iron in normoblasts suggested that a defect in these cells plays a major role in

the development of anemia. As a result of this defect, iron cannot be incorporated into hemoglobin and, instead, accumulates as nonhemoglobic iron.

In 1969, Moore [147] suggested an inverse relationship between vitamin
A and copper metabolism. He suggested that in the latter stages of human
gestation there is a decrease in maternal blood vitamin A but an increase
in copper, and that the fetal livers of animals usually have low levels of
vitamin A while concentrating high levels of copper. He further suggested
that studies of vitamin A and copper metabolism might give interesting results. Underwood [1] reviewed the role of copper in keratinization of
wool, although vitamin A was not mentioned. Subsequently, Moore et al. [48]
reported that experimental chronic copper poisoning in sheep was accompanied
by reduced plasma concentrations of retinol.

COPPER METABOLISM

Copper is absorbed from the small intestine of most species. Generally,
less than 30% of the copper consumed is absorbed in any species, and its
absorption and retention are greatly affected by the chemical forms in which
the metal is ingested, by the dietary levels of other minerals and organic
substances, and by the acidity of the intestinal contents in the absorptive
area [1]. Little is known, however, about the mechanism of copper absorption, although it is generally believed that a copper-protein binding or a
copper-metal or organic complex may be formed.

Copper entering the blood plasma from the intestine and from body tissues becomes loosely bound to serum albumin and is distributed widely to
the tissues and the erythrocytes. Copper in ceruloplasmin does not appear
to be so readily available for exchange or for transfer [1].

When copper reaches the liver, the primary organ of its metabolism,
it is incorporated into the mitochondria, microsomes, nuclei, and soluble
fraction of the parenchymal cells [49,50]. The copper is either stored in
these sites or released for incorporation into erythrocuprein, ceruloplasmin, or the various copper-containing enzymes. Hepatic copper is also secreted into the bile and excreted into the intestine. Smaller amounts of
plasma copper are also excreted into the urine [1,51]. The modes of action
of copper at the cellular level have been the subject of many basic re-
searches. Many copper-protein compounds have been isolated from living
tissues, many of which are enzymes with oxidative functions. Probably no
metal ion is more versatile than copper as a catalyst of enzymic reactions.

Tyrosinase, lactase, ascorbic acid oxidase, uricase, monoamine oxidase, δ-aminolevulinic acid dehydrase, dopamine-B-hydroxylase, and cytochrome oxidase have all been identified as copper enzymes [1]. The diminished activity of cytochrome oxidase is a sensitive indicator of copper deficiency [52]. Gallagher and Reeve [53] have suggested that an uncomplicated copper deficiency in the rat causes two major biochemical dysfunctions and that one is directly caused by the other. One, the loss of cytochrome oxidase activity, leads to the other, depressed synthesis of phospholipids by liver mitochondria, by interfering with the provision of endogenous ATP to maintain an optimal rate of synthesis. Copper undoubtedly is involved in many other biochemical functions, but many of the signs of deficiency apparently result from insufficient cytochrome oxidase. Copper is important to the formation and maintenance of myelin.

The mechanism(s) responsible for the copper-molybdenum-sulfate interaction is not entirely understood. There is good evidence that copper and molybdenum form an in vivo complex having a molar ratio of 4:3 [20,21,54] but that this complex may not prevent the intestinal absorption of copper. It is more likely that the copper-molybdenum complex inhibits copper and perhaps molybdenum utilization. As a tentative explanation of the mechanism of this interference, either an impairment of copper uptake by liver cells or a primary intracellular metabolic disturbance in the synthesis of copper-protein compounds, including ceruloplasmin or both, are postulated by Marcilese et al. [23]. Regardless of the mechanism(s), this phenomenon is dependent upon the presence of inorganic sulfate [16,23]. The copper-molybdenum-sulfate interactions also strongly affect urinary excretion of these elements. Dick [12] reported that increased urinary molybdenum excretion was associated with increased dietary inorganic sulfate, and Marcilese et al. [55] reported that increased dietary levels of molybdenum and sulfate resulted in increased urinary excretion of copper.

Buck [56] and Buck et al. [57] have suggested that copper added to sheep diets in the form of the sulfate is less toxic than when added as other salts such as oxide, carbonate, acetate, gluconate, iodide, chloride, orthophosphate, and pyrophosphate. This may be because the sulfate radical is available to prevent absorption of the copper or facilitate its excretion in the presence of molybdenum. Work by Todd et al. [58] demonstrated that the acetate salt of copper was more toxic to sheep than was the sulfate salt. If these clinical and experimental observations that copper sulfate

is less toxic to sheep than are other copper salts are indeed fact, then
much of the copper toxicity data reported in the literature should be reevaluated since most reports are based on studies with the sulfate form of
copper.

COPPER DEFICIENCIES

A variety of disorders in animals and man have been associated with
copper deficiencies. When discussing the deficiency of copper, however,
one should keep in mind that an important consideration is the relative
concentration of copper in relation to other elements and molecules such as
molybdenum, zinc, iron, and sulfate. Thus, the ratio of copper to these
dietary factors is just as significant as the actual level of copper in the
diet.

Disorders that have been associated with a relative copper deficiency
in various animal species include anemia, depressed growth, bone disorders,
depigmentation of hair and wool, abnormal wool growth, neonatal ataxia, impaired reproductive performance, heart failure, cardiovascular defects, and
gastrointestinal disturbances [1]. Many factors influence the severity of
these dysfunctions, especially species, age, dietary interrelationships,
environment, sex, and even breed or strain characteristics.

Anemia associated with copper deficiency occurs in most species and is
characteristic of the anemia associated with iron deficiency (see section
on interrelationships).

Bone abnormalities associated with copper deficiency have been reported
in rabbits, mice, chicks, dogs, pigs, foals, sheep, and cattle [1]. In ruminants, osteoporosis and spontaneous bone fractures are usually associated
with excess dietary molybdenum and thus a relative copper deficiency. Suttle
et al. [59] have presented evidence, however, of the development of osteoporosis in the offspring of ewes given copper-deficient diets.

Sheep suffering from simple copper deficiency and/or excess molybdenum
also develop depigmentation of dark wool together with loss of crimp and
quality of their fine wool [1,14]. In Australia a syndrome called enzootic
ataxia and in the United Kingdom a condition termed swayback are thought to
be due to copper deficiency. Ewes become anemic with stringy wool, which
corresponds with neurological signs in their lambs in enzootic ataxia;
whereas the typical case of swayback is more acute in lambs with the ewes'
wool being normal. These diseases are noted in lambs under 1 month of age.

Lambs are severely incoordinated, ataxic, and usually blind. Death is the result of starvation, exposure, or pneumonia [1,52]. Cordy [60] has also reported that this condition occurs in the United States.

Postmortem changes associated with excess molybdenum or deficient copper include harsh, stringy wool and hair; achromotrichia; microcytic, hypochromic anemia; emaciation, osteoporosis; bone rarification and fractures; and hemosiderosis. Lesions associated with enzootic ataxia and swayback in lambs are characterized by lysis of the white matter of the cerebrum and degeneration of the motor tracts of the spinal cord. The destruction of the white matter varies from microscopic foci to massive subcortical destruction. There is often neuronal degeneration as well as demyelination [52,60].

The first evidence of cardiovascular disorders in copper deficiency emerged from studies by Bennetts and co-workers [61-63] studying a disorder in cattle known as "falling disease." The primary lesion is progressive atrophy of the myocardium with replacement fibrosis. Sudden deaths characteristic of the disease were attributed to heart failure, usually after exercise or excitement. The disease can be prevented by copper supplementation. This condition has not been reported in sheep or horses but has occurred in pigs and chickens [64,65]. There is a derangement of the elastic tissue in major blood vessels resulting in spontaneous ruptures. The tensile strength of the aorta becomes markedly reduced and the myocardium becomes friable. The primary biochemical lesion has been described by Hill et al. [66] as a reduction in amine oxidase activity of the aorta, a copper-containing enzyme. This reduction in enzymic activity results in reduced capacity for deaminating lysine in elastin, which, in turn, results in less lysine being converted to desmosine, a cross-linkage group of elastin, and therefore lessened elasticity of the aorta and other major vessels. Rucker et al. [67] reported that reduced bone cytochrome oxidase activity was also associated with copper deficiency in chicks.

Cattle apparently are more susceptible than sheep to excess molybdenum and deficient copper in their diet [1,68,69]. When the ratio of copper to molybdenum in feed drops below 2:1, molybdenum poisoning can be expected in cattle [70]. This syndrome is manifested by emaciation, liquid diarrhea full of gas bubbles, swollen genitalia, anemia, and achromotrichia. Poor weight gains and death from prolonged purgation may occur. The average morbidity is about 80% [68]. Osteoporosis and bone fractures have been reported in prolonged cases of molybdenosis [1]. Copper deficiency-like syndromes have resulted from feeding of forages and grains grown on soils naturally high in molybdenum and/or low in copper [71]. In the United States, such soils have been found in California, Oregon, Nevada, and Florida [1,68,69]. Cattle grazing pastures on muck or shale soils in England, Ireland, New Zealand, and Holland have suffered severe molybdenosis [1].

Miltimore and Mason [70] apparently have made the first extensive report of molybdenum and copper concentrations and copper/molybdenum ratios in ruminant feeds. The overall mean copper/molybdenum ratio of all feeds in the area of Canada studied (legume hay, grass hay, sedge hay, oat forage, corn silage, and grains) was 5.7:1. The copper/molybdenum ratio in sedge hays was 2.1:1, near the critical ratio of 2:1. The mean ratio of other hays was 4.4:1, and the ratio for other feeds was 5:1 or higher. They reported that 19% of all samples had ratios below 2:1; the lowest copper/molybdenum ratio was 0.1:1 and the highest was 52.7:1. Molybdenum levels were generally low, 35% of all samples being below 1 ppm. Only 1% was above 8.0 ppm, and the highest molybdenum concentration was 9.9 ppm. Copper concentrations were generally low, 95% being below 10 ppm with legumes having 7.5 ppm and grass hays 3.3 ppm.

When the copper levels of feed or forages are in the normal range of 8 to 11 ppm, cattle can be poisoned on levels of molybdenum above 5-6 ppm and sheep on levels above 10-12 ppm. When the dietary copper level falls much below 8-11 ppm or the sulfate ion level is high, even 1-2 ppm molybdenum may be toxic to cattle. Increasing the copper level in the diet even 5 ppm above normal (13-16 ppm) will protect cattle against 150 ppm dietary molybdenum [1,14].

Copper deficiency in the nonruminant results in anemia, bone deformation and reduced calcification, cerebral edema and cortical necrosis, achromotrichia, fetal absorption, and aortic rupture [1]. The levels of ceruloplasmin and copper in serum and the levels of cytochrome oxidase and copper in tissues decrease in animals fed a copper-deficient diet [72-75].

Nonruminants are more tolerant of excess levels of molybdenum than ruminants. Pigs appear to be the most tolerant of the nonruminants since Davis [76] reported that a diet containing 1000 ppm molybdenum for a period of 3 months had no ill effects on pigs. In pigs, the storage of copper in the liver does not appear to be influenced by the level of molybdenum in the diet [28,77,78].

Molybdenum excess in rats can result in symptoms that are similar to copper deficiency. The level of molybdenum that is required to produce toxicity depends upon the copper status of the rat. Nielands et al. [79] and Gray and Daniel [31] have demonstrated that growth and hemoglobin levels of rats can be reduced by feeding 100 ppm of molybdenum when a diet low in copper is fed. When the diet contains adequate amounts of copper, 500-1000 ppm of molybdenum are required to cause such effects. In rats, the blood and liver copper level tends to increase when the dietary level of molybdenum is increased [80].

Supplemental ascorbic acid has been demonstrated to accentuate copper deficiency in chicks, swine, and rabbits. Hunt et al. [75] found that 0.5% dietary ascorbic acid reduced liver copper levels and increased mortality associated with the rupture of the aorta. Voelker and Carlton [81] have reported that 2.5% ascorbic acid in the diet of swine resulted in intensification of copper-deficiency symptoms. Ascorbic acid is thought to interfere with the absorption of copper.

COPPER TOXICITY

Ruminant Animals

When discussing the deficiency or toxicity of copper or molybdenum in ruminant animals, such as cattle and sheep, it should be understood that the interaction of these elements in the presence of inorganic sulfate makes it impossible to delineate between the toxicity of one and a deficiency of the other [82]. Thus, copper deficiency is manifested by the same syndrome as chronic molybdenum poisoning; and chronic copper poisoning is almost identical to molybdenum deficiency.

The various conditions under which copper poisoning (molybdenum deficiency) may occur are as follows:

1. Consumption of plants contaminated by copper-containing pesticides.

2. Copper sulfate used for the control of helminthiasis and foot rot.

3. Contamination of pasture in the vicinity of mines.

4. Consumption of pasture plants containing an imbalance of copper and molybdenum as in Australia where Trifolium subterraneum is used extensively to raise the nitrogen level of soils. Under adverse climatic conditions associated with an early autumn break, this clover may grow

abundantly.   Under these conditions, the plants contain little or no molybdenum [83,84].   Sheep grazing the plants store high levels of copper in the liver and become predisposed to hemolytic crisis of chronic copper poisoning.

5. Sheep in western Australia grazing on pastures containing various species of _Lupinus_, a plant well adapted to copper-deficient soils, appear to be predisposed to lupine toxicosis as a result of high levels of liver copper [148,149].   The practice of fertilizing soils with copper resulted in marked increases in liver copper storage when the lupine plants were consumed.

6. Chronic hepatogenous poisoning in sheep commonly occurs in Australia and New Zealand.   Plants of the _Heliotropium_, _Echium_, and _Senecio_ genera contain pyrrolizidine alkaloids that cause hepatic necrosis with resulting inability to metabolize and excrete normal dietary levels of copper [1,85].

7. Sheep kept under confinement conditions are more susceptible to copper toxicity [6,22].   This is also supported by the findings of Bracewell [86], who found that the typical syndrome in housed sheep receiving a small copper supplement in their ration with no molybdenum and sulfate continued up to 11 months after copper supplement had been withdrawn. Todd [82] has reported the occurrence of copper poisoning in housed sheep being fed oats and barley having low-normal levels of copper, 3.9 and 4.9 ppm, respectively, but with extremely low levels of molybdenum, less than 0.025 ppm [82].

8. Providing imbalanced mineral mixtures to sheep and cattle [9,17,18].

9. Consumption of complete feeds with a copper-molybdenum imbalance [18, 56,87,88].   This is a man-made problem in the United States and possibly other countries.   Copper is recognized as a safe and necessary ingredient for animal feeds by federal regulations.   Molybdenum is not recognized as safe and necessary and, by law, cannot be incorporated into animal feeds [56].   Mineral-vitamin premixes are routinely combined with other ingredients in the commercial formulation of complete feeds for cattle and sheep.

    Failure to recognize molybdenum as safe and necessary is based partly on the fact that molybdenum is accumulated in tissues and eliminated in milk when given in excessive amounts in the absence of copper and inorganic sulfate, thus creating a potential public health hazard. Apparently ignored, however, is the fact that under present regulations,

animals consuming a ration containing supplemental copper but no molybdenum may have copper concentrations of 2000 ppm or greater in the liver, which could also be hazardous [22].

If sheep, for instance, are fed a diet containing a normal concentration of copper (8-11 ppm) but with no molybdenum, copper toxicity
may result [89]. Therefore, when a vitamin-mineral preparation containing copper but no molybdenum is added to the ration, the copper
concentration of the ration may be elevated to 25-30 ppm or greater;
and since the natural molybdenum concentration in feed is usually low
(1-2 ppm), copper poisoning may occur. Over 20 such episodes were
encountered in Iowa during the period from June 1968 to June 1970 [56,
90]. Many episodes involved feeder lambs, show lambs, or ram lambs
being tested for weight gain and feed efficiency. Usually from 1 to 5%
of the flock was affected, and over 75% of those clinically affected
died.

10. Other sources of copper poisoning unrelated to feed mineral imbalance
include the use of copper formulations for pesticides and medicinal
agents. Bordeaux mixture ($CuSO_4$), 1-3%, has been used to spray orchards, resulting in poisoning of grazing animals. $CuSO_4$ as an anthel-
mintic and foot rot treatment has occasionally resulted in acute poisoning. In countries where copper deficiency problems are frequently
encountered in sheep, the use of CuCaEDTA as an injectable source of
copper has resulted in copper poisoning [91-93].

In acute copper poisoning, copper chloride ($CuCl_2$) is two to four times
more toxic than $CuSO_4$. Cattle are poisoned by 200-800 mg $CuSO_4$/kg body
weight, and sheep are poisoned by 20-100 mg/kg single dose.

When an animal consumes small but excessive amounts of copper over a
period of weeks to months, particularly when the copper-to-molybdenum ratio
is greater than 10:1, no toxic signs are manifested until a critical level
of copper is reached in the liver. The clinical signs are sudden, and the
course is usually 24-48 hr. The animal becomes weak, trembles, and is anoretic. Hemoglobinuria, hemoglobinemia, and icterus are usually present.
Occasionally, an animal will show only pale mucous membranes without icterus
and hemoglobinuria. Although the morbidity is usually less than 5%, the
mortality is usually over 75%.

In acute poisoning by large oral doses of a copper formulation, vomition, excessive salivation, abdominal pain, and diarrhea (greenish-tinged
fluid feces) are usual signs. Collapse and death follow within 24-48 hr.

Physiopathology. A sudden hemolytic crisis accounts for nearly all of the signs of chronic copper poisoning, although not all animals that die of chronic copper poisoning have a hemolytic crisis and jaundice [94]. Reduced hemoglobin and PCV with increased WBC and unchanged platelet counts are characteristic hematologic findings. Concomitant with the release of copper from the liver into the plasma, there is an elevation of plasma bilirubin and decreased liver function [95].

One of the important features of chronic copper poisoning in sheep and cattle is that blood copper concentration remains within normal range during the period of accumulation of copper and increases very markedly and abruptly 24-48 hr before clinical signs appear [96-99]. McCosker [95] describes an intermediate stage, however, when blood copper levels are slightly increased before the stage of acute hemolytic crisis.

Normal levels of copper in the blood range from 75 to 135 μg/100 ml (0.75-1.35 ppm); but at the onset of the hemolytic crisis, concentrations may be much higher [100].

Postmortem changes are usually characteristic and include generalized icterus (occasionally absent, however); greatly enlarged gunmetal-colored kidneys that sometimes have hemorrhagic mottling; slightly enlarged, friable, yellowish liver (may also be small, firm, and pale); gallbladder distended with thick, greenish bile; and enlarged spleen with a brown to black parenchyma of blackberry jam consistency.

Histologically, the hepatocytes may have cytoplasmic vacuolation and necrosis. All lobules may contain clusters of dead cells with the nucleus fragmented and the cytoplasm acidophilic. Fibrosis begins early and is of portal distribution [52]. The kidneys are clogged with hemoglobin and accompanying degeneration and necrosis of the tubular and glomerular cells (acute toxic nephritis). The spleen is crowded with fragmented erythrocytes. A status spongiosus in the white matter of the CNS has been reported [101].

Changes in serum enzyme levels have been recorded 6-8 weeks before the hemolytic crisis, and it has been thought that some of these indicate changes in the liver [98,101,102]. Increases in the activity of the following enzymes have been reported: serum glutamic oxaloacetic transaminase (SGOT) [98,102-105], lactic dehydrogenase (LDH) [98,99,103], sorbitol dehydrogenase, arginase, and glutamate dehydrogenase [101,102]. Elevations in activities of these enzymes, especially SGOT, have been used as indicators of chronic copper poisoning 4-6 weeks in advance of the acute hemolytic crisis. A reduction in activity of the hydrolytic enzymes adenosine triphosphatase [106] and nonspecific esterase and of the oxidative enzyme succinic dehydrogenase at the time of hemolytic crisis was reported by Ishmael et al. [101].

Sheep are more susceptible than cattle to excess copper and deficient molybdenum in their diet [107]. This has been shown by numerous accounts in the literature. Young calves are susceptible; but as they mature, their tolerance increases [6]. The liver metabolizes considerable copper without ill effects, provided molybdenum and sulfate are present. The sheep liver stores copper more readily than that of other species of animals, and a copper concentration of 10-50 ppm on a wet-weight basis (30-150 ppm dry weight) is normally present. When it reaches about 150 ppm (450 ppm dry weight) or more, the animal is predisposed to the characteristic hemolytic crisis of copper poisoning. Poisoning is brought about by the sudden release into the bloodstream of copper which has been stored in the liver. This sudden release of copper by the liver may be spontaneous or may be associated with stress such as reduced food intake, unaccustomed handling, or strenuous exercise [6]. The hepatic storage of copper may occur as a result of (1) intake of copper contaminated feed; (2) the consumption of diets containing improper levels of copper, molybdenum, and sulfate; and (3) liver damage affecting the copper metabolism of the hepatocyte.

Todd and Thompson [98,108] reported a marked reduction in blood glutathione concentration and an accumulation of methemoglobin associated with the hemolytic crisis of copper toxicosis. They postulated that the effect of high blood copper levels is to initiate a chain of events which results in the premature death of the red cell. Death may result from blockage of the kidneys by hemoglobin and resulting kidney failure.

Studies with rats and mice injected with copper compounds have shown that copper accumulates in liver lysosomes. Some workers have postulated that acid hydrolysases, which are capable of producing cellular injury by hydrolyzing cellular structures and constituents, are release from lysosomes following copper accumulation, thus causing hepatic and other cellular damage [109-115].

Nonruminant Animals

There is a marked difference between species in their ability to tolerate high levels of copper. Levels that are toxic to ruminants (30-50 ppm) are well tolerated by nonruminants. Dietary levels in excess of 250 ppm are required to produce toxicity in swine and rats [35,116,117]. Copper

toxicity in nonruminants may not result in the rapid destruction of red
blood cells (hemolytic crisis), although jaundice is observed in pigs fed
toxic levels of copper [1,117]. Milne and Weswig [50] have shown that sheep
accumulate copper in the liver in proportion to the dietary intake, while
rats maintain normal liver copper levels until a high dietary level is
reached (1000 ppm). This may partially explain the difference in the effects
of high levels of copper between the ruminants and nonruminants.

Copper levels of 125-250 ppm have been noted to increase the rate of
gain and feed efficiency of swine [118-121]. Such dietary levels also increase the unsaturation of depot fat, resulting in soft fat [121-123]. The
levels of zinc and iron are important when high levels of copper are fed
since deficient levels of these minerals accentuate the toxic properties of
copper [34,121,124]. Gipp et al. [27], however, reported that liver storage
of copper in pigs on a diet containing 150 ppm copper oxide was unaffected
by dietary levels of zinc, molybdenum, or sulfate. They concluded that
there was no conclusive evidence to support any copper-molybdenum-sulfate
interrelationship or any copper-zinc interaction.

Poultry may be more resistant to copper toxicity than most mammals
except rats. Smith [125] fed copper sulfate to day-old chicks for 25 days
at 0, 100, 200, or 350 ppm copper in a basal ration containing 10 ppm copper.
Those chicks on the 100-ppm copper diet had a slight increase in daily gain
while those on 350 ppm had a slight but statistically significant reduced
weight gain. These differences were attributed only to feed consumption.
Goldberg et al. [126] gave copper acetate to adult chickens (weighing 1.8
± 0.25 kg) via capsule at a rate of 50 mg copper per chicken/day for 1 week,
75 mg/day for a second week, and 100 mg/day until anemia or toxicity appeared or death occurred. After 2-6 weeks of copper administration the
birds became weak, anorexic, and lethargic. Eight of 23 chickens developed
anemia concomitant with toxicosis. They suggested that erythrocytes are
detroyed in the liver as a consequence of copper exposure. McGhee et al.
[127] reported iron and copper interacted in the diet of chicks. They
found that copper at 80 ppm or above depressed growth in young chicks when
iron was fed at the level of 40 ppm or above for a duration of 4 weeks. No
signs of toxicosis were observed, however, even in levels up to 160 ppm
copper and 1600 ppm iron. Turkey poults have been reported to tolerate up
to 676 ppm dietary copper (as copper sulfate) for 21 days, but growth depression occurred when fed 910 or 1000 ppm copper. Signs of toxicosis occurred at 1620 ppm [128]. Wiederanders [129] tried to produce copper toxi-

cosis in turkeys by injecting copper sulfate subcutaneously. He injected
0.5 mg per bird for 84 days followed by 5 mg per bird for an additional 17
days without producing copper toxicosis. He concluded that turkeys and
perhaps other fowl have metabolic and excretory pathways for copper different from those of mammals because there was no increase in ceruloplasmin in
the copper-loaded turkeys.

Extensive acute and chronic copper toxicity studies in chickens, pigeons,
and ducks were conducted by Pullar [130,131]. He found the MLD of copper
sulfate for chickens on a mg/kg body weight basis to be as follows: single
crystal, 900; powdered, 300-500; 4% solution, 1000-1500; mixed with twice
its weight in sodium chloride, 300-500. The MLD of copper carbonate to
chickens was found to be 900 mg/kg single dose. The MLD of copper fed to
pigeons on a mg/kg basis were as follows: single copper sulfate crystal,
1000-1500; copper carbonate, 1000-1500. The single MLDs of a single copper
sulfate crystal to domestic mallards and muscovy ducks were 400 and 600 mg/kg
body weight, respectively. The maximum daily intake of copper carbonate
tolerated by chickens was 60 mg/kg body weight and by mallards was 29 mg/kg.
It was not possible to produce poisoning in chickens given copper sulfate in
drinking water at a level of 250 ppm (1:4000 dilution of copper sulfate in
drinking water), and no obvious signs of copper poisoning were observed in
mallards consuming 250 ppm copper sulfate in their feed.

Numerous studies in rats and mice have been conducted in an effort to
learn more about hepatolenticular degeneration (Wilson's disease) in humans
[109-111,113-115,133-135]. As with other animals, copper seems to have a
propensity for the liver. Prolonged daily intraperitoneal injections of
as little as 0.3 mg Cu/kg will result in elevated hepatic levels [135].
Both hepatic and renal necrosis is observed in rats and mice associated
with increased copper levels [133,134]. However, there is no apparent deposition of copper in the brain, skeletal and cardiac muscle, or skin and
only transient elevations of copper in bone following copper exposure [134].

A form of hepatic cirrhosis in Bedlington terrier dogs that has striking
similarities to Wilson's disease has been described by Hardy et al. [132].
This fatal disease was associated with hepatic copper concentrations in
excess of 10,000 ppm dry weight.

Fish

Contamination of water by copper is commonly associated with mining
and industrial operations. Also, the widespread use of copper sulfate to

control aquatic vegetation may result in significant copper levels in water. A frequently used minimum application of 2.7 lb of copper sulfate per acre-foot of water provides a concentration of 300 ppb of copper (0.3 mg/l) [136]. The application rate is much higher when rapid control of plant growth is desired, sometimes over 5 lb of copper sulfate per acre-foot of water.

Fish are apparently susceptible to copper toxicosis, and trace concentrations, 1/10 to 1/20 of accepted standards for drinking water, can be lethal to fish in regions where surface water is very soft [137]. The 48-hr LC_{50} (lethal concentration 50) in rainbow trout (Salmo gairdneri) was found to be 750 μg/l (0.75 mg/l ranging from 0.67 to 0.84) [138]. The acute (10-day) lethal concentration of copper for brook trout (Salvelinus fontinalis) was about 50 μg/l (0.05 mg/l) [137]. The 96h TL_m (median threshold limit) of copper for bluegills (Lepomis macrochirus) is considered to be 240 μg/l (0.24 mg/l) [136]. Levels above 10 μg/l (0.01 mg/l), however, have been reported to alter oxygen consumption rates of bluegills [136]. Chronic studies using fathead minnows (Pimephales promelas) exposed to copper sulfate for 11 months indicated the fraction or percentage of the TL_m that does not affect growth and reproduction lies between 3 and 7% of the TL_m of 430 μg/l [139]. The absolute copper concentration that is toxic depends upon physical and chemical characteristics of the water, especially pH and calcium concentration. The toxicity of copper to fish has been reviewed by Doudoroff and Katz [140] and McKee and Wolf [141].

Certain metal chelating agents, nitrilotriacetic acid (NTA) and ethylenediaminetetraacetic acid (EDTA) have been shown to protect fish from poisoning by copper and zinc [137]. The ratio of NTA to copper must be at least 3:1 for protection of most fish and is effective against concentrations up to 300-400 times the toxic level. About six times as much EDTA as metal is required for protection, and this chemical is more expensive than NTA. These chelating agents are not cure-alls, however, since they are biodegradable in natural waters in a few days. It has been suggested that they are effective as temporary treatment for metal pollution or to carry a slug of pollution harmlessly past a critical section of river.

ANALYTICAL TECHNIQUES

Analyses for copper for the purpose of assessing tissue and feed levels in domestic and wild animals are best accomplished by studying liver, kidney, serum (or whole blood), and feed and water specimens. The tissues and feed specimens should be ashed at $500°$ C or digested by adding 1 g of specimen to 20 ml nitric and 1 ml sulfuric acid. Heat slowly with a Bunsen burner until the mixture becomes frothy and then add 3 parts nitric to 1 part perchloric acid dropwise, taking care not to char the sample. After digesting, dilute to 50 ml with water and analyze by atomic absorption spectrophotometry [142-145]. Serum or whole blood should be extracted with ammonium pyrollidine dithiocarbamate into methylisobutylketone and analyzed by atomic absorption spectrophotometry [144,145]. The official method of the Association of Official Analytical Chemists for analysis of copper in animal feeds involves the extraction of the specimen with tetraethylenepentamine and measuring the color concentration in a colorimeter or spectrophotometer [146].

REFERENCES

1. E. J. Underwood, Trace Elements in Human and Animal Nutrition, Academic Press, New York, 1971, pp. 57-106.

2. C. B. Ammerman, Symposium: Trace Minerals. Recent Developments in Cobalt and Copper in Ruminant Nutrition: A Review, J. Dairy Sci., 53 (8), 1097-1107 (1970).

3. A. S. Prasad, D. Oberleas, and G. Rajasekaran, Essential Micronutrient Elements, Amer. J. Clin. Nutr., 23 (5), 581-591 (1970).

4. R. P. Dowdy, Copper Metabolism, Amer. J. Clin. Nutr., 22 (7), 887-892 (1969).

5. Site of Copper Toxicity: Microsomal Membrane Adenosinetriphosphatase, Nutr. Rev., 25 (7), 213-215 (1967).

6. J. R. Todd, Chronic Copper Poisoning in Farm Animals, Vet. Bull., 32, 573-580 (1962).

7. I. H. Scheinberg and I. Sternlieb, Copper Metabolism, Pharmacol. Rev., 12, 355 (1960).

8. H. R. Marston, Cobalt, Copper and Molybdenum in the Nutrition of Animals and Plants, Physiol. Rev., 32, 66-121 (1952).

9. I. B. Boughton and W. T. Hardy, Chronic Copper Poisoning in Sheep, Tex. Agr. Exp. Sta. Bull., 499, 32 (1934).

10. W. S. Ferguson, A. H. Lewis, and S. J. Watson, Action of Molybdenum in Nutrition of Milking Cattle, Nature (London), 141, 553 (1938).

11. A. T. Dick and L. B. Bull, Some Preliminary Observations on the Effect of Molybdenum on Copper Metabolism in Herbivorous Animals, Aust. Vet. J., 21, 70-72 (1945).

12. A. T. Dick, The Effect of Inorganic Sulfate on the Excretion of Molybdenum in Sheep, Aust. Vet. J., 29, 18-26 (1953).

13. A. T. Dick, The Control of Copper Storage in the Liver of Sheep by Inorganic Sulfate and Molybdenum, Aust. Vet. J., 29, 1953 (1953).

14. A. T. Dick, Preliminary Observations on the Effect of High Intakes of
     Molybdenum and of Inorganic Sulfate on Blood Copper and on Fleece
     Character in Crossbred Sheep, Aust. Vet. J., 30, 196-202 (1954).

15. R. E. Pierson and W. A. Aanes, Treatment of Chronic Copper Poisoning
     in Sheep, J. Amer. Vet. Med. Ass., 133, 307-311 (1958).

16. I. J. Cunningham, K. G. Hogan, and B. M. Lawson, The Effect of Sulfate
     and Molybdenum on Copper Metabolism in Cattle, N. Z. Agr. Res., 2, 145-
     152 (1959).

17. T. Kowalczyk, A. L. Pope, and D. K. Sorenson, Chronic Copper Poisoning
     in Sheep Resulting from Free-Choice Trace Mineral-Salt Ingestion, J.
     Amer. Vet. Med. Ass., 141, 362-366 (1962).

18. T. Kowalczyk, A. L. Pope, K. C. Berger, and B. A. Muggenburg, Chronic
     Copper Toxicosis in Sheep Fed Dry Feed, J. Amer. Vet. Med. Ass., 145,
     352-357 (1964).

19. R. D. Goodrich and A. D. Tillman, Copper, Sulfate, and Molybdenum In-
     terrelationships in Sheep, J. Nutr., 90, 76-80 (1966).

20. R. P. Dowdy and G. Matrone, Copper-Molybdenum Interactions in Sheep
     and Chicks, J. Nutr., 95, 191-196 (1968).

21. R. P. Dowdy and G. Matrone, A Copper-Molybdenum Complex: Its Effects
     and Movement in the Piglet and Sheep, J. Nutr., 95, 197-201 (1968).

22. A. H. Adamson and D. A. Valks, Copper Toxicity in Housed Lambs, Vet.
     Rec., 85, 368-369 (1969).

23. N. A. Marcilese, C. B. Ammerman, R. M. Valsecchi, B. G. Dunavant, and
     G. K. Davis, Effect of Dietary Molybdenum and Sulfate Upon Copper Me-
     tabolism in Sheep, J. Nutr., 99, 177-183 (1969).

24. J. R. Todd, Chronic Copper Toxicity of Ruminants, Symp. Nutr. Disorders
     Ruminants, 28, 189-197 (1969).

25. D. B. Ross, The Effect of Oral Ammonium Molybdate and Sodium Sulfate
     Given to Lambs with High Liver Copper Concentrations, Res. Vet. Sci.,
     2, 295-297 (1970).

26. J. T. Huber, N. O. Price, and R. W. Engel, Response of Lactating Dairy
     Cows to High Levels of Dietary Molybdenum, J. Animal Sci., 32, 364-367
     (1971).

27. W. F. Gipp, W. G. Pond, and S. E. Smith, Effects of Level of Dietary
     Copper, Molybdenum, Sulfate, and Zinc on Body Weight Gain, Hemoglobin,
     and Liver Storage of Growing Pigs, J. Animal Sci., 26, 727-730 (1967).

28. W. V. Hays and R. D. Kline, Copper-Molybdenum-Sulfate Interrelation-
     ships in Growing Pigs, Feedstuffs, 41 (44), 18 (1969).

29. S. E. Dale, Effect of Molybdenum and Sulfate on Copper Metabolism in
     Young Growing Pigs, M.S. Thesis, Iowa State University, Ames, 1971.

30. G. L. Cromwell, Copper, Molybdenum, Sulfate, and Sulfide Interrela-
     tionships in Swine, An. Nutr. Health, 26 (12), 5-7 (1971).

31. L. F. Gray and L. J. Daniel, Effect of the Copper Status of the Rat
     on the Copper-Molybdenum-Sulfate Interaction, J. Nutr., 84, 31 (1964).

32. T. J. Hanrahan and J. F. O'Grady, Copper Supplementation of Pig Diets.
     The Effect of Protein Level and Zinc Supplementation on the Response
     to Added Copper, Animal Prod., 10, 423-432 (1968).

33. H. D. Ritchie, R. W. Luecke, B. V. Baltzer, E. R. Miller, O. E. Ulrey, and J. A. Hoefer, Copper and Zinc Interrelationships and Parakeratosis, J. Nutr., 79 (2), 117-123 (1963).

34. N. F. Suttle and C. F. Mills, Studies of the Toxicity of Copper to Pigs. I. Effects of Oral Supplements of Zinc and Iron Salts on the Development of Copper Toxicosis, Brit. J. Nutr., 20, 135-147 (1966).

35. N. F. Suttle and C. F. Mills, Studies of the Toxicity of Copper to Pigs. 2. Effect of Protein Source and Other Dietary Components on the Response to High and Moderate Intakes of Copper, Brit. J. Nutr., 20, 149-161 (1966).

36. R. Van Reen, Effects of Excessive Dietary Zinc in the Rat and the Interrelationship of Copper, Arch. Biochem. Biophys., 46, 337-344 (1953).

37. S. E. Smith and E. J. Larson, Zinc Toxicity in Rats; Antagonistic Effects of Copper and Liver, J. Biol. Chem., 163, 29-38 (1946).

38. P. D. Whanger and P. H. Weswig, Effect of Supplementary Zinc on the Intracellular Distribution of Hepatic Copper in Rats, J. Nutr., 101, 1093-1098 (1971).

39. A. MacPherson and R. G. Hemingway, Effects of Protein Intake on the Storage of Copper in the Liver of Sheep, J. Sci. Food Agr., 16, 220-227 (1965).

40. E. C. C. Parris and B. E. McDonald, Effect of Dietary Protein Source on Copper Toxicity in Early-Weaned Pigs, Can. J. Animal Sci., 49, 215-222 (1968).

41. Copper Toxicity, Nutr. Rev., 24, 305-308 (1966).

42. G. Gregoriadis and T. L. Sourkes, Role of Protein in Removal of Copper from the Liver, Nature, 218, 290-291 (1968).

43. M. E. Lahey, C. J. Gubler, M. S. Chase, G. E. Cartwright, and M. M. Wintrobe, Studies on Copper Metabolism. II. Hematologic Manifestations of Copper Deficiency in Swine, Blood, 7, 1053-1074 (1952).

44. G. E. Cartwright, C. J. Gubler, J. A. Bush, and M. M. Wintrobe, Studies on Copper Metabolism. XVII. Further Observations on the Anemia of Copper Deficiency in Swine, Blood, 11, 143-153 (1956).

45. G. R. Lee, G. E. Cartwright, and M. M. Wintrobe, Heme Biosynthesis in Copper Deficient Swine, Proc. Soc. Exp. Biol. Med., 127 (4), 977-981 (1968).

46. M. S. Chase, C. J. Gubler, G. E. Cartwright, and M. M. Wintrobe, Studies on Copper Metabolism. V. Storage of Iron in Liver of Copper-Deficient Rats, Proc. Soc. Exp. Biol. Med., 80, 749-750 (1952).

47. G. R. Lee, S. Nacht, J. M. Lukens, and G. E. Cartwright, Iron Metabolism in Copper-Deficient Swine, J. Clin. Invest., 47, 2058-2069 (1968).

48. T. Moore, I. M. Sharman, J. R. Todd, and R. H. Thompson, Copper and Vitamin A Concentrations in the Blood of Normal and Copper-Poisoned Sheep, Brit. J. Nutr., 28, 23-29 (1972).

49. G. Gregoriadis and T. L. Sourkes, Intracellular Distribution of Copper in the Liver of the Rat, Can. J. Biochem., 45, 1841-1851 (1967).

50. P. B. Milne and P. H. Weswig, Effect of Supplementary Copper on Blood and Liver Copper-Containing Fractions in Rats, J. Nutr., 95, 429 (1968).

51. G. E. Cartwright and M. M. Wintrobe, Copper Metabolism in Normal Subjects, Amer. J. Clin. Nutr., 14, 224-232 (1964).

52. K. V. F. Jubb and P. C. Kennedy, Pathology of Domestic Animals, Vol. I, Academic Press, New York, 1970, pp. 304-308.

53. C. H. Gallagher and Vivienne E. Reeve, Copper Deficiency in the Rat. Effect on Synthesis of Phospholipids, Aust. J. Exp. Biol. Med. Sci., 49, 21-31 (1971).

54. J. Huisingh and G. Matrone, Copper-Molybdenum Interactions with the Sulfate-Reducing System in Rumen Microorganisms, Proc. Soc. Exp. Biol. Med., 139, 518-521 (1972).

55. N. A. Marcilese, C. B. Ammerman, R. M. Valsecchi, B. G. Dunavant, and G. K. Davis, Effect of Dietary Molybdenum and Sulfate Upon Urinary Excretion of Copper in Sheep, J. Nutr., 100, 1399-1406 (1970).

56. W. B. Buck, Diagnosis of Feed-Related Toxicoses, J. Amer. Vet. Med. Ass., 156, 1434-1443 (1970).

57. W. B. Buck, G. D. Osweiler, and G. A. Van Gelder, Copper-Molybdenum, in Clinical and Diagnostic Veterinary Toxicology, Kendall-Hunt, Dubuque, 1973.

58. J. R. Todd, J. F. Gracey, and R. H. Thompson, Studies on Chronic Copper Poisoning. I. Toxicity of Copper Sulphate and Copper Acetate in Sheep, Brit. Vet. J., 118, 482-491 (1962).

59. N. F. Suttle, K. W. Angus, D. E. Nisbet, and A. C. Fields, Osteoporosis in Copper-Depleted Lambs, J. Comp. Pathol., 82, 93-97 (1972).

60. D. R. Cordy, Enzootic Ataxia in California Lambs, J. Amer. Vet. Med. Ass., 158, 1940-1942 (1971).

61. H. W. Bennetts, A. B. Beck, and R. Harley, The Pathogenesis of "Falling Disease," Aust. Vet. J., 24, 237-244 (1948).

62. H. W. Bennetts and H. T. B. Hall, "Falling Disease" of Cattle in the South-west of Western Australia, Aust. Vet. J., 15, 152-159 (1939).

63. H. W. Bennetts, R. Harley, and S. T. Evans, Studies on Copper Deficiency of Cattle: The Fatal Termination ("Falling Disease"), Aust. Vet. J., 18, 50-63 (1942).

64. G. S. Shields, W. H. Carnes, G. E. Cartwright, and M. M. Wintrobe, Vascular Lesions in Copper-Deficient Swine, Fed. Proc., Fed. Amer. Soc. Exp. Biol., 20, 118 (1961).

65. B. L. O'Dell, B. C. Hardwich, Genevieve Reynolds, and J. E. Savage, Connective Tissue Defect in the Chick Resulting from Copper Deficiency, Proc. Soc. Exp. Biol. Med., 108, 402-405 (1961).

66. C. H. Hill, B. Strarcher, and C. Kim, Role of Copper in the Formation of Elastin, Fed. Proc., Fed. Amer. Soc. Exp. Biol., 26, 129-133 (1967)

67. R. B. Rucker, H. E. Parker, and J. C. Rogler, Effect of Copper Deficiency on Chick Bone Collagen and Selected Bone Enzymes, J. Nutr., 98, 57-63 (1969).

68. J. W. Britton and H. Goss, Chronic Molybdenum Poisoning in Cattle, J. Amer. Vet. Med. Ass., 108, 176-178 (1946).

69. C. E. Fleming, J. A. McCormick, and W. B. Dye, The Effects of Molybdenosis on a Breeding Experiment, Nevada Agr. Exp. Sta. Bull., 220, 15 (1961).

70. J. E. Miltimore and J. L. Mason, Copper to Molybdenum Ratio and Molybdenum and Copper Concentrations in Ruminant Feeds, Can. J. Animal Sci., 51, 193-200 (1971).

71. I. Thornton, G. F. Kershaw, and M. G. Davies, An Investigation into Sub-clinical Copper Deficiency in Cattle, Vet. Rec., 90, 11-12 (1972).

72. C. A. Owen and J. B. Hazelrig, Copper Deficiency and Copper Toxicity in the Rat, Amer. J. Phys., 215, 334 (1968).

73. R. P. Dowdy, G. A. Kunz, and H. E. Sauberlich, Effect of a Copper-Molybdenum Compound Upon Copper Metabolism in the Rat, J. Nutr., 99, 491 (1969).

74. H. A. Ragen, S. Nacht, G. R. Lee, C. R. Bishop, and G. E. Cartwright, Effect of Ceruloplasmin on Plasma Iron in Copper-Deficient Swine, Amer. J. Phys., 217, 1320 (1969).

75. C. E. Hunt, J. Landesman, and P. M. Newberne, Copper Deficiency in Chicks: Effects of Ascorbic Acid on Iron, Copper, Cytochrome Oxidase Activity, and Aortic Mycopolysaccharides, Brit. J. Nutr., 24, 607 (1970).

76. G. K. Davis, in Symposium on Copper Metabolism (W. D. McElroy and B. Glass, eds.), Johns Hopkins Press, Baltimore, 1950, p. 216.

77. R. D. Kline, V. W. Hays, and G. L. Cromwell, Effects of Molybdenum, Sulfate, and Sulfide on Copper Store of Pigs, J. Animal Sci., 31, 205 (1970).

78. R. D. Kline, V. W. Hays, and G. L. Cromwell, Effects of Copper, Molybdenum and Sulfate on Performance, Hematology and Copper Stores of Pigs and Lambs, J. Animal Sci., 33 (4), 771-779 (1971).

79. J. B. Nielands, F. M. Strong, and C. A. Elvehjem, Molybdenum in the Nutrition of the Rat, J. Biol. Chem., 172, 431 (1948).

80. R. Compere, A. Burny, A. Riga, E. Francois, and S. Vanuytrecht, Copper in the Treatment of Molybdenosis in the Rat: Determination of the Toxicity of the Antidote, J. Nutr., 87, 412 (1965).

81. R. W. Voelker and W. W. Carlton, Effect of Ascorbic Acid on Copper Deficiency in Miniature Swine, Amer. J. Vet. Res., 30, 1825 (1969).

82. J. R. Todd, Copper, Molybdenum, and Sulphur Contents of Oats and Barley in Relation to Chronic Copper Poisoning in Housed Sheep, J. Sci. Comb., 79, 191-195 (1972).

83. L. B. Bull, H. E. Albiston, G. Edgar, and A. T. Dick, Toxaemic Jaundice of Sheep: Phytogenous Chronic Copper Poisoning, Heliotrope Poisoning, and Hepatogenous Chronic Copper Poisoning, Aust. Vet. J., 32, 220-236 (1956).

84. A. B. Beck and H. W. Bennetts, Copper Poisoning in Sheep in Western Australia, J. Roy. Soc. West. Aust., 46, 5-10 (1963).

85. T. D. St. George-Grambauer and R. Rac, Hepatogenous Chronic Copper Poisoning in Sheep in South Australia Due to the Consumption of Echium plantagineum (Salvation Jane), Aust. Vet. J., 38, 288-293 (1962).

86. C. D. Bracewell, A Note on Jaundice in Housed Sheep, Vet. Rec., 70, 342-344 (1958).

87. W. B. Buck and R. M. Sharma, Copper Toxicity in Sheep, I. S. U. Vet., 31, 4-8 (1969).

88. T. Suveges, F. Ratz, and G. Salyi, Pathogenesis of Chronic Copper Poisoning in Lambs, Acta Vet. Hung., 21 (4), 383-391 (1971).

89. K. G. Hogan, D. F. L. Money, and Annette Blayney, The Effect of a Molybdate and Sulphate Supplement on the Accumulation of Copper in the Livers of Penned Sheep, N. Z. J. Agr. Res., 11, 435-444 (1968).

90. W. B. Buck, unpublished data, 1970.

91. J. Ishmael, C. Gopinath, and J. McC. Howell, Studies with Copper Calcium E.D.T.A., J. Comp. Pathol., 81, 279-291 (1971).

92. J. Ishmael and C. Gopinath, Blood Copper and Serum Enzyme Changes Following Copper Calcium E.D.T.A. Administration to Hill Sheep of Low Copper Status, J. Comp. Pathol., 81, 455-461 (1971).

93. J. Ishmael and C. Gopinath, Effect of a Single Small Dose of Inorganic Copper on the Liver of Sheep, J. Comp. Pathol., 82, 47-57 (1972).

94. M. D. Sutter, D. C. Rawson, J. A. McKeown, and A. R. Hashell, Chronic Copper Toxicosis in Sheep, Amer. J. Vet. Res., 19, 890-892 (1958).

95. P. J. McCosker, Observations on Blood Copper in the Sheep. II. Chronic Copper Poisoning, Res. Vet. Sci., 9, 103-116 (1968).

96. H. D. Albiston, L. B. Bull, A. T. Dick, and J. C. Keast, A Preliminary Note on the Aetiology of Enzootic Jaundice, Toxaemic Jaundice, or "Yellows," of Sheep in Australia, Aust. Vet. J., 16, 233-243 (1940).

97. P. J. Barden and A. Robertson, Experimental Copper Poisoning in Sheep, Vet. Rec., 74, 252-256 (1962).

98. J. R. Todd and R. H. Thompson, Studies on Chronic Copper Poisoning. II. Biochemical Studies on the Blood of Sheep During the Haemolytic Crisis, Brit. Vet. J., 119, 161-173 (1963).

99. J. R. Todd and R. H. Thompson, Studies on Chronic Copper Poisoning. IV. Biochemistry of the Toxic Syndrome in the Calf, Brit. Vet. J., 121, 90-97 (1965).

100. A. B. Beck, The Copper Content of the Liver and Blood of Some Vertebrates, Aust. J. Zool., 4, 1 (1955).

101. J. Ishmael, C. Gopinath, and J. McC. Howell, Experimental Chronic Copper Toxicity in Sheep. Histological and Histochemical Changes During the Development of the Lesions in the Liver, Res. Vet. Sci., 12, 358-366 (1971).

102. J. Ishmael, C. Gopinath, and J. McC. Howell, Experimental Chronic Copper Toxicity in Sheep. Biochemical and Haematological Studies During the Development of Lesions in the Liver, Res. Vet. Sci., 13, 22-29 (1972).

103. P. W. Van Adrichem, Changes in the Activity of Serum-enzymes and in the LDH Iso-enzymes Pattern in Chronic Copper Intoxication in Sheep, Tydschr. Diergeneesk, 90 (20), 1371-1381 (1965).

104. D. B. Ross, The Diagnosis, Prevention, and Treatment of Chronic Copper Poisoning in Housed Lambs, Brit. Vet. J., 122, 279-284 (1966).

105. A. MacPherson and R. G. Hemingway, The Relative Merits of Various Blood Analyses and Liver Function Tests in Giving an Early Diagnosis of Chronic Copper Poisoning in Sheep, Brit. Vet. J., 125, 213-220 (1969).

106. R. A. Peters, M. Shorthouse, and J. M. Walshe, The Effect of Cu^{2+} on the Membrane ATPase and Its Relation to the Initiation of Convulsions, J. Physiol., 181, 27 (1965).

107. I. J. Cunningham, The Toxicity of Copper to Bovines, N. Z. J. Sci. Tech., 27 (5), 372-376 (1946).

108. J. R. Todd and R. H. Thompson, Studies on Chronic Copper Poisoning. III. Effect of Copper Acetate Injected Into the Blood Stream of Sheep, J. Comp. Pathol., 74 (4), 542-551 (1964).

109. Richard R. Lindquist, Studies of the Pathogenesis of Hepatolenticular Degeneration. I. Acid Phosphatase Activity in Copper-Loaded Rat Livers, Amer. J. Pathol., 51, 471-481 (1967).

110. Richard R. Lindquist, Studies on the Pathogenesis of Hepatolenticular Degeneration. III. The Effect of Copper on Rat Liver Lysosomes, Amer. J. Pathol., 53, 903-927 (1968).

111. M. Anthony Verity, J. K. Gambell, A. R. Reith, and W. Jann Brown, Subcellular Distribution and Enzyme Changes Following Subacute Copper Intoxication, Lab. Invest., 16 (4), 580-590 (1967).

112. Sidney Goldfischer, Demonstration of Copper and Acid Phosphatase Activity in Hepatocyte Lysosomes in Experimental Copper Toxicity, Nature, 215, 74-75 (1967).

113. Eugene N. McNatt, Wallace G. Campbell, Jr., and Brenda C. Callahan, Effects of Dietary Copper Loading on Livers of Rats. I. Changes in Subcellular Acid Phosphatases and Detection of an Additional Acid. p-Nitrophenylphosphatase in the Cellular Supernatant During Copper Loading, Amer. J. Pathol., 64, 123-144 (1971).

114. S. Lal and T. L. Sourkes, Deposition of Copper in Rat Tissues: The Effect of Dose and Duration of Administration of Copper Sulfate, Toxicol. Appl. Pharmacol., 20, 269-283 (1971).

115. T. Barka, P. J. Scheuer, Fenton Schaffner, and H. Popper, Structural Changes of Liver Cells in Copper Intoxication, Arch. Pathol., 78, 331-349 (1964).

116. R. Boyden, V. R. Potter, and C. A. Elvehjem, Effect of Feeding High Levels of Copper to Albino Rats, J. Nutr., 15, 397 (1938).

117. H. D. Wallace, J. T. McCall, B. Boss, and G. E. Combs, High Level Copper for Growing-Finishing Swine, J. Animal Sci., 19, 1153 (1960).

118. R. J. Bowler, R. Braude, R. C. Campbell, J. N. Craddock-Turnbull, H. F. Fieldsend, E. K. Grifiths, I. W. M. Lucas, K. G. Mitchell, N. J. D. Nickalls, and J. H. Taylor, High Copper-Mineral Mixtures for Fattening Pigs, Brit. J. Nutr., 9, 358 (1955).

119. R. S. Barber, R. Braude, K. G. Mitchell, J. A. F. Rook, and J. G. Rowell, Further Studies on Antibiotic and Copper Supplements for Fattening Pigs, Brit. J. Nutr., 11, 70 (1957).

120. R. G. Bunch, J. T. McCall, V. C. Speer, and V. W. Hays, Copper Supplementation for Weanling Pigs, J. Animal Sci., 24, 995 (1965).

121. L. W. De Goey, R. C. Wahlstrom, and R. J. Emerick, Studies of High Level Copper Supplementation to Rations for Growing Swine, J. Animal Sci., 33, 52 (1971).

122. M. Taylor and S. Thomke, Effect of High-Level Copper on the Depot Fat of Bacon Pigs, Nature, 201, 246 (1964).

123. J. I. Elliot and J. P. Bowland, Effects of Dietary Copper Sulfate and Protein on the Fatty Acid Composition of Porcine Fat, J. Animal Sci., 30, 923 (1970).

124. R. G. Bunch, V. C. Speer, V. W. Hays, and J. T. McCall, Effects of High Levels of Copper and Chlorotetracycline on Performance of Pigs, J. Animal Sci., 22, 56 (1963).

125. M. S. Smith, Responses of Chicks to Dietary Supplements of Copper Sulphate, Brit. Poultry Sci., 10, 97-108 (1969).

126. A. Goldberg, C. B. Williams, R. S. Jones, Mitsue Yamagita, G. E. Cartwright, and M. M. Wintrobe, Studies on Copper Metabolism. XXII. Hemolytic Anemia in Chickens Induced by the Administration of Copper, J. Lab. Clin. Med., 48, 442-453 (1956).

127. Flin McGhee, C. R. Creger, and J. R. Couch, Copper and Iron Toxicity, Poultry Sci., 44, 310-312 (1965).

128. Pran Vohra and F. H. Kratzer, Zinc, Copper and Manganese Toxicities in Turkey Poults and Their Alleviation by E.D.T.A., Poultry Sci., 47, 699-704 (1968).

129. R. E. Wiederanders, Copper Loading in the Turkey, Proc. Soc. Exp. Biol. Med., 128, 627-629 (1968).

130. E. Murray Pullar, The Toxicity of Various Copper Compounds and Mixtures for Domesticated Birds, Aust. Vet. J., 16, 147-162 (1940).

131. E. Murray Pullar, The Toxicity of Various Copper Compounds and Mixtures for Domesticated Birds, Aust. Vet. J., 16, 203-213 (1940).

132. R. M. Hardy, J. B. Stevens, and C. M. Stowe, Chronic Progressive Hepatitis in Bedlington Terriers Associated with Elevated Liver Copper Concentrations, Minnesota Vet., 15 (2), 13 (1975).

133. F. Stephen Vogel, Nephrotoxic Properties of Copper Under Experimental Conditions in Mice With Special Reference to the Pathogenesis of the Renal Alterations in Wilson's Disease, Amer. J. Pathol., 36, 699-711 (1960).

134. Sheldon M. Wolff, Copper Deposition in the Rat, A. M. A. Arch. Pathol., 69, 217-223 (1960).

135. S. Lal and T. L. Sourkes, Intracellular Distribution of Copper in the Liver During Chronic Administration of Copper Sulfate to the Rat, Toxicol. Appl. Pharmacol., 18, 562-572 (1971).

136. James O'Hara, Alterations in Oxygen Consumption by Bluegills Exposed to Sublethal Treatment with Copper, Water Res., 5, 321-327 (1971).

137. John B. Sprague, Promising Antipollutant: Chelating Agent NTA Protect Fish from Copper and Zinc, Nature, 220, 1345-1346 (1968).

138. V. M. Brown and R. A. Dalton, The Acute Lethal Toxicity to Rainbow Trout of Mixtures of Copper, Phenol, Zinc, and Nickel, J. Fish Biol., 2, 211-216 (1970).

139. Donald I. Mount, Chronic Toxicity of Copper to Fathead Minnows (Pimephales promelar, Rafinesque), Water Res., 2, 215-223 (1968).

140. P. Doudoroff and M. Katz, Critical Review of Literature on the Toxicity of Industrial Wastes and Their Components to Fish. II. The Metals, as Salts, Sewage Ind. Wastes, 25, 802-839 (1953).

141. J. E. McKee and H. W. Wolf, Water Quality Criteria, 2nd Ed., The Re-
     sources Agency of California, State Waste Quality Control Board, Pub-
     blication No. 3A, 1963, p. 548.

142. Analytical Methods for Atomic Absorption Spectrophotometry, Perkin
     Elmer Corporation, Norwalk, September 1968.

143. Committee Report on Metallic Impurities in Organic Matter, Analyst,
     84, 127-128 (1957).

144. W. B. Buck, Laboratory Toxicologic Tests and Their Interpretation, J.
     Amer. Vet. Med. Ass., 155, 1928-1941 (1969).

145. E. Berman, Application of Atomic Absorption Spectrophotometry to the
     Determination of Copper in Serum, Urine, and Tissue, Atomic Absorp-
     tion Newsletter, 4, 296 (1965).

146. W. Horwitz (ed.), Official Methods of Analysis of the Association of
     Official Analytical Chemists, 11th Ed., Ass. Offic. Anal. Chem.,
     Washington, D. C., 1970, p. 134.

Periodic Table of the Elements

IA	IIA	IIIB	IVB	VB	VIB	VIIB	VIII			IB	IIB	IIIA	IVA	VA	VIA	VIIA	Zero
1 H																1 H	2 He
3 Li	4 Be											5 B	6 C	7 N	8 O	9 F	10 Ne
11 Na	12 Mg											13 Al	14 Si	15 P	16 S	17 Cl	18 Ar
19 K	20 Ca	21 Sc	22 Ti	23 V	24 Cr	25 Mn	26 Fe	27 Co	28 Ni	29 Cu	30 Zn	31 Ga	32 Ge	33 As	34 Se	35 Br	36 Kr
37 Rb	38 Sr	39 Y	40 Zr	41 Nb	42 Mo	43 Tc	44 Ru	45 Rh	46 Pd	47 Ag	48 Cd	49 In	50 Sn	51 Sb	52 Te	53 I	54 Xe
55 Cs	56 Ba	57 *La	72 Hf	73 Ta	74 W	75 Re	76 Os	77 Ir	78 Pt	79 Au	80 Hg	81 Tl	82 Pb	83 Bi	84 Po	85 At	86 Rn
87 Fr	88 Ra	89 #Ac															

*LANTHANIDE
SERIES

58 Ce	59 Pr	60 Nd	61 Pm	62 Sm	63 Eu	64 Gd	65 Tb	66 Dy	67 Ho	68 Er	69 Tm	70 Yb	71 Lu

#ACTINIDE
SERIES

90 Th	91 Pa	92 U	93 Np	94 Pu	95 Am	96 Cm	97 Bk	98 Cf	99 Es	100 Fm	101 Md	102 No	103 Lr